Da Ruan, Guoqing Chen, Etienne E. Kerre, Geert Wets (Eds.)

Intelligent Data Mining

T0191885

Studies in Computational Intelligence, Volume 5

Editor-in-chief
Prof. Janusz Kacprzyk
Systems Research Institute
Polish Academy of Sciences
ul. Newelska 6
01-447 Warsaw
Poland
E-mail: kacprzyk@ibspan.waw.pl

Further volumes of this series
can be found on our homepage:
springeronline.com

Da Ruan
Guoqing Chen
Etienne E. Kerre
Geert Wets
(Eds.)

Intelligent Data Mining

Techniques and Applications

 Springer

Professor Dr. Da Ruan

Belgian Nuclear Research
Center (SCK· CEN)
Boeretang 200, 2400 Mol
Belgium
E-mail: druan@sckcen.be

Professor Dr. Etienne E. Kerre

Department of Applied Mathematics
and Computer Science
Ghent University
Krijgslaan 281 (S9), 9000 Gent
Belgium
E-mail: etienne.kerre@ugent.be

Professor Dr. Guoqing Chen

School of Economics
and Management, Division MIS
Tsinghua University
100084 Beijing
The People's Republic of China
E-mail: chengq@mail.tsinghua.edu.cn

Professor Dr. Geert Wets

Limburg University Centre
Universiteit Hasselt
3590 Diepenbeek
Belgium
E-mail: geert.wets@uhasselt.be

ISSN print edition: 1860-949X
ISSN electronic edition: 1860-9503

ISBN 978-3-642-06576-7   e-ISBN 978-3-540-32407-2

Springer is a part of Springer Science+Business Media
springeronline.com
© Springer-Verlag Berlin Heidelberg 2010
Printed in The Netherlands

Preface

In today's information-driven economy, companies may benefit a lot from suitable information management. Although information management is not just a technology-based concept rather a business practice in general, the possible and even indispensable support of IT-tools in this context is obvious. Because of the large data repositories many firms maintain nowadays, an important role is played by data mining techniques that find hidden, non-trivial, and potentially useful information from massive data sources. The discovered knowledge can then be further processed in desired forms to support business and scientific decision making.

Data mining (DM) is also known as Knowledge Discovery in Databases. Following a formal definition by W. Frawley, G. Piatetsky-Shapiro and C. Matheus (in AI Magazine, Fall 1992, pp. 213–228), DM has been defined as "The nontrivial extraction of implicit, previously unknown, and potentially useful information from data." It uses machine learning, statistical and visualization techniques to discover and present knowledge in a form that is easily comprehensible to humans. Since the middle of 1990s, DM has been developed as one of the hot research topics within computer sciences, AI and other related fields. More and more industrial applications of DM have been recently realized in today's IT time.

The root of this book was originally based on a joint China-Flanders project (2001–2003) on methods and applications of knowledge discovery to support intelligent business decisions that addressed several important issues of concern that are relevant to both academia and practitioners in intelligent systems. Extensive contributions were made possible from some selected papers of the 6th International FLINS conference on Applied Computational Intelligence (2004).

Intelligent Data Mining – Techniques and Applications is an organized edited collection of contributed chapters covering basic knowledge for intelligent systems and data mining, applications in economic and management, industrial engineering and other related industrial applications. The main objective of this book is to gather a number of peer-reviewed high quality contributions in the relevant topic areas. The focus is especially on those chapters that provide theoretical/analytical solutions to the problems of real interest in intelligent techniques possibly combined with other traditional tools, for data mining and the corresponding applications to engineers and managers of different industrial sectors. Academic and applied researchers and research students working on data mining can also directly benefit from this book.

The volume is divided into three logical parts containing 24 chapters written by 62 authors from 10 countries[1] in the field of data mining in conjunction with intelligent systems.

Part 1 on *Intelligent Systems and Data Mining* contains nine chapters that contribute to a deeper understanding of theoretical background and methodologies to be used in data mining. Part 2 on *Economic and Management Applications* collects six chapters that dedicate to the key issue of real-world economic and management applications. Part 3 presents nine chapters on *Industrial Engineering Applications* that also point out the future research direction on the topic of intelligent data mining.

We would like to thank all the contributors for their kind cooperation to this book; and especially to Prof Janusz Kacprzyk (Editor-in-chief of Studies in Computational Intelligence) and Dr Thomas Ditzinger of Springer for their advice and help during the production phases of this book. The support from the China Flanders project (grant No. BIL 00/46) is greatly appreciated.

April 2005   *Da Ruan*
                                                        Guoqing Chen
                                                     Etienne E. Kerre
                                                         Geert Wets

---

[1]   Australia, Belgium, Bulgaria, China, Greece, France, Turkey, Spain, the UK, and the USA.

Contents

Intelligent Systems and Data Mining

Part I

Intelligent Systems and Data Mining

Some Considerations
in Multi-Source Data Fusion

Ronald R. Yager

Machine Intelligence Institute, Iona College, New Rochelle, NY 10801
yager@panix.com

Abstract. We introduce the data fusion problem and carefully distinguish it from a number of closely problems. Some of the considerations and knowledge that must go into the development of a multi-source data fusion algorithm are described. We discuss some features that help in expressing users requirements are also described. We provide a general framework for data fusion based on a voting like process that tries to adjudicate conflict among the data. We discuss various of compatibility relations and introduce several examples of these relationships. We consider the case in which the sources have different credibility weight. We introduce the idea of reasonableness as a means for including in the fusion process any information available other than that provided by the sources.

Key words: Data fusion, similarity, compatibility relations, conflict resolution

1 Introduction

An important aspect of data mining is the coherent merging of information from multiple sources [1, 2, 3, 4]. This problem has many manifestation ranging from data mining to information retrieval to decision making. One type of problem from this class involves the situation in which we have some variable, whose value we are interested in supplying to a user, and we have multiple sources providing data values for this variable. Before we proceed we want to carefully distinguish our particular problem from some closely related problems that are also important in data mining. We first introduce some useful notation. Let Y be some class of objects. By an attribute A we mean some feature or property that can be associated with the elements in the set Y. If Y is a set of people then examples of attributes are age, height, income and mother's name. Attributes are closely related to the column headings used in a table in a relational data base [3]. Typically an attribute has a domain X in which the values of the attribute can lie. If Y is an element from Y we denote the value of the attribute A for object Y as $A[y]$. We refer to $A[y]$ as a variable. Thus if John is a member of Y the Age [John] is a variable. The value of the

Ronald R. Yager: *Some Considerations in Multi-Source Data Fusion*, Studies in Computational Intelligence (SCI) **5**, 3–XIII (2005)
www.springerlink.com

variable $A[y]$ is generally a **unique** element from the domain X. If $A[y]$ takes on the value x we denote this as $A[y] = x$. One problem commonly occurring in data mining is the following. We have the value of an attribute for a number of elements in the class Y, $(A[y_1] = x_1, A[y_2] = x_2, A[y_3] = x_3, \ldots, A[y_q] = x_q)$ and we are interested in finding a value $x* \in x$ as a representative or summary value of this data. We note since each of the $A[y_k]$ is different variables there is no inherent conflict in the fact the values associated with these variables are different. We emphasize that the summarizing value $x*$ is not associated with any specific object in the class Y. It is a value associated with a conceptual variable. At best we can consider $x*$ the value of a variable $A[Y]$. We shall refer to this problem of attaining $x*$ as the **data summarization problem**. A typical example of this would if Y are the collection of people in a city neighbor and A is the attribute salary. Here then we are interested in getting a representative value of the salary of the people in the neighborhood. The main problem we are interested in here, while closely related, is different. Here again we have some attribute A. However instead of being concerned with the class Y we are focusing on one object from this class y_q and we are interested in the value of the variable $A[y_q]$. For example if A is the attribute age and y_q is Osama bin Laden then our interest is in determining Osama bin Laden's age. In our problem of concern the data consists of $(A[y_q] = x_1, A[y_q] = x_2, A[y_q] = x_2, \ldots, A[y_q] = x_n)$. Here we have a number of observations provided by different sources on the value of the variable $A[y_q]$ and we are interested in using this to obtain "a value of the variable $A[y_q]$." We shall call this the **data fusion problem**. While closely related there exists differences. One difference between these problems is that in the fusion problem we are seeking the value of the attribute of a real object rather than the attribute value of some conceptual object. If our attribute is the number of children then determining then the summarizing value over a community is 2.6 may not be a problem, however if we are interested in the number of children that bin Laden has, 2.6 may be inappropriate. Another distinction between these two situations relates to the idea of conflict. In the first situation since $A[y_1]$ and $A[y_2]$ are different variables the fact that $x_1 \neq x_2$ is not a conflict. On the other hand in the second situation, the data fusion problem, since all observations in our data set are about the same variable $A[y_q]$ the fact that $x_a \neq x_b$ can be seen as constituting a conflict. One implication of this relates to the issue of combining values. For example consider the situation in which A is the attribute salary in trying to find the representative (summarizing) value of salaries within a community averaging two salaries such as $5,000,000 and $10,000 poses no conceptual dilemma. On the hand if these values are said by different sources to be the salary of some specific individual averaging them would be questionable.

Another problem very closely related to our problem is the following. Again let A be some attribute, y_q be some object and let $A[y_q]$ be a variable whose value we are trying to ascertain. However in this problem $A[y_q]$ is some variable whose value has not yet been determined. Examples of this would be tomorrow's opening price for Microsoft stock or the location of the next terrorist attack or how many nuclear devices North Korea will have in two years. Here our collection of data $(A[y_q] = x_1, A[y_q] = x_2, A[y_q] = x_2, \ldots, A[y_q] = x_n)$ is such that $A[y_q] = x_j$ indicates the jth source or experts conjecture as to the value of $A[y_q]$. Here we are interested in using this data to predict the value of the future variable $A[y_q]$. While formally almost the same as our problem we believe the indeterminate nature of the future variable introduces some aspects which can effect the mechanism we use to fuse the individual data. For example our tolerance for conflict between $A[y_q] = x_1$ and $A[y_q] = x_2$ where $x_1 \neq x_2$ may become greater. This greater tolerance may be a result of the fact that each source may be basing their predictions on different assumptions about the future world.

Let us now focus on our problem the multi-source data fusion problem. The process of data fusion is initiated by a users request to our sources of information for information about the value of the variable $A[y_q]$. In the following instead using $A[y_q]$ to indicate our variable of interest we shall more simply refer to the variable as V. We assume the value of V lies in the set X. We assume a collection $S_1, S_2, \ldots, S_q$ of information sources. Each source provides a value which we call our data. The problem here becomes the fusion of these pieces of data to obtain a value appropriate for the user's requirements. The approaches and methodologies available for solving this problem depend upon various considerations some of which we shall outline in the following sections. In Fig. 1 we provide a schematic framework of this multi-source data fusion problem which we use as a basis for our discussion.

Our fusion engine combines the data provided by the information sources using various types of knowledge it has available to it. We emphasize that the fusion process involves use of both the data provided by the sources as well as other knowledge. This other knowledge includes both context knowledge and user requirements.

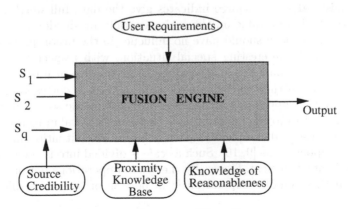

Fig. 1. Schematic of Data Fusion

2 Considerations in Data Fusion

Here we discuss some considerations that effect the mechanism used by the fusion engine. One important consideration in the implementation of the fusion process is related to the form, with respect to its certainty, with which the source provides its information. Consider the problem of trying to determine the age of John. The most certain situation is when a source reports a value that is a member of X, John's age is 23. Alternatively the reported value can include some uncertainty. It could be a linguistic value such as John is "young." It could involve a probabilistic expression of the knowledge. Other forms of uncertainty can be associated with the information provided. We note that fuzzy measures [5, 6] and Dempster-Shafer belief functions [7, 8] provide two general frameworks for representing uncertainty information. Here we shall assume the information provided by a source is a specific value in the space X.

An important of the fusion process is the inclusion of source credibility information. Source credibility is a user generated or sanctioned knowledge base. It associates with the data provided by a source a weight indicating its credibility. The mechanism of assignment of credibility weight to the data reported by a source can be involve various degrees of sophistication. For example, degrees of credibility can be assigned globally to each of the sources. Alternatively source credibility can be dependent upon the type of variable involved. For example, one source may be very reliable with information about ages while not very good with information about a person's income. Even more sophisticated distinctions can be made, for example, a source could be good with information about high income people but bad about income of low people.

The information about source credibility must be at least ordered. It may or may not be expressed using a well defined bounded scale. Generally when the credibility is selected from a well defined bounded scale the assignment of the highest value to a source indicates give the data full weight. The assignment of the lowest value on the scale generally means don't use it. This implies the information should have no influence in the fusion process.

There exists an interesting special situation, with respect to credibility where some sources may be considered as disinformative or misleading. Here the lowest value on the credibility scale can be used to correspond to some idea of taking the "opposite" of the value provided by the source rather than assuming the data provided is of no use. This somewhat akin to the relationship between false and complementation in logic. This situation may require the use of a bipolar scale [9, 10]. Such a scale is divided into two regions separated by a neutral element. Generally the type of operations performed using values from these bipolar depend on from portion of the scale which it was drawn.

Central to the multi-source data fusion problem is the issue of conflict and its resolution. The proximity and reasonableness knowledge bases shown in Fig. 1 play important roles in the handling of this issue.

One form of conflict arises when we have multiple values of a variable which are not the same or even compatible. For example one source may say the age of Osama Bin Laden is 25 another may say he is 45 and another may say he is 85. We shall refer to this as data conflict. As we shall subsequently see the proximity knowledge base plays an important role in issues related to the adjudication of this kind of conflict.

There exists another kind of conflict, one that can occur even when we only have a single reading for a variable. This happens when a sources reported value conflicts with what we know to be the case, what is reasonable. For example, if in searching for the age of Osama Bin Laden, one of the sources reports that he is eighty years old. This conflicts with what we know to be reasonable. This is information which we consider to have a higher priority than any information provided by any of the sources. In this case our action is clear: we discount this observation. We shall call this a context conflict, it relates to a conflict with information available to the fusion process external to the data provided by the sources. The repository of this higher priority information what we have indicated as the knowledge of reasonableness in Fig. 1. This type of a priori context or domain knowledge can take many forms and be represented in different ways.

As an illustration of one method of handling this type of domain knowledge we shall assume our reasonableness knowledge base in the form of a mapping over the domain of V. More specifically a mapping $R : X \rightarrow T$ called the **reasonableness mapping**. We allow this to capture the information we have, external to the data, about the possibilities of the different values in X being the actual value of V. Thus for any $x \in X, R(x)$ indicates the degree of reasonableness of x. T can be the unit interval $I = [0, 1]$ where $R(x) = 1$ indicates that x is a completely reasonable value while $R(x) = 0$ means x is completely unreasonable. More generally T can be an ordered set $T = \{t_1, \ldots, t_n\}$. We should point out that the information contained in the reasonableness knowledge base can come from a number of modes. It can be directly related to object of interest. For example from picture of bin Laden in a newspaper dated 1980, given that we are now in 2004, it would clearly be unreasonable to assume that he is less than 24. Historical observations of human life expectancy would make it unreasonable to assume that bin Laden is over 120 years old. Commonsense knowledge applied to recent pictures of him can also provide information regarding the idea reasonableness regarding bin Laden's age. In human agents their use of a knowledge of reasonableness plays fundamental role in distinguishing high performers from lesser. With this in mind it is noted that the need for tools for simply developing and applying these types of reasonableness knowledge bases is paramount.

The reasonableness mapping R provides for the inclusion of information about the context in which we are performing the fusion process. Any data provided by a source should be acceptable given our external knowledge about the situation. The use of the reasonableness type of relationship clearly provides a very useful vehicle for including intelligence in the process.

In the data fusion process, this knowledge of reasonableness often interacts with the source credibility in an operation which we shall call reasonableness qualification. A typical application of this is described in the following. Assume we have a source that provides a data value a_i and it has credibility t_i. Here we use the mapping R to inject the reasonableness, $R(a_i)$, associated with the value a_i and then use it to modify t_i to give us z_i, the support for data value a_i that came from source S_i. The process of obtaining z_i from t_i and $R(a_i)$ is denoted $z_i = g(t_i, R(a_i))$, and is called **reasonableness qualification**. In the following we shall suppress the indices and denote this operator as $z = g(t, r)$ where $r = R(a)$. For simplicity we shall assume t and r are from the same scale.

Let us indicate some of the properties that should be associated with this operation. A first property universally required of this operation is monotonicity, $g(t_1, r_1) \geq g(t_2, r_2)$ if $t_1 \geq t_2$ and $r_1 \geq r_2$. A second property that is required is that if either t or r is zero, the lowest value on the scale, then $g(t, r) = 0$. Thus if we have no confidence in the source or the value it provides is not reasonable, then the support is zero. Another property that may be associated with this operation is symmetry, $g(t, r) = g(r, t)$. Although we may necessarily require this of all manifestations of the operation.

The essential semantic interpretation of this operation is one of saying that in order to support a value we desire it to be reasonable and emanating from a source in which we have confidence. This essentially indicates this operation is an "anding" of the two requirements. Under this situation a natural condition to impose is the $g(t, r) \leq \text{Min}[t, r]$. More generally we can use a t-norm [11] for g. Thus we can have $g(t, r) = \text{Min}[t, r]$ or using the product t-norm $g(t, r) = tr$.

Relationships conveying information about the congeniality[1] between values in the universe X in the context of their being the value of V play an important role in the development of data fusion systems. Generally these types of relationships convey information about the compatibility and interchangeability between elements in X and as such are fundamental to the resolution and adjudication of internal conflict. Without these relationships conflict can't be resolved. In many applications underlying congeniality relationships are implicitly assumed, a most common example is the use of least squared based methods. The use of linguistic concepts and other granulation techniques are based on these relationships [12, 13]. Clustering operations require these relationships. These relationships are related to equivalence relationships and metrics.

The **proximity relationship** [14, 15] is an important example of these relations. Formally a proximity relationship on a space X is a mapping Prox:

---

[1] We use this term to indicate relationships like proximity, similarity, equivalence or distance.

$X \times X \to T$ having the properties: 1. $\text{Prox}(x,x) = 1$ (**reflexive**) and 2. $\text{Prox}(y,x) = \text{Prox}(x,y)$ (**symmetric**). Here T is an ordered space having a largest and smallest element denoted 1 and 0. Often T is the unit interval. Intuitively the value $\text{Prox}(x,y)$ is some measure of degree to which the values x and y are compatible and non-conflicting with respect to context in which the user is seeking the value of V. The concept of metric or distance is related in an inverse way to the concept of proximity.

A closely related and stronger idea is the concept of similarity relationship as introduced by Zadeh [16, 17]. A similarity relationship on a space X is a mapping $\text{Sim}:X \times X \to T$ having the properties: **1)** $\text{Sim}(x,x) = 1$, **2)** $\text{Sim}(x,y) = \text{Sim}(y,x)$ & **3)** $\text{Sim}(x,z) \geq \text{Sim}(x,y) \wedge \text{Sim}(y,z)$. A similarity relationship adds the additional requirement of transitivity. Similarity relationships provide a generalization of the concept of equivalent relationships.

A fundamental distinction between proximity and similarity relationships is the following. In a proximity relationship x and y can be related and y and z can be related without having x and z being related. In a similarity relationship under the stated premise a relationship must also exist between x and z.

In situations in which V takes its value on a numeric scale then the bases of the proximity relationship is the absolute difference $|x - y|$. However the mapping of $|x - y|$ into $\text{Prox}(x,y)$ may be highly non-linear.

For variables having non-numeric values a relationship of proximity can be based on relevant features associated with the elements in the variables universe. Here we can envision a variable having multiple proximity relationships. As an example let V be the country in which John was born, its domain X is the collection of all the countries of the world. Let us see what types of proximity relationship can be introduced on X in this context. One can consider the continent in which a country lies as the basis of a proximity relationship, this would actually generate an equivalence relationship. More generally, the physical distance between the country can be the basis of a proximity relationship. The spelling of the country's name can be the basis of a proximity relationship. The primary language spoken in a country can be the basis of a proximity relationship. We can even envision notable topographic or geographic features as the basis of proximity relationships. Thus many different proximity relationships may occur. The important point here is that the association of a proximity relationship over the domain over a variable can be seen as a very creative activity. More importantly the choice of proximity relationship can play a significant role in the resolution of conflicting information.

A primary consideration that effects the process used by the fusion engine is what we shall call the compositional nature of the elements in the domain X of V. This characteristic plays an important role in determining the types of operations that are available in the fusion process. It determines what types of aggregations we can perform with the data provided by the sources. We shall distinguish between three types of variables with respect to this characteristic. The first type of variable is what we shall call celibate or nominal. These are variables for which the composition of multiple values is meaningless. An example of this type of variable is a person's name. Here the process of combining names is completely inappropriate. Here fusion can be based on matching and counting. A next more structured type of variable is an ordinal variable. For these types of variables these exists some kind of meaningful ordering of the members of the universe. An example of this is a variable corresponding to size which has as its universe {small, medium, large}. For these variables some kind of compositional process is meaning, combining small and large to obtain medium is meaningful. Here composition operations must be based on ordering. The most structured type of variable is a numeric variable. For these variables in addition to ordering we have the availability of all the arithmetic operators. This of course allows us a great degree of freedom and we have a large body of compositional operators.

3 Expressing User Requirements

The output of any fusion process must be guided by the needs, requirements and desires of the user. In the following we shall describe some considerations and features that can be used to define or express the requirements of the user.

An important consideration in the presentation of the output of the fusion process is the users level of conflict tolerance. Conflict tolerance is related to the **multiplicity** of possible values presented to the user. Does the user desire one unique value or is it appropriate to provide him with a few solutions or is the presentation of all the multi source data appropriate?

Another different, although closely related, issue focuses on the level of granulation of the information provided to the user. As described by Zadeh [18] a granule is a collection of values drawn together by proximity of various types. Linguistic terms such as cold and old are granules corresponding to a collection of values whose proximity is based on the underlying temperature scale. In providing information we must satisfy the user's required level of granularity for the task for which he is requiring the information. Here we are not referring to the number of solutions provided but the nature of each solution object. One situation is that in which each solution presented to the user must be any element from the domain X. Another possibility is one in which we can provide, as a single solution, a subset of closely related values. Presenting ranges of values is an example of this. Another situation is where use a vocabulary of linguistic terms to express solutions. For example if the task is to determine what jacket to wear being told that it is cold is sufficient. Using $a > b$ to indicate that a has larger granularity than b if we consider providing information where somebody lives we see

country > region > state > city.> building address > floor in building
    > apartment on floor.

Recent interest in ontologies [19] involves many aspects related to granulation.

Another issue related to the form of the output is whether output values presented to the user are required to be values that correspond to one supplied by a source as the input or can we blend source values using techniques such as averaging to construct new values that didn't appear in the input. A closely related issue is the reasonableness of the output. For example consider the attempt to determine the number of children that John has. Assume one source says 8 and another says 7, taking the average gives us 7.5. Well, clearly it is impossible for our John to have 7.5 children. For some purposes this may be an appropriate figure. In addition we should note the that sometimes the requirement for reasonableness may be different for the output than input.

Another feature of the output revolves around the issue of qualification. Does the user desire qualifications associated with suggested values or does he prefer no qualification? As we indicated data values inputted to a fusion system often have attached values of credibility, this being due to the credibility of the source and the reasonableness of the data provided. Considerations related to the presentation of this credibility arise regarding the requirements of the user. Are we to present weights of credibility with the output or present it without these weights? In many techniques, such as weighted averaging, the credibility weight gets subsumed in the fusion process.

In most cases the fusion process should be deterministic, a given informational situation should always result in the same fused value. In some cases we may allow for a non-deterministic, random mechanism in the fusion process. For example in situations in which some adversary may have some role in effecting the information used in the fusion process we may want to use randomization to blur and confuse the influence of their information.

4 A Framework for Multi-Source Data Fusion

Here we shall provide a basic framework in which to view and implement the data fusion process. We shall see that this framework imposes a number of properties that should be satisfied by a rational data fusion technology.

Consider a variable of interest V having an underlying universe X. Assume we have as data a collection of q assessment of this variable, $\{V = a_1, V = a_2, V = a_3, \ldots, V = a_q\}$ Each assessment is information supplied by one of our sources. Let a_i be the value provided by the source S_i. Our desire here is to fuse these values to obtain some value $\tilde{a} \in X$ as the fused value. We denote this as a $\tilde{a} = \text{Agg}(a_1, \ldots, a_n)$. The issue then becomes that of obtaining the operator Agg that fuses these pieces of data. One obvious requirement of such an aggregation operator is idempotency, if all $a_i = a$ then $\tilde{a} = a$.

In order to obtain acceptable forms for Agg we must conceptually look at the fusion process. At a meta level multi-source data fusion is a process in which the individual sources must agree on a solution that is acceptable to each of them, that is compatible with the data they each have provided.

Let a be a proposed solution, some element from X. Each source can be seen as "voting" whether to accept this solution. Let us denote $\text{Sup}_i(a)$ as the support for solution a from source i. We then need some process of combining the support for a from each of the sources. We let

$$\text{Sup}(a) = F(\text{Sup}_1(a), \text{Sup}_2(a), \ldots, \text{Sup}_q(a))$$

be the total support for a. Thus F is some function that combines the support from each of the sources. The fused value $\tilde{a}$ is then obtained as the value $a \in X$ that maximizes $\text{Sup}(a)$. Thus $\tilde{a}$ is such that $\text{Sup}(\tilde{a}) = \text{Max}_{a \in X}[\text{Sup}(a)]$. In some situations we may not have to search through the whole space X to find an element $\tilde{a}$ having the property $\text{Sup}(\tilde{a}) = \text{Max}_{a \in X}[\text{Sup}(a)]$.

We now introduce the ideas of solution set and minimal solution set which may be useful We say that a subset G of X is a **solution set** if all a s.t. $\text{Sup}(a) = \text{Max}_{a \in X}[\text{Sup}(a)]$ are contained in G. The determination of G is useful in describing the nature of the type of solution we can expect from a fusion process. We shall say that a subset H of X is a **minimal solution set** if there always exists one element $a \in H$ s.t. $\text{Sup}(a) = \text{Max}_{a \in X}[\text{Sup}(a)]$. Thus a minimal solution set is a set in which we can always find an acceptable fused value. The determination of a minimal solution set can help reduce the task of searching.

Let us consider some properties of F. One natural property associated with F is that the more support from the individual sources the more overall support for a. Formally if a and b are two values and if $\text{Sup}_i(a) \geq \text{Sup}_i(b)$ for all i then $\text{Sup}(a) \geq \text{Sup}(b)$. This requires that F be a monotonic function, $F(x_1, x_2, \ldots, x_q) \geq F(y_1, y_2, \ldots, y_q)$ if $x_i \geq y_i$ for all i. A slightly stronger requirement is **strict monotonicity**. This requires that F be such that if $x_i \geq y_i$ for all i and there exists at least one i such that $x_i > y_i$ then $F(x_i, \ldots, x_q) > F(y_1, \ldots, y_q)$.

Another condition we can associate with F is a symmetry with respect to the arguments. That is the indexing of the arguments should not affect the answer. This symmetry implies a more expansive situation with respect to monotonicity. Assume $t_1, \ldots, t_q$ and $\hat{t}_1, \ldots, \hat{t}_q$ are two sets of arguments of F, $\text{Sup}_i(a) = t_i$ and $\text{Sup}_i(\hat{a}) = \hat{t}_i$. Let perm indicate a permutation of the arguments, where $\text{perm}(i)$ is the index of the ith element under the permutation. Then if there exists some permutation such that $t_i \geq \hat{t}_{\text{perm}(i)}$ for all i we get

$$F(t_1, \ldots, t_q) \geq F(\hat{t}_1, \ldots, \hat{t}_q).$$

Let us look further into this framework. A source's support for a solution, $\text{Sup}_i(a)$, should depend upon the degree of compatibility between the proposed solution a and the value provided by the source, a_i. Let us denote $\text{Comp}(a, a_i)$ as this compatibility. Thus $\text{Sup}_i(a)$ is some function of the compatibility between a_i and a. Furthermore, we have a monotonic type of relationship. For any two values a and b if $\text{Comp}(a, a_i) \geq \text{Comp}(b, a_i)$ then $\text{Sup}_i(a) \geq \text{Sup}_i(b)$.

The compatibility between two objects in X is based upon some underlying proximity relationship. The concept of a proximity relationship, which we introduced earlier, has been studied is the fuzzy set literature [20]. Here then we shall assume a relationship Comp, called the compatibility relationship, which has at least the properties of a proximity relationship. Thus Comp: $X \times X \rightarrow T$ in which T is an ordered space with greatest and least elements denoted 1 and 0 and having the properties: 1) $\text{Comp}(x,x) = 1$ and 2) $\text{Comp}(x,y) = \text{Comp}(y,x)$. A suitable although not necessary, choice for T is the unit interval.

We see that this framework imposes an idempotency type condition on the aggregation process. Assume $a_i = a$ for all i. In this case $\text{Comp}(a, a_i) = 1$ for all i. From this it follows that for any $b \in X$ $\text{Comp}(a, a_i) \leq \text{Comp}(b, a_i)$ hence $\text{Sup}_i(a) \geq \text{Sup}_i(b)$ for all b thus $\text{Sup}(a) \geq \text{Sup}(b)$ for all a. Thus there can never be a better solution than a. Furthermore, if F is assumed strictly monotonic and Comp is such that $\text{Comp}(a,b) \neq 1$ for $a \neq b$ then we get a strict idempotency.

5 Compatibility Relationships

What is important to emphasize here is that by basing our fusion process on the idea of the compatibility relationship we can handle, **in a unified manner,** the fusion of variables whose values are drawn from sets (universes) having widely different properties. Consider the variables John's age and John's city of residence. These variables take their values from sets of a completely different nature. Age is drawn from a purely mathematical set possessing all the structure that this affords, we can add or subtract or multiply elements. The city of residence has none of these properties. Its universe is of a completely different nature. What is also important to emphasize is that in order to use this approach on a variable V we must be able to obtain an appropriate context sensitive compatibility relation over its domain X. It is in this process of obtaining the compatibility relationship that we make use of the nature, the features and properties, of the elements in X. The construction of the compatibility relationship is often an extremely subjective task and greatly effects the end result. While in the numeric variables the basic feature used to form $\text{Comp}(a,b)$ is related to the difference $|a - b|$ this may be very complicated. For example the compatibility between salaries of 20 million and 30 million may be greater then the compatibility between salaries of 30 thousand and 50 thousand. While in the case numeric variables where the only feature of the elements in the domain useful for constructing the compatibility relationship is the numeric value in the case of other variables such as the country of residence the elements in the domain X have a number of features that can be used as the basis of an underlying compatibility relationship. This leads to the possibility of having multiple available compatibility relationships in our fusion process. While in the remainder of our work we shall assume the fusion process is based on one well defined compatibility relationship we would like to describe one generalization related to the situation of having the availability of multiple compatibility relations over the domain of the variable of interest. Earlier we indicated that the fused value is $\tilde{a}$ such $\mathrm{Sup}(\tilde{a}) = \mathrm{Max}_{a \in X}[\mathrm{Sup}(a)]$. In the case of multiple possible compatibility relations C_k for $k = 1$ to m then if we let $\mathrm{Sup}(a)/k$ indicate the Sup for a under compatibility relation C_k the process of obtaining the fused value may involve finding $\tilde{a}$ and compatibility relation C_{k*} such that $\mathrm{Sup}(\tilde{a})/k* = \mathrm{Max}_k[\mathrm{Max}_{a \in X}[\mathrm{Sup}(a)/k]]$.

At a formal level compatibility relations are mathematical structures that well studied and characterized. We now look at some very important special examples of compatibility relationships. We particularly focus on the properties of the solution sets that can be associated with relations. This helps us understand the nature of the fused values we may obtain. In the following discussion we shall let B be the set of all the values provided by the sources, $B = \{a_j \mid V = a_j \text{ for some source}\}$.

First we consider a very strict compatibility relation. We assume $\mathrm{Comp}(a, b) = 1$ if $a = b$ and $\mathrm{Comp}(a, b) = 0$ if $a \neq b$. This is a very special kind of equivalence relationship, elements are only equivalent to themselves. It can be shown under the condition of monotonicity of F the minimal solution set is the set B. This means the fused value for this type of compatibility relation must be one the data points provided by the sources.

Consider now the case where Comp is an equivalence relationship, $\mathrm{Comp}(a, b) \in \{0, 1\}$ and $\mathrm{Comp}(a, a) = 1$, $\mathrm{Comp}(a, b) = \mathrm{Comp}(b, a)$ and if $\mathrm{Comp}(a, b) = 1$ and $\mathrm{Comp}(b, c) = 1$ the $\mathrm{Comp}(a, c) = 1$. It can be shown [21] in this case that B also provides a minimal solution set, no solution can be better than some element in B.

We turn to another type of compatibility relationship, one in which there exists some **linear ordering** on the space X which underlies the compatibility relation. Let L be a linear ordering on X where $x \underset{L}{\geq} y$ indicates that x is larger than y in the ordering. Let Comp be a compatibility relationship on X which in addition to being reflexive and symmetric is such that the closer two elements are in the ordering L the more compatible they are. More formally we assume that if $x \underset{L}{\geq} y \underset{L}{\geq} z$ then $\mathrm{Comp}(x, y) \geq \mathrm{Comp}(x, z)$. We say this connection between ordering and compatibility is strict if $x \underset{L}{\geq} y \underset{L}{\geq} z$ implies $\mathrm{Comp}(x, y) > \mathrm{Comp}(x, z)$. Again let B be the set of data values provided by the sources. Let a^* be the largest element in B with respect to the underlying ordering $\underset{L}{\geq}$ and let a_* be the smallest element in B with respect to the ordering. It can be shown that the subset H of H where $H = \{a \mid a_* \underset{L}{\leq} a \underset{L}{\leq} a^*\}$ is a minimal solution set. Thus under this type of compatibility relationship only requiring only that F is monotonic leads to the situation which our fused value will be found in the "interval of X" bounded by a_* and a^*. This is a very interesting and deep result. Essentially this is telling us that if we view the process of obtaining the fused value as an aggregation of the data, $a = \mathrm{Agg}(a_1, a_2, \ldots, a_q\}$ then Agg is a mean like operation.

6 Additional Requirement on F

We described the process of determining the fused value to a data collection $\langle a_1, \ldots, a_q \rangle$ as to be conceptually implemented by the following process:

(1) For any $a \in X$ obtain $\mathrm{Sup}_i(a) = \mathrm{Comp}(a, a_i)$
(2) Evaluate $\mathrm{Sup}(a) = F(\mathrm{Sup}_i(a), \ldots, \mathrm{Sup}_q(a))$
(3) Select as fused value the $\tilde{a}$ such that $\mathrm{Sup}(\tilde{a}) = \mathrm{Max}_{a \in X}[\mathrm{Sup}(a)]$

We explicitly made two assumptions about the function F, we assumed that F was symmetric the indexing of input information is not relevant and F is monotonic. An implicit assumption we made about F was an assumption of pointwiseness.

There exists another property we want to associate with F, it is closely related to the idea of self-identity discussed by Yager and Rybalov [22]. Assume that we have a data set $\langle a_1, \ldots, a_q \rangle$ and using our procedure we find that $\tilde{a}$ is the best solution $\mathrm{Sup}(\tilde{a}) \geq \mathrm{Sup}(x)$ for all x in X. Assume now that we are provided an additional piece of data a_{q+1} such that $a_{q+1} = \tilde{a}$, the new data suggests $\tilde{a}$ as its value. Then clearly $\tilde{a}$ should still be the best solution. We shall formalize this requirement. In the following we let $\tilde{a}$ and $\hat{a}$ be two possible solutions and let $\tilde{c}_i = \mathrm{Comp}(\hat{a}, a_i)$ and $\hat{c}_i = \mathrm{Comp}(\hat{u}, a_i)$. We note that if $a_{q+1} = \tilde{a}$ then $\tilde{c}_{q+1} \geq \hat{c}_{q+1}$ since

$$\tilde{c}_{q+1} = \mathrm{Comp}(\tilde{a}, a_{q+1}) = \mathrm{Comp}(\tilde{a}, \tilde{a}) = 1 \geq \mathrm{Comp}(\hat{a}, \tilde{a})$$
$$\geq \mathrm{Comp}(\hat{a}, a_{q+1}) = \hat{c}_{q+1}$$

Using this we can more formally express our additional requirement on F. If

$$F(\tilde{c}_1, \ldots, \tilde{c}_q) \geq F(\hat{c}_i, \ldots, \hat{c}_q)$$

and if $\tilde{c}_{q+1} \geq \hat{c}_{q+1}$ then we require that

$$F(\tilde{c}_1, \ldots, \tilde{c}_q, \tilde{c}_{q+1}) \geq F(\hat{c}_i, \ldots, \hat{c}_q, \hat{c}_{q+1}) \,.$$

We note that this last condition is not exactly a standard monotonicity condition. We call this property stepwise monotonicity. We now have specified four conditions on F: pointwise, monotonicity, symmetry and stepwise monotonicity.

Let us now consider the issue of providing some formulations for F that manifest the conditions we require. Before we do this we must address the measurement of compatibility. In our work so far we have assumed a very general formulation for this measurement. We have defined Comp: $X \times X \to T$ in which T is an ordered space with greatest and least elements denoted 1 and 0. Let us consider the situation in which T has only an ordering. In this case one form for F is that of a Max operator. Thus $F(t_1, t_2, \ldots, t_q) = \mathrm{Max}_i[C_i]$ satisfies all the conditions required. We also note that the Min operator satisfies our conditions.

If we consider the situation in which the compatibility relation takes its values in the unit interval, $[0, 1]$ one formulation for F that meets all our required conditions is the sum or totaling function, $F(x_1, x_2, \ldots x_q) = \sum_{i=1}^{q} x_i$. Using this we get $\text{Sup}(a) = \sum_{i=1}^{q} \text{Sup}_i(a) = \sum_{i=1}^{q} \text{Comp}(a, a_i)$. Thus our fused value is the element that maximizes the sum of its compatibilities with the input.

7 Credibility Weighted Sources

In the preceding we have implicitly assumed all the data had the same credibility. Here we shall consider the situation in which each data has a credibility weight w_i. Thus now our input is q pairs of (w_i, a_i). We also note that the weight w_i must be drawn from a scale that has at least an ordering. In addition we assume this scale has minimal and maximal elements denoted 0 and 1.

Again in this situation for any $a \in X$ we calculate $\text{Sup}(a) = F(\text{Sup}_1(a), \ldots, \text{Sup}_q(a))$ where $\text{Sup}_i(a)$ is the support for **a** from the data supplied by source $i, (w_i, a_i)$. However in this case, $\text{Sup}_i(a)$ depends upon two components. The first being the compatibility of a with a_i, $\text{Comp}(a, a_i)$ and the second being the weight or strength of credibility source i. Thus in this case

$$\text{Sup}_i(a) = g(w_i, \text{Comp}(a, a_i))$$

Ideally we desire that both w_i and $\text{Comp}(a, a_i)$ be drawn from the same scale, which has at least an ordering. For the following discussion we shall not implicitly make this assumption. However, we shall find it convenient to use 0 and 1 to indicate the least and greatest element on each of the scales. We now specify the properties that are required of the function g. A first property we require of g is monotonicity with respect to both of the arguments: $g(x, y) \geq g(z, y)$ if $x > z$ and $g(x, y) \geq g(x, w)$ if $y > w$. Secondly we assume that zero credibility or zero compatibility results in zero support: $g(x, 0) = g(0, y) = 0$ for all x and y. We see that g has the character of an "and" type operator. In particular at a semantic level we see that we are essentially saying is "source i provides support for solution if the source is credible and the solution is compatible with the sources data".

With this we see that $g(1, 1) = 1$ and $g(x, y) \neq 0$ if $x \neq 0$ and $y \neq 0$. We must make one further observation about this process with respect to source credibility. Any source that has zero credibility should in no way effect the decision process. Thus if $((w_1, a_1), \ldots, (w_q, a_q))$ has as its fused value $\tilde{a}$ then the data $((w_1, a_1), \ldots, (w_q, a_q), (w_{q+1}, a_{q+1}))$ where $w_{q+1} = 0$ should also have the same result. With this understanding we can discard any source with zero credibility. In the following we shall assume unless otherwise stated all sources have non-zero credibility.

8 Including Reasonableness

In an early part we introduced the idea of a **R**easonableness **K**nowledge **B**ase (RKB) and indicated its importance in the data fusion process. Formally we use this structure to introduce into the fusion process any information we have about the value of the variable exclusive of the data provided by the sources. The information in the reasonableness knowledge base will affect our proposed fusion process in at least two ways. First it will interact with the data provided by the sources. In particular, the weight (credibility) associated with a source providing an unreasonable input value should be diminished. This results in our giving the data less importance in the fusion process. Secondly some mechanism should be included in the fusion process to block unreasonable values from being provided as the fused value.

A complete discussion of the issues related to the construction of the RKB and those related to formal methods for the interaction of the RKB with the data fusion process is complex and beyond our immediate aim as well as well being beyond our complete understanding at this time. In many ways the issue of reasonableness goes to the very heart of intelligence. Here we shall focus on the representation of a specific type of knowledge effecting what are reasonable values for a variable and suggest a method for introducing this in the fusion process.

We shall distinguish between two types of information about the value of a variable with the terms intimate and collective knowledge. Before making this distinction we recall a variable V is formally denoted as $A(y)$ where A is an attribute and y is a specific object. For example if the variable is *John's age* then age is the attribute and John is the object. By intimate knowledge we mean information directly about the variable whose value we are trying obtain. Knowing that John was born after Viet Nam war or that Mary lives in Montana are examples of intimate knowledge. By collective knowledge we mean information about the value of the attribute for a class of objects in which our object of interest lies. Knowing that Singaporeans typically are college graduates is collective knowledge while knowing that Min-Sze has a PhD is intimate knowledge. Generally intimate knowledge has a possibilistic nature while collective knowledge has a probabilistic nature. (The preceding statement is an example of collective knowledge). Another type of knowledge related to reasonableness is what has been called default (commonsense) knowledge [23, 24]. This knowledge is such that while we have not been given intimate knowledge that xyz is the value of a variable we can act as if this is the case unless we have some overriding intimate knowledge saying that this is not the case. One view of default knowledge is that it is collective knowledge that is so pervasively true from a pragmatic point of view it is more economical to act as if it is categorical, holds for all objects, and deal with exceptions as they are pointed out.

Here we consider only the situation in which our knowledge about reasonableness is intimate and can be expressed by fuzzy subset, a mapping

$R : X \to T$. As pointed out by Zadeh [25] this kind of knowledge induces a constraint on the values of the variable and has a possibilistic nature [26]. Here for any $x \in X, R(x)$ indicates the reasonableness (or possibility) that x is the value of the variable V. For example, if our interest is to obtain John's age and before soliciting data from external sources we know from our personal interview that John is *young* then we can capture this information using the fuzzy subset R corresponding to and thus constrain the values that are reasonable.

Let us see how we can include this information into our data fusion process. In the following we assume that T is a linear ordering having maximal and minimal elements, usually denoted 1 and 0. Assume the data provided by source i is denoted a_i and w_i is the credibility assigned to source i. We assume these credibilities are measured on the same scale as the reasonableness, T. In the fusion process the importance weight, u_i, assigned to the data a_i should be a function of the credibility of the source, w_i, **and** the reasonableness of the data, $R(a_i)$. An unreasonable value, whatever the credibility of the source, should not be given much significance in the fusion. Similarly a piece of data coming from a source with low credibility, whatever the reasonableness of its value, should not be given much significance in the fusion. Using the Min to implement this "anding" we obtain $u_i = \text{Min}[R(a_i), w_i]$ as the importance weight assigned to the data a_i coming from this source. In this environment the information that goes to the fusion mechanism is the collection $\langle (u_1, a_1), \ldots, (u_q, a_q) \rangle$.

As in the preceding the overall support for a proposed fused value a should be a function its support from each of the sources, $\text{Sup}(a) = F(\text{Sup}_i(a), \ldots, \text{Sup}_q(a))$. The support provided from source i for solution a should depend on the importance weight u_i assigned to data supplied by source i as well as the compatibility of the data a_i and the proposed fused value, $\text{Comp}(a, a_i)$. In addition we should also include information about the reasonableness of the proposed solution a. Here then for a solution a to get support from source i it should be compatible with the data a_i and compatible with what we consider to be reasonable, $\text{Comp}(a, R)$. Here then we let $\text{Comp}_i(a) = \text{Comp}(a, a_i) \wedge \text{Comp}(a, R)$. Furthermore $\text{Comp}(a, R) = R(a)$ hence $\text{Comp}_i(a) = \text{Comp}(a, a_i) \wedge R(a)$. In addition, as we have indicated, the support afforded any solution by source i should be determined in part by the importance weight assigned i. Taking these considerations into account we get $\text{Sup}_i(a) = g(u_i, \text{Comp}_i(a))$. Substituting our values we get

$$\text{Sup}_i(a) = g(w_i \wedge R(a_i), \text{Comp}(a, a_i) \wedge R(a))$$

What is clear is that g should be monotonically increasing in both its arguments and be such that if any of the arguments are 0 then $\text{Sup}_i(a) = 0$. In the case where we interpret g as implementing an *anding* and using the Min operator as our *and* we get $\text{Sup}_i(a) = w_i \wedge R(a_i) \wedge R(a) \wedge \text{Comp}(a, a_i)$. Here we observe that the support afforded from source i to any proposed fused solution is related to the credibility of the source, the reasonableness of value provided by the source, the reasonableness of the proposed fusion solution and the compatibility of the data and solution.

Earlier we looked at the form of solution set for the fused value under different assumptions about the underling compatibility relationship. Let us now investigate how the introduction of reasonableness affects our results about boundedness and minimal solution sets. For simplicity neglect the issue of source credibility, we assume all sources are fully credible.

Consider the case in which our underlying compatibility relationship is very strict, $\text{Comp}(x, y) = 1$ iff $x = y$ and $\text{Comp}(x, y) = 0x \neq y$. Let B be the set of data values and let $\hat{B}$ be the subset of B such that $b \in \hat{B}$ if $R(b) \neq 0$, it is the set of reasonable data values. If $a \notin B$ then $\text{Comp}(a, a_i) = 0$ for all a_i and hence $\text{Sup}_i(a) = 0$ for all i. Let $d \in B - \hat{B}$, here $R(d) = 0$ and again we get that $\text{Sup}_i(d) = 0$ for all i. On the other hand for $b \in \hat{B}$ then $R(b) \neq 0$ and $b = a_j$ for some j and hence $\text{Sup}_j(b) > 0$. Thus we see that we will always find our solution in the space $\hat{B}$, the set of data values that are not completely unreasonable. Actually in this case for each $b \in \hat{B}$ its overall support is the number of sources that provided this value.

Consider now the case in which Prox is an ordinary equivalence relation. Again let $\hat{B}$ be our set of input data which have some degree of reasonableness. Let E_i be the equivalence class of a_i, for all $y \in E_i$, $\text{Prox}(y, a_i) = 1$. Let $E = \bigcup_i E_i$, the union of all equivalence classes that have input value. If $a \notin E$ then $\text{Prox}(a, a_i) = 0$ for all i. From this we see that if $a \notin E$ then $\text{Sup}_i(a) = 0$ for all i and hence we can always find at least as good a solution in E. We can obtain a further restriction on the minimal solutions. Let $D_i \subseteq E_i$ be such that $d_i \in D_i$ if $R(d_i) = \text{Max}_{x \in F_i}(R(x))$. Thus D_i is the subset of elements that are equivalent to a_i and are most reasonable. For any $d_i \in D_i$ and any $e_i \in E_i$ we have that for all input data a_j $\text{Comp}(e_i, a_j) = \text{Comp}(d_i, a_j)$. Since $R(d_i) \geq R(e_i)$ we see that $\text{Sup}_j(d_i) \geq \text{Sup}_j(e_i)$ for all j. Hence d_i is always at least as good a fused value as any element in E_i. Thus we can always find a fused solution in $D = \bigcup_i D_i$. Furthermore if x and $y \in D_i$ then $R(x) = R(y)$ and $\text{Comp}(x, z) = \text{Comp}(y, z)$ for all z. Hence $\text{Sup}_i(x) = \text{Sup}_i(y)$. Thus $\text{Sup}(x) = \text{Sup}(y)$. The result is that we can consider any element in D_i. Thus all we need consider is the set $\tilde{D} = \bigcup_i \{\tilde{d}_i\}$ where $\tilde{d}_i$ is any element from D_i. We note that if $a_i \in D_i$ then this is of course the preferred element.

We now consider the case where the proximity relationship is based on a linear ordering L over space X. Let B be the set of data values provided by the sources. Let x^* and x_* be the maximal and minimal elements in B with respect to the ordering L. Let H be the set of x_j so that $x^* \underset{L}{\geq} x_j \underset{L}{\geq} x_*$. In the preceding we showed that we can always find a fused value element a in H. We now show that the introduction of reasonableness removes this property.

In the preceding we indicated that for any proposed fused value we get that $\text{Sup}_i(a) = g(u_i, \text{Comp}_i(a))$ where g monotonic in both the arguments, $u_i = w_i \wedge R(a_i)$ and $\text{Comp}_i(a) = R(a) \wedge \text{Comp}(a, a_i)$. We shall now show that here we can have an element $a \notin H$ in which $\text{Sup}_i(a) \geq \text{Sup}_i(b)$ for all $b \in H$.

This implies that we can't be guaranteed of finding the fused value in H. Consider now the case in which there exists $b \in H$ for which $R(b) \leq \alpha$. In this case $\operatorname{Sup}_i(b) = g(u_i, R(b) \wedge \operatorname{Comp}(b, a_i)) \leq g(u_i, \alpha)$. Let $a \notin H$ be such that $R(a) > \alpha$. For this element we get $\operatorname{Sup}_i(a) = g(u_i, R(a) \wedge \operatorname{Comp}(a, a_i))$. If $\operatorname{Comp}(a, a_i) > \alpha$ then $R(a) \wedge \operatorname{Comp}(a, a_i) = \beta$ then $\beta > \alpha$ and hence $\operatorname{Sup}_i(a) = g(u_i, \beta) \geq g(u_i, \alpha) = \operatorname{Sup}_i(b)$ and then it is not true we can eliminate a as a solution. Thus we see that the introduction of this reasonableness allows for the possibility of solutions not bounded by the largest and smallest of input data.

An intuitive boundary condition can be found in this situation. Again let H be the subset of X bounded by our data: $H = \{x \mid x^* \succeq_L x \succeq_L x_*\}$ where let $\alpha^* = R(x_*)$ and let $\alpha_* = R(x^*)$. Let $H^* = \{x \mid x \succ_L x^*$ and $R(x) > R(x^*)\}$ and let $H_* = \{x \mid x \prec_L x_*$ and $R(x) > R(x_*)\}$. Here we can restrict ourselves to looking for the fused value in the set $\hat{H} = H \cup H_* \cup H^*$. We see that as follows. For any $x \succeq_L x^*$ we have, since the proximity relationship is induced by the ordering, that $\operatorname{Comp}(x, a_i) \leq \operatorname{Comp}(x^*, a_i)$ for all data a_i. If in addition we have that $R(x) \leq R(x^*)$ then $\operatorname{Sup}_i(x) = g(u_i, R(x) \wedge \operatorname{Comp}(x, a_i)) \leq \operatorname{Sup}_i(x^*) = g(u_i, R(x^*) \wedge \operatorname{Comp}(x^*, a_i))$ for all i and hence $\operatorname{Sup}(x) \leq \operatorname{Sup}(x^*)$. Thus we can eliminate all $x \succeq_L x^*$ having $R(x) \leq R(x^*)$. Using similar arguments we can eliminate $x \prec_L x_*$ which have $R(x) \leq R(x_*)$.

9 Conclusion

We presented a general view of the multi-source data fusion process and described some of the considerations and information that must go into the development of a data fusion algorithm. Features playing a role in expressing users requirements were also discussed. We introduced a general framework for data fusion based on a voting like process which made use of compatibility relationships. We described several important examples of compatibility relationships. We showed that our formulation resulted in specific bounding conditions on the fused value depending on the underlying compatibility relationships. We noted the existence of these bounding conditions essentially implied that the fusion process has the nature of a mean type aggregation. We presented the concept of reasonableness as a means for including in the fusion process any information available other then that provided by the sources. We considered the situation in which we allowed our fused value to be granular objects such as linguistic terms or subsets.

References

1. Berry, M. J. A. and Linoff, G., Data Mining Techniques, John Wiley & Sons: New York, 1997.

2. Dunham, M., Data Mining, Prentice Hall: Upper Saddle River, NJ, 2003.
3. Han, J. and Kamber, M., Data Mining: Concepts and Techniques, Morgan Kaufmann: San Francisco, 2001.
4. Mitra, S. and Acharya, T., Data Mining: Multimedia. Soft Computing and Bioinformatics, New York: Wiley, 2003.
5. Murofushi, T. and Sugeno, M., "Fuzzy measures and fuzzy integrals," in Fuzzy Measures and Integrals, edited by Grabisch, M., Murofushi, T. and Sugeno, M., Physica-Verlag: Heidelberg, 3–41, 2000.
6. Yager, R. R., "Uncertainty representation using fuzzy measures," IEEE Transaction on Systems, Man and Cybernetics 32, 13–20, 2002.
7. Shafer, G., A Mathematical Theory of Evidence, Princeton University Press: Princeton, N.J., 1976.
8. Yager, R. R., Kacprzyk, J. and Fedrizzi, M., Advances in the Dempster-Shafer Theory of Evidence, John Wiley & Sons: New York, 1994.
9. Yager, R. R. and Rybalov, A., "Uninorm aggregation operators," Fuzzy Sets and Systems 80, 111–120, 1996.
10. Yager, R. R., "Using a notion of acceptable in uncertain ordinal decision making," International Journal of Uncertainty, Fuzziness and Knowledge-Based Systems 10, 241–256, 2002.
11. Klement, E. P., Mesiar, R. and Pap, E., Triangular Norms, Kluwer Academic Publishers: Dordrecht, 2000.
12. Zadeh, L. A., "Toward a theory of fuzzy information granulation and its centrality in human reasoning and fuzzy logic," Fuzzy Sets and Systems 90, 111–127, 1997.
13. Lin, T. S., Yao, Y. Y. and Zadeh, L. A., Data Mining, Rough Sets and Granular Computing, Physica-Verlag: Heidelberg, 2002.
14. Kaufmann, A., Introduction to the Theory of Fuzzy Subsets: Volume I, Academic Press: New York, 1975.
15. Bouchon-Meunier, B., Rifqi, M. and Bothorol, S., "Towards general measures of comparison of objects," Fuzzy Sets and Systems 84, 143–153, 1996.
16. Zadeh, L. A., "Similarity relations and fuzzy orderings," Information Sciences 3, 177–200, 1971.
17. Yager, R. R., Ovchinnikov, S., Tong, R. and Nguyen, H., Fuzzy Sets and Applications: Selected Papers by L. A. Zadeh, John Wiley & Sons: New York, 1987.
18. Zadeh, L. A., "Toward a logic of perceptions based on fuzzy logic," in Discovering the World with Fuzzy Logic, edited by Novak, W. and Perfilieva, I., Physica-Verlag: Heidelberg, 4–28, 2001.
19. Gomez-Perez, A., Fernandez-Lopez, M. and Corcho, O., Ontological Engineering, Springer: Heidelberg, 2004.
20. Shenoi, S. and Melton, A., "Proximity relations in fuzzy relational databases," Fuzzy Sets and Systems 31, 287–298, 1989.
21. Yager, R. R., "A framework for multi-source data fusion," Information Sciences 163, 175–200, 2004.
22. Yager, R. R. and Rybalov, A., "Noncommutative self-identity aggregation," Fuzzy Sets and Systems 85, 73–82, 1997.
23. Reiter, R., "A logic for default reasoning," Artificial Intelligence 13, 81–132, 1980.
24. McCarthy, J., "Applications of circumscription to formalizing common sense knowledge," Artificial Intelligence 28, 89–116, 1986.

25. Zadeh, L. A., "Outline of a computational theory of perceptions based on computing with words," in Soft Computing and Intelligent Systems, edited by Sinha, N. K. and Gupta, M. M., Academic Press: Boston, 3–22, 1999.
26. Zadeh, L. A., "Fuzzy sets as a basis for a theory of possibility," Fuzzy Sets and Systems 1, 3–28, 1978.

Granular Nested Causal Complexes

Lawrence J. Mazlack

Applied Computational Intelligence Laboratory, University of Cincinnati,
Cincinnati, Ohio 45221-0030
mazlack@uc.edu

Abstract. Causal reasoning occupies a central position in human reasoning. In many ways, causality is granular. This is true for: perception, commonsense reasoning as well as for mathematical and scientific theory. At a very fine-grained level, the physical world itself may be made up out of granules. Knowledge of at least some causal effects is imprecise. Perhaps, complete knowledge of all possible factors might lead to a crisp description of whether an effect will occur. However, in the commonsense world, it is unlikely that all possible factors can be known. Commonsense understanding of the world deals with imprecision, uncertainty and imperfect knowledge. In commonsense, every day reasoning, we use approaches that do not require complete knowledge. Even if the precise elements of the complex are unknown, people recognize that a complex collection of elements can cause a particular effect. They may not know what events are in the complex; or, what constraints and laws the complex is subject to. Sometimes, the details underlying an event can be known to a fine level of detail, sometimes not. Usually, commonsense reasoning is more successful in reasoning about a few large-grain sized events than many fine-grained events. Perhaps, a satisficing solution would be to develop large-grained solutions and then only go to the finer-grain when the impreciseness of the large-grain is unsatisfactory. An algorithmic way of handling causal imprecision is needed. Perhaps fuzzy Markov models might be used to build complexes. It may be more feasible to work on a larger-grained size. This may reduce the need to learn extensive hidden Markov models, which in computationally expensive.

Key words: Causality, commonsense, causal complex, granularity, satisficing

1 Introduction

Causal reasoning occupies a central position in human reasoning. It plays an essential role in human decision-making. Considerable effort has been spent examining causation. For thousands of years, philosophers, mathematicians, computer scientists, cognitive scientists, psychologists, economists, and others have formally explored questions of causation. Whether causality exists at all or can be recognized has long been a theoretical speculation of scientists and

Lawrence J. Mazlack: *Granular Nested Causal Complexes*, Studies in Computational Intelligence
(SCI) **5**, 23–22 (2005)
www.springerlink.com   © Springer-Verlag Berlin Heidelberg 2005

philosophers. At the same time, people operate on the commonsense belief that causality exists.

In many ways, causality is granular. This is true for commonsense reasoning as well as for more formal mathematical and scientific theory. At a very fine-grained level, the physical world itself may be granular. Our commonsense perception of causality is often large-grained while the underlying causal structures may be described in a more fine-grained manner.

Causal relationships exist in the commonsense world; for example:

When a glass is pushed off a table and breaks on the floor

it might be said that

Being pushed from the table *caused* the glass to break.

Although,

Being pushed from a table is not a **certain** *cause* of breakage; sometimes the glass bounces and no break occurs; or, someone catches the glass before it hits the floor.

Counterfactually, usually (but not always),

Not falling to the floor prevents breakage.

Sometimes,

A glass breaks when an errant object hits it, even though it does not fall from the table.

Positive causal relationships can be described as: *if α then β* (or, $\alpha \rightarrow \beta$). For example:

When an automobile driver fails to stop at a red light and there is an accident it can be said that the failure to stop was the accident's *cause*.

However, negating the causal factor does not mean that the effect does not happen; sometimes effects can be *overdetermined*. For example:

An automobile that did not fail to stop at a red light can still be involved in an accident; another car can hit it because the other car's brakes failed.

Similarly, simple negation does not work; both because an effect can be overdetermined and because negative statements are weaker than positive statements as negative statements can become *overextended*. It cannot be said that $\neg\alpha \rightarrow \neg\beta$, for example:

Failing to stop at a red light is not a **certain** *cause* of no accident occurring; sometimes no accident at all occurs.

Some describe events in terms of *enablement* and use counterfactual implication whose negation is implicit; for example [22]:

Not picking up the ticket *enabled* him to miss the train.

There is a multiplicity of definitions of enable and not-enable and how they might be applied. To some degree, logic notation definitional wars are involved. It is not in the interests of this paper to consider notational issues.

Negative causal relationships are less sure; but often stated; for example, it is often said that:

Not walking under a ladder prevents bad luck.

Or, usually (but not always),

Stopping for a red light avoids an accident.

In summary, it can be said that the knowledge of at least some causal effects is imprecise for both positive and negative descriptions. Perhaps, complete knowledge of all possible factors might lead to a crisp description of whether an effect will occur. However, it is also unlikely that it may be possible to fully know, with certainty, all of the elements involved. Consequently, the extent or actuality of missing elements may not be known. Additionally, some well described physics as well as neuro-biological events appear to be truly random [5]; and some mathematical descriptions randomly uncertain. If they are, there is no way of avoiding causal imprecision.

1.1 Complexes of Elements

In applying commonsense causal reasoning, it may be recognized that a complex collection of elements can be involved causally in a particular effect, even if the precise elements of the complex are unknown. It may not be known what events are in the complex; or, what constraints and laws the complex is subject to. Sometimes, the details underlying an event are known to a fine level of detail, sometimes not. Events are rarely known to the finest possible grain size. Furthermore, there appears to be limits as the grain size at which events are knowable.

Nested granularity may be applied to causal complexes. A complex may be several larger-grained elements. In turn, each of the larger-grained elements may be a complex of more fine-grained elements. Recursively, in turn, these elements may be made up still finer-grained elements. In general, people are more successful in applying commonsense reasoning to a few large-grain sized events than to many fine-grained elements that might make up a complex.

When using large-grained commonsense reasoning, people do not always need to know the extent of the underling complexity. This is also true for situations not involving commonsense reasoning; for example:

When designing an electric circuit, designers are rarely concerned with the precise properties of the materials used; instead, they are concerned with the devices functional capabilities and take the device as a largergrained object.

Complexes often may be best handled on a black-box, large-grained basis. It may be recognized that a fine-grained complex exists; but it is not necessary need to deal with the details internal to the complex.

1.2 Satisficing

People do things in the world by exploiting commonsense *perceptions* of cause and effect. Manipulating perceptions has been explored [44] but is not the focus of this paper. The interest here is how perceptions affect commonsense causal reasoning, granularity, and the need for precision.

When trying to precisely reason about causality, complete knowledge of all of the relevant events and circumstances is needed. In commonsense, every day reasoning, approaches are used that do not require complete knowledge. Often, approaches follow what is essentially a *satisficing* [32] paradigm. The use of non-optimal mechanisms does not necessarily result in ad hocism; [7] states:

> "Zadeh [43] questions the feasibility (and wisdom) of seeking for optimality given limited resources. However, in resisting naive optimizing, Zadeh does not abandon the quest for justifiability, but instead resorts to modifications of conventional logic that are compatible with linguistic and fuzzy understanding of nature and consequences."

Commonsense understanding of the world tells us that we have to deal with imprecision, uncertainty and imperfect knowledge. This is also the case with scientific knowledge of the world. An algorithmic way of handling imprecision is needed to computationally handle causality. Models are needed to algorithmically consider causes and effects. These models may be symbolic or graphic. A difficulty is striking a good balance between precise formalism and commonsense imprecise reality.

2 Complexes

When events happen, there are usually other related events. The entire collection of events can be called a complex. The events can be called the elements of the complex.

A "mechanism" [34] or a "causal complex" [11, 12] is a collection of events whose occurrence or non-occurrence results in a consequent event happening. Hobbs' causal complex is the *complete* set of events and conditions necessary for the causal effect (consequent) to occur. Hobbs suggests human casual reasoning that makes use of a causal complex does not require precise, complete knowledge of the complex. (Different workers may use the terms "mechanism and "causal complex" differently; I am using them as these author's use them.)

Each complex, taken as a whole, can be considered to be a granule. Larger complexes can be decomposed into smaller complexes; going from large-grained to small-grained. For example, when describing starting an automobile, A large-grained to small-grained, nested causal view would start with

> When an automobile's ignition switch is turned on, this *causes* the engine to start.

But, it would not happen if a large system of other nested conditions were not in place.

> There has to be available fuel. The battery has to be operational. The switch has to be connected to the battery so electricity can flow through it. The wiring has to connect the switch to the starter and ignition system (spark plugs, etc.). The engine has to be in good working order; and so forth.

Turning the ignition switch on is one action in a complex of conditions required to start the engine. One of the events might be used to represent the collection of equal grain sized events; or, a higher-level granule might be specified with the understanding that it will invoke a set of finer-grained events. In terms of nested granules, the largest grained view is: turning on the switch is the sole causal element; the complex of other elements represents the finer-grains. These elements in turn could be broken down into still finer-grains; for example, "available fuel" could be broken down into:

> fuel in tank, operating fuel pump, intact fuel lines, and so forth.

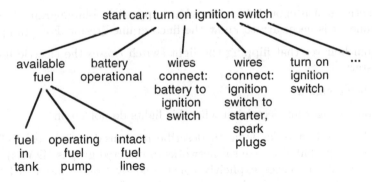

Fig. 1. Nested causal complex

Sometimes, it is enough to know what happens at a large-grained level; at other times it is necessary to know the fined grained result. For example, if

> Bill believes that turning the ignition key of his automobile *causes* the automobile to start.

It is enough if

> Bill engages an automobile mechanic when his automobile does not start when he turns the key on.

However,

> The automobile mechanic needs to know a finer-grained view of an automobile's causal complex than does Robin.

Instead of being concerned with all of the fined grained detail, a better approach may be to incorporate granulation using rough sets and/or fuzzy sets to soften the need for preciseness. And then accept impreciseness in the description. Each complex can be considered to be a granule. Larger complexes can be decomposed into smaller complexes. Thus, going from large-grained to small-grained.

Others are less explicitly top-down, large-grain to small-grain; although the examples have similar elements. Reference [11] uses the example of turning on an electric light. He says:

> "It is natural to say that when you flip a light switch, you cause the light to go on. But it would not happen if a whole large system of other conditions were not in place. The wiring has to connect the switch to the socket, and be intact. The light bulb has to be in good working order. The switch has to be connected to a system for supplying electricity. The power plant in that system has to be operational. And so on. Flipping the light switch is only the last small move in a large-scale systems of actions and conditions required for the light to go on."

Sometimes, it is enough to know what happens at a large-grained level; at other times it is necessary to know the fined grained result. For example, if

> Robin believes that flipping the light switch *causes* the electric light to start.

It may be enough if

> Robin engages an electrician when his lights do not go on.

Hobbs uses first order logic to describe his causal complexes. Reference [26] develops probabilistic causal networks of directed graphs (DAGs).

The causal complexes explicitly considered by Hobbs and Pearl have a required structure that may be overly restrictive for commonsense causal understanding, namely:

- If *all* of the events in the causal complex appropriately happen, then the effect will occur
- There is nothing in the causal complex that is irrelevant to the effect.

These requirements are probably too precise and extensive to be realized in a commonsense world. Sometimes, only some of the events need to happen; for example:

Someone may be able to save more money:

- If their taxes are lowered or
- If they earn more money.

Either even may lead to greater savings. However,

Neither may result in increased savings if they also have to pay a large divorce settlement.

So, if all of the events happen, the effect may happen. If some of the events happen, the effect may happen. In the commonsense world, we rarely whether all of the events are in a complex are necessary. For example,

A man may want to attract the attention of a woman. He may do a large number of things (e.g., hair, clothes, learn to dance, etc.). If he does attract the woman, he may never know which things were relevant and which were not

An issue is how to distinguish between what is in a complex and what is not. Another issue is how to distinguish between the things that deserve to be called "causes" and those that do not. Hobbs suggests that a consideration of causal complexes can be divided into:

- Distinguishing what events are in a causal complex from those outside of it. [18, 23, 26, 37, 38]
- Within a causal complex, recognizing the events that should be identified as causes from those that are not [33].

A major question concerning complexes is: To what extent can we increase the causal grain size and still have useful causal information. Conversely, can we start with a large-grained causal event and then derive the finer-grained structure? Can we measure and/or control the imprecision involved in changing grain size? If we start with a large-grained structure and resolve it, will our computational complexity burdens be reduced?

3 Defining Causality

Coming to a precise description of what is meant by causality is difficult. There are multiple and sometimes conflicting definitions. Reference [45] suggested that a precise, formal definition might not be possible. To this, [27] suggested: "For me, the adequacy of a definition lies not in abstract argumentation but in whether the definition leads to useful ways of solving concrete problems." Regardless, we do have a commonsense belief that there are causal relationships. Satisfactorily and explicitly specifying them is difficult. This paper is not dependent on precise notation; it argues from a commonsense view and occasionally comments on more formal notations.

Reference [18] defined causality depends on one-way, time ordered conception of causality. In contrast, [37, 38] provides an analysis of causality that does not rely on time order. Some believe [10] 1 that causal relations are mostly indicated by asymmetric relationships. An abbreviated version of the relationships that Hausman lists is:

- *Time-order:* Effects do not come before causes – This corresponds with commonsense understanding. Unfortunately, it is at variance with Einsteinium space-time. This variance raises the interesting question: If there is a committal to commonsense reasoning, how is it handled when commonsense reasoning differs from scientific understanding?
- *Probabilistic Independence*
- *Agency* or *manipulability:* Causes can be used to manipulate their effects, but effects cannot be used to manipulate their causes. Effects of a common cause cannot be used to manipulate one another.
- *Counterfactual dependence:* Effects counterfactually depend on their causes, while causes do not counterfactually depend on their effects.
- *Overdetemination:* Effects over determine their causes, while causes rarely over determine their effects
- *Invariance:* Dependent variables in an equation are effects of the independent variables.
- *Screening-off:* Causes screen off their effects
- *Robustness:* The relationship between cause and effect is invariant with respect to the frequency of the cause.
- *Connection dependence:* If the connection between cause and effect is broken, only the effect would be affected.

Reference [6] argues that any cause that we isolate is never the whole cause and that every direct cause itself has its own direct causes, so that networks of causation spread synchronically across the economy and diachronically back into the mists of time. If this is true, granules must necessarily be imprecise as separation trough truncation from a network would be required.

4 Recognizing Causality Is of Interest In Many Domains

Recognizing causality is of interest in many areas. Of particular interest to this paper are areas where the analysis is non-experimental. The world is taken as it is and not subject to experimentation. In the computational sciences, data mining is of concern. An area not well known to people working in the computational sciences is economics.

Perhaps, the applied area that has the greatest history of attempting to deal with causality and non-observational data is economics. Econometrics is distinguished from statistics by econometrics interest in establishing causation [13]. How and if causality can be recognized has been a significant area of discussion. Some of this discussion mirrors discussion that has gone on in the computational sciences. Hoover provides a good entry to the discussion of causality in economics.

IF Age < 20
  THEN Income < $10,000
    with {belief = 0.8}

IF Age is old
  THEN vote frequency is: often
    with {belief = high}

Fig. 2. Conditional rules

Customers who
  buy **beer** and **sausage**
  also tend to buy **hamburger**
    with {confidence = 0.7}
    in {support = 0.2}

Customers who buy **strawberries**
  also tend to buy **whipped cream**
    with {confidence = 0.8}
    in {support = 0.15}

Fig. 3. Association rules

Reference ([15], p. 165), as a philosopher, suggested that causal statements are really about constant conjunction and time-ordering. However,
when speaking as an economist, ([14], p. 304) was less insistent on causal
ordering: "it is of consequence to know the principle whence any phenomenon
arises, and to distinguish between a cause and a concomitant effect." The issue of causal ordering is also often of importance to those modeling causality
in data discovery.

Data mining analyzes data previously collected; it is non-experimental.
There are several different data mining products. The most common are *conditional rules* or *association rules*. Conditional rules are most often drawn
from induced trees while association rules are most often learned from tabu-
lar data.

At first glance, association rules seem to imply a causal or cause-effect
relationship. That is:

A customer's purchase of both sausage and beer *causes* the customer
to also buy hamburger.

But, all that is discovered is the *existence* of a statistical relationship between
the items. They have a degree of joint occurrence. The *nature* of the relationship is not identified. Not known is whether the presence of an item or sets of
items causes the presence of another item or set of items; or the converse, or
some other phenomenon causes them to occur together.

Purely accidental relationships do not have the same decision value, as do causal relationships. For example,

IF it is true that buying both **beer** and **sausage** somehow causes someone to *buy* **beer**,
- THEN: A merchant might profitably put **beer** (or the likewise associated **sausage**) on sale
- AND at the same time: Increase the price of **hamburger** to compensate for the sale price.

On the other hand, knowing that

Bread and **milk** are often purchased together.

may not be useful information as both products are commonly purchased on every store visit. What might be of interest is discovering if there is a causal relationship between the purchase of *bananas* and something else. (It turns out that *bananas* are the most frequently purchased food item at Wal-Mart [21].)

When typically developed, rules do not *necessarily* describe causality. The confidence measure is simply an estimate of conditional probability. Support indicates how often the joint occurrence happens (the joint probability over the entire data set). The joint occurrence count is symmetric; that is, it does not matter what we count first. Also, the strength of any causal dependency may be very different from that of a possibly related association value. In all cases

confidence ≥ causal dependence

All that can be said is that associations describe the strength of joint co-occurrences.

Sometimes, the association might be causal; for example, if

Someone eats salty peanuts and then drinks beer.

or

Someone drinks beer and then becomes inebriated.

there may be a causal relationship. On the other hand, if

A rooster grows and then the sun rises.

or

Someone wears a "lucky" shirt and then wins a lottery.

there may not be a causal relationship. Recognizing true causal relationships would greatly enhances the decision value of data mining results.

The most popular market basket association rule development method identifies rules of particular interest by screening for joint probabilities (associations) above a specified threshold. Conditional probabilities (confidences) may well be asymmetric; whether this asymmetry indicates causality is a matter for speculation. For example, in the market basket data shown in Fig. 4,

$$p(c \mid a) > p(a \mid c) .$$

This does not mean that there is a causal relationship $(a \to c)$, it may simply be an artifact of there being more occurrences of c than a.

Perhaps, if the conditional probability is greater than the joint probability, it might indicate the relationship should be further explored for causality. An indicator if whether a discovered association might be causal is whether the joint probability is greater than the combined independent probabilities; i.e.,

$$p(a)p(c) < p(a, c)$$

A more specific threshold test could be placed on the conditional probabilities; i.e.,

  IF conditional probability > joint probability
  THEN conditional probability – joint probability = causal effect

	a	b	c
t_1	1	0	0
t_2	1	1	1
t_3	1	0	1
t_4	0	1	1
t_5	0	1	1

Fig. 4. Binary market basket data with imbalanced confidences

4.1 Association Rules without Causality Can Lead to Naive Decisions

Association rules are used is to aid in making retail decisions. However, simple association rules may lead to errors. Errors might occur; either if causality is recognized where there is no causality; or if the direction of the causal relationship is wrong [20, 35]. Errors might occur; either if causality is recognized where there is no causality; or if the direction of the causal relationship is wrong. For example, if

  A study of past customers shows that 94% are sick.

- Is it the following rule?

  Our customers are sick, so they buy from us.

- Is it the following complementary rule?

  If people use our products, they are likely to become sick.

- Is the relationship coincidental; and, consequently, there is no causal rule?

From a decision-making viewpoint, it is not enough to know that

> People both buy our products and are sick.

what is needed is knowledge of what causes what; if anything at all.

If causality is not recognized, the naive application of association rules can result in bad decisions [35]. This can be seen in an example from [20]:

Example: At a particular store, a customer buys:

- *hamburger* 33% of the time
- *hot dogs* 33% of the time
- both *hamburger* and *hot dogs* 33% of the time
- *sauerkraut** only if *hot dogs* are also purchased

This would produce the transaction matrix:

	hamburger	hot dog	sauerkraut
t_1	1	1	1
t_2	1	0	0
t_3	0	1	1

This would lead to the associations:

- $(hamburger,\ hot\ dog) = 0.5$
- $(hamburger,\ sauerkraut) = 0.5$
- $(hot\ dog,\ sauerkraut) = 1.0$

If the merchant:

- Reduced price of *hamburger* (as a sale item)
- Raised price of *sauerkraut* to compensate (as the rule *hamburger fi sauerkraut* has a high confidence.
- The offset pricing compensation would not work, as the sales of sauerkraut would not increase with the sales of *hamburger*. Most likely, the sales of *hot dogs* (and consequently, *sauerkraut*) would likely decrease as buyers would substitute *hamburger* for *hot dogs*.

4.2 Inherently Uncertain Recognition

Recognizing many things with absolute certainty is problematic. As this is the case, our causal understanding is based on a foundation of inherent uncertainty and incompleteness. Consequently, causal reasoning models must accommodate inherent ambiguity. Reference [19] lists:

---

* Sauerkraut is a form of pickled cabbage. Some people greatly enjoy using sauerkraut as a garnish with sausages. However, it is rarely consumed as a garnish with hamburger. For more about sauerkraut, see: *http://www.sauerkraut.com/*

- *Quantum Physics:* In particular, Heisenberg's uncertainty principle
- *Observer Interference:* Knowledge of the world might never be complete because we, as observers, are integral parts of what we observe. As observers, we necessarily affect what we observe.
- *Gödel's Theorem:* Showed that in any logical formulation of arithmetic that there would always be statements whose validity was indeterminate. This strongly suggests that there will always be inherently unpredictable aspects of the future.
- *Turing Halting Problem:* Turning showed that any problem solvable by a step-by-step procedure could be solved using a Turing machine. However, there are many routines where you cannot ascertain if the program will take a finite, or an infinite number of steps. Thus, there is a curtain between what can and cannot be known mathematically.
- *Chaos Theory:* Chaotic systems appear to be deterministic; but are computationally irreducible. If nature is chaotic at its core, it might be fully deterministic, yet wholly unpredictable ([9], 139).
- *Space-Time:* Space-time impacts the idea that causality can be described by a time separation (timeline) between cause and effect. The malleability of Einstein's space-time that has the effect that what is "now" and "later" is local to a particular observer; another observer may have contradictory views. This causes problems with the common definition of causality that cause precedes effect. If we cannot use a time line to determine what precedes what, what can be used?
- *Arithmetic Indeterminism:* Arithmetic itself has random aspects that introduce uncertainty as to whether equations may be solvable. References [1, 2] discovered that Diophantine equations may or may not have solutions, depending on the parameters chosen to form them. Whether a parameter leads to a solvable equation appears to be random.

It may well be that a precise and complete knowledge of causal events is not possible or at least uncertain. On the other hand, we have a commonsense belief that causal effects exist in the real world. If we can develop models tolerant of imprecision, it would be useful. Also, to some degree, the degree of importance that some of these items have decreases as grain size increases.

5 Granular Space-Time

One of the key principles of space-time is that of *background independence*. This principle says that the geometry of space-time is not fixed. Instead, the geometry is an evolving, dynamical quantity. A closely related principle is *diffeomorphism invariance*. This principle implies that unlike theories prior to general relativity, one is free to choose any set of coordinates to map space-time and express the equations. A point in space-time is defined only by what physically happens at it, not by its location according to some special set of coordinates (no coordinates are special).

Modern physics has developed a theory that entails that space and time are granular [41]. This is an extension of quantum theory. Quantum mechanics require that certain quantities, such as the energy of an atom, can only come in specific, discrete units. Over the last few years, theory has evolved concerning quantum gravity and quantum space-time. This area of endeavor is sometimes called *loop quantum gravity*. (The term *loop* arises from how some computations in the theory involve small loops marked out in space-time.) The work is concerned with quantum theory of the structure of space-time at the smallest size scales.

What concerns us in this paper is that there are apparently limits on fine grain size. These limits apply to areas, volumes, and time [41]. There is a non-zero minimum area (about one square Planck length, or 10^{-66} square centimeter) and a discrete series of allowed quantum areas. Similarly, there is a non-zero absolute minimum volume (about one cubic Planck length, or 10^{-99} cubic centimeter) and it restricts the set of larger volumes to a discrete series of numbers. Time is also discrete; it comes in "clicks" of 10^{-43} seconds (approximately the Planck time). Time does not exist between the clicks; there is no "in between," in the same way that there is no water between adjacent molecules of water.

This information should influence how we think about causality. If the universe is fundamentally granular, causal descriptions need to somehow deal with granularity. How to do this is unclear. Rough sets might be the best way of handling the granularity of causal complexes. Similarly, they seem to be a good tool to initially approach the granularity of space-time.

6 Causality Recognition

Various causality descriptions and discovery tools have been suggested. It may eventually turn out that different subject domains may have different methodological preferences.

Different aspects of causality have been examined. The idea of "positive" causation ($\alpha \rightarrow \beta$) is at the core of commonsense causal reasoning. Often a positive causal relationship is represented as a network of nodes and branches [19].

Fig. 5. Diagram indicating that α is causally dependent on β

Negation or counterfactuals ($\neg\alpha \rightarrow \neg\beta$) also have a place; although it may result in errors in reasoning. For example, the rule:

If a person drinks *wine*, they may become inebriated.

cannot be simply negated to

If a person **does not** drink *wine*, they will **not** become inebriated.

One reason is that effects can be *overdetermined;* that is: more than one item can cause an effect. If so, eliminating one cause does not necessarily eliminate the effect. In this case:

A person may also drink *beer* or *whiskey* to excess and become inebriated.

Events that do *not* happen can similarly be overdetermined. From a commonsense reasoning view, it is more likely that things do not happen than they do. For example, [22] says that it is not true that

His closing the barn door caused the horse not to escape.

because the horse might not have attempted to escape even if the door was open. Therefore, a false counterfactual is:

If he had not closed the barn door, the horse would have escaped.

Similarly, for example, the rule

If a person smokes, they will get cancer.

cannot be simply negated to

If a person *does not* smoke, they will *not* get cancer.

Again, effects can be overdetermined. In this case,

People who do not smoke may also get cancer.

Another idea that is sometimes involved in causal reasoning is *causal uncorrelatedness* [31] where if two variables have no common cause they are causally uncorrelated. This occurs if there are no single events that cause them to both change.

Similarly, [4] focuses on the negative; i.e., when α goes not affect β. Dawid speaks in terms of *unresponsiveness* and *insensitivity*. If β is *unresponsive* to α if whatever the value of α might be set to, the *value* of β will be unchanged. In parallel, if β is *insensitive* to α if whatever the value α may be set, the *uncertainty* about β will be unaffected. Along the same vein, [33, 34] distinguishes between *causing, enabling,* and *preventing*. The enabling factor is often considered to be a causal factor. Shoham distinguished between background (enabling) conditions and foreground conditions. The background (enabling) conditions are inferred by default. For example [34]:

"If information is present that the key was turned and nothing is mentioned about the stated about the state of the battery, then it is inferred that the motor will start, because the battery is assumed, by default to be alive.

Given this distinction, causing is taken to refer to the foreground conditions where enabling and preventing refer to the background conditions (in this example, turning the key causes the motor to start, the live battery enables it, the dead battery prevents it)."

Other ideas that are sometimes involved in causal reasoning are *causal uncorrelatedness* [31] where if two variables share no common cause they are causally uncorrelated. This occurs if there are no single events that cause them to both change. Similarly, causal independence occurs when speaking about probabilities.

6.1 Causality Recognition Through Statistical Testing

Statistics is the traditional tool used to discover causality when handling experimental (observational) data. However, the data of greatest interest in the computational sciences is non-observational. In this domain, traditional statistical methods are either not useful an/or are often too computationally complex.

The standard method in the experimental sciences of recognizing causality is to perform randomized, controlled experiments. This produces observational data. Depending on their design, randomized experiments may remove reasons for uncertainty whether or not a relationship is casual. However, large data sets are typically the subject of data mining. Even if some experimentation is possible, the amount of experimentation in contrast to the amount of data to be mined will be small. This said, some work has been done using chi-squared testing to reduce the search space [34].

6.2 Directed Graphs

Various graph based Bayesian based methods have been suggested to recognize causality. Probably the best known is the class of methods based on Directed Acyclic Graphs (DAGs). The most fully developed approach is [26]. Reference [34] followed a similar approach.

References [24] and [42] claim that it is possible to infer causal relationships between two variables from associations found in observational (nonexperimental) data without substantial domain knowledge. Spirtes claims that directed acyclic graphs can be used if (a) the sample size is large and (b) the distribution of random values is faithful to the causal graph. Reference [29] argues that their argument is incorrect. Lastly, [30] only claims that in some situations will it be possible to determine causality. Their discussion is tangential to the focus of this paper; going deeply into their discussion is outside this paper's scope. It is enough to note that these methods are possibly the most thoroughly developed methods of computational causal analysis.

From the commonsense causal reasoning view, the various directed graph methods have similar liabilities, specifically:

Discrete or Continuous Data Must be Reduced to Boolean Values

Objection: This is an early technique that was and is used in data mining when analyzing market basket data. However, it is essentially flawed. Quantities do matter; some data co-occurrences are conditioned on there being a sufficiency of a co-occurring attribute. Also, some relationships may be non-linear based on quantity [20].

Example: Situation: *Customers frequently buy either wine or beer for them-selves in varying amounts. However, when buying for a party, they often pur-chase both beer and wine and they usually purchase in larger quantities.*

Actual basket:		Binary basket:	
Beer	Wine	Beer	Wine
6	0	1	0
0	1	0	1
12	0	1	0
0	3	0	1
24	4	1	1
24	5	1	1
48	2	1	1

Missed rule: When at least 24 beers purchased, wine also purchased; *other-wise, there is no relationship between beer and wine.*

Naively constructing an association rule would find a rule that misleadingly represents the situation; i.e.,

Misleading rule: When beer is purchased, wine is also purchased {*confidence = 0.6*} {*support = 0.43*}

This rule is misleading. To the naive, it implies that purchase probabilities are uniform, in fact they are not. Under one set of conditions, *beer* and *wine* are *never* purchased together under one set of conditions; and, under another set of conditions *always* purchased together.

There is no Missing Data

Objection: This is at variance with day-to-day experience. There is almost always missing data of some sort. Data collection is rarely fully representative and complete. Incremental data is often acquired that is at variance with previously acquired data. What is needed is a methodology that is not brittle in the face of incompleteness.

Causal Relationships are not Cyclic, Either Directly or Indirectly (Through Another Attribute)

Objection: This is at variance with our commonsense understanding of the world.

For example: I tell Jane that I love her. Then, she tells me that she loves me. Then, I tell Jane that I love her more than before; then, she ... and so forth and so forth. Clearly, the cyclic reinforcement would be substantial.

Another example is shown in the following figure.

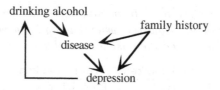

Fig. 6. Cyclic relationship

Another form of a cycle is *joint mutual dependency*. It is possible that there might be mutual dependencies [19]; i.e., $\alpha \rightarrow \beta$ is well as $\beta \rightarrow \alpha$. It seems to be possible that they do so with different strengths. They can be described as shown in the following figure where $S_{i,j}$ represents the strength of the causal relationship from i to j. It would seem that the strengths would be best represented by an approximate belief function, either quantitatively or verbally.

Fig. 7. Cyclic relationship: Mutual unequal dependency. Reference [19]

There would appear to be two variations: differing causal strengths for the same activity; and, different causal strengths for symmetric activities occurring at different times.

Alternative: Different Causal Strengths for the Same Activity, Occurring at the Same Time

Some argue that causality should be completely asymmetric and if it appears that items have mutual influences it is because there is another cause that causes both. A problem with this is that it can lead to eventual regression to a first cause. Whether this is true or not, it is not useful for commonsense representation. In contrast, [40] and [34] identify cases where causality is simultaneous.

It is also our commonsense experience. For example, in the preceding figure, α could be *short men* and β could be *tall women*. If $S_{\alpha,\beta}$ meant the strength of desire for a social meeting that was caused in *short men* by the sight of *tall women*, it might be that $S_{\alpha,\beta} > S_{\beta,\alpha}$.

Alternative: Different Causal Strengths for Symmetric Activities, Occurring at Different Times

It would seem that if there were causal relationships in market basket data, there would often be imbalanced dependencies. For example, if

A customer first buys strawberries, there may be a reasonably good chance that they will then buy whipped cream.

Conversely, if

They first buys whipped cream, the subsequent purchase of strawberries may be less likely.

This situation could be represented by the previous figure.

However, the issue of time sequence would be poorly represented. A graph representation could be used that implies a time relationship. Nodes in a sequence closer to a root could be considered to be earlier in time than those more distant from the root. Redundant nodes would have to be inserted to capture every alternate sequence. For example, one set of nodes for when strawberries are bought before whipped cream and another set when whipped cream is bought before strawberries.

However, this representation is less elegant and not satisfactory when a time differential is not a necessary part of causality. It also introduces multiple nodes for the same object (e.g., strawberries, whipped cream); which at a minimum introduces housekeeping difficulties.

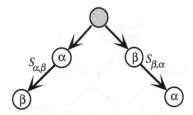

Fig. 8. Alternative time sequences for two symmetric causal event sequences where representing differing event times are necessary for representing causality. Nodes closer to the root occur before nodes more distant from the root. Causal strengths may be different depending on sequence. Reference [19]

Markov Stationary Condition Holds: Probabilities are Time Independent

Objection: This does not correspond to our commonsense understanding of the world. If one event is dependent on two other causal events, if one causal event happens much earlier (or later) than the other causal event, there may well be a different result.

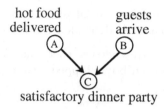

satisfactory dinner party

Fig. 9. Case where differing times in causal events affects probability of causal result

The Markov Condition Holds: Memoryless States

The *Markov Condition* is defined as: Let A be a node in a causal Bayesian network, and let B be any node that is not a descendant of A in the network. Then the Markov (Markoff) condition holds if A and B are independent, conditioned on the parents of A. The intuition of this condition is: If A and B are dependent, then B must either be (a possibly indirect) cause of A or (possibly indirectly) caused by A. In the second case, B is a descendant of A, while in the first B is an ancestor of A and has no effect on A once A's immediate parents are fixed.

This makes sense in the example shown in the following figure.

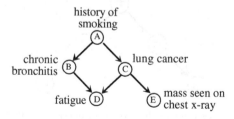

Fig. 10. "Memoryless" Markov condition holds

Objection: However, not all of our commonsense perceptions of causality work this way. Often, we believe that history matters as in the example shown in the following figure.

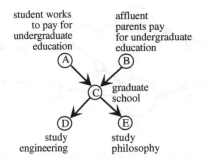

student works
to pay for
undergraduate
education

affluent
parents pay
for undergraduate
education

graduate
school

study
engineering

study
philosophy

Fig. 11. Causality where memory play a part

7 Markov Models

One possible way of describing the causal relationships is with first order Markov Models. They treat a system as a series of states with specific, constant rate transitions between them. At all times, the system is in exactly one state. (Transitions are considered to be instantaneous.) The only information available is the current state, the allowed transitions, and the probability of these transitions. This means that the system is totally characterized by its current state. None of the past states or transitions have any effect on the transitions out of the current state.

Quantitatively describing the relationships between the nodes can be complex. One possibility is an extension of the random Markov model, shown in Fig. 7. The state value is 1/0 as an event either happens or does not.

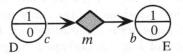

Fig. 12. Random Markov model: $c = P(D)$, m = the possibility/probability that when D is present, the causal mechanism brings about E, b = the probability that some other (unspecified) causal mechanism brings about E

Perhaps, fuzzy sets and fuzzy Markov algorithms in particular might be a good way of approaching causality represented by Markov models. What we must work with is inherently imprecise and incomplete.

Fuzzy sets are useful for dealing with any situation where the exact value of a variable is unknown. Instead of a guess of the value of the variable (which can easily be wrong), or a distribution of its possible values (which is usually unknown, so this problem reduces to a guess), fuzzy logic deals with the *possibility* of the variable taking on a set of values. In this way, it assumes less, and shows explicitly both what is and is not known.

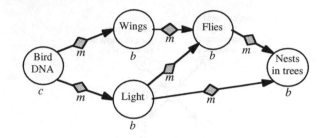

Fig. 13. Markov model applied to a "bird" example [28]

Reference [3] shows how the probabilities in the classic Markov model may be extended to fuzzy possibilities. The generalized hidden Markov Model can be defined with the same type of model parameters, but with a different mathematical basis than the classic one. Fuzzy logic replaces probability theory and this leads to a new definition of the model variables. The structure in terms of states and observations, remains the same.

There has been a fair amount of work in fuzzy Markov models [16, 17, 43]. A fair amount of the work has a control theory or medical diagnostic orientation. Whether it can be extended to representing causality is to be seen.

7.1 Increasing Grain Size Through Nested Models

Depending on the goal of the work, it may be more useful to work on a larger-grain size. This may reduce the need to learn extensive hidden Markov models, which in computationally expensive. Perhaps, a satisficing [39] solution would be to develop large-grained solutions and then only go to the finer-grain when the impreciseness of the large-grain is unsatisfactory. For example, if instead of a complex in the form of Fig. 13, perhaps a diagram similar to Fig. 14, would be sufficient. This form of representation is widely used in automata design, its extension to causal reasoning and imprecise reasoning would seem to be reasonable.

The greater grain size would be more useful in commonsense reasoning. It also might be more computationally tractable in the discovery process. Rough sets might be the best way of handling the granularity of causal complexes. Similarly, they seem to be a good tool to initially approach the granularity of space-time.

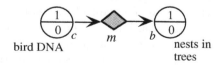

Fig. 14. Greater grain size (higher level) abstraction of Fig. 13

8 Epilogue

Commonsense causal reasoning occupies a central position in human reasoning. It plays an essential role in human decision-making. Considerable effort has been spent examining causation. Philosophers, mathematicians, computer scientists, cognitive scientists, psychologists, and others have formally explored questions of causation beginning at least three thousand years ago with the Greeks.

Whether causality can be recognized at all has long been a theoretical speculation of scientists and philosophers. At the same time, in our daily lives, we operate on the commonsense belief that causality exists.

Commonsense understanding of the world tells us that we have to deal with imprecision, uncertainty and imperfect knowledge. This is also the case of our scientific knowledge of the world. A difficulty is striking a good balance between precise formalism and commonsense imprecise reality.

Causal relationships exist in the commonsense world. Our commonsense perception of causality is often granular. Knowledge of at least some causal effects is imprecise. Perhaps, complete knowledge of all possible factors might lead to a crisp description of whether an effect will occur. However, in our commonsense world, it is unlikely that all possible factors can be known. In commonsense, every day reasoning, we use approaches that do not require complete knowledge.

People recognize that a complex collection of elements causes a particular effect, even if the precise elements of the complex are unknown. They may not know what events are in the complex; or, what constraints and laws the complex is subject to. Sometimes, the details underlying an event are known to a fine level of detail, sometimes not. At a very fine-grained level, the physical world itself may be composed of granules. In general, people are more successful in reasoning about a few large-grain sized events than many fine-grained events. Perhaps, this can transfer over to computational models of causality.

A lack of complete, precise knowledge should not be discouraging. People do things in the world by exploiting our commonsense *perceptions* of cause and effect. When trying to precisely reason about causality, we need complete knowledge of all of the relevant events and circumstances. In commonsense, every day reasoning, we use approaches that do not require complete knowledge. Often, approaches follow what is essentially a *satisficing* paradigm.

Instead of being concerned with all of the fined grained detail, a better approach may be to incorporate granulation using rough sets and/or fuzzy sets to soften the need for preciseness. And then accept impreciseness in the description. Each complex can be considered to be a granule. Larger complexes can be decomposed into smaller complexes. Thus, going from large-grained to small-grained.

Perhaps Markov models might be used to build these complexes. Depending on the goal of the work, it may be more useful to work on a larger-grain size. This may reduce the need to learn extensive hidden Markov models, which in computationally expensive. Perhaps, a satisficing solution would be to develop large-grained solutions and then only go to the finer-grain when the impreciseness of the large-grain is unsatisfactory.

References

1. G. Chatin [1987] *Algorithmic Information Theory*, Cambridge University Press, Cambridge, United Kingdom.
2. G. Chatin [1990] "A Random Walk In Arithmetic," New Scientist 125, n 1709 (March, 1990), 44–66.
3. A. Cheok, S. Chevalier, M. Kaynak, K. Sengupta, K. Chung [2002] "Use of a Novel Generalized Fuzzy Hidden Markov Model for Speech Recognition," *Proceedings. IEEE International Conference on Fuzzy Systems*, December 2–5, 2002 Melbourne, Australia.
4. A. Dawid [1999] "Who Needs Counterfactuals" in *Causal Models and Intelligent Data Management* (ed) A. Gammerman) Springer-Verlag, Berlin.
5. W. Freeman [1995] *Societies of Brains*, Lawrence Erlbaum, 1995.
6. M. Friedman [1949] "The Marshallian Demand Curve," *Journal of Political Economy*, v 57, 463–495.
7. M. Goodrich, W. Stirling, E. Boer [2000] "Satisficing Revisited," *Minds and Machines*, v 10, 79–109.
8. C. Granger [1969] "Investigating Causal Relations By Econometric Models and Cross-Spectral Methods," *Econometrica*, v 37, 424–438.
9. P. Halpern [2000] *The Pursuit of Destiny*, Perseus, Cambridge, Massachusetts.
10. D. Hausman [1998] *Causal Asymmetries*, Cambridge University Press, Cambridge, United Kingdom.
11. J. Hobbs [2001] "Causality," *Proceedings, Common Sense 2001, Fifth Symposium on Logical Formalizations of Commonsense Reasoning*, New York University, New York, May, 145–155.
12. J. Hobbs [2003] "Causality and Modality: The Case of 'Would'," to appear in *Journal of Semantics*.
13. K. Hoover [2003] "Lost Causes," *HES Conference*, Presidential Address, Durham, North Carolina.
14. D. Hume [1742/1985] *Essays: Moral, Political, and Literary*, Eugene Miller (ed.), Liberty Classics, Indianapolis, 1985.
15. D. Hume [1777/1902] "An Enquiry Concerning Human Understanding," in L. Selby-Bigge (ed.) *Enquiries Concerning Human Understanding and Concerning the Principles of Morals*, 2nd edition, Clarendon Press, Oxford, 1902.
16. L. Jeanpierre, F. Charpillet [2002] "Hidden Markov Models for Medical Diagnosis," *International Workshop on Enterprise Networking and Computing in Health Care Industry, HEALTHCOM'2002*, Nancy, France, June 2002.
17. M. Leuschen, I. Walker, J. Cavallaro [1998] "Robot Reliability Through Fuzzy Markov Models," *Proceedings IEEE Annual Reliability and Maintainability Symposium*, Anaheim, CA, January 1998, 209–214.
18. D. Lewis [1973] *Counterfactuals*, Harvard University Press, Cambridge University Press.

19. L. Mazlack [2003a] "Commonsense Causal Modeling in the Data Mining Context," IEEE ICDM Proceedings, Melbourne, Florida, November 19–22, 2003.

20. L. Mazlack [2003b] "Causality Recognition for Data Mining In An Inherently Ill Defined World," 2003 BISC FLINT-CIBI International Joint Workshop On Soft Computing for Internet and Bioinformatics, December, 2003.

21. E. Nelson [1998, October 6] "Why WalMart sings, 'Yes, we have bananas'," The Wall Street Journal, B1.

22. C. Ortiz [1999a] "A Commonsense Language for Reasoning about Causation and Rational Action," Artificial Intelligence, v 108, n 1–2, pp. 125–178.

23. C. Oritz [1999b] "Explanatory Update Theory: Applications of Counterfactual Reasoning to Causation," Artificial Intelligence, v 108, n 1–2, 125–178.

24. J. Pearl, T. Verma [1991] "A Theory of Inferred Causation," *Principles of Knowledge Representation and Reasoning: Procccdings of the Second International Conference,* Morgan Kaufmann, 441–452.

25. J. Pearl, T. Verma [1991] "A Theory of Inferred Causation," *Principles of Knowledge Representation and Reasoning: Proceedings of The Second International Conference,* Morgan Kaufmann, 441–452.

26. J. Pearl [2000] *Causality,* Cambridge University Press, New York, NY.

27. J. Pearl [2001] From a web page supporting Pearl's 2000 book, in reply to a question, Date: September 28, 2001; From: Sampsa Hautaniemi, NIH; Subject: Zadeh's "CAUSALITY IS UNDEFINABLE", http://bayes.cs.ucla.edu/BOOK-2K/hautaniemi.html.

28. B. Rehder [2002] "A Causal-Model Theory of the Conceptual Representation of Categories," FLAIRS 2002 Workshop on Causality, Pensacola, May.

29. R. Robins, L. Wasserman [1999], "On the Impossibility of Inferring Causation from Association without Background Knowledge," in (eds) C. Glymour, G. Cooper, *Computation, Causation, and Discovery* AAAI Press/MIT Press, Menlo Park, 305–321.

30. R. Scheines, P. Spirtes, C. Glymour, C. Meek [1994] *Tetrad II: Tools for Causal Modeling,* Lawrence Erlbaum, Hillsdale, NJ.

31. G. Shafer [1999] "Causal Conjecture," in *Causal Models and Intelligent Data Management* (ed) A. Gammerman) Springer-Verlag, Berlin

32. II. Simon [1955] "A Behavior Model of Rational Choice," Quarterly Journal of Economics, v 59, 99–118.

33. Y. Shoham [1990] "Nonmonotonic Reasoning and Causation," *Cognitive Science,* v 14, 213–252.

34. Y. Shoham [1991] "Remarks on Simon's Comments," *Cognitive Science,* v 15, 301–303.

35. C. Silverstein, S. Brin, R. Motwani [1998a] "Beyond Market Baskets: Generalizing Association Rules to Dependence Rules," Data Mining and Knowledge Discovery, v 2, 39–68.

36. C. Silverstein, S. Brin, R. Motwani, J. Ullman [1998b] "Scalable techniques for mining causal structures," *Proceedings 1998 VLDB Conference,* New York, NY, August 1998, 594–605.

37. H. Simon [1952] "On the Definition of the Causal Relation," *The Journal of Philosophy,* v 49, 517–528. Reprinted in Herbert A. Simon, *Models of Man,* John Wiley, New York, 1957.

38. H. Simon [1953] "Causal Ordering and Identifiability," Reprinted in Herbert A. Simon, *Models of Man,* John Wiley, New York, 1957.

39. H. Simon [1955] "A Behavioral Model of Rational Choice," *Quarterly Journal Economics,* v 59, 99–118.
40. H. Simon [1991] "Nonmonotonic Reasoning and Causation: Comment," *Cognitive Science,* v 15, 293–300.
41. L. Smolin [2004] "Atoms of Space and Time," *Scientific American,* January, 2004, 66–75.
42. P. Spirtes, C. Glymour, R. Scheines [1993] *Causation, Prediction, and Search,* Springer-Verlag, New York.
43. L. Zadeh [1998] "Maximizing Sets and Fuzzy Markov Algorithms," *IEEE Transactions on Systems, Man, and Cybernetics – Part C: Applications and Reviews,* v 28, 9–15.
44. L. Zadeh [1999] "From computing with Numbers to Computing with Words – From Manipulation of Measurements to Manipulation of Perceptions," *IEEE Transactions on Circuits and Systems,* v 45, n 1, 108–119.
45. L. Zadeh [2001] "Causality Is Undefinable," Abstract of a Lecture Presented at the BISC Seminar, University of California, Berkeley, reported to: *Fuzzy Distribution List,* fuzzymail@ dbai.tuwien.ac.at, January 16, 2001 (on line only), also available at: http://www.cs.berkeley.edu/~nikraves/zadeh/Zadeh2.doc.

Gene Regulating Network Discovery

Yingjun Cao[1], Paul P. Wang[1], and Alade Tokuta[2]

[1] ECE Department, 130 Hudson Hall, P.O. Box 90291, Duke University, Durham,
NC 27708, USA
[2] {yc23, ppw}@ee.duke.edu
Dept. of Math. & Computer Sci., NC Central University, 1801 Fayetteville St.,
Durham, NC 27707, USA
atokuta@nccu.edu

Abstract. Gene regulation has been an important research topic for the past
30 years and data processing models based on different assumptions have been proposed by many researchers with different methodologies. In this paper, a novel gene
regulation data processing algorithm has been investigated in detail. Based on fuzzy
logic technique and the philosophy of mutual information, the algorithm is intuitive
and is believed to be robust. The key motivation for this algorithm is that genes with
regulation relationship should have more mutual information and this feature can
be represented as the smaller accumulated distance so far as the available time sequence of microarray data lasting a period of time intervals. One unique feature of
this algorithm is that it makes very limited prior assumptions concerning the modeling; hence the algorithm is a data-driven algorithm. Therefore an algorithm that
is robust should be entirely based on data only unless some priori information has
been firmly established. As a non-parametric model, the algorithm is very useful
when the limited prior knowledge is known about the target microarray. Another
helpful characteristic of this algorithm is that the microarray data have been treated
as a gene regulation network so it is fuzzified into fuzzy sets which make this algorithm agreeable to reality. Finally, the algorithm has been exclusively tested on
the simulated microarray data under different conceivable situations (such as prior
knowledge availability, the regulation time span, number of chips in microarray,
noise level etc). In conclusion, we believe this algorithm will be proven to be quite
promising in practical applications.

Key words: gene regulation, microarray, fuzzy logic, hamming distance, reverse
engineering, data driven

1 Introduction

The commonly accepted definition of the so called "gene regulation" is that
DNA and protein interactions in a gene that determine the temporal and
spatial modes of expression as well as the amplitude of expression [8]. Each cell

Yingjun Cao et al.: *Gene Regulating Network Discovery*, Studies in Computational Intelligence
(SCI) **5**, 49–48 (2005)
www.springerlink.com

in the human body contains the same DNA molecule and a gene is a sequence of nucleotides in DNA that is coded for a function product. The process from DNA to the protein is based on transcription and translation. First, the DNA codes are deciphered by mRNA (messenger RNA). Then mRNA translates those information and carries them to ribosome. The ribosome uses the information to produce proteins. This transcription process is then also affected by the feedback mechanism from the protein. The whole procedure is detailed in Fig. 1. This biological feedback represents the spatial-temporal conditions in gene regulating and makes genes in DNA "on/off" for different times.

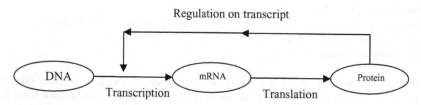

Fig. 1. Gene regulation mechanism

The actual gene regulation process is much more complicated and it involves many chemical reactions. The most well-known chemical equations were established by Gibson in 1999 and the following equations include the whole transcription and translation processes.

$$RNAP + DNA \xrightarrow{k_1} RNAP \bullet DNA_0$$
$$RNAP \bullet DNA_n \xrightarrow{k_2} RNAP \bullet DNA_{n+1}$$
$$RNAP \bullet DNA_{MAX} \xrightarrow{k_3} RNAP + DNA + mRNA$$
$$Ribosome + mRNA \xrightarrow{k_4} Ribosome \bullet mRNA_0$$
$$RNase + mRNA \xrightarrow{k_5} RNase \bullet mRNA$$
$$RNase \bullet mRNA \xrightarrow{k_6} RNase$$
$$Ribosome \bullet mRNA_n \xrightarrow{k_7} Ribosome \bullet mRNA_{n+1}$$
$$Ribosome \bullet mRNA_{MAX} \xrightarrow{k_8} Ribosome + mRNA + protein$$
$$protein \xrightarrow{k9}$$

In these equations, the k_s, where $s = 1, 2, \ldots, 9$, are the rate constants which represent the speed or probability of the reactions.

Historically, a lot of attention has been drawn on the transcription part; however, there is growing recognition that many regulation mechanism happens after the RNA has completely synthesized (post-transcriptional gene regulation). Normally, gene regulation takes place when the cell is experiencing environmental changes and this metabolism usually prevents the cells from starvation or injury. But when the regulation malfunctions, the regulation can results in diseases, such as cancer. In this case, the cell becomes a "cancerous cell". So if the regulatory network is completely understood, new drugs can be developed to prevent such regulating errors, at least in theory.

It has been widely known to the molecular biologist that some gene is regulated by some other genes. Unfortunately, the number of genes participating in the regulation process and the manners they regulate, usually AND, OR, or COMPLEMENT, are not fully known yet. On the other hand, this problem presents quite a challenge to the bioinformatics researchers and computer scientists, theoretical and experimental alike. This problem has become widely known in recent years as the "reverse engineering" in molecular biology community and as the "system identification problem" by the dynamic control engineers. Arguably, the most acceptable theory for the gene regulation can be represented in a Boolean network format. A.S. Kauffman proposed the NK Boolean Network format in the 1960's and interesting results about regulating mechanism have been found [18]. For example, Gene A and Gene B interact during some instant t and their interactions will determine or regulate the status of another gene C at time instant $t + 1$. According to A.S. Kauffman's theory, their regulating mechanism can be modeled by logic function stochastically as follows: $C(t + 1) = Logic(A(t), B(t))$. This means the current state of A and B at time t will regulate the next state $(t + 1)$ of gene C through the intrinsic logic function. Should numerous genetic regulations occur simultaneously, will the participating genes altogether constitute a gene regulation network. The network will self-evolve in synchronized fashion and will eventually reach certain final states either as attractors or limit cycles. In the so called NK network, N is the total number of participating genes in the network and K is the maximum number of regulating genes. According to Kauffman, the NK network fits the biological systems phenomena the best when K equals to 2 and the mutations and evolution phenomena will result from the chaotic dynamics of the system eventually. Due to the limitations of crisp Boolean values, however, the NK network is not practical in real gene regulation analysis. So other approaches were considered in the mathematical modeling of the gene regulation mechanism, such as differential equations [12], Bayesian networks [10]. With the distinctive properties in processing real-life uncertainties, the gene regulation analysis based on fuzzy logic theory did emerge in 1995 [17] and good fuzzy analysis results have been shown recently. Unfortunately, all above models have to deal with vast amount of data in order to understand the principles which govern the true cellular functioning [1, 4, 6, 7, 9, 24].

2 Algorithm

There are two key concepts in the course of developing our "knowledge mining" algorithm. One concept is the measurement of "distance" between two patterns. The second concept is the "templates matching". With these two concepts as our basic building blocks, we then discover much more complex relationships of multiple genes according to all possible logic combinations with the other genes in the network. The distances represent the mutual information between different genes in a reverse relationship. If some genes have regulation relationship, their mutual information must be stronger than others. This implies a smaller distance. In addition, the algorithm has been based on fuzzy set modeling which allows it to capture the non-linear relationship in gene regulation processes and the algorithm under investigation is quite robust to report genetic activities under different situations. In real gene regulation analysis, some genes are more important than others and they are preferred to be analyzed first. In this paper, we name those important genes as our "target genes". So the algorithm is first investigated under the situation that target genes are known. This first class of simulation is important in the sense that the historical fact about biology science is that some genes are known biologically and they are very important in governing the behavior of the whole biological system. A case in point is that for a song bird where the male or female songbirds sing [5]. Secondly, the algorithm is simulated for the case that the target gene is unknown which means a blind search. It is safe to say that a robust algorithm should be capable of revealing a complex relationship among gene regulations.

In practice, however, noise is a noticeable phenomenon in the microarray chips that one must deal with and further research is definitely needed here to understand the basic statistical characteristics of the noise. Since the field of bioinformatics is still at its infancy, a great deal of work lies ahead. To name two examples, one is that the microarray chips makers, in the order of fifty to sixty companies who all exhibit their chips products with prominent noise presence. Most companies in this industry chose to keep the characteristics of noise in secrecy due to the propriety of their products. It is nearly inconceivable that research works on this line has not been extensively carried out so far. The second example is the lack of standards in bioinformatics. Unlike the professional society such as IEEE where the standard committees exist for about every major issue, there is no formal standards committee, to our knowledge, in its existence in most of the bioinformatics corporations.

The noise level is usually rather high in microarray as compared with the consumer electronics industry. It is not unusual that sometimes the noise can reach as big as 20%. The actual noise in microarray varies from different experiment methods and there is no established noise estimation method, nor the design of noise filters for microarrays. So in this paper, the noise is assumed to be truncated white Gaussian noise and the noise levels are assumed to be less than 10%. To accommodate and to tolerate the noise, parameters in the algorithm should be adjusted accordingly and the analysis of inference results helps to improve the results themselves. In this paper, the noise tolerance for both known and unknown target genes has been investigated, respectively.

2.1 Development of the Algorithm

The principle idea for this knowledge mining algorithm is intuitive and it is data driven. However the development of this fuzzy microarray data processing algorithm is not as straight forward as its outside appearance. Rather, it is the result of a number of different theoretical and experimental explorations. The evolutionary process of our final proposal is briefly described as below.

The Hardship of a Reverse Engineering Approach

One fact stands out as agreed by most researchers in the area of bioinformatics research is that the reverse engineering problem is very difficult. Yet most publications in the open literature nearly always end up with treating genetic network discovery problem as a reverse engineering problem. Although intuitively appealing, the reverse engineering takes an approach by maintaining a methodology of searching and synthesizing. Initially, we have also tried to reduce the computational complexity as our primary goal. Some paper indeed took such reduction approaches which end up with differential equations or equivalent linear form as their models. However, we soon realized the only possible solution is via dictionary-like approaches in a two-gene network [19]. We finally give up this approach because it is very difficult to have any major break through such a straight forward strategy. The Boolean network always ends up with NP-hard in computation complexity for most attempts. Alternative approaches of belief network, artificial neural network, ordinary differential equations or other hybrid approaches end up with at least NP-complete. No substantial advantage in terms of computation complexity has been observed yet which is capable of resulting in appealing results.

Data Driven and Data Mining Approach

In due time, the problem of processing microarray chips is anticipated to become a major challenging one because of the needs to process large amount of microarray data which is estimated to yield vast amount of information about biological systems. So far, the manufacturer's detailed designs are not readily available, at least in an open literature field and the characteristics of the associated parameters are also unavailable. One thing is certain though which appeared to be available is the huge amount of microarray data. Parameters which are introduced into models must have practical meanings and unnecessarily assumed parameters will make the problem unnecessarily more complex and more ineffective. A case in point is the approach of belief network which by definition is using a number of probabilities where the source of their estimation is highly questionable. Furthermore, we claim that the discovery of the genetic network can be the most efficient if the problem is being treated a data mining problem.

It is highly possible that many parameters will affect the outcome of the gene regulations. This is not a question of design philosophy but rather related to the issue of the suitability. More fundamentally, there is a real need for a more rigorous and robust definition for gene regulation. The fundamental question to be asked is that "what do you mean a gene is regulated by a number of other genes?" Further more, the parameters to model gene regulation should be robust. In the course of our studies, the "accumulated distances" over a finite time span has been adopted as an important measurement. The reason why the finite length of time span must be considered is due to the fact of the limitation of the data obtainable from biological phenomena. Let us consider one extreme condition where the time span is "ultra short" and this will probably yield useless results because the sample size is too small, stochastically speaking. As it turned out, the genetic regulation network discovery algorithm is quite versatile, flexible and productive for needed information.

2.2 Algorithm Structure

The algorithm consists of 3 major parts: Logic function iteration, Distance computation and Result inference. In Fig. 2, logic function iteration and distance computation generate a distance matrix whose property will strongly influence the final inference part. The analysis of inference results will be associated with the distance distribution in the results and this part will be addressed in detail in simulation results.

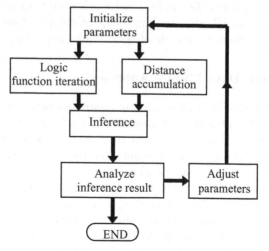

Fig. 2. Algorithm structure

Logic Function Iteration

As mentioned in introduction, the number of regulating genes is usually assumed to be 2 genes or less than 2 genes according to A.S. Kauffman's theory and concurrent biology phenomena. So in the logic function iteration part, the maximum regulating genes are assumed to be 1 or 2 and this assumption includes the simple promoter and repressor case. Suppose two genes in the microarray are labeled as G_1 and G_2 and they regulate another gene G_3 in the microarray. All possible regulating logics can be deducted from Karnaught map and the self regulating cases are eliminated because this kind of regulation has no associated biological meaning. As listed below, there are totally $2^{2^2} - 2$ different logics for the two gene case.

$$G_1 \Rightarrow G_3 \qquad G_2 \Rightarrow G_3 \qquad \bar{G}_1 \Rightarrow G_3 \qquad \bar{G}_2 \Rightarrow G_3$$
$$G_1 \wedge G_2 \Rightarrow G_3 \quad \bar{G}_1 \wedge G_2 \Rightarrow G_3 \quad G_1 \wedge \bar{G}_2 \Rightarrow G_3 \quad \bar{G}_1 \wedge \bar{G}_2 \Rightarrow G_3$$
$$G_1 \vee G_2 \Rightarrow G_3 \quad G_1 \vee \bar{G}_2 \Rightarrow G_3 \quad \bar{G}_1 \vee G_2 \Rightarrow G_3 \quad \bar{G}_1 \vee \bar{G}_2 \Rightarrow G_3$$
$$(G_1 \wedge G_2) \vee (\bar{G}_1 \wedge \bar{G}_2) \Rightarrow G_3 \qquad (\bar{G}_1 \wedge G_2) \vee (G_1 \wedge \bar{G}_2) \Rightarrow G_3$$

In this formula, $\wedge$ means logic AND, $\vee$ means logic OR and $^-$ means logic NEGATE.

In gene regulations, all these logical functions must take place in a time series which means the states of the regulating genes at time t will affect the state of the regulated gene at time $t + 1$. Besides, all logical functions are based on fuzzy logic MAX-MIN composition as described in appendix I. The function to apply gene regulations is termed as "Fuz_logic" for the ease of communication.

Distance Accumulation

The distance used in this algorithm is the modified Hamming distance which shows the similarities between two time series. One of two time series is the actual time series in microarray data, the other is the "pseudo" time series generated by logic function iterations emulating the behavior according to the logic functions. The comparison of two time series is defined as the accumulated distance at different time instances. Suppose the actual dynamics of a gene expression is $G = \{c_1, c_2, c_3, \ldots, c_{n-1}, c_n\}$, where $c_i \in [0, 1]$ and the index represents different time slots (n slots in this case). Suppose the data series generated by two regulating genes and one of the logics are $S = \{s_1, s_2, s_3, \ldots, s_{n-1}, s_n\}$. The distance between the simulated sequence and the real data sequence is $d = \sum_{i=2}^{n} H(s_i, c_i)$ where the Hamming distance is computed as $H(s_i, c_i) = \{ {1 \text{ when } |s_i - c_i| > threshold \atop 0 \text{ when} |s_i - c_i| \leq threshold}$. The distance is then computed starting from the 2nd time point in the microarray because we assume that regulation does not take place in time 1. The threshold in the formula is called "Hamming threshold" and its value is determined from the noise level and data. In general, the accuracy of data is based on the experiment methods and post-experiment data processing which are beyond the scope of this paper. As for the noise level, if the estimated noise level is high, the threshold should be increased in order to tolerate the noise and vice versa.

The algorithm computes all the distances for all possible gene combinations under all possible logical functions. So the minimum accumulated distance is 0 when the true regulation exists in all time series which is unlikely to happen in nature; on the other hand, the maximum distance should be the number of chips in the microarray which means either no regulation exists at all between the selected genes or the selected logical function turns out to be wrong. Distance in between can be classified into 2 categories. Category 1 is the distance that is smaller than maximum but is not small enough to infer the existence of true regulations. Category 2 is the distance that is very small with respect to others and hence it should represent the true regulations. Differentiation of the category 2 distances from that of the category 1 turns out to be the major goal of this algorithm. And through the adjustment of parameters, two categories of distance can be recognized accurately.

Inference

After logical function iterations and the distance accumulations, the algorithm will generate a distance matrix D which is the result of all possible logic combinations with respect to all genes in the microarray. Figure 3 shows the histogram of matrix D from 20-genes, 500-chips simulated microarray. In this microarray, only four regulations exist and their regulations are not equally strong. Two of them are stronger regulations which means the regulation exists in most of the time spots while the other two are relatively weaker regulations. It shows that most of the distances are near the maximum distance because most of the data do not show the visible regulation relationship. While there exist some distances that are much smaller than maximum and the smaller the distances are, the smaller number of cases do exist.

As one can see in Fig. 3, the actual regulations are not very easy to be inferred because the distribution of smaller distances does not have a clear "separation" from category 1 to that of the category 2 distances. Although there exists two extremely small distance (associated with the two strong regulations) in the histogram, the other two true regulations cannot discern themselves with respect to their distances. The reason for this phenomenon is that they are not strong enough so their distances are not distinctively smaller with respect to others. However, the histogram can help to eliminate most of the cases and only a small number of possible regulations are left as potential candidates for further investigation. When the number of microarray chips increases and the regulation activity becomes stronger, the distance distribution will yield a good inference threshold to separate category 1 and category 2 distances. Besides, the adjustment of Hamming threshold also may

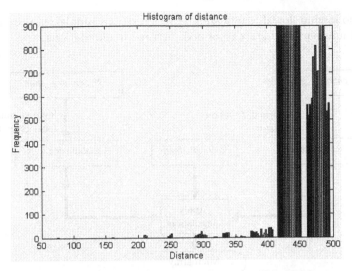

Fig. 3. Histogram of distance (20-gene, 500-chips)

help us to reduce the possible errors of recognizing category 1 distances as category 2 distances.

2.3 Algorithm Flowchart

Microarray Data Simulator

The data formula for this algorithm is a microarray chip presented as a matrix whose rows are different genes and columns represent time series (i.e. each column in the matrix represents the dynamics of one gene). The simulated microarray data are first randomly generated (uniform distribution) and then the designated regulations are applied to the microarray. Here are the designated parameters for the microarray data simulator.

N_genes: number of genes in the microarray
N_chips: number of timeslots in the microarray
N_reg: number of regulations existing in the microarray
Positions: the indexes for the regulating genes and the regulated gene
Logics: the regulation logical functions
Reg_span: the regulation time span for each regulation

In reality, the genetic regulation phenomenon is a relatively ambiguous concept because the gene regulations can take many forms and shapes. Furthermore, the logical functions, or the rules of regulation, can be highly time-varying. This means the rules can change as a function of time in true biological behavior. Hence, for each regulated gene, the regulating time spots are randomly chosen and the sum of regulating times is in accordance with the predefined Reg_span.

The following flow diagram summarizes the data simulation procedure and the final microarray matrix is named "Array" in simulations.

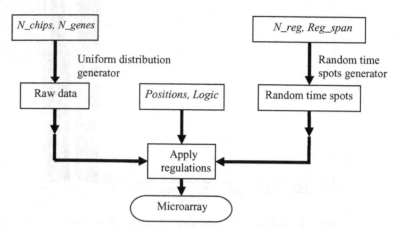

Fig. 4. Data simulation diagram

Flowchart

The algorithm under investigation has been constructed to deal with two major different situations, namely whether the target genes are known or not known. If we want to know whether a particular gene is being regulated or not, this gene is considered to be known. In this situation, the inference is relatively easy. We only need to compare the target gene with all the possible combinations of genes and logical functions. The inference step is just to figure out the minimum distance by which the regulation logic and regulating genes will be located. The flow chart for this part of the algorithm with known target genes is shown in Fig. 5.

In the search for each regulation, all possible cases for logic and gene combinations in a microarray are $\frac{N\_genes \cdot (N\_genes-1)}{2} \cdot 14$ and the minimum distance represents the most significant regulation that is presented in the microarray. If there are several small distances in the final distance matrix D, the Hamming threshold should be decreased in order to sharpen the difference. However, suppose noise exists in the microarray, the threshold cannot be decreased to a level that is smaller than the noise level because the noise will pop up in the distance computation and hence degrade the inference results.

For the case when target genes are unknown, no prior knowledge is given and the iteration and comparison parts in Fig. 5 should be adjusted. All possible combinations should be checked and any gene in the microarray should be compared. A new threshold, inference threshold, is then needed in order to

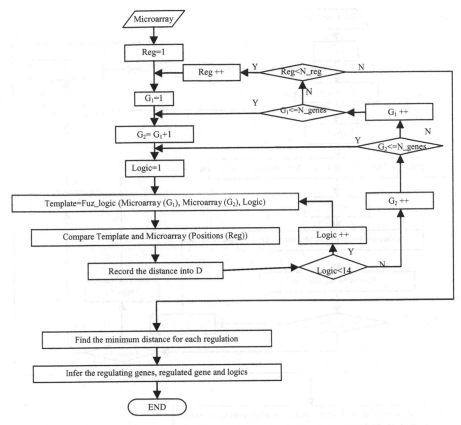

Fig. 5. Flowchart of the algorithm for known target genes

differentiate the category 1 distance the category 2 distances. The modified algorithm is shown in Fig. 6.

All possible cases in a microarray are $\frac{N\_genes \cdot (N\_genes - 1)}{2} \cdot N\_genes \cdot 14$ and those cases are the only source to infer all true regulations. This is the major difference between known targets and unknown targets. In this situation, the algorithm can not guarantee to find all regulations, especially for the weaker ones, because the inference threshold has to be empirically chosen. If the inference threshold is chosen to be too small, the inference might miss some weak regulations. On the contrary, if the inference threshold is too big, the inference would select some erroneous regulations as part of the final results. So for good inference results, the choice of inference threshold should be combined with known biology knowledge.

It is worthwhile to point out that the blind search algorithm exhibit good "robustness" in the proposed algorithm. In a sense, one can come up with any number of regulating genes and with a particular functional logic combination. This particular combination indeed would be a template. What this

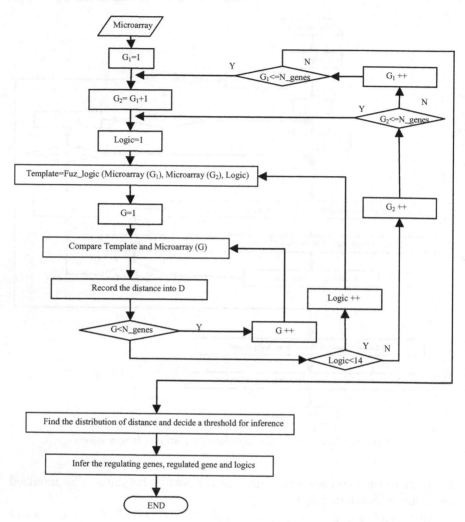

Fig. 6. Flowchart of the algorithm for unknown target genes

really implies is that we can always find out the answer to the important question, "is this part of the game?" Should we ask enough of such questions, can the genetic regulating network always be discovered through the means of some comparative study. The major difference lying between different templates hence is the computational complexity which must be tackled by further research.

3 Simulation Results

The simulations of this algorithm are based on two different situations (i.e. known target genes and the unknown target genes) and both situations are investigated with additive noise. A special gene regulation network is designed to test the algorithm. The hypothesized gene regulation network is shown in Fig. 7.

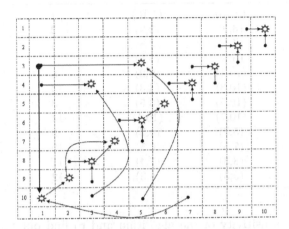

Fig. 7. Hypothesized two-gene network

In Fig. 7, $\star$ represents the target genes and $\rightarrow$ shows the regulation route which starts at regulating gene and ends at the regulated gene with an arrow. In the 100-gene microarray, every gene is labeled with an ordered pair of column and row numbers. 12 regulations are hypothesized in this microarray and the regulating genes are no more than 2 (namely, we assume $k = 2$ in A.S. Kauffman's network). In these 12 regulations, eight regulations are simple 2-gene regulations; two regulations are single-gene promoter or repressor cases; the other two regulations are somewhat more complicated regulations. Besides, the regulating logic is designed to cover all possible combinations of logical functions and they are listed explicitly as follows:

$$G_{1,10} \Leftarrow \overline{G_{1,9}} \wedge \overline{G_{2,10}} \qquad\qquad G_{5,6} \Leftarrow G_{6,5}$$
$$G_{2,9} \Leftarrow \overline{G_{2,8}} \wedge G_{3,9} \qquad\qquad G_{6,5} \Leftarrow G_{6,4} \vee \overline{G_{7,5}}$$
$$G_{3,5} \Leftarrow G_{3,1} \wedge \overline{G_{10,5}} \qquad\qquad G_{7,4} \Leftarrow \overline{G_{9,2}} \vee G_{8,3}$$
$$G_{3,8} \Leftarrow G_{3,7} \wedge G_{4,8} \qquad\qquad G_{8,3} \Leftarrow \overline{G_{8,2}} \vee G_{9,3}$$
$$G_{4,3} \Leftarrow \overline{(G_{4,1}} \wedge G_{10,3}) \vee (G_{4,1} \wedge \overline{G_{10,3}}) \quad G_{9,2} \Leftarrow \overline{G_{10,1}}$$
$$G_{4,7} \Leftarrow \overline{G_{4,6}} \vee G_{5,7} \qquad\qquad G_{10,1} \Leftarrow (G_{1,3} \wedge G_{10,7}) \vee (\overline{G_{1,3}} \wedge \overline{G_{10,7}})$$

3.1 Simulation for the Known Target Genes

In this situation, the algorithm was first tested via the designed microarray without additive noise. The regulation time spans of those 12 regulations are assumed to be strong. The simulation result gives 100% correct recognition correctness. As one would the following table shows the recognition results by the algorithm.

Table 1. Results of identification

Reg Gene 1	Reg Gene 2	Target Gene	Logic Number	Distance
(1,9)	(2,10)	(1,10)	1	5
(2,8)	(3,9)	(2,9)	2	7
(3,7)	(4,8)	(3,8)	4	7
(6,5)	(6,5)	(5,6)	5	5
(10,1)	(10,1)	(9,2)	8	5
(4,1)	(10,3)	(4,3)	13	5
(3,1)	(10,5)	(3,5)	3	6
(1,3)	(10,7)	(10,1)	14	5
(4,6)	(5,7)	(4,7)	9	7
(6,4)	(7,5)	(6,5)	10	6
(9,2)	(8,3)	(7,4)	11	6
(8,2)	(9,3)	(8,3)	12	6

To test the sensitivity of the algorithm and to find out the relationship between parameters and inference quality, the algorithm has been simulated with varying regulation time span, varying number of chips in the microarray as well as the noise tolerance for the known target genes situation.

Regulation Time Span

The regulation time span is defined as the percentage of total time spans that the regulation activities are going on in the whole microarray time span. If the regulation time span is high, it is supposed to be strong and vice versa. The simulation has been based upon a 10-gene simulated microarray "cut" from the designed array. The number of chips is intentionally set to be small (10 chips) in our experiments. For one random selected regulation in those 12 regulations as mentioned before, the regulation time span is varied from 10% to 90%. The simulation result is in Fig. 8.

In Fig. 8, the accuracy has been computed as the degree of accuracy or the percentage of correct recognitions (correctness means correct recognition of both regulating genes and regulation logical functions) for running the algorithm 100 times. As can be seen, when the regulation time span is low, the recognition is inaccurate because the microarray does not provide sufficient information to recognize such weak mutual information. However, as the regulation time durations become longer, the accuracy rises up as it is supposed to be. When the regulation time span is high enough, the accuracy almost stays at 100% which is as expected for the algorithm. It means the algorithm

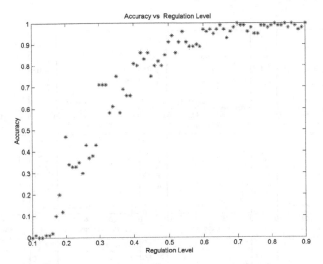

Fig. 8. Accuracy vs. Time span (or regulation level) for known targets

guarantees to find the strong regulations even the time span of the microarray is short. What is really significant here is the convergence phenomenon observed.

Number of Chips

Another important relationship is between the recognition accuracy and the number of chips in the available microarray. If the number of chips in the microarray is small, the total time span presented does not provide enough gene regulation information. While if the microarray covers a longer time span, more regulating information can be garnished in it. Consequently, the performance of the algorithm will improve due to the availability of chips.

This simulation is based on the special designed microarray and each regulation is assigned a fixed regulation time span (range from 10% to 90%). The number of chips varies from 10 to 50 and the accuracy is defined as before but by running the algorithm 30 times. The result is shown in Fig. 9. As expected, the accuracy of recognition increases as the time span becomes larger in the microarray. This result is also intuitively correct and hence appealing.

As can be seen from Fig. 9, when the number of chips reaches above 50, the accuracy reaches 100% for all regulation time spans. For strong regulations, we observe the accuracy remains 100% all the time. But regulations with low regulation time spans need more chips in order to reach 100% correct recognition. This is because when chip numbers are smaller, the accuracy of low regulation time spans is badly affected for lacking of information. So the algorithm of known targets can guarantee to locate all regulations with a moderate requirement of chips in available microarray.

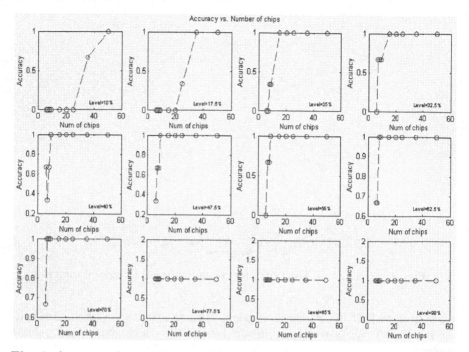

Fig. 9. Accuracy of recognition as a function of number of chips for known targets

Noise Tolerance

It has been tested that with small noise (smaller than 5%) and a large number of chips (300), the algorithm can find both strong and weak regulations in the designed microarray. In general, if noise exists in the microarray, the Hamming threshold should be increased in order to tolerate the noise effectively. To further investigate the general robustness of the algorithm with respect to combating the noise, the Hamming threshold is set as a fixed value and the noise level is varied. This simulation is based on a 50-gene, 20-chips microarray and only one 90% regulation exists in it. The accuracy is then computed for running the algorithm 30 times. The noise level (0.01 to 0.1) is the standard deviation of the white Gaussian noise. The result is shown in the Fig. 10.

As one can see, the accuracy drops sharply as the noise level increases. Not surprisingly, if the noise is too big, the algorithm cannot find the regulation at all. However, this sensitivity to the noise level can be improved if the Hamming threshold is increased accordingly in order to make the algorithm more robust to relatively large noise. The adjustment of Hamming threshold to tolerate noise should involve prior noise estimations. This strategy will be discussed especially for unknown target gene cases.

Beside, if the microarray has more chips, it gives more regulation information and the noise tolerance would be better. The relationship between noise tolerance and number of chips are simulated based on the previous microarray

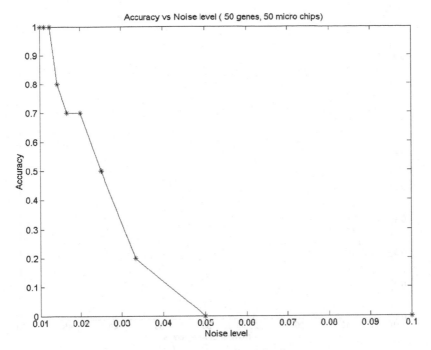

Fig. 10. Accuracy vs. Noise level for known targets

and the noise strength is set to increase from 3% to 9% as the number of chips becomes larger. The simulation result is shown in Fig. 11.

As can be seen, although the noise level rises, the accuracy gets better as the chip number becomes large. So if the microarray data are available for a long time span series, the algorithm can eventually guarantee to find the regulations even under the condition of large noise disturbance.

3.2 Simulation for the Unknown Target Genes

We have studied extensively with prior knowledge for the preferred genes in the investigated microarray. We now shift to the issue when no priori information is known. In other words, the algorithm has to search blindly. Now to ensure good recognition results, the Hamming threshold and inference threshold consequently become very crucial because the distance between category 1 and category 2 are more difficult to be separated under this situation.

One special case of the unknown target genes is that although the exact regulated gene is unknown, but the number of regulations in the microarray is known. Hence it can be taken as a special case between the known targets and unknown targets. To find the regulation mechanism in this situation, we can use the search method for unknown targets and then use the inference technique for searching the known targets (i.e. only the minimum distances

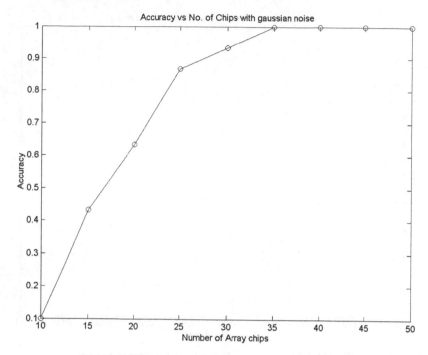

Fig. 11. Accuracy vs. Number of chips with noise

are needed for inference). The simulation is based on the 100-genes designed microarray with one random selected 90% time span of the regulation and the number of chips varies from 10 to 100. The accuracy is plotted on Fig. 12 with the computations of 10 runs each in Monte Carlo style.

This simulation result shows that the algorithm can handle the special situation and when compared with that of the Fig. 9, the number of chips required for 100% accuracy becomes larger. Not surprisingly, it is because this special case has less prior information as compared with known target genes. So the algorithm needs more data and information from the available microarray data (i.e. more chips) to achieve good inference accuracy results.

For the generally unknown target genes situation, the algorithm has been tested with varying inference threshold, varying number of chips, varying Hamming threshold as well as the noise levels.

Inference Threshold

The simulation is based on the designed 100-gene microarray with 12 regulations and the upper limit of the number of chips is fixed at 500. The regulation time spans ranges from 55% to 90%. The inference threshold varies from 100 to 250 (i.e. 20% to 50% of the maximum distance). Because the inference results in unknown target genes situation can not guarantee that the number

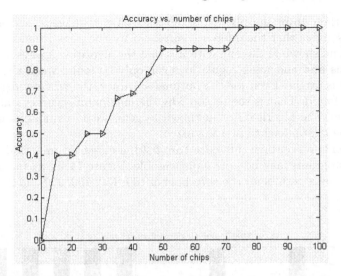

Fig. 12. Accuracy vs. Number of chips for the special case

of inference results equals to the number of true regulations, the accuracy consequently is defined as the percentage of correct recognitions within the total findings, $accuracy = \frac{correct\ recognition}{\max(total\ findings, N\_reg)}$. Suppose the algorithm found 16 regulations but there are only 6 are true regulations in the microarray, the accuracy hence to be determined as $\frac{6}{16}$ which is 37.5% correct recognition.

In Fig. 13, the accuracy first rises with the increase of inference threshold. However, after reaching a peak (where inference threshold is around 200), the

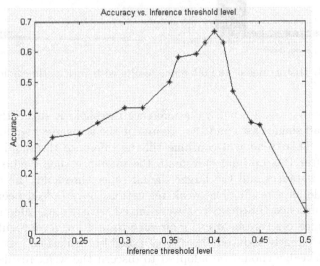

Fig. 13. Accuracy vs. Inference threshold for unknown target genes

accuracy then drops sharply as the inference threshold keeps increasing. It is quite reasonable that the inference threshold for the peak is indeed the optimal. A trend shown in this simulation is that the algorithm always finds strong regulations first and weak regulations can only be found when the inference threshold is larger. Even for the optimal accuracy, the weakest regulation is yet to be found. That is the reason why the maximum accuracy only reaches 68%. So for the algorithm of unknown targets, weak regulations in general can not be guaranteed to be recognized.

To decide an optimal inference threshold, a closer examination of the resulted histograms may become indispensable. Figure 14 shows the histogram of the inference results for the threshold of 100, 150, 190, 220 and 250, respectively.

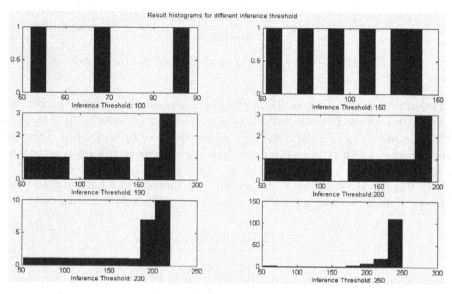

Fig. 14. Histograms of the inference results with various threshold values

As shown in Fig. 14, when the inference threshold is small, the inference results are all single cases and they represent the strongest regulations in the microarray. This trend will continue till the inference threshold reaches its optimal. After this optimal threshold, the inference then begins to involve category 1 distances and the larger the inference threshold, the more errors will be made as a result. The weak regulations have indeed been found by the large inference threshold but were mixed with wrong inference results. However, since big threshold can guarantee to find low level regulations, the results have greatly reduced the size of possible regulations. At this point, biology knowledge should be adopted to discern the weak regulations from errors. This shows that we can never emphasize more for the importance of biological facts and knowledge. Besides, a good inference threshold should not result in an inference that has too many distances in one range. On the other hand, this threshold should not be too small in order to ensure the findings of the weak regulation. This is the criteria for the inference result analysis which is related to Fig. 2. With this criterion, other parameters should also be adjusted accordingly with respect to the trend of their influence on accuracy for an improved result.

Number of Chips

When the number of chips in microarray is large, it implies that more information is inherently hidden. Through simulation, however, the algorithm still cannot locate low level (<20%) regulation without introducing large erroneous recognition results even when large chips are now provided.

To investigate how the number of chips will affect the accuracy, the algorithm has been simulated on the same microarray as in the previous part except the inference threshold is now being fixed at 41% of chips and the Hamming threshold has been fixed at 0.09. The simulation result is shown in the Fig. 15.

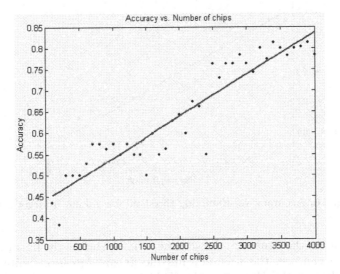

Fig. 15. Accuracy vs. Number of chips for unknown targets

Based upon Fig. 15, the accuracy rises as the number of chips increases. The line is a linear regression fit which clearly shows the linear relationship between the accuracy and the number of chips. Besides, accuracy ascendance levels off during a range of chips and climb up again. It means that only when the number of chips reaches a certain level, will the additional chips then provide enough information for a good inference result. In this simulation, the regulation time span is being set from 55% to 90% which makes 2 weak regulations "invisible" to the 41% inference threshold. So the maximum accuracy will be $\frac{12-2}{12} = 83.3\%$. This has also been confirmed by the simulation result. The maximum accuracy in Fig. 15 is 81.25% and the accuracy stays around 80% despite the increase of chips. Mathematically speaking, if the number of microarray chips approaches infinity, the accuracy should converge to optimal accuracy.

Hamming Threshold

This simulation is based on the special microarray has 500 chips and the regulation time span varies from 55% to 90%. The inference threshold is fixed at the optimal 205. The Hamming threshold is varied from 0.0001 to 0.1 and the simulation result has been shown in Fig. 16.

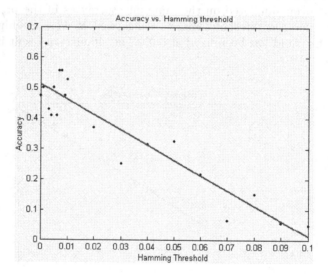

Fig. 16. Accuracy vs. Hamming threshold for unknown target genes

The dots in the figure are the accuracy and the line is the linear regression fit. As one can see, the accuracy fits the linear regression quite well and it shows that as the Hamming threshold increases, the accuracy decreases linearly accordingly. The reason is that big Hamming threshold cannot tell the actual difference in the microarray data discriminately. If noise does not exist in the microarray (which is not impractical) the Hamming threshold should be set as small as possible. However, big Hamming threshold makes the algorithm more robust and less sensitive to noise. So an optimal Hamming threshold is also needed for the algorithm.

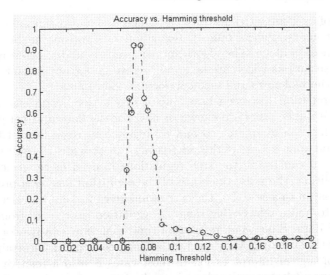

Fig. 17. Accuracy vs. Hamming threshold with noise disturbance

Noise Tolerance

To test the noise tolerance for the unknown target genes situation, only the case of varying the Hamming threshold is investigated because it is closely related to the issue of noise tolerance. The simulation has been based on the 100-gene microarray with 12 strong regulations. The number of chips is once again fixed at 500. As for the white Gaussian noise, the standard deviation is now set to be 10% of the data standard deviation. The simulation result is presented in Fig. 17.

A closer examination of Fig. 17 reveals that the accuracy is zero with small Hamming threshold because noise has disguised the true regulations and small Hamming threshold can no longer discern the true regulations. However, when the Hamming threshold reaches to the level of noise, good inference once again emerges and a spike of accuracy almost reaches 100%. As expected, large Hamming distance introduce more errors into the inference results and the accuracy drops sharply after Hamming threshold cross its optimal value. The lesson can be learned from this simulation is the optimal Hamming threshold should be set around the noise level. So a prior estimation of noise is absolutely necessary in order to ensure good inference.

4 Conclusion

This paper proposed a novel data driven algorithm with a main goal of knowledge discovery for gene regulation microarray data processing. The algorithm has been proven to be quite robust. The main idea of the algorithm is the definition of "accumulated distance" as a measurement of mutual information among the genes. Hence, built-in regulation mechanism can be recognized based upon data alone. The algorithm has been simulated extensively under numerous situations which are believed to have important semantic interpretation for the biologists and medical doctors' viewpoints.

The simulation has been based upon two major conditions concerning whether the target genes are known or not. For the known target genes case, the algorithm can guarantee to find all regulations regardless of the length of regulation time spans. If the microarray can provide a larger time span, the algorithm can locate the true regulations in even more efficiently manner. In addition, the noise tolerance of the algorithm has been proven to be rather robust in this case. For the case of unknown targets, the algorithm can guarantee to find a relative big range of regulation time spans from available microarray data. Since gene regulation time span is a relative concept, it is really, strictly speaking, very much of a fuzzy concept of uncertainties. The semantic biological interpretation in most cases is also a relative and comparative concept because the absolute and crispy concept does not appear to work when dealing with biological problems. As for the weak regulations, the algorithm can be adjusted to find a range of weak regulations and biological knowledge should be combined with the inference results to find the true weak regulations. Besides, criteria to analyze inference results are also provided. As for noise issues in this case, the algorithm can still locate most strong regulations and a criterion is addressed for best inference under noise disturbance. The major accomplishment of the gene regulation network really is to "discover" the relative degree of the regulation which is expected to be the very nature of the biological phenomena. Another conclusion can be made is the template approach will allow us to discover the number of genes participating regulation mechanism and the modes of the regulation through the idea of repeatedly comparison of the accumulated distances.

In a sense, this paper formalizes some of the concept dealing with genetic network discovery. The definition of "what is gene regulation" has been explained in details at various locations in this paper. We still maintain that the microarray chips are the most important available access for different models. The problem here, of course, is the quality of data. In addition, the characteristics of noise arise from the technological design is also very crucial for the performance of the algorithm.

5 Future Research

We believe that this algorithm has opened a door to a new methodology in microarray data processing and genetic network discovery. But this algorithm is only tested on the simulated microarray data. So we will focus our future research in the following aspects:

1. Test the algorithm with the real dynamic microarray data which will be available in due time. From the data processing point of view, a detailed study of normalization method should also be advised in order to avoid large normalization noise.
2. Studies concerning the noise properties in the microarray should be conducted as soon as the true regulation data is available because noise issues are of great importance in microarray data processing. It is nearly impossible to improve the performance without a good handling of noise characteristics.
3. Modify the algorithm to make it more computational efficient. A good heuristic search method will be proposed.
4. It is somewhat anticipated that the proposed algorithm eventually will be called upon to handle huge amount of microarray data. Hence the running of this algorithm will most likely use supercomputer in the near future.

Appendix I

Fuzzy Logic Theory

Fuzzy logic was first proposed by Lotfi A. Zadeh in 1965 as a new way of quantifying the uncertainties in real life. Fuzzy logic incorporates the traditional Boolean logic by generalize it into fuzzy logic and uses partial membership to model the nature of human cognitive process. For example, one would say that "I am not comfortable when other people stand too close to me" rather than "I am not comfortable when people stand 17.521 inches from me." The latter is too precise to be implemented and it is only reasonable to think that the computational models should be able to recognize fuzzy uncertainties [2].

The fuzzy set theory generalizes the membership definition of Boolean network which are only 2 memberships in the set $\{0, 1\}$. In the classical set theory, a subset B of a set S can be defined as a mapping from elements of S to set $\{0, 1\}$;

$$B : S \rightarrow \{0,1\}$$

On the other hand, this mapping can be represented as a set of ordered pairs with exactly one ordered pair associated with each element in B. For instance,

$$S = \{s_1, s_2, s_3\}, B = \{s_1, s_4\}$$
$$B : S \rightarrow \{0, 1\} \Leftrightarrow \{(s_1, 1), (s_4, 0)\}$$

So in Boolean network, the statement of whether an element belongs to a set is only true or false. In parallel comparison, the fuzzy subset F of a set S is also mapping from the elements of S to the closed set $[0, 1]$. So each element in F is also associated with exactly one ordered pair and the last element in each pair describes "the extent" that this element belongs to the set. For the example of weather temperature,

$$S = \{High\ temperature\}, F = \{78F^o, 100F^o, 30F^o\}$$
$$F : S \to [0,1] \Leftrightarrow \{(78F^o, 0.7), (100F^o, 1), (30F^o, 0.05)\}$$

The degrees of membership are presented by the membership functions and membership function is usually defined by the expertise's knowledge. When dealing with gene regulation data, the situation becomes complicated because no prior expertise exists for determining what is "high" or "low". Hence good normalizing methodology based on meaningful biology sense must be employed.

Most recent research concerning gene regulation model has been based on a stochastic model. The difference between fuzzy logic and probability in handling uncertainty is that the fuzziness deals with the type of uncertainty arising when the boundaries of a class of objects are not sharply defined. Membership is just a matter of relative degree rather than the certainty has to be one way or the other. So the degree of gene control mechanism is a very basic fuzzy implication in this model and it makes perfectly sense from human cognitive mind viewpoint. Although the probability and the degrees of membership are distinctive things, they can be implemented synergistically together sometimes. Probabilistic fuzzy logic has already emerged and various researches concerning it are currently under taken by.

To get inference from microarray data, the fuzzy inference format should be determined in advance. Fuzzy inference differs from the traditional inference considerably. It is done through several methods such as MAX-MIN composition, Max-Star composition, Max-product composition or more generally, the concept of so called "T-norm" lately emerged. In this paper, the Max-Min composition is adopted without losing any generality and all possible 2-gene logic combinations are implemented. The MAX-MIN fuzzy logic inference is defined as:

$$IF\ Gene1\ is\ A_1\ AND\ Gene2\ is\ A_2,\ THEN\ MIN(A_1, A_2)$$
$$IF\ Gene1\ is\ A_1\ OR\ gene2\ is\ A_2,\ THEN\ MAX(A_1, A_2)$$
$$IF\ Gene1\ is\ A_1\ NEG(Gene1)\ is\ A_1,\ THEN\ 1 - A_1$$

Fuzzy logic has widely used in control systems and expert systems with many applications from washing machines to auto focus cameras. There are advantages in applying Fuzzy logic in gene regulation data analysis [27]. Firstly, there are high levels of noise in practical microarray data and Fuzzy logic is quite robust in handling the accumulated noise because it emphasizes on the trend, rather than the exact value in which the data driven algorithm is supposed to behave. In comparison to other algorithms such as polynomial fits, fuzzy logic produces the prediction based on real life cognition. Hence the second advantage of using the Fuzzy logic analysis is that its results are easier to interpret by human. Furthermore, facing the availability of vast amount of data from microarrays in the near future, the employment of fuzzy logic techniques are also quite capable of achieving the computational efficiency.

Appendix II

Microarray Technology

A microarray is a small analytical device that allows the measurement of the genomic exploration with high speed and adequate precision unprecedented in the history of biological research. The microarray technique was first developed by Schena and his co-workers in the early 1990s and it is an ordered Matrix like array of microscopic elements on a planar substrate which allows the specific binding of genes or gene products [13]. Its striking characteristic is that it can acquire vast amount of data from a relatively shorter experiment. Through detailed, snapshot-like images, it allows researchers to see how cells use their genetic information in their minute to minute functions and hence to identify the patterns in some unhealthy cells. This experimental methodology to get a microarray is to use biochemistry process to spot mRNA in the fluorescent samples and therefore the intensity of that spot represents the levels of gene activities. The process by which proteins are made from the "instructions" as encoded in a DNA is so called the "gene expression". Each cell contains a blueprint to build the whole biological body of which the proteins are the very basic building blocks. Gene expression provides control over the body's blueprint by selectively producing specific types of proteins in our cells [26]. Therefore, modern microarray technology provides the laboratory technique to learn which genes in the cell are making which products. More importantly, the knowledge of how large number of genes interact with each other and how a regulatory network controls over such a vast number of genes in a stochastically manner is of some paramount importance.

At present, there are mainly two different kinds of microarray in the market. One is called the spot microarray. The base pairing of a sequence of DNA to metaphase chromosomes on a microscope slide is called "In Situ Hybridization". This is a technique that researchers use to determine the presence of a particular DNA or RNA sequence in a cell. By hybridizing a probe for a specific genetic sequence, we are able to look in the microscope and see the place in the cell where that probe binds. This approach is useful in the mapping of genes and hence in the study of gene expressions. As it turned out, it is also a key process in DNA fingerprinting. Researchers then attach fluorescent labels (usually $Cy5$ and $Cy3$) to the DNA (PCR product) & mRNA samples from the cell they are studying. The labeled probes are allowed to bind the complementary DNA stands on the slides. Then the slides are put into a scanning microscope that can measure the relative brightness of each fluorescent dot; brightness reveals how much of a specific DNA fragment is present, an indicator of how active that fragment is. Based on the color and brightness of the reference spots, the microarray data in term of brightness must be processed into some raw data for further refined data processing. This image scanning and intensity measurement can be done using the some commercial software. This normalization process can be done based on statistical normalization ($mean\ (Cy3) = mean\ (Cy5)$) or using spike control normalization. The second kind of microarray is the so called "Affymetrix microarray" which is based on oligonuleotide hybridization. The oligonuleotide are single-stranded molecules, allowing hybridization to complementary probe molecules without denaturing or concerns about reannealing. The "Affymetrix microarray" has been developed by Affymetrix Inc. and the main advantages of the oligonuleotide targets are increased specificity. The capacity to work directly from the sequence database information can be viewed as a plus advantage. The following table shows the most important difference between the two different kinds of microarrays.

Table 2. Comparison of spot microarray and Affymetrix microarray

Array	Advantage	Disadvantage
Spot	Flexibility	Variation
	Price	Optimization
Affymetrix	Optimization	Organism specific
	Fast results	Large subsets only
	Internal controls	Price

For gene regulation analysis, the dynamic data or time series microarray data must be acquired as our target data set. This dynamic process holds the key to our understanding of the gene regulation; consequently the proteins products. It is also generally recognized that a microarray data is quite suitably represented by a fuzzy set over a range of [0, 1] because one can always normalize a set of number as fuzzy set once the definition of the "regulation" is crystallized. In practice, the time intervals in microarray may not always achieve synchronization. However, in the data processing period, an assumption has been made in this paper that different time intervals are treated as the same. This assumption implies the situation that the regulation processes at different time intervals are equally effective.

References

1. Akutsu, T. et al., *Algorithm for inferring qualitative models of biological networks*, Pacific Symposium on Biocomputing, 2000.
2. Barro, S. et al., *Fuzzy logic in medicine*, Physica-Verlag, 2002.
3. Carvalho, J. et al., *Automatic Implementation and Simulation of Dynamic Qualitative Systems using Fuzzy Boolean Networks and Fuzzy Rule-Based Cognitive Maps*, International Conference on Computational Science, pp. 217–220. 2001.
4. Chen, T., *Modeling gene expression with differential equations*, Pacific Symposium on Biocomputing, 2000.
5. Jarvis, E. et al., *Integrating the Songbird Brain*, (under revision for *Physiology Journal*), 2002.
6. Friedman, N. et al., *Learning Bayesian Network Structure of Dynamic Probabilistic Network*, Proceeding of the Fifteenth Conference on Uncertainty in Artificial Intelligence 206–215. 1999.
7. Gibson, M. et al., *Modeling the activity of single genes, in computational modeling of genetic and biochemical networks*, Bower, J. M. and Bolouri, H. (Eds) MIT Press, Cambridge MA, pp. 1–48, 2001.
8. Harford, J. et al., *mRNA metabolism & post-transcriptional gene regulation*, Wiley-Liss, 1997.
9. Jong, H., *Modeling and simulation of genetic regulatory systems: A literature review*, J. Computational Biology. 9(1): 67–103, 2002.
10. Murphy, K. et al., *Modeling gene expression data using dynamic Bayesian Networks*, Technical Report, University of California, Berkeley, 1999.
11. Pasanen, T., *Formatting gene regulatory patterns with fuzzy logic*, 3rd international conference on system biology, 2002.
12. Plahte, E. et al., *Global analysis of steady points for systems of differential equations with sigmoid interactions*, Dyn. Stab. Syst. 9(4):275–291, 1994.
13. Schena, M., *Microarray analysis*, Wiley-Liss, 2002.
14. Smith, V. et al., *Evaluating Functional Network Inference Using Simulations of Complex Biological Systems*, Proceeding of the 10th international conference on Intelligent Systems for Molecular Biology, 2002.
15. Sokhansanj, B., *Building and testing scalable fuzzy models of genetic regulation in bacteria: analysis of genomic data*, International conference on computational nanoscience and nanotechnology, 2002.
16. Sokhansanj, B. et al., *Applying URC Fuzzy Logic to model Complex Biological Systems in the Language of Biologists*, 2nd international conference on system biology, 2001.
17. Thieffey, D. et al., *Dynamical behavior of biological regulatory networks*II. Immunity control in bacteriophage lambda*, Bull. Math. Biol. 57, 277–297, 1995.
18. Vladimir. G., *Kauffman's NK Boolean network*, http://pespmel.vub.ac.be/BOOLNETW.html, 2000.
19. Wang, P. et al., *A study of two gene network- the simplest special case of SORE (Self Organizable & Regulating Engine)*, Proceedings of 7th JCIS joint conference, 2003.
20. Wang, P. et al., *SORE- an example of a possible building block for a "Biologizing" control system*, Proceedings of 7th JCIS joint conference, 2003.
21. Wang, P. et al., *What is SORE (Self Organizable & Regulating Engine)*, Proceedings of 7th JCIS joint conference, 2003.

22. Wang, P. et al., *A novel method of error correcting code generation based on SORE*, Proceedings of 7th JCIS joint conference, 2003.
23. Wang, P. et al., *SORE, Self Organizable & Regulating Engine-A powerful classifier*, Proceedings of 7th JCIS joint conference, 2003.
24. Weaver, D. et al., *Modeling regulatory networks with weight matrices*, Pacific Symposium on Biocomputing, 4, 112–123, 2002.
25. Wen, X. et al., *Large scale temporal gene expression mapping of central nervous system development*, Proc. Natl. Acad. Sci. USA, 95, 1 334–339, 1998.
26. Witherly, J. et al., *An A to Z of DNA science*, Cold spring harbor laboratory press, 2001.
27. Woolf, P. et al., *A fuzzy logic approach to analyzing gene expression data*, Physio. Genomics: 9–15, 2000.
28. Zhu, J. et al., *Cluster, function, and promoter: analysis of yeast expression array*, Pacific Symposium on Biocomputing, 5, 476–487, 2000.

Semantic Relations and Information Discovery

D. Cai and C.J. van Rijsbergen

Department of Computing Science, University of Glasgow, G12 8RZ, Scotland
{caid;keith}@dcs.gla.ac.uk

Abstract. The treatment of semantic relations between terms is essential in information retrieval (IR). Each term in a thesaurus might have classes of synonymous, narrower, broader, or related terms. The ability to express formally the semantic relations is a core issue in applying mathematical tools to IR. In the application of evidential theory, particularly, the problems become more apparent since evidential theory is based on set theory and individual terms have to be expressed as subsets of the frame of discernment. Four basic questions are to be faced: How to establish the frame of discernment using a thesaurus? How to express terms using subsets of the frame? How to apply evidential functions to represent documents or queries using the term subset-expressions? What are appropriate agreement measures for ranking documents against a given query? This study attempts to answer these questions.

1 Introduction

The objective of an IR system is to identify latent useful information in response to user information needs. The effectiveness of an IR model depends mainly on three central issues: document representation, query representation and agreement (similarity) measurement. Many important studies focus on the three issues, and some good formal methods have been developed.

In IR, each document is characterized by a set of *index terms* that appear in the document. There exist complex semantic relations between index terms. Generally, a *weighting* function, which maps each index term to a numerical quantity related to a given document, is used to represent the importance of the index term concerning the document. It should be pointed out that to arrive at a precise representation of a document by means of a set of weights of index terms is difficult. This is because it is very hard to obtain sufficient statistical data for estimating the importance of index terms. It is also very hard to explicate the complicated semantic relations between index terms. Extensive studies on document representation can be found in, for instance [1, 11, 17, 20, 28, 29, 30, 32, 33, 34].

D. Cai and C.J. van Rijsbergen: *Semantic Relations and Information Discovery*, Studies in Computational Intelligence (SCI) **5**, 79–78 (2005)
www.springerlink.com

Query representation is also an obstacle to developing an effective retrieval system. In practice, the original queries are usually imprecise and incomplete descriptions of information needs. A retrieval system cannot thus be expected to produce ideal retrieval results by using a poor query representation. Some thorough investigations into query representation and query expansion can be found in, for instance [4, 6, 7, 8].

An agreement measure determines the degree to which individual documents are relevant to the query. To be successful, the determination should be performed in such a way that retrieval output and actual outcome are, on average, in close agreement. The choice of agreement measure is essential for effective retrieval. The relevance problem has been studied by many researchers, for instance [2, 9, 15, 23].

The treatment of semantic relations between terms has long been a significant subject of interest in IR. Terms in a thesaurus might have a class of synonymous terms, a class of narrower terms, a class of broader terms, or a class of related terms. For example, semantically, terms *cows* and *goats* are narrower than term *mammals*. In a semantic net, "term t_1 is narrower than term t_2" is expressed by an arrow from t_1 to t_2 to impose a specification/generalization relation on terms. An example for related terms is that terms *ducks*, *geese*, *hens* might be related to terms *eggs*, *feathers*. It is necessary to normalize a thesaurus for treating term semantic relations in IR applications. This study presents a method for the normalization.

The ability to express formally the semantic relations of terms is a core issue in IR. Terms are not mutually exclusive, and naive probabilistic methods may not be adequate for handling the issue. The evidential method appears to be more convenient than the usual probabilistic methods [32] to express the semantic relations, to represent *objects* (i.e., documents and queries), and to rank documents against a given query. Some retrieval methods based on evidential theory (Dempster-Shafer's theory of evidence [25]) have been proposed [17, 18, 24, 26]. However, in applications of evidential theory, the construction of a σ-algebra and a base for this σ-algebra in order to construct a probability space is an arduous task [26]. This study proposes a method to establish the frame of discernment, and to express each term by a subset of the frame.

In this study, we are also concerned with the application of evidential theory to some practical IR problems: we formally discuss the representations of objects by means of evidential functions; we introduce agreement measures over the evidential representations for ranking documents against the query.

This paper is organized as follows. In Sect. 2, we suggest a method for normalizing a thesaurus semantically. In Sect. 3, we propose a method for establishing a frame of discernment and for expressing individual terms as subsets of the frame. In Sect. 4, after introducing basic concepts of evidential theory, a novel method for representing objects based on evidential functions is proposed, and the agreement measures for ranking documents based on the evidential representations are introduced.

2 Normalization of a Thesaurus

This section concentrates on how to normalize a thesaurus. We introduce notation for the normalization. The notation is used to describe the semantic relations found in a general thesaurus. An example thesaurus $\aleph$, taken from [26], is used throughout this paper. It is given in Table 1. We will denote T as the set of 28 terms contained in $\aleph$.

2.1 Treatment of Synonymous Terms

We first introduce two pieces of notation, $\Leftarrow$ and $\Rightarrow$.

☞ "$\Leftarrow$" is used to denote "use-for".
   "$t_1 \Leftarrow t_2, t_3, \ldots, t_l$" means that t_1 and $t_2, t_3, \ldots, t_l$ are synonymous terms, and that t_1 is their representative.
☞ "$\Rightarrow$" is used to denote "use".
   "$t_1 \Rightarrow t_2$" means that t_1 and t_2 are synonymous terms, and that t_2 is their representative.

The *synonymous* relation between terms is an equivalence relation. That is, it has the following properties:

reflexive: $t \Leftarrow t$;
symmetric: if $t_1 \Leftarrow t_2$, then $t_2 \Leftarrow t_1$;
transitive: if $t_1 \Leftarrow t_2$, and $t_2 \Leftarrow t_3$, then $t_1 \Leftarrow t_3$.

Removing the tuples with relations $\supset$, $\subset$ and $\cap$ from the thesaurus, and then with symmetry, further removing relation $\Rightarrow$, we can obtain a *synonym-normalized* thesaurus, denoted $\mathcal{S}_{(k,t)}$. For the example thesaurus $\aleph$, we obtain Table 2.

In contrast to thesaurus $\aleph$, we can see that terms that do not have relations $\Leftarrow$ or $\Rightarrow$, but have relations $\supset$, $\subset$ and $\cap$ remain in the first column of thesaurus $\mathcal{S}_{(k,t)}$ (e.g., terms "domestic-birds", "eggs", etc.). In other words, only those terms that have only relation $\Rightarrow$ are removed from the first column of $\aleph$ (e.g., terms "barnyard-birds", "farm-animals", etc.).

In what follows, we will call the terms listed in the first column of thesaurus $\mathcal{S}_{(k,t)}$ *key-terms*, and denote $K = \{k_1, k_2, \ldots, k_n\}$ as the set of the key-terms. Obviously, $K \subseteq T$ and $n = |K| \leq |T|$. For instance, for thesaurus $\aleph$, we have $n = 15$ key-terms.

2.2 Treatment of Ordering Terms

Let us introduce two further pieces of notation $\supset$ and $\subset$.

☞ "$\supset$" is used to denote "narrower-terms".
   "$t_1 \supset t_2, t_3, \ldots, t_l$" means that term t_1 has narrower terms $t_2, t_3, \ldots, t_l$. The *narrower* relation between terms is a partial ordering relation. That is, it has the following properties:

Table 1. An example thesaurus ℵ

Term	Relation	Term(s)
animals	⇐	animal
	⊃	birds, domestic-animals, mammals
barnyard-birds	⇒	poultry
birds	⇐	bird
	⊂	animals
	⊃	domestic-birds, poultry
	∩	eggs, feathers
cows	⇐	cow
	⊂	domestic-mammals
domestic-animals	⇐	farm-animals
	⊂	animals
	⊃	domestic-birds, domestic-mammals, poultry
domestic-birds	⊂	birds, domestic-animals
	⊃	poultry
domestic-mammals	⊂	domestic-animals, mammals
	⊃	cows, goats
ducks	⇐	duck
	⊂	poultry
eggs	∩	birds, poultry
farm-animals	⇒	domestic-animals
farmyard-birds	⇒	poultry
feathers	∩	birds, poultry
geese	⇐	goose
	⊂	poultry
goats	⇐	goat
	⊂	domestic-mammals
hens	⇐	chick, chicken-cock, hen
	⊂	poultry
mammals	⇐	mammal
	⊂	animals
	⊃	domestic-mammals
	∩	milk
milk	∩	mammals
poultry	⇐	barnyard-birds, farmyard-birds
	⊂	birds, domestic-animals, domestic-birds
	⊃	ducks, geese, hens
	∩	eggs, feathers

Table 2. A synonym-normalized thesaurus $\mathcal{S}_{(k,t)}$

Key-Term	Relation	Term(s)
animals	$\Leftarrow$	animal
birds	$\Leftarrow$	bird
cows	$\Leftarrow$	cow
domestic-animals	$\Leftarrow$	farm-animals
domestic-birds	$-$	
domestic-mammals	$-$	
ducks	$\Leftarrow$	duck
eggs	$-$	
feathers	$-$	
geese	$\Leftarrow$	goose
goats	$\Leftarrow$	goat
hens	$\Leftarrow$	chick, chicken-cock, hen
mammals	$\Leftarrow$	mammal
milk	$-$	
poultry	$\Leftarrow$	barnyard-birds, farmyard-birds

irreflexive: $t \not\supset t$;

asymmetric: if $t_1 \supset t_2$ then $t_2 \not\supset t_1$;

transitive: if $t_1 \supset t_2$ and $t_2 \supset t_3$, then $t_1 \supset t_3$.

☞ "$\subset$" is used to denote "broader-terms".

"$t_1 \subset t_2, t_3, \ldots, t_l$" means that term t_1 has broader terms $t_2, t_3, \ldots, t_l$. The *broader* relation between terms is a partial ordering relation. That is, it has the following properties:

irreflexive: $t \not\subset t$;

asymmetric: if $t_1 \subset t_2$ then $t_2 \not\subset t_1$;

transitive: if $t_1 \subset t_2$ and $t_2 \subset t_3$, then $t_1 \subset t_3$.

Clearly, term t_i has a narrower term t_j if and only if term t_j has a broader term t_i. This implies that we can use only one of these two ordering relations in a thesaurus to obtain an *ordering-normalized* thesaurus, denoted $\mathcal{O}_{(k_i,k_j)}$.

For the example thesaurus $\aleph$, removing the tuples with relations $\Leftarrow, \Rightarrow$ and $\cap$, and then further removing relation $\subset$, we obtain Table 3.

The narrower and broader relations of terms, forming a hierarchical structure, are given in Fig. 1. Generally, in IR, a *hierarchical structure*, represented using a *directed acyclic graph* (using arrows), is a tree-like structure, which embeds relations $\supset$ or $\subset$ in a form such that no key-term appears on a level below that of its narrower key-term.

Notice that all terms given in the first and third columns of thesaurus $\mathcal{O}_{(k_i,k_j)}$ (i.e., on the hierarchical structure) are key-terms. In what follows, we will denote the set of all key-terms on the hierarchical structure as $hie(K)$. Obviously, $hie(K) \subseteq K$, and it may be that $K - hie(K) \neq \emptyset$. For instance, from our example, we can see that $K - hie(K) = \{eggs, feathers, milk\}$.

Table 3. An ordering-normalized thesaurus $\mathcal{O}_{(k_i, k_j)}$

Key-Term	Relation	Key-Term(s)
animals	⊃	birds, domestic-animals, mammals
birds	⊃	domestic-birds, poultry
domestic-animals	⊃	domestic-birds, domestic-mammals, poultry
domestic-birds	⊃	poultry
domestic-mammals	⊃	cows, goats
mammals	⊃	domestic-mammals
poultry	⊃	ducks, geese, hens

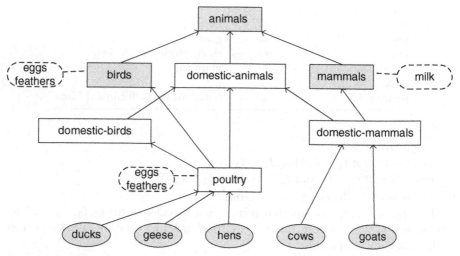

Fig. 1. The narrower and broader relations of key-terms on a hierarchical structure

We denote the hierarchical structure itself as $H(K)$. A key-term k is said to be on structure $H(K)$, denoted by $k \vdash H(K)$, if there exists one node of $H(K)$, such that it is used to represent k. Clearly, $k \in hie(K)$ if $k \vdash H(K)$.

2.3 Treatment of Related Terms

We need to introduce another piece of notation, ∩.

☞ "∩" is used to denote 'related-terms'.

"$t_1 \cap t_2, t_3, \ldots, t_l$" means that term t_1 has related terms $t_2, t_3, \ldots, t_l$. The *related* relation between terms has the following properties:

reflexive: $t \cap t$;

symmetric: if $t_1 \cap t_2$, then $t_2 \cap t_1$;

transitive: if $t_1 \cap t_2$, and $t_2 \cap t_3$, then $t_1 \cap t_3$ may not hold.

For the example thesaurus ℵ, deleting the tuples with relations ⇐, ⇒, ⊃ and ⊂, and then with symmetry, we can obtain Table 4.

Table 4. A related-normalized thesaurus $\mathcal{R}_{(k_i, k_j)}$

Key-Term	Relation	Key-Term(s)
birds	$\cap$	eggs, feathers
mammals	$\cap$	milk
poultry	$\cap$	eggs, feathers

The related relation of terms linked to the hierarchical structure (using dashed lines) is given in Fig. 1.

Notice that all terms listed in the first and third columns of thesaurus $\mathcal{R}_{(k_i, k_j)}$ are key-terms. For two related key-terms $k_i, k_j \in K$, we always have $k_i \cap k_j$ and $k_j \cap k_i$. However, we use the following rules to write their related relation:

- if $k_i \in hie(K)$ and $k_j \in K - hie(K)$, write $k_i \cap k_j$;
- if $k_i \in K - hie(K)$ and $k_j \in hie(K)$, write $k_j \cap k_i$;
- if $k_i, k_j \in hie(K)$, simply ignore them as their related relation has been implied by their narrower or broader relations (see Sect. 2.4);
- if $k_i, k_j \notin hie(K)$, simply ignore them as their related relation is not linked to the hierarchical structure.

That is, the right side of $\cap$ is always a key-term in $hie(K)$, i.e., in the first column of the related-normalized thesaurus; the left side of $\cap$ is always a key-term in $K - hie(K)$, i.e., in the third column of the related-normalized thesaurus.

In what follows, we will denote the set of the key-terms linked to the hierarchical structure (i.e., listed in the third column of thesaurus $\mathcal{R}_{(k_i, k_j)}$) as $rla(K)$. Obviously, $rla(K) \subseteq K$. For our examples, we can see that $rla(K) = \{eggs, feathers, milk\}$.

Notice also that, for an arbitrary key-term k_j in $K - hie(K)$, we have three and only three cases:

→ There exists at least one $k_i \vdash H(K)$, such that, $k_i \cap k_j$; in this case, $k_j \in rla(K)$.
→ There exists one $k_{i_1} \vdash H(K)$ and $k_{i_2}, \ldots, k_{i_\lambda} \nvdash H(K)$ ($\lambda \geq 2$), such that, $k_{i_1} \cap k_{i_2}, k_{i_2} \cap k_{i_3}, \ldots, k_{\lambda-1} \cap k_\lambda, k_\lambda \cap k_j$; in this case, $k_j \notin rla(K)$, as the related relation does not satisfy transitivity.
→ There exists no $k_i \vdash H(K)$, such that, $k_i \cap k_j$; in this case, $k_j \notin rla(K)$.

2.4 Superiority and Inferiority

Having discussed the properties of the ordering and related relations we can further discuss the properties between these two relations. The properties which are useful for an insight into the semantic relations are:

superior: if $t_1 \cap t_2$ and $t_3 \supset t_2$, then $t_1 \cap t_3$;
inferior: if $t_1 \cap t_2$ and $t_2 \supset t_3$, then $t_1 \cap t_3$ may not hold.

From the superiority property we obtain the other two properties:

– If $t_1 \supset t_2$, then $t_1 \cap t_2$.
  In fact, from reflexivity we have $t_2 \cap t_2$. Now $t_1 \supset t_2$. So $t_2 \cap t_1$ by superiority, i.e., $t_1 \cap t_2$ by symmetry.
– If $t_1 \subset t_2$, then $t_1 \cap t_2$.
  In fact, from reflexivity we have $t_1 \cap t_1$. Now $t_1 \subset t_2$. So $t_1 \cap t_2$.

By the superiority property, we can further infer the related relations between key-terms. For instance, from $milk \cap mammals$ and $animals \supset mammals$, we have $milk \cap animals$.

However, the inferiority property may not always hold. For instance, from $milk \cap animals$ and $animals \supset birds$, an inferred result $milk \cap birds$ obviously makes no sense.

In what follows, we will call $\mathcal{S}_{(k,t)}$, $\mathcal{O}_{(k_i,k_j)}$, $\mathcal{R}_{(k_i,k_j)}$ together the *semantically normalized thesaurus*, and denote it by

$$\aleph_{SOR} = \left[\aleph; \ \mathcal{S}_{(k,t)} | \mathcal{O}_{(k_i,k_j)} | \mathcal{R}_{(k_i,k_j)} \right] .$$

Our aim is to formally express the semantic relations between terms, that is, to establish the frame of discernment, and then to express all key-terms as subsets of the frame. The next section attempts to discuss this core issue.

Before discussing the core issue, we first point out that the normalized thesaurus is a precondition for our method to be used. However, in the real world, it is very likely that a thesaurus will not satisfy the conditions of normalization. The problem of normalizing a thesaurus is a pressing one, and may necessitate much effort. It is beyond the scope of this paper to discuss such a problem, and will be treated as a significant subject for further study. Thus, in what follows, we always assume that thesaurus $\aleph$ has been normalized to thesaurus $\aleph_{SOR}$.

3 Subset-Expressions of Key-Terms

As is known, in the application of evidential theory to IR, the frame of discernment, in which elements are exclusive and exhaustive, must first be established, and each key-term must be expressed as the subset of the frame. This section attempts to give a method to establish the frame and to derive the subset-expressions of key-terms.

To do so, we need to introduce two further pieces of notation, $\rightleftharpoons$ and $\leftleftarrows$.

- "$\rightleftharpoons$" is used to denote "expressed-by".
  "$k_1 \rightleftharpoons \{k_2, k_3, \ldots, k_l\}$", where $k_2, k_3, \ldots, k_l$ are atomic-terms, means that key-term k_1 is expressed by the set of atomic-terms $\{k_2, k_3, \ldots, k_l\}$.
- "$\leftleftarrows$" is used to denote "equivalent-to".
  "$k_1 \leftleftarrows k_2, k_3, \ldots, k_l$", where $k_1, k_2, \ldots, k_l$ have the same subset-expression, means that $k_1, k_2, k_3, \ldots, k_l$ are equivalent key-terms, and k_1 is their representative.

3.1 The Sub-Frame of Discernment Θ'

The derivation of all atomic-terms is the starting point for establishing the frame of discernment.

Atomic-terms can be derived from the hierarchical structure $H(K)$. In order to generate $H(K)$, we arrange key-terms from narrower to broader (or from specific to general) by using an *arrow* pointing to a "parent" node (representing a key-term) k_i from a "child" node k_j; the arrow denotes the relation $k_i \supset k_j$. The oldest node of $H(K)$, such as key-term *animals*, is called the root of the hierarchical structure (see Sect. 3.8).

The nodes to which no arrows point, are called *terminal nodes*. All terminal nodes are regarded as *atomic-terms*. The set of all atomic-terms, denoted by Θ', is called the *sub-frame of discernment*. Obviously, $\Theta' \subseteq hie(K)$. From Fig. 1, we can see that only five key-terms are atomic-terms in thesaurus $\aleph_{SOR}$:

$$\Theta' = \{cows, ducks, geese, goats, hens\} .$$

Once all atomic-terms on $H(K)$ are derived, we are able to further express general key-terms using a subset of the sub-frame, called a *subset-expression*.

Each atomic-term can be expressed by itself. That is, for an arbitrary $a' \in \Theta'$, $a' \rightleftharpoons \{a'\}$. Thus, atomic-terms are always pairwise unrelated since the intersection of their subset-expressions is always empty.

3.2 Key-Terms in Set $hie(K)$

The subset-expressions of general key-terms can also be derived from the hierarchical structure $H(K)$.

First, consider all key-terms that are narrower or broader than at least one other key-term. For an arbitrary $k \in hie(K) \subseteq K$, a key point of the derivation is to find all possible narrower key-terms, and then traverse downward to atomic-terms. The subset of atomic-terms narrower than k is used as the subset-expression of k. For instance, in Fig. 1, we see that key-term *birds* has narrower atomic-terms *ducks*, *geese* and *hens*, we can thus express *birds* by a subset $\{ducks, geese, hens\}$. Further, with the symbol $\rightleftharpoons$, we can write subset-expressions for all key-terms in $hie(K)$ as shown in Table 5.

3.3 Key-Terms in Set $rla(K)$

Next, consider all key-terms that are not narrower or broader than any other key-terms, but are directly related to at least one other key-term on $H(K)$. For an arbitrary $k \in rla(K) \subseteq K$, link k to the hierarchical structure using a *dashed line* between k and the key-terms to which k is related; check the *youngest* one among the key-terms; express k using the same subset-expression as the youngest one. For instance, in Fig. 1, we see key-term *eggs* linked to the hierarchical structure by dashed lines between it and two key-terms *birds* and

Table 5. Subset-expressions for key-terms in set $hie(K)$

Key-Term	Relation	Subset-Expression
animals	$\rightleftharpoons$	$\{cows, ducks, geese, goats, hens\}$
birds	$\rightleftharpoons$	$\{ducks, geese, hens\}$
cows	$\rightleftharpoons$	$\{cows\}$
domestic-animals	$\rightleftharpoons$	$\{cows, ducks, geese, goats, hens\}$
domestic-birds	$\rightleftharpoons$	$\{ducks, geese, hens\}$
domestic-mammals	$\rightleftharpoons$	$\{cows, goats\}$
ducks	$\rightleftharpoons$	$\{ducks\}$
geese	$\rightleftharpoons$	$\{geese\}$
goats	$\rightleftharpoons$	$\{goats\}$
hens	$\rightleftharpoons$	$\{hens\}$
mammals	$\rightleftharpoons$	$\{cows, goats\}$
poultry	$\rightleftharpoons$	$\{ducks, geese, hens\}$

poultry. The youngest of the two key-terms is *poultry* with subset-expression $\{ducks, geese, hens\}$, we can thus express *eggs* by the same subset *poultry* has. Subset-expressions for all key-terms in $rla(K)$ are shown in Table 6.

Table 6. Subset-expressions for key-terms in set $rla(K)$

Key-Term	Relation	Subset-Expression
eggs	$\rightleftharpoons$	$\{ducks, geese, hens\}$
feathers	$\rightleftharpoons$	$\{ducks, geese, hens\}$
milk	$\rightleftharpoons$	$\{cows, goats\}$

Notice that the choice of the youngest key-term among the key-terms related to a given key-term is essential: it ensures that the related relation is able to satisfy the superiority property (but does not ensure it satisfies the inferiority property).

3.4 Representatives in Set $rep(K)$

Often, while all key-terms in $hie(K) \cup rla(K)$ are expressed by subset-expressions, some key-terms have the same subset-expressions.

Two key-terms $k_i, k_j \in hie(K) \cup rla(K)$ are said to be *equivalent*, denoted by $k_i \leftleftarrows k_j$, if they have the same subset-expression. The set of equivalent key-terms is called an *equivalent class*, denoted by $equ(k)$, where k is a representative of the equivalent class. Denote $rep(K)$ as the set of all the representatives and, obviously, $rep(K) \subseteq (hie(K) \cup rla(K))$.

We can choose a *representative* for each equivalent class by taking the *oldest* key-term of the class. For instance, from Fig. 1 and from Tables 5 and 6, we see that key-terms *birds*, *domestic-birds*, *eggs*, *feathers* and *poultry* have the same subset-expression, and that *birds* is the oldest one, we can thus take *birds* as the representative of the equivalent class *birds*, *domestic-birds*, *eggs*, *feathers* and *poultry*. Further, with the symbol $\Leftarrow$, we can write all representatives for key-terms in $hie(K) \cup rla(K)$ as shown in Table 7.

Table 7. Equivalent classes' representatives in set $rep(K)$

Representative	Relation	Key-Term(s)
animals	$\Leftarrow$	domestic-animals
birds	$\Leftarrow$	domestic-birds, eggs, feathers, poultry
mammals	$\Leftarrow$	domestic-mammals, milk

Notice that the choice of the oldest key-term as the representative of an equivalent class is immaterial: this is done only for the purpose that the root of the hierarchical structure can be chosen as a representative.

It is worth mentioning, similar to the synonym relation $\Leftarrow$ on the term set T, that relation $\Leftarrow$ is an equivalent relation on the key-term set K with the following properties:

reflexive: $k \Leftarrow k$ for an arbitrary $k \in K$;
symmetric: if $k_i \Leftarrow k_j$ then $k_j \Leftarrow k_i$, where $k_i, k_j \in K$;
transitive: if $k_i \Leftarrow k_j$ and $k_j \Leftarrow k_l$ then $k_i \Leftarrow k_l$, where $k_i, k_j, k_l \in K$.

It may seem odd that we have relation *birds* $\Leftarrow$ *eggs* and *mammals* $\Leftarrow$ *milk*, etc. Nevertheless, it is mathematically reasonable. Due to the limitations of thesaurus $\aleph_{sor}$, key-terms may not be entirely distinguished from each other. Semantically, it might be doubtful that key-terms *birds* and *eggs* are "the same" according to some knowledge. Mathematically, key-terms *birds* and *eggs* are both equal to $\{ducks, geese, hens\}$ according to the sub-frame of discernment Θ', which has only five atomic-terms totally. Semantically, two key-terms are "the same" if they are synonymous. Mathematically, two key-terms are equivalent if they contain the same atomic-terms.

3.5 Isolated-Terms in Set $iso(K)$

Also, we need consider some isolated key-terms. A key-term k' is said to be an *isolated-term* if there exists no key-term $k_i \in hie(K)$, such that, $k_i \supset k'$ or $k_i \cap k'$. That is, if key-term k' is isolated, then k' is neither on the hierarchical structure, nor (directly) related to any key-term which is on the hierarchical structure. Denote

$$iso(K) = K - hie(K) - rla(K) .$$

Then, $k' \in iso(K)$ is an isolated-term. For instance, from Tables 2, 5 and 6, we can see that $iso(K) = \emptyset$.

Like atomic-terms, each isolated-term can be expressed by itself. That is, for an arbitrary $k' \in iso(K)$, $k' \rightleftharpoons \{k'\}$. Thus, isolated-terms are also pairwise unrelated as the intersection of their subset-expressions is always empty.

However, in practice, isolated-terms may themselves be semantically related to one another. Further study on how to treat the related relation of isolated-terms is needed.

3.6 The Frame of Discernment Θ

Generally, a *frame of discernment*, denoted by Θ, can immediately be established after the sub-frame of discernment and the set of isolated-terms are given:

$$\Theta = \Theta' \cup iso(K) = \{a_1, a_2, \ldots, a_{|\Theta|}\} .$$

Clearly, $|\Theta| = |\Theta'| + |iso(K)|$ as $\Theta' \cap iso(K) = \emptyset$.

3.7 The Evidence Sub-Space K'

Finally, we form an evidence sub-space K'. An *evidence sub-space*, denoted by K', is a subset of the power set of the frame of discernment Θ, over which the evidence functions can be defined:

$$K' = \Theta \cup rep(K) = \{k'_1, k'_2, \ldots, k'_m\} ,$$

where the dimensionality of the sub-space is Θ, and the size of the sub-space satisfies: $m = |K'| \le |\Theta| + |rep(K)|$ as it may be that $\Theta \cap rep(K) \neq \emptyset$ [5].

The key-terms in K' are called *kernel-terms*. From the above discussion, it is clear that each kernel-term in $K' \subseteq K$ can be expressed by a subset of Θ. For thesaurus $\aleph_{SOR}$, we can write subset-expressions for all kernel-terms as shown in Table 8.

Table 8. Subset-expressions for kernel-terms in set K'

Kernel-Term	Relation	Subset-Expression
animals	$\rightleftharpoons$	$\{cows, ducks, geese, goats, hens\}$
birds	$\rightleftharpoons$	$\{ducks, geese, hens\}$
cows	$\rightleftharpoons$	$\{cows\}$
ducks	$\rightleftharpoons$	$\{ducks\}$
geese	$\rightleftharpoons$	$\{geese\}$
goats	$\rightleftharpoons$	$\{goats\}$
hens	$\rightleftharpoons$	$\{hens\}$
mammals	$\rightleftharpoons$	$\{cows, goats\}$

It is important to understand that representatives (kernel-terms) in $rep(K)$ may be related to each other as the intersection of their subset-expressions over Θ' may not be empty. For instance, representatives *birds* and *mammals* are not related to one another, but both of them are related to representatives *animals*. Consequently, kernel-terms in $K' \supseteq rep(K)$ may be related to each other, and may not partition the frame of discernment $\Theta \supseteq \Theta'$.

3.8 Multiple Subset-Expressions

In the above discussion, we gained insight into the concept of term semantic relations by means of a normalized thesaurus, and a hierarchical structure generated from the normalized thesaurus. From thesaurus $\aleph_{SOR}$, for instance, we generated a hierarchical structure $H(K)$ with a root *animals*. In a real world application, however, there may exist many roots with respect to a given normalized thesaurus.

A key-term k is called a *root*, if

- there exists no key-term k_i, such that, $k_i \supset k$;
- there exists at least one key-term k_j, such that, $k \supset k_j$.

Suppose there are s key-terms that are roots, and denote the set of roots as $R(K) = \{r_1, r_2, \ldots, r_s\}$, where $0 \leq s < n$. Each root will generate a hierarchical structure or, more precisely, a *hierarchical sub-structure*.

Denote $H_{r_i}(K)$ as the hierarchical sub-structure corresponding to root r_i, $hie_{r_i}(K)$ as the set of key-terms on $H_{r_i}(K)$ (including root r_i), Θ_{r_i} as the set of atomic-terms corresponding to root r_i. Clearly, $hie_{r_i}(K) \supset \Theta_{r_i}$ and so $|hie_{r_i}(K)| > |\Theta_{r_i}| \geq 1$.

For two arbitrary roots $r_i, r_j \in R(K)$, it is likely that $hie_{r_i}(K) \cap hie_{r_j}(K) \neq \emptyset$. That is, a key-term may be on, both sub-structures $H_{r_i}(K)$ and $H_{r_j}(K)$. In particular, it may be that $\Theta_{r_i} \cap \Theta_{r_j} \neq \emptyset$, that is, an atomic-term may be on both $H_{r_i}(K)$ and $H_{r_j}(K)$.

With the above notation, the hierarchical structure is an assembly of the individual hierarchical sub-structures: $H_R(K) = \left[H_{r_1}(K), H_{r_2}(K), \ldots, H_{r_s}(K)\right]$. The set of key-terms on at least one hierarchical sub-structure is denoted by $hie(K) = hie_{r_1}(K) \cup hie_{r_2}(K) \cup \cdots \cup hie_{r_s}(K)$. The sub-frame of discernment is denoted by $\Theta' = \Theta_{r_1} \cup \Theta_{r_2} \cup \cdots \cup \Theta_{r_s}$.

In the context of IR, each root r_i should be regarded as referring to one specific topic, and other key-terms on $H_{r_i}(K)$ are its narrower key-terms under the same topic. Thus, all roots should be regarded as pairwise unrelated.

While key-term k is on, or linked to, more than one hierarchical sub-structure, it would have multiple subset-expressions corresponding to the individual roots. This usually happens when key-term k is polysemous (multiple-meaning). For instance, key-term "phoenix" may have several meanings:

"the capital and largest city of Arizona", or
"a bird in Egyptian mythology", or

"a constellation in the Southern Hemisphere near Tucana and Sculptor".

Thus, "phoenix" may be on, three sub-structures.

Since each root is considered to refer to only one topic, the multiple subset-expressions will be treated as different from root to root. In particular, when $a \in \Theta_{r_i} \cap \Theta_{r_j}$, we denote $a \rightleftharpoons \{a\}_{r_i}$ and $a \rightleftharpoons \{a\}_{r_j}$, and regard a as having different (semantic) meanings, corresponding to roots r_i and r_j, respectively.

In an extreme case, there is no root: $s = 0$. Thus, the sub-frame of discernment $\Theta' = \emptyset$, and $n = |K|$ key-terms are all isolated-terms. We can then immediately obtain $\Theta = iso(K) = K = \{k_1, k_2, \ldots, k_n\}$. That is, all the key-terms are merged into the frame of discernment (i.e., they are treated as "atomic-terms"), and considered unrelated to each other. In many existing IR models, key-terms are dealt with in this way.

We give a detailed algorithm for establishing a frame of discernment and for expressing key-terms as subsets of the frame in [5].

3.9 Thesaurus Classes and Query Expansion

Our method reduces terms (i.e., $t \in T$) to their thesaurus classes: terms are replaced by representatives either synonymous (i.e., key-term $k \in K$), or equivalent (i.e., kernel-term $k' \in K'$). In order to clarify how thesaurus classes can be used to represent objects at a later stage, we introduce three further pieces of notation: $equ(k')$, $syn(k)$ and $cla(k')$.

For an arbitrary key-term $k \in K$, denote

$$syn(k) = \{t \,|k \Leftarrow t, t \in T\}$$

as the set of terms which are synonymous with key-term k.

For an arbitrary kernel-term $k' \in K'$, denote

$$equ(k') = \{k \,|k' \rightleftharpoons k, k \in K\}$$

as an *equivalent class* of key-terms which are equivalent to kernel-term k'.

Also, denote

$$cla(k') = \bigcup_{k \in equ(k')} syn(k) = \bigcup_{k \in equ(k')} \{t \,|k \Leftarrow t, t \in T\}$$
$$= \{t \,|k \Leftarrow t, t \in T; k' \rightleftharpoons k, k \in K\}$$

as a *thesaurus class* of terms either synonymous with, or equivalent to, kernel-term k'.

Let us consider the following example.

Example 1. Suppose a user enters a query: $q = $ "*birds*". Notice that term *birds* $\in K'$. Thus, from Table 7, we can write an equivalent class

$$equ(birds) = \{birds, domestic\text{-}birds, eggs, feathers, poultry\} \,,$$

Also, from thesaurus $\mathcal{S}_{(k,t)}$ given in Table 2, we have

$$syn(birds) = \{bird, birds\},$$
$$syn(domestic\text{-}birds) = \{domestic\text{-}birds\},$$
$$syn(eggs) = \{eggs\},$$
$$syn(feathers) = \{feathers\},$$
$$syn(poultry) = \{barnyard\text{-}birds, farmyard\text{-}birds, poultry\}.$$

Hence, we can expand query q from a single term birds to a thesaurus class

$$cla(birds) = \{bird, birds, domestic\text{-}birds, eggs, feathers,$$
$$barnyard\text{-}birds, farmyard\text{-}birds, poultry\}. \quad \Diamond$$

Query expansion, an important component in a retrieval system, has long been an effective technique to improve retrieval performance [4, 6, 8, 16, 19, 21, 31, 35]. Some good reviews of query expansion methods can be found in [10, 13].

In a practical IR environment, document collections are processed, and documents matching the user's query are displayed in real time. Documents that do not have any term matching the query are disregarded. Users of an IR system employing term matching as a basis for retrieval are faced with the challenge of expressing their queries with terms in the vocabulary of the documents they wish to retrieve. This difficulty is especially severe in extremely large, wide-ranging, full-text collections containing many different terms describing the same concept. The problem of term-mismatch has long been serious in IR.

The problem is more pronounced for short queries consisting of just a few terms related to the subject of interest: this can best be illustrated through the scenario of information search on the World Wide Web where users tend to enter very short queries. The shorter the query is, the less chance for important terms to co-occur in both relevant documents and the query. Hence a good way of matching terms is urgently needed. Query expansion is a process that modifies the original query representation so as to more precisely express the information needs.

Since our method can reduce different terms to their thesaurus classes, retrieval systems can achieve the effect of automatically expanding objects with thesaurus classes of the original query terms. The expansion may be expected to improve performance as it greatly increases matching between relevant document terms and query terms.

Consider several examples. If a user describes his information need as "aviation school", then relevant information indexed by terms aeronautical engineering institute might meet with retrieval failure – term mismatch arises from synonymous terms. In a retrieval based on a term "cow", the user might be also interested in the documents containing term *mammal* – term mismatch arises from narrower terms. A user enters a term *"planet"*, he might be thinking of something like *Mercury* or *Venus* – term mismatch arises from broader terms. A user tries terms *"crime"* and *"murder"* when she desires to find some thrillers – term mismatch arises from the related terms. Thus, if the reader traces through all discussions given in this paper, it should become clear that the structure of the thesaurus classes embodies intuitive meaning, and this structure can be expected to resolve these term mismatching problems.

4 Retrieval Based on Evidential Theory

So far, we have concentrated on developing an effective method for tackling the problem of expressing key-terms as subsets of the frame of discernment. Before seeing how to apply our knowledge of expression to practical IR problems, we need to introduce evidential theory, which underpins the formal method proposed in this paper.

4.1 Evidential Theory

Evidential theory is by now a familiar one for many IR researchers. A detailed account of it can be found in [25]. Some general definitions applied in this study can be written as follows.

Let $\Theta = \{y_1, y_2, \ldots, y_{|\Theta|}\}$ denote a frame of discernment. Then the power set of Θ can be represented as $2^\Theta = \{Y_1, Y_2, \ldots, Y_{2^{|\Theta|}}\} = \{Y_i \mid Y_i = \cup_{j=1}^{m_i}\{y_{i_j}\}, 1 \leq m_i \leq |\Theta|, 1 \leq i \leq 2^{|\Theta|}\}$, that is, each element $Y_i \in 2^\Theta$ is a subset of Θ.

A function $m : 2^\Theta \to [0, 1]$ is a *mass function* if there is a random subset variable Y over the *evidence space* 2^Θ, such that $m(Y)$ satisfies (1) $m(\emptyset) = 0$ and, (2) $\sum_{Y_i \subseteq \Theta} m(Y_i) = 1$. In evidential theory, masses are assigned to only those propositions (subsets) that are supported by evidence.

A function $bel : 2^\Theta \to [0, 1]$ is a *belief function* if there is a random subset variable Y on Θ such that $bel(Y)$ satisfying: (1) $bel(\emptyset) = 0$, (2) $bel(\Theta) = 1$ and, (3) for any collection $A_1, A_2, \ldots, A_k$ ($k \geq 1$) of subsets of Θ, $bel(A_1 \cup A_2 \cup \ldots \cup A_k) \geq \sum_{I \subseteq \{1,2,\ldots,k\}, I \neq \emptyset} (-1)^{|I|+1} bel(\cap_{i \in I} A_i)$. If we suppose $bel(Y) = \sum_{A \subseteq Y} m(A)$, then it is not difficult to verify that $bel(Y)$ is a belief function. We call $bel(Y)$ the belief function corresponding to mass function $m(Y)$.

Also, suppose $pls(Y) = \sum_{A \subseteq \Theta, A \cap Y \neq \emptyset} m(A)$, and call it the *plausibility function* corresponding to mass function $m(Y)$.

It can be verified that (1) $pls(Y) = 1 - bel(\Theta - Y)$, (2) $bel(Y) = 1 - pls(\Theta - Y)$ and, (3) $bel(Y) \leq pls(Y)$. Thus, $bel(Y)$ is also referred to as the *lower probability function*, and $pls(Y)$ as the *upper probability function*. The interval $[bel(Y), pls(Y)]$ is referred to as the belief interval. Here value $bel(Y)$ gives the degree to which the current evidence supports subset Y. The degree to which Y remains plausible is given by value $pls(Y) = 1 - bel(\Theta - Y)$. The difference $pls(Y) - bel(Y)$ represents the residual ignorance, $ign(Y) = pls(Y) - bel(Y)$, and is called the *ignorance function* corresponding to mass function $m(Y)$.

4.2 Object Representations and Mass Function

Having introduced the evidential functions, we move on to two other important issues – defining the representations of objects, and introducing agreement measures for ranking documents against a given query. We discuss the first in this subsection and the following; the second is discussed in Sect. 4.4.

In the IR context, we can simulate:

(a) the frame of discernment by $\Theta = \{a_1, a_2, \ldots, a_{|\Theta|}\}$;
(b) the evidence space by the evidence sub-space $K' = \{k'_1, k'_2, \ldots, k'_m\}$;
(c) an evidence by an object (i.e., by the statistical information within the object, more precisely, by term weights obtained from the statistical information within the object);
(d) a proposition by a statement "kernel-term k' appears".

Thus, masses are assigned to only those kernel-terms that are supported by the object. By kernel-term k' supported by an object x (i.e., a document $x = d$ or query $x = q$), we mean here that it or, term(s) of thesaurus class $cla(k')$, appears in x.

Putting the above simulations together is equivalent to saying that each object x can be represented by a mass function $m_x(k')$ over sub-space K':

$$\left[m_x(k')\right]_{1 \times m} = \left[m_x(k'_1), m_x(k'_2), \ldots, m_x(k'_m)\right],$$

where component $m_x(k')$ can be interpreted as indicating the strength of kernel-term k' or, a thesaurus class $cla(k')$, when supported by x. For instance, from Table 8, we have

$$\left[m_x(k')\right]_{1 \times 8} = \big[m_x(animal), m_x(birds), m_x(cows), m_x(ducks),$$
$$m_x(geese), m_x(goats), m_x(hens), m_x(mammals)\big].$$

In order to estimate the strength of kernel-term k' supported by object x, suppose that term *weights*, $w_x(t)$, have been obtained, which are considered to reflect the importance of terms t $(\in V^x \subseteq T)$ concerning x. Thus, the mass, $m_x(k')$, can be estimated from the weights of (i) k', (ii) synonymous terms of k', (iii) equivalent key-terms of k' and (iv) the synonymous terms of equivalent key-terms of k', in object x.

More specifically, for an arbitrary kernel-term $k' \in K'$, the mass is defined:

$$m_x(k') = \frac{\psi_x(k')}{N_x} = \frac{\psi_x(k')}{\sum_{k' \in (V^x \cap K')} \psi_x(k')},$$

where V^x is the set of terms appearing in object x; N_x, is the *normalization factor* of object x; function

$$\psi_x(k') = \sum_{t \in cla(k')} w_x(t) = \sum_{k \in equ(k')} \left(\sum_{t \in syn(k)} w_x(t) \right).$$

It can be seen that function $\psi_x(k')$ is the sum of weights, $w_x(t)$, of terms in the thesaurus class $cla(k')$. Thus, mass $m_x(k')$ is proportional to the sum. It is therefore evident that the design of weighting function $w_x(t)$ is crucial in determining retrieval performance. The effectiveness of the weighting function for reflecting the statistical importance of a term in respect to individual objects, has been investigated extensively in the literature [1, 4, 12, 14, 20, 22, 27, 28, 29, 31].

Let us see an example below, which may help to clarify the above idea and assist in understanding the computation involved in function $m_x(k')$.

Example 2. Let us return to Example 1. Suppose we are given the weights of terms in document d as follows.

$w_d(birds) = .646, \quad w_d(bird) = .421,$

$w_d(domestic\text{-}birds) = .0,$

$w_d(eggs) = .285,$

$w_d(feathers) = .17,$

$w_d(barnyard\text{-}birds) = .01, \quad w_d(farmyard\text{-}birds) = .0, \quad w_d(poultry) = .0,$

and so on. Suppose also that the normalization factor of d is $N_d = 6.52$. Then we arrive at the mass for kernel-term $k' = birds$ or, the thesaurus class $cla(birds)$:

$$m_d(birds) = \frac{.646 + .421 + .0 + .285 + .17 + .01 + .0 + .0}{6.52} \approx .235 . \quad \diamondsuit$$

Notice that all terms in a given object x may not be partitioned into distinct thesaurus classes (that is, each term $t \in V^x$ may be classified into at least one thesaurus class), and that the normalization factor in the estimation of $m_x(k')$ is therefore given by $N_x = \sum_{k' \in (V^x \cap K')} \psi_x(k')$, rather than simply by $N_x = \sum_{t \in V^x} w_x(t)$.

Notice also that there is no necessity to design a weighting function in advance for estimating mass $m_x(k')$. The estimation can simply be made using occurrence frequencies of terms. In this case, $w_x(t) = f_x(t)$.

4.3 Object Representations and Other Evidential Functions

Objects can also be represented by the belief and plausibility functions. In order to compute functions *bel* and *pls*, the narrower relation and related relation between kernel-terms are involved, and the relations take their mathematical meanings: for two arbitrary kernel-terms $k'_i, k'_j \in K'$, we can consider their semantic relations $k'_i \subset k'_j$ and $k'_i \cap k'_j$ by the set relations and operations of their corresponding subset-expressions. We can clarify this idea by considering the example below.

Example 3. Suppose that the mass functions for documents d_1, d_2, d_3, d_4 and queries q_1, q_2 are obtained. These are given in Table 9.

Table 9. Mass functions

Functions	Kernel-Terms	d_1	d_2	d_3	d_4	q_1	q_2
mass	animals	0.125	0.000	0.133	0.300	0.000	0.200
functions	birds	0.560	0.445	0.867	0.000	1.000	0.600
	mammals	0.315	0.555	0.000	0.700	0.000	0.200

and $m_x(cows) = m_x(ducks) = m_x(geese) = m_x(goats) = m_x(hens) = 0$ for $x = d_1, d_2, d_3, d_4, q_1, q_2$.

Then, the corresponding belief and plausibility functions are calculated, and results are given in Table 10.

Table 10. Belief and plausibility functions

Functions	Kernel-Terms	d_1	d_2	d_3	d_4	q_1	q_2
belief	animals	1.000	1.000	1.000	1.000	1.000	1.000
functions	birds	0.560	0.445	0.867	0.000	1.000	0.600
	mammals	0.315	0.555	0.000	0.700	0.000	0.200
plausibility	animals	1.000	1.000	1.000	1.000	1.000	1.000
functions	birds	0.685	0.445	1.000	0.300	1.000	0.800
	mammals	0.440	0.555	0.133	1.000	0.000	0.400

For instance, for kernel-term $k' = birds$ in document d_1, we have,

$$bel(birds) = \sum_{k' \subseteq birds} m_{d_1}(k') = m_{d_1}(birds) = 0.560 ;$$

$$pls(birds) = \sum_{k' \subseteq \Theta; k' \cap birds \neq \emptyset} m_{d_1}(k') = m_{d_1}(animals) + m_{d_1}(birds)$$

$$= 0.125 + 0.560 = 0.685 ,$$

where, from Table 5, all kernel-terms satisfying $k' \subseteq birds$ are

$$birds \rightleftharpoons \{ducks, geese, hens\},$$

$$ducks \rightleftharpoons \{ducks\}, \quad geese \rightleftharpoons \{geese\}, \quad hens \rightleftharpoons \{hens\};$$

and all kernel-terms satisfying $k' \cap birds \neq \emptyset$ are

$$animals \rightleftharpoons \{cows, ducks, geese, goats, hens\},$$
$$birds \rightleftharpoons \{ducks, geese, hens\},$$
$$ducks \rightleftharpoons \{ducks\}, \quad geese \rightleftharpoons \{geese\}, \quad hens \rightleftharpoons \{hens\}. \qquad \Diamond$$

4.4 Agreement Measures

In the foregoing, we discussed the definition of the representations of objects based on evidential functions. In this subsection, for ranking documents against a given query, we further discuss agreement (similarity) measures, over the evidential representations.

Suppose that document d and query q can be represented by $m_d(k')$ and $m_q(k')$, respectively. Our method involves computing belief $bel(m_d = m_q)$ and plausibility $pls(m_d = m_q)$. A study on the computation can be found in [3]. The *agreement measures* between $m_d(k')$ and $m_q(k')$ are defined by

$$bel(m_d = m_q) = \sum_{k' \subseteq K'} m_d(k') m_q(k'),$$

$$pls(m_d = m_q) = \sum_{k_i', k_j' \subseteq K'; \; k_i' \cap k_j' \neq \emptyset} m_d(k_i') m_q(k_j')$$

$$= \sum_{k_j' \subseteq K'} m_q(k_j') \left(\sum_{k_i' \subseteq K'; \; k_i' \cap k_j' \neq \emptyset} m_d(k_i') \right).$$

Obviously, the *lower agreement*, $bel(m_d = m_q)$, measures the belief that document kernel-terms and query kernel-terms are equal (i.e., their subset-expressions are the same); the *upper agreement*, $pls(m_d = m_q)$, measures the plausibility that document kernel-terms and query kernel-terms are related (i.e., their subset-expressions have non-empty intersections).

Further, let us consider the belief interval

$$\left[bel(m_d = m_q), \; pls(m_d = m_q) \right].$$

As mentioned, $bel(m_d = m_q)$ gives the degree to which the current evidence supports $m_d(k') = m_q(k')$. The degree to which evidence $m_d(k') = m_q(k')$ remains plausible is given by $pls(m_d = m_q) = 1 - bel(m_d \neq m_q)$.

The following example illustrates the computation involved.

Example 4. Let us return to Example 3. We there gave mass functions for documents d_1, d_2, d_3, d_4 and queries q_1, q_2. Thus, for d_1 and q_1, we have

$$bel(d_1 = q_1) = m_{d_1}(animals)m_{q_1}(animals) + m_{d_1}(birds)m_{q_1}(birds)$$
$$+ m_{d_1}(mammals)m_{q_1}(mammals)$$
$$= 0.125 \times 0 + 0.560 \times 1 + 0.315 \times 0 = 0.560 \; ;$$
$$pls(d_1 = q_1) = m_{q_1}(birds)\big(m_{d_1}(animals) + m_{d_1}(birds)\big)$$
$$= 1 \times \big(0.125 + 0.560\big) = 0.685 \; ,$$

also, we have

$$bel(d_2 = q_1) = 0.445, \quad bel(d_3 = q_1) = 0.687, \quad bel(d_4 = q_1) = 0 \; ;$$
$$pls(d_2 = q_1) = 0.445, \quad pls(d_3 = q_1) = 1, \quad pls(d_4 = q_1) = 0.3 \; .$$

Similarly, for d_1 and q_2, we have

$$bel(d_1 = q_2) = m_{d_1}(animals)m_{q_2}(animals) + m_{d_1}(birds)m_{q_2}(birds)$$
$$+ m_{d_1}(mammals)m_{q_2}(mammals)$$
$$= 0.125 \times 0.2 + 0.560 \times 0.6 + 0.315 \times 0.2 = 0.424 \; ;$$
$$pls(d_1 = q_2) = m_{q_2}(animals)\big(m_{d_1}(animals) + m_{d_1}(birds)$$
$$+ m_{d_1}(mammals)\big)$$
$$+ m_{q_2}(birds)\big(m_{d_1}(animals) + m_{d_1}(birds)\big)$$
$$+ m_{q_2}(mammals)\big(m_{d_1}(animals) + m_{d_1}(mammals)\big)$$
$$= 0.2 \times \big(0.125 + 0.560 + 0.315\big) + 0.6 \times \big(0.125 + 0.560\big)$$
$$+ 0.2 \times \big(0.125 + 0.315\big) = \; 0.699 \; ,$$

also, we have

$$bel(d_2 = q_2) = 0.378, \quad bel(d_3 = q_2) = 0.5468, \quad bel(d_4 = q_2) = 0.2;$$
$$pls(d_2 = q_2) = 0.578, \quad pls(d_3 = q_2) = 0.8266, \quad pls(d_4 = q_2) = 0.58 \; .$$

Finally, ranking documents by the belief interval, we respond to users' queries as follows.

For query q_1, the response is
$$d_3[0.687, 1.0] \succ d_1[0.560, 0.685] \succ d_2[0.445, 0.445] \succ d_4[0, 0.3] \; .$$
For query q_2, the response is
$$d_3[0.5468, 0.8266] \succ d_1[0.424, 0.699] \succ d_2[0.378, 0.578] \succ d_4[0.2, 0.58] \; ,$$

where $d_i[bel_i, pls_i] \succ d_j[bel_j, pls_j]$ is explained as "document d_i is more in agreement with the query than document d_j". $\Diamond$

Conclusion and Further Work

The ability to formally express term semantic relations is a core issue in IR. The problems for the expression are how to establish the frame of discernment, and how to express key-terms as subsets of the frame. The problems lead to many other IR problems as pointed out repeatedly in the literature. Solution of the problems is a technical barrier to applying mathematical tools, especially evidential theory, to IR. In this study, we focus on the problems, and present a method for establishing the frame of discernment and for deriving subset-expressions of key-terms. Then, we propose a novel method for representing documents and queries based on evidential functions, and for ranking documents against a given query. A central aim of this study is to treat the semantic relations between terms and incorporate the relations into the retrieval strategies for more effective retrieval.

A key-term, if polysemous, may be expressed by different frame subsets corresponding to different roots. In IR, it is difficult to automatically determine which meaning is being used in the context. Almost all existing IR methods suffer from the same problem. Thus, it is hard to determine into which thesaurus class a polysemous term should be placed. This paper does not deal with how the class is determined; it is left as a significant subject for further study.

Thesaurus class methods can be regarded as *recall*[1] improving devices. The thesaurus classes of query terms may be expected to retrieve more relevant documents because extra "related" terms are added to the query when the thesaurus classes are assigned to the query instead of single terms. However, if terms included in a thesaurus class have high *document frequencies*[2], then the addition of these terms would be likely to lead to unacceptable losses in *precision*[3]. For this reason, some studies suggest that thesaurus classes should be formed only from those terms which have low document frequencies, [36] for instance. This interesting issue needs to be investigated in further work.

We intend to develop an experimental investigation into the performance of our method.

Acknowledgements

This research was supported in part by both EPSRC and Microsoft.

References

1. G. Amati and C. J. Van Rijsbergen. Probabilistic models of information retrieval based on measuring the divergence from randomness. *ACM Transactions on Information Systems*, 20(4):357–389, 2002.
2. A. Bookstein. Relevance. *Journal of the American Society for Information Science*, 30:269–273, 1979.

---

[1] The proportion of relevant documents actually retrieved in answer to a query.

[2] The number of documents in a collection in which a term appears.

[3] The proportion of retrieved documents actually relevant to the query.

3. D. Cai. Extensions and applications of evidence theory. In *Soft Computing for Risk Evaluation and Management: Applications in Technology, Environment and Finance*, volume 76, pp. 73–93, New York, 2001. Physica-Verlag, Heidelberg.

4. D. Cai. *IfD – Information for Discrimination*. PhD thesis, University of Glasgow, Glasgow, Scotland, 2004.

5. D. Cai and C. J. Van Rijsbergen. An algorithm for modelling key-terms. Technical Report TR-2005-190, Department of Computing Science, University of Glasgow, 2005.

6. D. Carmel, E. Farchi, Y. Petruschka, and A. Soffer. Automatic query refinement using lexical affinities with maximal information gain. In *Proceedings of the 25th Annual International ACM-SIGIR Conference on Research and Development in Information Retrieval*, pp. 283–290, 2002.

7. C. Carpineto, R. Mori, and G. Romano. Informative term selection for automatic query expansion. In *The 7th Text REtrieval Conference (TREC-7)*, pp. 363–369. NIST Special Publication, 1998.

8. C. Carpineto, R. D. Mori, G. Romano, and B. Bigi. An information-theoretic approach to automatic query expansion. *ACM Transactions on Information Systems*, 19(1):1–27, 2001.

9. W. S. Cooper. A definition of relevance for information retrieval. *Information Storage and Retrieval*, 7:19–37, 1971.

10. E. N. Efthimiadis. Query expansion. *Annual Review of Information Systems and Technology*, 31:121–187, 1996.

11. N. Fuhr. Optimum polynomial retrieval functions based on the probability ranking principle. *ACM Transactions on Information Systems*, 7(3):183–204, 1989.

12. N. Fuhr. Probabilistic models in information retrieval. *The Computer Journal*, 35(3):243–255, 1992.

13. S. Gauch, J. Wang, and S. M. Rachakonda. A corpus analysis approach for automatic query expansion and its extension to multiple databases. *ACM Transactions on Information Systems*, 17(3):250–269, 1999.

14. D. Harman. An experimental study of factors important in document ranking. In *Proceedings of the 9th Annual International ACM-SIGIR Conference on Research and Development in Information Retrieval*, pp. 186–193, 1986.

15. S. P. Harter. Psychological relevance and information science. *Journal of the American Society for Information Science*, 43(9):602–651, 1992.

16. M. A. Hearst. Improving full-text precision on short queries using simple constraints. In *Proceedings of the 5th Annual Symposium on Document Analysis and Information Retrieval*, 1996.

17. J. Lafferty and C. Zhai. Document language models, query models, and risk minimization for information retrieval. In *Proceedings of the 24th Annual International ACM-SIGIR Conference on Research and Development in Information Retrieval*, pp. 111–119, 2001.

18. M. Lalmas. Dempster-Shafer's theory of evidence applied to structured documents: modelling uncertainty. In *Proceedings of the 20th Annual International ACM-SIGIR Conference on Research and Development in Information Retrieval*, pp. 110–118, 1997.

19. M. Mitra, A. Singhal, and C. Buckley. Improving automatic query expansion. In *Proceedings of the 21st Annual International ACM-SIGIR Conference on Research and Development in Information Retrieval*, pp. 206–214, 1998.

20. S. E. Robertson and K. Sparck Jones. Relevance weighting of search terms. *Journal of the American Society for Information Science*, 27(3):129–146, 1976.
21. G. Salton and C. Buckley. Improving retrieval performance by relevance feedback. *Journal of the American Society for Information Science*, 41(4):288–297, 1990.
22. G. Salton and C. S. Yang. On the specification of term values in automatic indexing. *Journal of Documentation*, 29(4):351–372, 1973.
23. T. Saracevic. Relevance reconsidered '96. In *Proceedings of the 2nd International Conference on Conceptions of Library and Information Science*, pp. 201–218, 1996.
24. S. Schocken and Hummel. On the use of Dempster-Shafer model in information indexing and retrieval applications. *International Journal of Man-Machine Studies*, 39:843–879, 1993.
25. G. Shafer. *A Mathematical Theory of Evidence*. NJ: Princeton University, Princeton, 1976.
26. W. T. Silva and R. L. Milidiú. Belief function model for information retrieval. *Journal of the American Society for Information Science*, 44(1):10–18, 1993.
27. K. Sparck Jones. A statistical interpretation of term specificity and its application to retrieval. *Journal of Documentation*, 28(1):11–21, 1972.
28. H. Turtle and W. B. Croft. Inference networks for document retrieval. In *Proceedings of the 13th Annual International ACM-SIGIR Conference on Research and Development in Information Retrieval*, pp. 1–24, 1990.
29. C. J. Van Rijsbergen. A theoretical basis for the use of co-occurrence data in information retrieval. *Journal of Documentation*, 33(2):106–119, 1977.
30. C. J. Van Rijsbergen. A non-classical logic for information retrieval. *The Computer Journal*, 39:481–485, 1986.
31. C. J. Van Rijsbergen, D. J. Harper, and M. F. Porter. The selection of good search terms. *Information Processing & Management*, 17:77–91, 1981.
32. S. K. M. Wong and Y. Y. Yao. A probability distribution model for information retrieval. *Information Processing & Management*, 25(1):39–53, 1989.
33. S. K. M. Wong and Y. Y. Yao. A probabilistic inference model for information retrieval. *Information Systems*, 16(3):301–321, 1991.
34. S. K. M. Wong, W. Ziarko, and P. C. N. Wong. Generalised vector space model in information retrieval. In *Proceedings of the 8th Annual International ACM-SIGIR Conference on Research and Development in Information Retrieval*, pp. 18–25, 1985.
35. J. Xu and W. B. Croft. Query expansion using local and global document analysis. In *Proceedings of the 19th Annual International ACM-SIGIR Conference on Research and Development in Information Retrieval*, pp. 4–11, 1996.
36. C. T. Yu and G. Salton. Effective information retrieval using term accuracy. *Journal of the Association for Computing Machinery*, 20(3):135–142, 1977.

Sequential Pattern Mining*

Tian-Rui Li[1], Yang Xu[1], Da Ruan[2], and Wu-ming Pan[3]

[1] School of Science, Southwest Jiaotong University, Chengdu 610031, P. R. China
`{trli, xuyang}@swjtu.edu.cn`
[2] Belgian Nuclear Research Centre (SCK•CEN), Boeretang 200, 2400 Mol,
Belgium
`druan@sckcen.be`
[3] College of Software Engineering, Sichuan University, Chengdu 610065,
P.R. China
`fluedpan@hotmail.com`

Summary. Sequential pattern discovery has emerged as an important research topic in knowledge discovery and data mining with broad applications. Previous research is mainly focused on investigating scalable algorithms for mining sequential patterns while less on its theoretical foundations. However, the latter is also important because it can help to use existing theories and methods to support more effective mining tasks. In this chapter, we conduct a systematic study on models and algorithms in sequential pattern analysis, especially discuss the existing algorithms' advantages and limitations. Then, we build the relation between the closed sequential patterns and fixed point, which can serve as a theoretical foundation of sequential patterns. Finally, we discuss its applications and outline the future research work.

1 Introduction

Data mining is mainly concerned with methodologies for extracting patterns from large data repositories. Sequential pattern mining, since its introduction in [1], has become an active research topic in data mining, with broad applications [2, 3]. For example, consider a Web access database at a popular site, where a Web user and Web page are regarded as an object and attribute respectively. The discovered patterns are the sequences of most frequently accessed pages at that site. This kind of information can help to improve a system design such as better a hyperlinked structure between the correlated pages and lead to better marketing decisions like strategic advertisement placement [2]. There are many other domains where sequence mining has been applied, which include discovering customer buying patterns in retail stores,

---

* This work was partially supported by the National Natural Science Foundation of China (NSFC) under the grant No.60474022.

analysis of Web access databases, identifying plan failures, mining DNA sequences and gene structures, and finding network alarm patterns. Moreover, a deep understanding of efficient sequential pattern mining methods may also have strong implications on the development of efficient methods for mining frequent subtrees, lattices, subgraphs, and other structured patterns in large databases [4].

Many studies have been contributed to the efficient mining of sequential patterns in the literature, most of which was focused on developing efficient algorithms for finding all sequential patterns such as AprioriAll [1], GSP [5], SPADE [6], PrefixSpan [7] and so on. In addition, enormous sizes of available databases and possibly large number of mined sequential patterns demand efficient and scalable parallel algorithms. Therefore, several parallel algorithms such as HPSPM [8], pSPADE [9], TPF [10] were proposed and have good performance. Moreover, recent research on sequential pattern mining has progressed to closed sequential pattern mining, which can greatly reduce the number of frequent subsequences and improve the efficiency [4, 11].

Yet, the above work all assumes that the database is static, and a database updates requires rediscovering all the patterns by scanning the entire old and new database. Then, there is a need for efficient algorithms to update, maintain and manage the information discovered. Some incremental mining algorithms of sequential patterns were proposed, e.g. [12, 13, 14, 15, 16, 17], and perform significantly better than the naïve approach of mining the whole updated database from scratch.

However, a major common thread that runs through the vast majority of earlier work is the lack of user-controlled focus in the pattern mining process and then it demonstrates the need for novel pattern mining solutions. Recent feasible solutions are to incorporate user-specified constraints in sequential pattern mining, which enable the incorporation of user-controlled focus in the mining process and avoid overwhelming volume of potentially useless results [18, 19, 20].

Other work contributes on an extension of the problem of sequential pattern mining like mining cyclically repeated patterns [21], approximate mining of consensus sequential patterns [22], mining multidimensional sequential pattern [23], mining top-k closed sequential patterns [24], mining frequent Max sequential patterns [25], mining long sequential patterns in a noisy environment [26], mining hybrid sequential patterns and sequential rules [27], etc.

In this chapter, the models in sequential pattern mining are presented in Sect. 2. A systematic analysis on algorithms for mining sequential patterns is conducted in Sect. 3. In Sect. 4, a theoretical foundation of sequential pattern mining is provided. Its applications are discussed in Sect. 5. Our current study is summarized and some future research issues is also pointed out in Sect. 6.

2 Models of Sequential Patterns Mining

2.1 Original Model of Sequential Patterns Mining

The original model of mining sequential patterns was proposed in [1]. It can be stated as follows:

Let $I = \{i_1, i_2, \ldots, i_k\}$ be a set of all items. A subset of I is called an *itemset*. A *sequence* $s = \langle s_1, s_2, \ldots, s_m \rangle (s_i \subseteq I)$ is an ordered list. The *size*, $|s|$, of a sequence s is the number of itemsets in the sequence. The *length*, $l(s)$, is the total number of items in the sequence, namely, $l(s) = \sum_{i=1}^{m} |s_i|$. A sequence $\alpha = \langle \alpha_1, \alpha_2, \ldots, \alpha_m \rangle$ is a *sub-sequence* of another sequence $\beta = \langle \beta_1, \beta_2, \ldots, \beta_n \rangle$, denoted as $\alpha \sqsubseteq \beta$, (if $\alpha \neq \beta$, written as $\alpha \sqsubset \beta$), if and only if $\exists k_1, k_2, \ldots, k_m$, such that $1 \leq k_1 < k_2 < \ldots < k_m \leq n$ and $\alpha_1 \subseteq \beta_{k_1}, \alpha_2 \subseteq \beta_{k_2}, \ldots, \alpha_m \subseteq \beta_{k_m}$. We also call β is a super-sequence of α and β contains α. A *sequence database*, $D = \{d_1, d_2, \ldots, d_k\}$, is a set of sequences. Each sequence is associated with an *id*. For simplicity, say the *id* of d_i is i. $|D|$ represents the number of sequences in the database D. The *support* of a sequence α in a sequence database D is the number of sequences in D which contain α, $support(\alpha) = |\{d|d \in D \text{ and } \alpha \sqsubseteq d\}|$. Given a minimum support threshold, *min_sup*, the set of *frequent sequential patterns*, FS, includes all the sequences whose support is no less than *min_sup*. Given a sequence database and a user-specified *min_sup*, the problem of mining sequential patterns is to find all the frequent subsequences in the database.

The set of the closed frequent sequential pattern is defined as follows, $CS = \{\alpha | \alpha \in FS \text{ and } \nexists \beta \in FS \text{ such that } \alpha \sqsubseteq \beta \text{ and } support(\alpha) = support(\beta)\}$. Since CS includes no sequence which has a super-sequence with the same support, we have $CS \subseteq FS$. The problem of the closed sequence mining is to find CS above a minimum support threshold.

Example 1. Table 1 is a sample sequence database, referred as D when the context is clear. The alphabetic order is taken as the default lexicographical order. The set of *items* in the database is $\{a, b, c, d, e, f\}$. Obviously, $\langle (ce)(a)(af)(c) \rangle$ is a *sequence* in the database and its *size* is equal to 4. This whole sequence contributes only one to the *support* of (a) although item a appears more than once in it. The *support* of (a) is 5 in the database. If *min_sup* = 3 (taken as default in this chapter), CS, FS are listed in support descending order (in the form of *sequence* : *support*) as below,

$$CS = \{(a) : 5, (c) : 4, (a)(a) : 3, (a)(c) : 3, (a)(b) : 3, (c)(a) : 3\} ;$$

$$FS = \{(a) : 5, (c) : 4, (a)(a) : 3, (a)(c) : 3, (a)(b) : 3, (c)(a) : 3, (b) : 3\} .$$

CS has the exact same information as FS, but includes much fewer patterns.

Table 1. A sample sequence database D

Identifier	Sequence
1	$\langle(ce)(a)(af)(c)\rangle$
2	$\langle(a)(a)(b)(d)\rangle$
3	$\langle(a)(bf)(b)(c)\rangle$
4	$\langle(b)(c)(a)(b)\rangle$
5	$\langle(a)(c)(a)(be)\rangle$

2.2 Related Models of Mining Sequential Pattern

First is an episode discovery approach presented by Mannila et al in [28]. Sequences of events describing the behavior and actions of users or systems can be collected in several domains. An episode is a collection of events that occur relatively close to each other in a given partial order. Once such episodes are known, one can produce rules for describing or predicting the behavior of the sequence. The problem of this model is to find all patterns that occur in some user-specified percentage of windows through moving a time window across the input sequence. It could provide such information like "events of type b, c, and d occur together in 8% of windows of 50 time units" or "event of type b is followed by event of type d 20 times in the event sequence".

Second is the generalized sequential patterns also proposed by Srikant and Agarwal in [5], motivated by real applications such as in a book club. Time constraints that specify a minimum and/or maximum time period between adjacent elements in a pattern are incorporated in the original model and the items to be presented in a set of transactions whose transaction-times are within a user-specified time window are allowed. Given a user-defined taxonomy on items, sequential patterns are also allowed to include items across all levels of the taxonomy. The problem of this model is to find all sequences whose support is greater than the user-specified minimum support, given a database D of data-sequences, a taxonomy T, user-specified min-gap and max-gap time constraints, and a user-specified sliding-window size.

Third is a universal formulation of sequential patterns proposed in [29], which can unify and generalize the above formulations. There are two novel concepts in this universal formulation. One is the directed acyclic graph representation of the structural and timing constraints of sequential patterns. The other is that this approach supplies several different ways in which support of a pattern can be defined, each of which can be suitable in specific applications, depending on the user's perception. By choosing specific combinations of structural constraints, timing constraints, and support counting methods, this formulation can be made identical to the above formulations. A sequential pattern is said to be interesting if it occurs *enough* number of times (support threshold) satisfying the given timing constraint (ms, ws, xg, ng), where ms, ws, xg, ng represent maximum span, event-set window size, maximum gap, minimum gap, respectively. The problem of this model is to find all interesting sequential patterns, given timing constraint (ms, ws, xg, ng) and support threshold.

In sum up, investigation of models is the first step to discover useful information from database, which helps to support more efficient decision making. It is an important and interesting problem still worthy of study in the future.

3 Analysis of Sequential Pattern Mining Algorithms

Due to its extensive applications, lots of sequential mining algorithms were developed to increase efficiency and effectiveness. They can be categorized into the following classes: BFS(Breadth-first Search)-based method, DFS(Depth-first Search)-based method, closed sequential pattern based method, parallel-based method, incremental-based method and constraint-based algorithm. Here we study several representative algorithms in those classes respectively.

3.1 BFS-Based Method

AprioriAll, AprioriSome and DynamicSome, these three algorithms are first proposed in the pioneering work of sequential pattern mining by Agrawal and Srikant [1]. The latter two algorithms were developed to discover only maximal sequential patterns. The first algorithm, AprioriAll, finds all patterns. It is regarded as a three-phase algorithm in brief. It first finds all frequent itemsets using Apriori, transforms the database so that each transaction is replaced by the set of all frequent itemsets contained in the transaction, and then makes multiple passes over the database to generate candidates, count the supports of candidates, and to discover sequential patterns. Its performance was shown better than or comparable to the other two algorithms. However, this approach nearly doubles the disk space requirement which could be prohibitive for large databases.

The GSP algorithm, proposed in [2], performs like the AprioriAll algorithm, but it does not need find all the frequent itemsets first. In addition, it allows for (1) time-gap constraints, placing bounds on the time separation between adjacent elements in a pattern, (2) sliding time windows, permitting the items in an element of a pattern to span a set of transactions within a user-specified time window, (3) item taxonomies, given a user-defined taxonomy (is-a hierarchy), enabling the discovery of patterns across different levels of a user-defined taxonomy. GSP is also designed to discover these generalized sequential patterns. It makes multiple passes over the database and finds out frequent k-sequences at the kth database scanning. In each pass, every data sequence is examined to update the support counts of the candidates contained in this sequence. Initially, each item is a candidate 1-sequence for the first pass. Frequent 1-sequences are determined after checking all the data sequences in the database. In succeeding passes, frequent $(k-1)$-sequences are self-joined to generate candidate k-sequences. Again, the supports of these candidate sequences are counted by examining all data sequences, and then those candidates having minimum supports become frequent sequences. This process terminates when there is no candidate sequence left. Empirical evaluation indicates that GSP is much faster than the AprioriAll algorithm by up to 20 times. However, because of the sliding window, minimum and maximum time gaps, it needs backtracking to check if the data sequence s contains each candidate sequence stored in the leaf, which leads to the high computational cost. In addition, its I/O cost may be very high since the number of I/O passes required is determined by the length of the longest frequent sequences.

MFS, proposed in [30], is a modified version of GSP. It tries to reduce the I/O cost needed by GSP. With GSP, every database scan discovers frequent sequences of the same length. MFS, on the other hand, takes a successive refinement approach. It first computes a rough estimate of the set of all frequent sequences as a suggested frequent sequence set, makes use of it and generalizes the candidate generation function of GSP to maintain the set of maximal frequent sequences known so far. Then, longer sequences can be generated and counted early, which is the major source of efficiency improvement of MFS over GSP. Experiment results show that MFS saves I/O cost significantly compared with GSP.

SPADE, proposed in [6], uses the observation that the subsequence relation induces a lattice which is *downward closed* on the support, i.e., if β is frequent, then all subsequences $\alpha\beta$ are also frequent. It decomposes the original lattice into smaller sub-lattices, so that each sublattice can be processed entirely in main-memory using a breadth-first or depth-first search (SPADE also belongs to DFS-based method) for frequent sequences. Starting with the frequent single items, during each step the frequent sequences of the previous level are extended by one more item. Before computing the support of a new sequence, a pruning step ensures that all its subsequences are also frequent, greatly reducing the search space. The experimental results show that SPADE is about twice as fast as GSP. In addition, if we do not count the cost of computing frequent 2-item sequences, SPADE outperforms GSP by an order of magnitude in most cases. The reason is that SPADE uses a more efficient support counting method based on the idlist structure. Furthermore, SPADE only scans the original database twice to generate frequent 1-item sequences and 2-item sequences respectively, and the remaining operations are solely performed on the idlist of each sequence, which keeps shrinking during the mining process and is much smaller than the original database. SPADE also shows a linear scalability with respect to the number of sequences.

3.2 DFS-Based Method

In [31], Han et al. proposed a projection-based algorithm called FreeSpan, which aims at reducing the generation of candidate subsequences. Its general idea is to use frequent items to recursively project sequence databases into a set of smaller projected databases based on the currently mined frequent sets, and grow subsequence fragments in each projected database respectively. This process partitions both the data and the set of frequent patterns to be tested, and confines each test being conducted to the corresponding smaller projected database. FreeSpan only needs to scan the original database three times, independent of the maximal length of the sequence. Performance study shows that FreeSpan mines the complete set of patterns and is efficient and runs considerably faster than the GSP algorithm. The major cost of FreeSpan is to deal with projected databases. Moreover, since a length-k subsequence may grow at any position, the search for length-$(k + 1)$ candidate sequences will need to check every possible combination, which is quite costly.

To solve this problem existing in FreeSpan, in [7], Pei et al. proposed another projection based algorithm called PrefixSpan. Its general idea is to examine only the prefix subsequences and project only their corresponding postfix subsequences into projected databases, instead of projecting sequence databases by considering all the possible occurrences of frequent subsequences. In each projected database, sequential patterns are grown by exploring only local frequent patterns. PrefixSpan mines the complete set of patterns and is efficient and runs considerably faster than both GSP algorithm and FreeSpan. However, similar to FreeSpan, the major cost of PrefixSpan is also the construction of projected databases. In the worse case, PrefixSpan needs to construct a projected database for every sequential pattern. If there are a large number of sequential patterns, the cost is non-trivial.

In [32], Sequential PAttern Mining using a bitmap representation (SPAM) was proposed by J. Ayres, et al. Based on a lexicographic tree of sequences, SPAM utilizes a depth-first traversal of the search space combined with a vertical bitmap representation to store each sequence, which allows for efficient support counting as well as significant bitmap compression. In addition, various pruning mechanisms are implemented to reduce the search space. SPAM is especially efficient when the sequential patterns in the database are very long. Moreover, a salient feature of this algorithm is that it incrementally outputs new frequent itemsets in an online fashion. Experimental results demonstrate that this algorithm outperforms SPADE and PrefixSpan on large datasets by over an order of magnitude. However, SPAM assumes that the entire database (and all data structures used for the algorithm) completely fit into main memory which is not suitable for mining sequential pattern from large databases and it consumes more space in comparison with SPADE and PrefixSpan.

3.3 Closed Sequential Pattern Based Method

Previous sequential pattern mining algorithms mine the full set of frequent subsequences satisfying a minimum support threshold in a sequence database. However, since a frequent long sequence contains a combinatorial number of frequent subsequences, such mining will generate an explosive number of frequent subsequences for long patterns, which is prohibitively expensive in both time and space. It is proved that for mining frequent patterns (for both itemsets and sequences), one should not mine *all* frequent patterns but the *closed* ones since the latter leads to not only more compact yet complete result set but also better efficiency, which can greatly reduces the number of frequent subsequences [4, 11].

CloSpan is such an algorithm for mining closed sequential patterns [4]. It divides the mining process into two stages. In the first stage, a candidate set is generated. The second stage helps eliminate non-closed sequences. CloSpan develops several efficient search space pruning methods and it is hash-based algorithm which can efficiently execute the search space optimization with negligible cost. The performance of CloSpan shows that it not only generates a complete closed subsequence set which is substantially smaller than that generated by PrefixSpan, but also runs much faster. However, it follows a *candidate maintenance-and-test* paradigm and results in a rather poor scalability in the number of frequent closed patterns because a large number of frequent closed patterns (or just candidates) will occupy much memory and lead to a large search space for the closure checking of new patterns, which is usually the case when the support threshold is low or the patterns become long.

BIDE is a more efficient algorithm to mine closed sequential patterns [11]. It avoids the curse of the *candidate maintenance-and-test* paradigm, prunes the search space more deeply and checks the pattern closure in a more efficient way while consuming much less memory in contrast to the previously developed closed pattern mining algorithms. It does not need to maintain the set of historic closed patterns, thus it scales very well in the number of frequent closed patterns. BIDE adopts a strict depth-first search order and can output the frequent closed patterns in an online fashion. A thorough performance study shows that BIDE consumes order(s) of magnitude less memory and runs over an order of magnitude faster than the previously developed frequent (closed) sequence mining algorithms, especially when the support is low. It also has linear scalability in terms of the number of sequences in the database. However, like other closed sequence mining algorithms, it will lose to some all-frequent-sequence mining algorithms with a high support threshold.

3.4 Parallel-Based Method

The most time consuming operation in the discovery process of sequential patterns is the computation of the frequency of the occurrences of interesting subsequences in the sequence database. However, the number of sequential patterns grows exponentially and the task of finding all sequential patterns requires a lot of computational resources, which make it an ideal candidate for parallel processing.

Three parallel algorithms, NPSPM, SPSPM, HPSPM, based on GSP for mining sequential patterns on a shared-nothing environment were presented in [8]. All three approaches partition the datasets into equal sized blocks among the nodes. In NPSPM, the candidate sequences are replicated on all the processors, and each processor gathers local support using its local database block. A reduction is performed after each iteration to get the global supports. Since NPSPM replicates the entire candidate set on each node, it can run into memory overflow problems for large databases. SPSPM partitions the candidate set into equal-sized blocks and assigns each block to a separate processor. While SPSPM utilizes the aggregate memory of the system, it suffers from excessive communication, since each processor's local database has to broadcast to all other processors to get global support. HPSPM uses a more intelligent strategy to partition the candidate sequences among the nodes using hash function, which eliminates the customer transaction data broadcasting and reduces the comparison workload. Among three algorithms HPSPM was shown to be the best approach through the experiments on an IBM SP2 distributed memory machine. However, the main limitation of all these parallel algorithms is that they make repeated passes over the disk-resident database partition, incurring high I/O overheads. Furthermore, the schemes involve exchanging the remote database partitions during each iteration, resulting high communication and synchronization overhead. They also use complicated hash structures, which entail additional overhead in maintenance and search, and typically also have poor cache locality [9].

pSPADE is a parallel algorithm based on SPADE for fast discovery of frequent sequences in large databases, targeting shared-memory systems [9]. It decomposes the original search space into small suffix-based classes. Each class can be solved in main-memory using simple join operations and efficient search techniques, which not only minimizes I/O costs by reducing database scans, but also computational costs. Further each class can be solved independently on each processor requiring no synchronization. It has good locality and little false sharing. Experiments on a 12 processor SGI Origin 2000 shared memory system show that pSPADE delivers good speedup and has excellent scale-up properties. However, like SPADE, the limitation of pSPADE is that it works on the assumption that each class and its intermediate idlists fit in main memory, which require lots of memory.

DPF, TPF are two different parallel algorithms for finding sequential patterns on distributed-memory parallel computers [10]. DPF decomposes the computation by exploiting data parallelism, whereas TPF utilizes task parallelism. The feature of TPF is the development of a static task decomposition scheme that uses a bipartite graph partitioning algorithm to simultaneously balance the computations and at the same time reduce the data sharing overheads, by minimizing the portions of the database that needs to be shared by different processors. Experiments on the 32-processor IBM SP2 parallel computer show that they incur small communication overheads, achieve good speedups, and can effective utilize the different processors, and that TPF outperformed DPF. However, as number of processors increased, the accuracy of estimated work-load decreased and the computation became increasingly un-balanced.

DTPF is an improved parallel algorithm of TPF, which uses task parallelism along with dynamic load balancing scheme that minimizes idle time in case when distributed workload is unbalanced [33]. Experiments on the 32-processor IBM SP2 parallel computer show that it is capable of achieving good speedups, substantially reducing the amount of the required work to find sequential patterns in large databases, and it outperforms static load balancing scheme algorithm, TPF.

3.5 Incremental-Based Method

The above studies all assume the database is static, and even a small change in the database will require the algorithms to run again to get the updated frequent sequences since previous sequential patterns would become irrelevant and new sequential patterns might appear. In practice, the content of a database changes over time and thus there is a need for efficient algorithms to update, maintain and manage the information discovered. If each time we have to rerun the mining algorithms from scratch, it will be very inefficient or infeasible. Incremental algorithm should be developed for sequential pattern mining so that mining can be adapted to frequent and incremental database updates, including both insertions and deletions. There are two cases in developing incremental algorithm: (1) Whole sequences are inserted into and/or removed from the old database called as Sequence Model; (2) Sequences in the old database are updated by appending new transactions at the end called as Transaction Model. The two models can model each other. Algorithms designed for one model can also work under the other model.

A work for incremental sequential pattern updating based on *SuffixTree* techniques was proposed in [12]. The structure used in that context acquires the data and builds up the frequent sequences in one scan by means of a *SuffixTree*. This method is thus very appropriate to an incremental sequence extraction, because it only has to continue the data reading after the update. The limitations are the complexity in space of this algorithm depends on the size of the database and the *SuffixTree* is very expensive for dynamic strings because of the sensitivity of the position to the update operation. Then, a modified work of that on incremental sequential pattern updating was proposed in [13]. The approach uses a dynamic *SuffixTree* structure, in which substrings are referenced by addresses rather than positions, for incremental mining in a single long sequence. The address reference restricts the impact of updates to a small part of the dynamic *SuffixTree*, making efficient update of the dynamic *SuffixTree* possible. Experiments showed that this incremental method performs substantially better than the non-incremental one for large and dynamic databases. However, those two algorithms only focus on incremental mining in a single long sequence.

FASTUP, proposed in [34], is in effect an enhanced GSP with improvement on candidate generation and support counting. It takes into account the previous mining result before generating and validating candidates using the generating-pruning method. Experiments show that the performance of this algorithm could be much faster than previous algorithms for the maintenance of sequential patterns. However, it suffers the same limitations as GSP.

Reference [14] developed an incremental mining algorithm ISM based on SPADE by maintaining a sequence lattice of an old database. The sequence lattice includes all the frequent sequences and all the sequences in the negative border, which is the collection of all sequences that are not frequent but both of whose generating subsequences are frequent. Compared with SPADE, the experiment results show that ISM is an improvement in execution time by up to several orders of magnitude in practice, both for handling increments to the database, as well as for handling interactive queries. However, maintaining negative border, the number of which can be very huge, is memory consuming and not well adapted for large databases. In addition, sequences in the negative border may be unlikely to become frequent in the updated database if they have low support. Moreover, ISM can only deal with the case of insertion.

Reference [16] developed two algorithms, GSP+ and MFS+, for incremental mining sequential patterns when sequences are inserted into or deleted from the original database: one based on GSP and the other based on MFS. These two algorithms can efficiently compute the updated set of frequent sequences given the set of frequent sequences obtained from mining the old database. Experiments show that GSP+ and MFS+ effectively reduce the CPU costs of their counterparts with only a small or even negative expense on I/O cost.

Reference [35] developed another incremental mining algorithm, ISE, using candidate generate-and-test approach, namely, using information collected during an earlier mining process to cut down the cost of finding new sequential patterns in the updated database. The main new feature of it is that the set of candidate sequences to be tested is substantially reduced. Furthermore, it incorporates some optimization techniques for improving the efficiency. Empirical evaluations show that the algorithm performs significantly faster than the approach, GSP, of mining the whole updated database from scratch. The limitation of this algorithm is the candidate set can be very huge, which makes the test-phase very slow and its level-wise working manner requires multiple scans of the whole database. This is very costly, especially when the sequences are long.

IncSP is another efficient incremental updating algorithm for up-to-date maintenance of sequential patterns after a nontrivial number of data sequences are appended to the sequence database [36]. Assume that the minimum support keeps the same. It utilizes the knowledge of previously computed frequent sequences, merges data sequences implicitly, prunes candidates early, and separately counts supports with respect to the original database and the newly appended database. Implicit merging ensures that IncSP employs correctly combined data sequences while preserving previous knowledge useful for incremental updating. Candidate pruning after updating pattern supports against the increment database further accelerates the whole process, since fewer but more promising candidates are generated by just checking counts in the increment database. Eventually, efficient support counting of promising candidates over the original database accomplishes the discovery of new patterns. IncSP both updates the supports of existing patterns and finds out new patterns for the updated database. The simulation performed shows that IncSP is several times faster than re-mining using the GSP algorithm, with respect to various data characteristics or data combinations. IncSP outperforms GSP with regard to different ratios of the increment database to the original database except when the increment database becomes larger than the original database.

IncSpan is developed in [37] for incremental mining over multiple database increments. Two novel ideas are introduced in the algorithm development. First is maintaining a set of "*almost frequent*" sequences as the candidates in the updated database, which has several nice properties and leads to efficient techniques. Second is that two optimization techniques, *reverse pattern matching* and *shared projection*, are designed to improve the performance. Reverse pattern matching is used for matching a sequential pattern in a sequence. Since the appended transactions are at the end of a sequence, reverse pattern matching can prune additional search space. Shared projection is designed to reduce the number of database projections for some sequences which share a common prefix. Performance study shows that IncSpan outperforms ISM and PrefixSpan on incrementally updated databases by a wide margin.

3.6 Constraint-Based Algorithm

Recent work has highlighted the importance of the paradigm of constraint-based mining. Not only the paradigm allows users to express their focus in mining, but also allows many kinds of constraints to be pushed deep inside mining, confining the search for patterns only to those of interest to the user, and therefore, improving performance. Constraint-based algorithms are close related to user-specified constraints because of the arising problem, namely, how to push kinds of constraints deep inside mining in order to improve performance.

The use of Regular Expressions (REs) was proposed in [18] as a flexible constraint specification tool that enables user-controlled focus to be incorporated into the pattern mining process. A family of novel algorithms, SPIRIT, was developed for mining frequent sequential patterns that also satisfy user-specified RE constraints. The main distinguishing factor among the proposed schemes is the degree to which the RE constraints are enforced to prune the search space of patterns during computation. The SPIRIT algorithms are illustrative of the tradeoffs that arise when constraints that do not subscribe to nice properties (like anti-monotonicity) are integrated into the mining process. Experimental results clearly validate the effectiveness of the approach, showing that speedups of more than an order of magnitude are possible when RE constraints are pushed deep inside the mining process and also illustrates the versatility of REs as a user-level tool for focusing on interesting patterns.

cSPADE was proposed in [38] for discovering the set of all frequent sequences with the following constraints: length and width restrictions, minimum and maximum gap between sequence elements, time window of occurrence of the whole sequence, item constraints for including or excluding certain items, and finding sequences distinctive of at least one class, i.e., a special attribute-value pair, that we are interested in predicting. The two main strengths of cSPADE are that it delivers performance far superior to existing approaches to constrained sequences, and that it incorporates the constraints with relative ease.

Prefix-growth is also one of constraint-based algorithms for mining sequential patterns developed in [19] to push prefix-monotone constraints, which covers many commonly used constraints such as all monotonic, anti-monotonic constraints and regular expression, deep into the mining process. Moreover, some tough constraints, like those involving aggregate $avg()$ and $sum()$, can also be pushed deep into prefix-growth with some minor extensions. Experimental results and performance study show that prefix-growth is efficient and scalable in mining large databases with various constraints compared with SPIRIT.

Moreover, there are many other kinds of algorithms for mining sequential patterns like DSG (Direct sequential pattern generation), a graph-based algorithm, proposed in [39]. Though the disk I/O cost of DSG is very low because it scans database only once, the related information may not fit in the memory when the size of the database is large.

All in all, the existing algorithms for mining sequential patterns depend heavily on massive computation that might cause high dependency on the memory size or repeated I/O scans for the data sets. Though they are very efficient, they are not sufficient for extremely large datasets and new solutions, that do not depend on repeated I/O scans and less reliant on memory size, still have to be found.

4 Theoretical Foundation of Sequential Pattern Mining

Previous work on sequential pattern discovery was mainly focused on studying scalable algorithms while less on its theoretical foundations, which is also important and makes it possible to use the existing theories or methods to support more effective sequential pattern mining tasks. In the following, we build the relation between the closed sequential patterns and the fixed point by using the theory of formal concept analysis [40, 41], which can serve as a theoretical foundation of sequential patterns [42].

Definition 1. *Reference [40] A triple (G, M, I) is called a context if G and M are sets and $I \subseteq G \times M$ is a binary relation between G and M. We call the elements of G objects, those of M attributes, and I the incidence of the context (G, M, I).*

For the object g and the attribute m, $(g, m) \in I$ or more commonly, gIm implies that 'the object g possesses the attribute m'.

Definition 2. *Reference [40] Let (G, M, R) be a context, then the following mappings are Galois connections between $P(G)$ and $P(M)$:*

$$s : G \mapsto M, s(X) = \{m \in M \,|(\forall g \in X)gRm\}\,,$$

$$t : M \mapsto G, t(Y) = \{g \in G|\,(\forall m \in Y)gRm\}\,,$$

where $P(G)$ and $P(M)$ are the power sets of G and M, respectively.

Obviously, $s \circ t$ and $t \circ s$ are closed operators, also call them as Galois closed operators. Let the identifier set ID, the sequential set D of a sequence database be G and M of a context (G, M, R), a binary relation between U and D be R, then (U, D, R) becomes a context. Their Galois connections are as follows:

$$t : D \mapsto U, t(\alpha) = \{g \in U|\,(\forall m \in \alpha)gRm\}$$
$$s : U \mapsto D, s(X) = \{m \in D|\,(\forall g \in X)gRm\}$$

Then, $t(\alpha)$ denotes all the id set that includes a sequence α. Moreover, $s \circ t$ and $t \circ s$ are closed operators, also call them as Galois closed operators.

Theorem 1. $\{\alpha \in P(D)\,|s \circ t(\alpha) = \alpha\}$ *is the set of all closed sequential patterns of sequence database D.*

Proof. (Sufficiency) Suppose that a sequence α, satisfying $s \circ t(\alpha) = \alpha$, is not a closed sequential pattern. Then there exists a sequence β, satisfying $\alpha \subset \beta$ and $support(\alpha) = support(\beta)$. Namely, $t(\alpha) = t(\beta)$ and thus $s \circ t(\beta) \supseteq \beta$. It is concluded that $\alpha = s \circ t(\alpha) = s \circ t(\beta) \supseteq \beta$, which contradicts that $\alpha \subset \beta$. Therefore, α is a closed sequential pattern.

(Necessity) Suppose that a sequence α is a closed sequential pattern of D. Since $s \circ t$ is a closed operator, $s \circ t(\alpha) \supseteq \alpha$. If $s \circ t(\alpha) \supset \alpha$, according to the definition of s and t, every sequence contains α also contains $s \circ t(\alpha)$. Thus, $support(s \circ t(\alpha)) = support(\alpha)$, which contradicts the assumption that α is a closed sequential pattern. Therefore, we have $s \circ t(\alpha) = \alpha$. □

Definition 3. *Reference [43] Let P be a partial order set, $\Phi:P \to P$ is a mapping, $a \in P$. If $\Phi(a) = a$, then call a is a fixed point of Φ.*

Then, every sequence in the set $\{\alpha \subseteq D\,|s \circ t(\alpha) = \alpha\}$ is a fixed point of the mapping $s \circ t$. To mine all frequent closed sequential patterns is equal to find all fixed points of $s \circ t$. The existence of fixed point of the mapping is confirmed by the following theorems.

Theorem 2. *Reference [43] Let P be a partial order set, $\Phi: P \to P$ is a mapping, satisfy that for every $a \in P, a \leq \Phi(a)$, then Φ has fixed points.*

Theorem 3. *[43] Let P be a partial order set, $\Phi : P \to P$ is an order-preserving mapping, then Φ has fixed points and the minimum fixed point.*

Theorem 4. $s \circ t : P(D) \to P(D)$ *has fixed points.*

Proof. It is obvious that $P(D)$ is a partial order set. Since $s \circ t : P(D) \to P(D)$ is a closed operator. Then, for every $\alpha \in P(D)$, $\alpha \subseteq s \circ t(\alpha)$ and $s \circ t$ is an order-preserving mapping. Therefore, there exist fixed points of $s \circ t$.

Since $(P(D), \subseteq)$ is a complete lattice and closed operator is order-preserving mapping, then a concrete fixed point(namely, closed sequential pattern) can be obtained by the fixed point theorem proposed by Knaster and Tarski [43]. □

Theorem 5. *Reference [43] Let P be a complete lattice, $\Phi : P \to P$ is order-preserving mapping, then $\vee \{ a \in P \,|\, a \leq \Phi(a) \}$ is a fixed point of Φ.*

Theorem 6. $(\{ \alpha \in P(D) \,|\, s \circ t(\alpha) = \alpha \}, \subseteq)$ *constitues a join semi-lattice, called as fixed point semi-lattice.*

Proof. It is obvious. □

Because the closed sequential patterns keep all the support information of all sequential patterns of sequence database, fixed point semi-lattice also keeps them. Therefore, to mine all frequent closed sequential patterns is equal to establish fixed point semi-lattice and mine all points on it that satisfy support constraint.

Example 2. Frequent fixed-point semi-lattice of Table 1 is as follows (Fig. 1). Every node's support in this lattice is above the minimum support threshold.

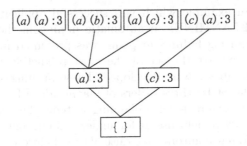

Fig. 1. Frequent fixed-point semi-lattice

To sum up, much work has been done to efficiently discovery sequential patterns while less work on its theoretical foundations. Based on the Galois closed operator, identifier set, sequential set in sequential database together with their binary relation constituted a context. The relation between the fixed point and closed sequential pattern was established. To mine all frequent closed sequential patterns is equal to build fixed point semi-lattice and mine all points on it that satisfy support constraint which serves as a theoretical foundation of sequential patterns and makes it possible to use the existing fixed point and lattice theories to support more effective sequential pattern mining tasks.

5 Applications of Sequential Pattern Mining

In the daily and scientific life, sequential data are available and used everywhere. Since discovering interesting patterns can benefit us by predicting coming activities, interpreting certain phenomena, extracting outstanding similarities and differences for close attention, etc. Methods for mining sequential patterns have successfully applied in many domains as follows.

- Retail industry: analysis of customer buying behavior.
- Medical treatment: discovering patterns in histories of medical records to improve level of diagnosis.
- DNA sequences and gene structures: discovery of motifs and tandem repeats in DNA sequences.
- Telecommunication: finding network alarm patterns, telephone calling patterns.
- Administration: identifying plan failures.
- Web service: discovering user access patterns.
- Information security: analysis of user's behavior.
- Natural disasters: earthquakes forecasting.
- Science & engineering processes: study of engineering & scientific processes such as experiment runs, satellite data streams, weather data.
- Stocks and markets: stock prices trend.

Following is an example which is a successful application of sequential pattern mining. PLANMINE, an automatic mining method that discovers event sequences causing failures in plans, was presented in [2]. Novel pruning techniques to extract the set of the most predictive rules from highly structured plan databases were developed. These pruning strategies reduced the size of the rule set by three orders of magnitude. PLANMINE has been fully integrated into two real-world planning systems. The rules discovered by PLANMINE were extremely useful for understanding and improving plans, as well as for building monitors that raise alarms before failures happen.

All in all, several cases should be carefully studied during the process of applications of sequential pattern mining: (1) Developing operations for data cleaning; (2) Integration of the algorithm into the real-world system; (3) Devising approaches to reduce the size of the rule set; (4) Visualization should also be carefully considered.

6 Conclusions

Sequential pattern mining is one of the most popular pattern-discovery approaches in the field of knowledge discovery and data mining. In this chapter, we conduct a systematic study on models and algorithms in sequential pattern analysis, build the relation between the closed sequential patterns and the fixed point and discuss its application domains. Despite recent advance in the problem of sequential pattern mining, several problems still need serious and immediate attention.

- Developing more efficient and scalable methods for mining sequential patterns, including incremental mining of closed sequential patterns, incorporating user-specified constraints in the mining of closed sequential patterns, developing methods to mine sequential patterns with other more complicated constraints, devising sampling-based methods and random access disk-based approaches like Inverted Matrix for efficient discovery of frequent itemsets [44, 45], etc.
- Proceeding to explore its application domains, including development of application-specific data mining system, invisible data mining (mining as built-in function) and integration of existing algorithms for other complicated structured patterns, etc.
- To continue studying theoretical foundations of sequential pattern mining like Codd's relational model [46] in order for us to apply them to the development of more efficient mining algorithms and methods.
- To establish a benchmark to evaluate sequential pattern mining algorithms like FIMI workshop for the problem of frequent itemset mining [47] in order to generate a very healthy and critical discussion on the state-of-affairs in sequential pattern mining implementations.
- To integrate efficient algorithms with database management systems, data warehouse systems, and Web database systems, etc. which can maximally benefit end-users.
- Development of techniques that incorporate privacy concerns in sequential pattern mining since data mining, with its promise to efficiently discover valuable, non obvious information from large databases, is particularly vulnerable to misuse [48, 49].

References

1. Agrawal R, Srikant R (1995) Mining sequential patterns. In Proc of the 11th Int Conf on Data Eng. Mar. 1995, Taipei, Taiwan. 3–14
2. Zaki MJ, Lesh N, Ogihara M (2000) PLANMINE: Sequence mining for plan failures. Artificial Intelligence Review, special issue on the Application of Data Mining 14(6): 421–446
3. Wu PH, Peng WC, Chen MS (2001) Mining sequential alarm patterns in a telecommunication database. In Proc of VLDB-01 Workshop on Databases in Telecommunications. Sept. 2001, Roma, Italy. 37–51

4. Yan X, Han J, Afshar R (2003) CloSpan: Mining Closed Sequential Patterns in Large Databases. In SDM'03. May 2003, San Francisco, CA. 166–177
5. Srikant R, Agrawal R (1996) Mining sequential patterns: generalizations and performance improvements. In Proc of the 5th Int Conf on Extending Database Technology. Mar. 1996, Avignon, France. 3–17
6. Zaki MJ (2001) An eficient agorithm for mining fequent sequences. Machine Learning J 42(1/2): 31–60
7. Pei J, Han J, Pinto H, Chen Q, Dayal U, Hsu MC (2001) PrefixSpan: Mining sequential patterns efficiently by prefix-projected pattern growth. In Proc of 2001 Int Conf on Data Eng. Apr. 2001, Heidelberg, Germany. 215–224
8. Shintani T, Kitsuregawa M (1998) Mining algorithms for sequential patterns in parallel: Hash based approach. In Proc of the Second Pacific-Asia Conf on Know Discovery and Data mining. Apr. 1998, Merburn, Australia. 283–294
9. Zaki MJ (2001) Parallel sequence mining on shared-memory machines. J of Parallel and Distributed Computing 61(3): 401–426
10. Guralnik V, Garg N, Karypis G (2001) Parallel tree projection algorithm for sequence mining. In Sakellariou R Keane J Gurd J Freeman L (eds) Proc of 7th European Conf on Parallel Computing. 310–320. Springer, Berlin Heidelberg New York
11. Wang J, Han J (2004) BIDE: Efficient mining of frequent closed sequences. In Proc. of 2004 Int. Conf. on Data Eng. Apr. 2004, Boston, MA. 79–90
12. Wang K, Tan J (1996) Incremental discovery of sequential patterns. In Proc of Workshop on Research Issues on Data Mining and Know Discovery. June 1996, Montreal, Canada. 95–102
13. Wang K (1997) Discovering patterns from large and dynamic sequential data. J of Intelligent Information Systems 9(1): 33–56
14. Parthasarathy S, Zaki MJ, Ogihara M, Dwarkadas S (1999) Incremental and interactive sequence mining. In Proc of the 8th Int Conf on Information and Know Management. Nov. 1999, Kansas, Missouri, USA. 251–258
15. Lee CH, Lin CR, Chen MS (2001) Sliding-Window filtering: An efficient algorithm for incremental mining. In Proc of the ACM 10th Int Conf on Information and Know Management. Oct. 2001, Atlanta, Georgia. 263–270
16. Zhang M, Kao B, Cheung D, Yip CL (2002) Efficient algorithms for incremental update of frequent sequences. In Proc of the 6th Pacific-Asia Conf on Know Discovery and Data Mining. May, 2002, Taipei, Taiwan. 186–197
17. Masseglia F, Poncelet P, Teisseire M (2003) Incremental mining of sequential patterns in large databases. Data Know. Eng 46: 97–121
18. Garofalakis MN, Rastogi R, Shim K (1999) SPIRIT: Sequential pattern mining with regular expression constraints. In Proc of the 25th Int Conf on Very Large Data Bases. Sept. 1999, Edinburgh, Scotland. 223–234
19. Pei J, Han J, Wang W (2002) Mining sequential patterns with constraints in large databases. In Proc of the 11th Int Conf on Information and Know Management. Nov. 2002, McLean, VA. 18–25
20. Pei J, Han J (2002) Constrained frequent pattern mining: A pattern-growth view. SIGKDD Explorations 4(1): 31–39
21. Toroslu IH, Kantarcioglu M (2001) Mining cyclically repeated patterns. In: Kambayashi Y, Winiwarter W, Arikawa M (eds): DaWaK 2001, LNCS 2114. 83–92

22. Kum HC, Pei J, Wang W, Duncan D (2003) ApproxMAP: Approximate mining of consensus sequential patterns. In Proc of the 2003 SIAM Int Conf on Data Mining. May 2003, San Francisco, CA. 311–315
23. Pinto H, Han J, Pei J, Wang K, Chen Q, Dayal U (2001) Multi-dimensional sequential pattern mining. In Proc of the 10th Int Conf on Information and Know Management. Nov. 2001, Atlanta, Georgia. 81–88
24. Tzvetkov P, Yan X, Han J (2003) TSP: Mining top-k closed sequential patterns. In Proc. 2003 Int. Conf. on Data Mining. Nov. 2003, Melbourne, FL. 347–354
25. Afshar R (2001) Mining frequent Max and closed patterns. MA thesis, Simon Fraser University, Canada
26. Yang J, Yu PS, Wang W, Han J (2002) Mining long sequential patterns in a noisy environment. In SIGMOD'02. June 2002, Madison, WI. 406–417
27. Chen YL, Chen SS, Hsu PY (2003) Mining hybrid sequential patterns and sequential rules, Information Systems 27: 345–362
28. Mannila H, Toivonen H, Verkamo AI (1995) Discovering frequent episodes in sequences. In Proc of the 1st Int Conf on Know Discovery and Data Mining. Aug. 1995, Montreal, Canada. 210–215
29. Joshi MV, Karypis G, Kumar V (1999) Universal formulation of sequential patterns. Technical Report Under Preparation. Department of Computer Science, University of Minnesota, Minneapolis
30. Zhang M, Kao B, Yip CL, Cheung D (2001) A GSP-based efficient algorithm for mining frequent sequences. In Proc. of IC-AI'2001. June 2001, Las Vegas, Nevada, USA
31. Han J, Pei J, Mortazavi-Asl B, Chen Q, Dayal U, Hsu MC (2000) FreeSpan: Frequent pattern-projected sequential pattern mining. In Proc of the 6th ACM SIGKDD int conf on Know discovery and data mining. Aug. 2000, Boston, MA. 355–359
32. Ayres J, Gehrke JE, Yiu T, Flannick J (2002) Sequential pattern mining using bitmaps. In Proc of the 8th ACM SIGKDD Int Conf on Know Discovery and Data Mining. July 2002, Edmonton, Alberta, Canada. 429–435
33. Guralnik V, Karypis G (2001) Dynamic load balancing algorithms for sequence mining. Technical Report 00-056, Department of Computer Science, University of Minnesota
34. Lin MY, Lee SY (1998) Incremental update on sequential patterns in large databases. In Proc of the 10th IEEE Int Conf on Tools with Artificial Intelligence. Nov. 1998, Taipei, Taiwan. 24–31
35. Masseglia F, Poncelet, Teisseire M (2003) Incremental mining of sequential patterns in large databases. Data & Know Eng 46: 97–121
36. Lin MY, Lee SY (2004) Incremental update on sequential patterns in large databases by implicit merging and efficient counting. Information Systems 29: 385-404
37. Cheng H, Yan X, Han J (2004) IncSpan: Incremental mining of sequential patterns in large database. In Proc. 2004 Int. Conf. on Know Discovery and Data Mining. Aug. 2004, Seattle, WA. 527–532
38. Zaki MJ (2000) Sequence mining in categorical domains: Incorporating constraints. In 9th Int Conf on Information and Know Management. Nov. 2000, Washington, DC. 422–429
39. Yen SJ, Chen ALP (1996) An efficient approach to discovering knowledge from large databases. In Proc of 4th Int Conf on Parallel and Distributed Information Systems. IEEE Computer Society. 8–18

40. Ganter B, Wille R (1999) Formal concept analysis: Mathematical foundations. Springer, Berlin Heidelberg, New York
41. Pasquier N, Bastide Y, Taouil R, Lakhal L (1999) Efficient mining of association rules using closed itemset lattices. Information Systems 24: 25–46
42. Li TR, Yang N, Ma J, Xu Y (2004) Theoretical foundations of sequential patterns. In The World Congress on Intelligent Control and Automation. June 2004, Hangzhou, China. 4241–4244
43. Smart DR (1974) Fixed point theorems. Cambridge University Press, Cambridge, UK
44. Lee S, Cheung D, Kao B (1998) Is sampling useful in data mining? A case in the maintenance of discovered association rule, Data Mining and Know Discovery 2(3): 233–262
45. EI-Hajj M, Zaiane OR (2003) Inverted Matrix: Efficient discovery of frequent items in large datasets in the context of interactive mining, In Proc. of the 9th ACM SIGKDD Int. Conf. on Knowledge Discovery and Data Mining. Aug. 2003, Washington, DC. 109–118
46. Mannila H (2000) Theoretical frameworks of data mining, SIGKDD explorations 1(2): 30–32
47. Goethals B, Zaki MJ (2003) Introduction: Advances in frequent itemset mining implementations. In Proc of the IEEE ICDM Workshop on Frequent Itemset Mining Implementations. Nov. 2003, Melbourne, Florida, USA. 1–13
48. Agrawal R, Srikant R (2000) Privacy-preserving data mining. In Proc of the ACM SIGMOD Conf on Management of Data. May 2000, Dallas, TX. 439–450
49. Zhan Z, Chang LW, Matwin S (2004) Privacy-preserving collaborative sequential pattern mining. In SIAM DM 2004 Workshop on Link Analysis, Counter-Terrorism & Privacy. Apr. 2004, Lake Buena Vista, Florida

Uncertain Knowledge Association Through Information Gain

Athena Tocatlidou[1], Da Ruan[2], Spiros Th. Kaloudis[1], and
Nikos A. Lorentzos[1]

[1] Agricultural University of Athens, Informatics Laboratory, Iera Odos 75, 11855
   Athens, Greece
[2] Belgian Nuclear Research Centre (SCK•CEN), 2400 Mol, Belgium

Abstract. The problem of entity association is at the core of information mining
techniques. In this work we propose an approach that links the similarity of two
knowledge entities to the effort required to fuse them in one. This is implemented
as an iterative updating process. It unites an evolving initial knowledge entity and
a piece of new information, which is repeatedly incorporated, until a convergence
state is reached. The number of updating repetitions can be used as an importance
index qualifying the new evidence.

Key words: Uncertain knowledge updating, Similarity, Information fusion, Fuzzy
sets.

1 Introduction

The present era is characterized by the enormous amount of available information in both raw and processed forms. The most important concern of the
modern knowledge engineer is to find the means to extract useful pieces of
knowledge as well as to add value and intelligence to computerized systems.
Therefore there is a need for efficient use of data, avoidance of over fitting and
high dimensionality solutions. This eventually helps enlarging applications of
computer models and extends the spectrum of information sources, departing from the exclusive use of numbers, and moving towards qualitative data,
which is a very important requirement for multimedia implementations.

Along another dimension, intelligent systems must possess the ability to
handle incomplete data, confront noise, to exploit and use missing information,
and effectively interact with the user [3, 6]. Data mining is a collective set of
techniques and methods towards these goals [1, 2].

Various methods such as neural networks, decision trees etc. have been
used at the core of knowledge management and information mining. In this
work a different approach is proposed: Two pieces of information are related

Athena Tocatlidou et al.: *Uncertain Knowledge Association Through Information Gain*, Studies
in Computational Intelligence (SCI) **5**, 123–122 (2005)
www.springerlink.com   © Springer-Verlag Berlin Heidelberg 2005

to the degree defined by the effort needed to fuse them in one piece of information. In a monotonic framework this results in an information gain of the initial knowledge. The implementation consists of a repeated updating of the initial information with values from the new evidence. This process converges after a number of repetitions, a number that is related to the *distance* of the two entities. Seen from another perspective, the selection of the number of updating steps can be the choice of the decision maker. Complementing this concept a probability distribution difference is defined. This difference detects the change in the shape of the initial information representation due to the inclusion of new elements.

2 Uncertain Information Representations

Consider a space X and a function $\mu_F(x) : X \rightarrow [0,1]$ such that for each $x : x \in X$ μ_F gives the degree to which the element x belongs to the concept described by F. The pairs $(x, \mu_F(x))$ define a fuzzy set F over the domain X [10]. This constitutes a knowledge representation scheme, which is able to handle uncertain information by allowing partial membership to a concept.

A discrete fuzzy set is the set defined over a discrete domain. Given a fuzzy set F, its support set is defined as the crisp set

$$S_F = \{x \in X : \mu_F(x) \neq 0\}$$

Consider another fuzzy set G, and let denote by set T the elements in X that belong to the support set of G, but not to the support set of F, i.e.,

$$T = \{x \in X : \mu_G(x) \neq 0 \quad \text{and} \quad \mu_F(x) = 0\}$$

A generalization or information updating algorithm will extend F to a new fuzzy set H, containing all the elements in F plus elements from T, such that

$$S_H = S_F \cup T' \quad \text{and} \quad \emptyset \neq T' \subseteq T$$

This process implements an information gain for the initial set F, which is monotonic in nature.

3 Description of the Knowledge Updating Algorithm

In this section we present a more systematic exposition of the knowledge updating algorithm and the notion information gain that it entails. Assuming that initial knowledge is represented by a fuzzy set, then this is realized by the production of a more general fuzzy set, enriched with elements from the new evidence. The resulting new information can be compared with the initial one within a probabilistic framework, since the probability distribution over the initial set of elements is enlarged to include new data. This captures the change in the shape of the fuzzy set involved.

More specifically, consider a piece of uncertain data, in the form of a discrete fuzzy set F, representing existing information, and the corresponding possibility distribution $\{F_i, m_i\}$ [7]. To the focal elements $F_1 \subseteq F_2 \subseteq F_3 \subseteq \cdots \subseteq F_n$ are assigned the basic probability masses $m_1, m_2, m_3, \ldots, m_n$, respectively [5].

Let now G be another discrete fuzzy set representing new evidence and $G_1 \subseteq G_2 \subseteq G_3 \subseteq \ldots \subseteq G_q$ be the family of focal elements of the corresponding possibility distribution. For the first focal G_1, containing the elements with a full membership in the fuzzy set G, we can locate a focal element F_k so that the two sets are disjoint.

More formally: there exists an index $k \in \{1, 2, \ldots, n\}$ such that for some F_k in $\{F_1, F_2, F_3, \ldots, F_n\}$ the following relations hold:

$$G_1 \cap F_k = \emptyset \quad \text{and} \quad G_1 \cap F_{k+1} \neq \emptyset .$$

Subsequently, we can create the following family of focals

$$\{F_1, F_2, \ldots, F_k \cup G_1, F_{k+1} \cup G_2, \ldots, F_n \cup G_q\}$$

resulting from the union of the focal elements of the fuzzy set G and the subset of the focals from the fuzzy set F starting from F_k. To this family of new focal elements we associate the basic probability masses $m_1, m_2, m_3, \ldots, m_n$ from the initial fuzzy set F.

In the case where $n - k \neq q - 1$, that is, when the families of focal elements are not aligned, two modifications are possible: Either the final focal element F_n of the fuzzy set F is repeated with its mass m_n accordingly divided, or the final focal G_q is repeated. In that way we can generate a new fuzzy set H. Figure 1 describes the steps involved in the algorithm.

This is an extension of F towards the support set of G that preserves the existing information in F. If by S_F, S_G, and S_H we denote the support sets of F, G, and H respectively, then

$$S_F \subseteq S_H \quad \text{and} \quad S_G \subseteq S_H$$

This outcome represents an information gain of the initial knowledge towards the new evidence represented by G. This process can be repeated, generalizing the resulting fuzzy set H again with fuzzy set G as new data. This process will eventually converge, reaching a point where no further updating will be possible.

The aboved defined number $k \in \{1, 2, \ldots, n\}$ indicates the number of permissible repetitions. The number of these iterations can be considered as an indication of the conceptual distance of the entities represented by the two fuzzy sets, in the sense that the greater the number of repetitions the more dissimilar are the two concepts.

Given a discrete fuzzy set F, representing existing information, let G be another discrete fuzzy set representing new evidence

I.   Arrange the two fuzzy sets in decreasing order of membership function

II.   From the two fuzzy sets derive the corresponding possibility distributions, i.e. the series of focal elements $F_1 \subseteq F_2 \subseteq F_3 \subseteq ... \subseteq F_n$, and the assigned basic probability masses $m_1, m_2, m_3, ..., m_n$ for set F, and $G_1 \subseteq G_2 \subseteq G_3 \subseteq ... \subseteq G_q$ and $r_1, r_2, r_3, ..., r_q$ for set G.

III. Find a focal F_k in $\{F_1, F_2, F_3, ..., F_n\}$ such that for the first focal element G_1 the following relations hold:
$$G_1 \cap F_k = \varnothing \qquad \text{and} \qquad G_1 \cap F_{k+1} \neq \varnothing .$$

IV.   Arrange the two series of focals in a way that G_1 is aligned to F_k. In the case where the last focal F_n does not correspond to the last focal G_q, then

   a. If the focals F_i after the alignment point, are fewer than the focals G_j, repeat the last focal element F_n and divide accordingly its mass m_n.

   b. If the focals G_j after the alignment point, are fewer than the focals F_i, repeat the last focal element G_q and divide accordingly its mass r_q.

V.   Create a series of new focals by taking the union: $\{F_1, F_2, ..., F_k \cup G_1, F_{k+1} \cup G_2, ...\}$ and associate with them the basic probability mass assignment of fuzzy set F (divided if necessary).

VI.   Construct a fuzzy set H from this family of focals and the associated basic probability masses.

Fig. 1. Steps of the generalization algorithm

A detailed exposition of the steps involved in realizing the algorithm is explained in the examples in the subsequent section.

Initial experiments with various configurations of fuzzy set pairs have indicated the following characteristics of the updating mechanism:

- It has a gradual updating behavior that depends on the relative position of the two fuzzy sets, the overlapping area, and the relation of the core sets.
- It is not based on the selection of a parameter. Instead the number of necessary iterations depends on the conceptual distance of the two entities.
- It can be applied to nominal and/or non-ordered data.
- It has a set theoretic basis, expanding the initial fuzzy set by enriching its constituent parts, i.e., the focal elements, departing from a point wise calculation.
- It preserves the normality of fuzzy sets.

Consider again the previously mentioned discrete fuzzy sets F and H. Over these sets a probability distribution can be defined as follows:

For each $x \in S_F$, let F_p be a focal element that contains x, i.e., such that

$$\{x\} \cap F_p \neq \emptyset$$

Since $\forall p \leq n \quad F_p \subseteq \ldots \subseteq F_n$, the value of a probability distribution at the point x can be defined as

$$p(x) = \sum_p^n \frac{m_p}{|F_p|}$$

$|F_p|$ being the cardinality of the focal F_p and m_p is the basic probability mass associated with F_p.

This defines a uniform probability distribution over each focal element. After the generalization of F, producing the set H, the value of the distribution at point x becomes

$$p'(x) = \sum_r^n \frac{m_r}{|H_r|}$$

where again H_r is a focal element such that $\{x\} \cap H_r \neq \emptyset$.

We can then take the difference of the values of these distributions at the point x_1 such that $\mu_F(x_1) = 1$, that is the core elements of the initial fuzzy set. The amount $\Delta pr = p(x_1) - p'(x_1)$ provides an indication of the degree of change imposed on the initial fuzzy set due the performed generalization. As such it can be used as a similarity index.

4 Examples

The above-described algorithm has been tested on a non-ordered, non-numeric domain. This is a sequence of qualitative values $\{LX, LY, HY, HE, CR, HX,$

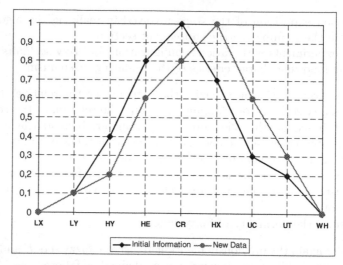

Fig. 2. Fuzzy sets representing initial and new information

$\{CR\}$: 0.2

$\{CR, HE\}$: 0.1

$\{CR, HE, HX\}$: 0.3

$\{CR, HE, HX, HY\}$: 0.1

$\{CR, HE, HX, HY, UC\}$: 0.1

$\{CR, HE, HX, HY, UC, UT\}$: 0.1

$\{CR, HE, HX, HY, UC, UT, LY\}$: 0.1

$\{HX\}$: 0.1

$\{HX, CR\}$: 0.3

$\{HX, CR, HE, UC\}$: 0.1

$\{HX, CR, HE, UC, UT\}$: 0.1

$\{HX, CR, HE, UC, UT, HY\}$: 0.1

$\{HX, CR, HE, UC, UT, HY, LY\}$: 0.1

Fig. 3. Possibility distributions for initial information and new data

$UC, UT, WH\}$, initially originated from coding conifer tree characteristics. In some experiments some additional arbitrary variables are used.

Consider the two fuzzy sets depicted in Fig. 2, representing Initial Information and New Data. We can generate the related possibility distributions given in Fig. 3.

After aligning the two focal element families (Fig. 4), according to the algorithm described in the previous section (in this case $k = 2$), we obtain a new possibility distribution as shown in Fig. 5.

This set of focal elements can result in a fuzzy set (Fig. 6) corresponding to the first generalization of the initial information when updated with the New Data fuzzy set.

The procedure can be repeated but using as initial information the new fuzzy set. The convergence stage is obtained after the second (final) updating (Fig. 7)

$\{CR\}:0.2$

$\{CR,HE\}\cup\{HX\}:0.1$

$\{CR,HE,HX\}\cup\{HX,CR\}:0.3$

$\{CR,HE,HX,HY\}\cup\{HX,CR,HE,UC\}:0.1$

$\{CR,HE,HX,HY,UC\}\cup\{HX,CR,HE,UC,UT\}:0.1$

$\{CR,HE,HX,HY,UC,UT\}\cup\{HX,CR,HE,UC,UT,HY\}:0.1$

$\{CR,HE,HX,HY,UC,UT,LY\}\cup\{HX,CR,HE,UC,UT,HY,LY\}:0.1$

Fig. 4. The union of the constituent possibility distributions

$\{CR\}:0.2$

$\{CR,HE,HX\}:0.1$

$\{CR,HE,HX\}:0.3$

$\{CR,HE,HX,HY,UC\}:0.1$

$\{CR,HE,HX,HY,UC,UT\}:0.1$

$\{CR,HE,HX,HY,UC,UT\}:0.1$

$\{CR,HE,HX,HY,UC,UT,LY\}:0.1$

Fig. 5. Generated possibility distribution

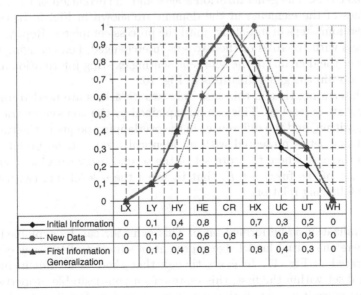

	LX	LY	HY	HE	CR	HX	UC	UT	WH
Initial Information	0	0,1	0,4	0,8	1	0,7	0,3	0,2	0
New Data	0	0,1	0,2	0,6	0,8	1	0,6	0,3	0
First Information Generalization	0	0,1	0,4	0,8	1	0,8	0,4	0,3	0

Fig. 6. First stage of generalization for initial information fuzzy set

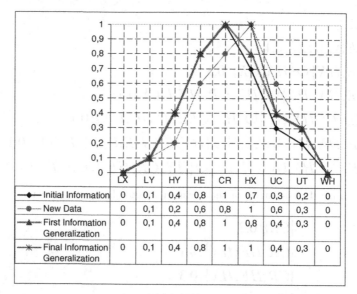

	LX	LY	HY	HE	CR	HX	UC	UT	WH
Initial Information	0	0,1	0,4	0,8	1	0,7	0,3	0,2	0
New Data	0	0,1	0,2	0,6	0,8	1	0,6	0,3	0
First Information Generalization	0	0,1	0,4	0,8	1	0,8	0,4	0,3	0
Final Information Generalization	0	0,1	0,4	0,8	1	1	0,4	0,3	0

Fig. 7. First and final generalization stages for initial information fuzzy set

The second index based on the probability distribution is defined over the support sets of the initial and the final fuzzy set is compute and its values over the corresponding domains are depicted in Fig. 8.

Additionally a second example is now given involving this time two trapezoidal fuzzy sets. The generalization stages and the evolution of the membership values of the elements in the domain are shown in Fig. 9. In this case there exist four steps of updating, until the process completes. Repeating now the process with the roles of the two sets interchanged, i.e., changing the position (over the domain) of the sets representing existing information and new data, we obtain the generalizations in Fig. 10.

It is interesting that in this case only two iterations are needed until convergence state is reached. This can be explained by the non-symmetric nature of the algorithm, and the differences in the overlapping points in the two cases.

If a fuzzy set is presented with an identical fuzzy set, no updating take place. In the case where the core set of the second fuzzy set G is a subset of the core set of the fuzzy set F, that is, when the α-level sets (with $\alpha = 1$) have the following relation

$$G_1 \subseteq F_1$$

the generalization is performed once. This behavior can be explained in the cases where the initial knowledge represents a superset of the new evidence, as illustrated in Fig. 11. In cases though that there is no full inclusion of the second set within the first, this represents a non-desirable property of the algorithm, since it cannot distinguish between various families of subsets.

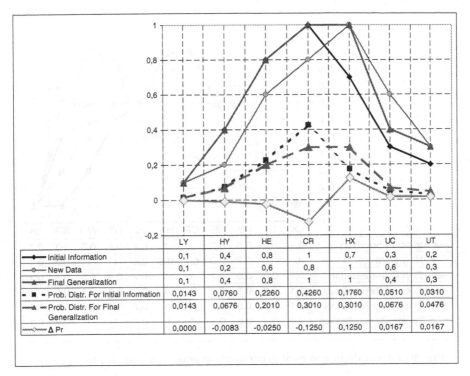

	LY	HY	HE	CR	HX	UC	UT
Initial Information	0,1	0,4	0,8	1	0,7	0,3	0,2
New Data	0,1	0,2	0,6	0,8	1	0,6	0,3
Final Generalization	0,1	0,4	0,8	1	1	0,4	0,3
Prob. Distr. For Initial Information	0,0143	0,0760	0,2260	0,4260	0,1760	0,0510	0,0310
Prob. Distr. For Final Generalization	0,0143	0,0676	0,2010	0,3010	0,3010	0,0676	0,0476
Δ Pr	0,0000	-0,0083	-0,0250	-0,1250	0,1250	0,0167	0,0167

Fig. 8. Probability distributions for initial and generalized fuzzy set

To this end the second proposed index, based on the generation of a probability distribution, and described implemented in the sequel, could be used.

The selection of updating steps implemented on the initial knowledge, given a new piece of evidence, can be considered as a number supplied by the user. In this context this number can be interpreted as an importance factor associated with this evidence. As such it can take a value smaller or equal to the total number of repetitions (the number k mentioned in the previous section), until the process converges. The relation of this user-supplied number to the total number of iterations allowed is an indication of the importance of the new information.

An example of this series of actions is illustrated in Fig. 12. In this example the fuzzy set representing initial knowledge is faced with three sets modeling new evidence, i.e., sets new data 1, 2 and 3. These sets are assigned different importance degrees, shown in Table 1. In this table the selected degree is shown along with the highest possible value it can take, due to the convergence limit. This is realized by implementing an equal number of repetitions of the associated fuzzy set.

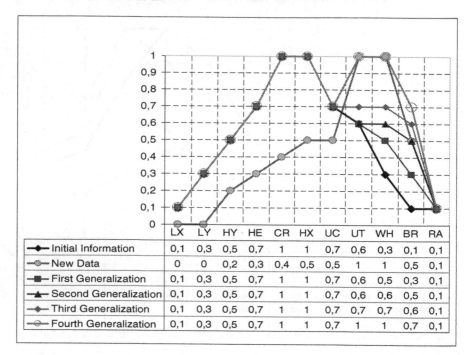

Fig. 9. Intermediate and final generalization stages for trapezoidal fuzzy sets

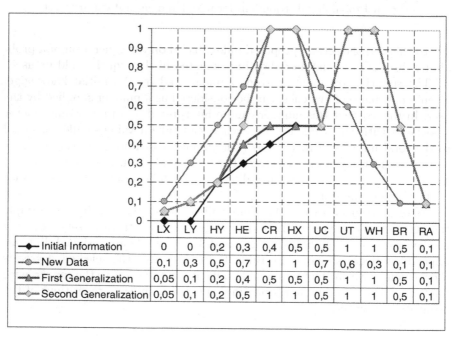

Fig. 10. Generalization stages for a second pair of trapezoidal fuzzy sets

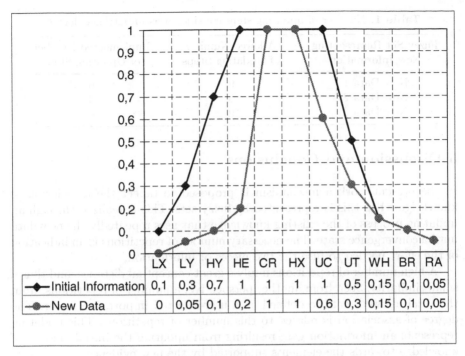

	LX	LY	HY	HE	CR	HX	UC	UT	WH	BR	RA
Initial Information	0,1	0,3	0,7	1	1	1	1	0,5	0,15	0,1	0,05
New Data	0	0,05	0,1	0,2	1	1	0,6	0,3	0,15	0,1	0,05

Fig. 11. Fuzzy set of initial information subsuming fuzzy set of new data

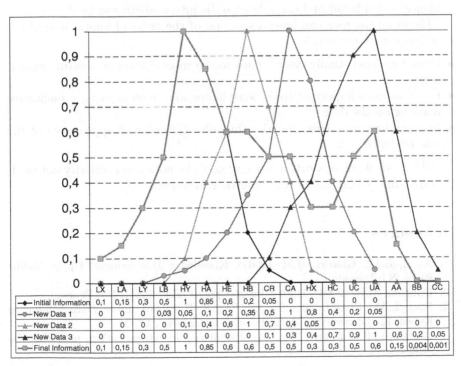

	LX	LA	LY	LB	HY	HA	HE	HB	CR	CA	HX	HC	UC	UA	AA	BB	CC
Initial Information	0,1	0,15	0,3	0,5	1	0,85	0,6	0,2	0,05	0	0	0	0	0			
New Data 1	0	0	0	0,03	0,05	0,1	0,2	0,35	0,5	1	0,8	0,4	0,2	0,05			
New Data 2	0	0	0	0	0,1	0,4	0,6	1	0,7	0,4	0,05	0	0	0	0	0	0
New Data 3	0	0	0	0	0	0	0	0	0,1	0,3	0,4	0,7	0,9	1	0,6	0,2	0,05
Final Information	0,1	0,15	0,3	0,5	1	0,85	0,6	0,6	0,5	0,5	0,3	0,3	0,5	0,6	0,15	0,004	0,001

Fig. 12. Initial information updated given various pieces of new data

Table 1. Number of updating steps considered as importance degree

Fuzzy Set Representing New Information	Allowed Number of Updating Steps	Implemented Number of Updating Steps
New Data 1	9	6
New Data 2	5	3
New Data 3	9	6

5 Discussion and Conclusions

In the present work a mechanism is proposed to detect the association of knowledge when this is represented by fuzzy sets. This is realized through an updating process of the existing knowledge that uses repeatedly the new data until a convergence state. The necessary number of repetitions is an indication of the effort needed to fuse the two pieces of information.

A high number of repetitions imply a great conceptual distance, and thus a small similarity. Seen from another perspective, a number of repetitions, from the total allowed, can be defined as a user-selected importance factor. The degree of association is related to this number of repetitions. This updating represents an information gain resulting from updating the initial amount of knowledge towards the elements supported by the new evidence.

Since the mechanism is not symmetric, the order of presentation of the new evidence is significant and can influence the intermediate and final outcomes.

Therefore the user can have a selection of the order of appearance of the new evidence, for example

- From the conceptually closest data to the more distant to existing knowledge
- From the most important to the least important, implementing a qualitative marking on the data
- In a random way, based on the chronological order of appearance of the new evidence

This framework supports a conception of the notion of similarity not modeled by a linear function or a symmetric relation [4, 8, 9].

References

1. Awad, E.M., Ghaziri, H.M. (2004), Knowledge Management, Upper Saddle River, NJ: Pearson.
2. Dunham, M.H. (2003), Data Mining, Upper Saddle River, NJ: Pearson.
3. Fayyad, U., Piatesky-Shapiro, G., Smyth, P. (1996), From Data Mining to Knowledge Discovery in Data Bases, A I Magazine, **17**, 37–54.
4. Goldstone, R.L. (1994), The role of similarity in categorization: providing the groundwork, Cognition, **52**, 125–157.

5. Klir, G.J., Yuan, B. (1995), Fuzzy Sets and Fuzzy logic: Theory and Applications, Upper Saddle River, NJ: Prentice Hall.
6. Mitra, S., Pal, S.K., Mitra, P. (2002), Data mining in soft computing framework: a survey, IEEE Transactions on Neural Networks, **13**, 3–14.
7. Shafer, G. (1976), A Mathematical Theory of Evidence, Princeton: Princeton University Press.
8. Tocatlidou, A. (1998), Learning based similarity for fuzzy sets, International Journal of Intelligent Systems, **13**, 193–220.
9. Tversky, A. (1977), Features of similarity, Psychological Review, **84,** 327–352.
10. Zadeh, L.A. (1965), Fuzzy sets, Information and Control, **8**, 338–353.

Data Mining for Maximal Frequent Patterns in Sequence Groups

J.W. Guan[1,2], D.A. Bell[1], and D.Y. Liu[2]

[1] School of Computer Science, The Queen's University of Belfast, Belfast BT7 1NN, Northern Ireland, UK
[2] College of Computer Science and Technology, Jilin University, 130012, Changchun, P.R. China

Abstract. In this paper, we give a general treatment for mining some kinds of sequences such as customer sequences, document sequences, and DNA sequences. Large collections of transaction, document, and genomic information have been accumulated in recent years, and embedded latently in it there is potentially significant knowledge for exploitation in the retailing industry, in information retrieval, and in medicine and the pharmaceutical industry, respectively. The approach taken here to the distillation of such knowledge is to detect strings in sequences which appear frequently, either within a given sequence (e.g. for a particular customer, document, or patient) or across sequences (e.g. from different customers, documents, or patients sharing a particular transaction, information retrieval, or medical diagnosis; respectively). Patterns are strings that occur very frequently. These concepts are generalisation of the concept of *motifs* for DNA sequences. There are interesting differences between the 3 applications.

We present basic theory and algorithms for finding very frequent and common strings. Strings which are maximally frequent are of particular interest and, having discovered such patterns we show briefly how to mine association rules by an existing rough sets based technique.

Key words: Rough Sets, Data Mining, Sales Data, Document Retrieval, DNA sequences/profiles, Bioinformatics

1 Introduction

Retail organisations collect and store massive amounts of sales data. Large amounts of data are also being accumulated for document retrieval, and for biological and genomic information systems. For example, Celera reportedly maintains a 70 Tbyte database which grows by 15–20 gbytes every day. Another organization in the pharmaceutical industry is pooling 1 Tbyte of data at each of 4 sites and the volume doubles every 8–9 months. Making full use of this data to gain useful insights into, for example, health issues, presents

J.W. Guan et al.: *Data Mining for Maximal Frequent Patterns in Sequence Groups*, Studies in Computational Intelligence (SCI) **5**, 137–135 (2005)
www.springerlink.com

a tremendous challenge and opportunity. For example, we can potentially inform diagnoses and treatments for the patient in a hospital by taking careful account of patterns in the DNA sequences in a group of a patient's genes. *Data mining* is the computer-based technique of discovering interesting, useful, and previously unknown patterns from massive databases [8] – such as those generated in gene expression. Exploiting the similarity between DNA sequences can lead to significant understanding in bioinformatics [11].

Patterns in a group of genes (DNA sequences) can be considered as phrases of a collection of documents [11]. This suggests that text mining techniques [5, 6, 7, 12] can be used for finding patterns and discovering knowledge about patterns, and ultimately ailments and treatments.

Mining sequential patterns is an attractive and interesting issue, there are various and extensive areas of exploration and application to which it is related. Some existing results for other applications can be extended to this area. For example, [14] have addressed and solved a general problem of finding maximal patterns from large datasets; [5, 6, 7] have investigated maximal association rules for mining for keyword co-occurrences in large document collection and proposed an integrated visual environment for text mining; etc. Here we present a general method to treat general sequences. It closely follows the format of [10] and is a generalisation of motifs to patterns.

We develop general theory and algorithms for discovering patterns and maximal patterns systematically. The paper is organised as follows.

Section 2 introduces what we need to know of sequences, and the definition of frequent patterns in a group of sequences is presented in Sect. 3. Theorems and Algorithm Appending for discovering patterns with certain support or levels of occurrence in the group of sequences are proposed in Sects. 4 and 5, respectively. Section 6 proposes Algorithm Checking for finding higher support patterns with lower computational cost. Section 7 is to find maximal patterns. Algorithm Printing is for finding 1-support maximal patterns (i.e., maximal patterns that occur in at least one sequence in the group of sequences) directly and Algorithm Comparing is for finding higher support (occurring in at least more than one sequences in the group) maximal patterns with lower cost. Sections 8 and 9 introduce the information table for maximal patterns and show how associations between patterns in the group of sequences can be discovered using rough sets techniques [3, 9, 13].

2 Sequences and Containing Relations

Sequences appear in various data and convey information to be mined. Customer, document, and DNA sequences are examples. Investigation of sequences is on the containing relation between sequences, patterns from sequences.

2.1 Customer, Document, and DNA Sequences

Generally, sequences can be defined as follows. Given a non-empty set $\mathcal{B}$, we call elements in it as items. A *sequence* s over $\mathcal{B}$ is an ordered list of non-empty subsets of $\mathcal{B}$, expressed as $s = A_1 A_2 \ldots A_n \ldots A_N$, where $\emptyset \subset A_n \subseteq \mathcal{B}$ for $n = 1, 2, \ldots, N$ and $N > 0$.

We call N the *length* of sequence s and denote $|s| = N$. Let us denote the set of sequences over $\mathcal{B}$ as $\mathcal{S}_\mathcal{B}$ and denote the set of sequences over $\mathcal{B}$ with length N as $\mathcal{S}_{\mathcal{B},N}$.

Example 1. (*Customer sequences*) Consider a large database of customer transactions. Each transaction consists of three fields: transaction Date, customer Id, transaction Items. The following example is given in [2].

Date	Id	Items		Id	Date	Items
10/6/93	2	10, 20		1	25/6/93	30
12/6/93	5	90		1	30/6/93	90
15/6/93	2	30		2	10/6/93	10, 20
20/6/93	2	40, 60, 70		2	15/6/93	30
25/6/93	4	30	or	2	20/6/93	40, 60, 70
25/6/93	3	30, 50, 70		3	25/6/93	30, 50, 70
25/6/93	1	30		4	25/6/93	30
30/6/93	1	90		4	30/6/93	40, 70
30/6/93	4	40, 70		4	25/7/93	90
25/7/93	4	90		5	12/6/93	90

This database can be expressed as a group of customer sequences as follows.

Customers i	Sequences s_i	Length s_i
1	{30}{90}	2
2	{10, 20}{30}{40, 60, 70}	3
3	{30, 50, 70}	1
4	{30}{40, 70}{90}	3
5	{90}	1

where $\mathcal{B} = \{10, 20, 30, 40, 50, 60, 70, 90\}$.

Example 2. (*Document sequences*) Consider a large collection of documents. Each document consists of several fields: document Id, term Categories (e.g., country names, topics, people names, organisations, stock exchanges, etc.). The following example is given by Feldman et al in their paper [5] to investigate maximal association rules and mining for keywords co-occurrences in document collection, where collection D consists of 10 documents $d_1, d_2, \ldots, d_{10}$; and Categories T_1, T_2 are *countries, topics*; respectively.

$D \backslash 2^{T_k}$	$T_1 = countries$	$T_2 = topics$
d_1, d_2	$\{Canada, Iran, USA\}$	$\{crude, ship\}$
d_3	$\{USA\}$	$\{earn\}$
d_4, d_5	$\{USA\}$	$\{jobs, cpi\}$
d_6	$\{USA\}$	$\{earn, cpi\}$
d_7	$\{Canada\}$	$\{sugar, tea\}$
d_8, d_9	$\{Canada, USA\}$	$\{trade, acq\}$
d_{10}	$\{Canada, USA\}$	$\{earn\}$

Example 3. (DNA sequences/Profiles) Let $\mathcal{B}$ be the set of nucleotides A,C,G, T. Then, a sequence formed by singletons (one element subsets) is a DNA sequence [4, 9, 11]. For example, let $\mathcal{U} = \{u_1, u_2, u_3, u_4\}$, where

$$u_1 = ACGTAAAAGTCACACGTAGCCCCACGTACAGT \, ,$$

$$u_2 = CGCGTCGAAGTCGACCGTAAAAGTCACACAGT \, ,$$

$$u_3 = GGTCGATGCACGTAAAATCAGTCGCACACAGT \, ,$$

$$u_4 = ACGTAAAAGTAGCTACCCGTACGTCACACAGT \, .$$

It is interesting that DNA sequences are usually used in scientific area of biology and medicine while DNA profiles are frequently used by journalists to report crime events.

Theorem 1. *The number of sequences over $\mathcal{B}$ with length N is* $|S_{\mathcal{B},N}| = (2^{|\mathcal{B}|} - 1)^N$.

Corollary 1. *The number of DNA sequences of length N is* 4^N.

Corollary 2. *The number of document sequences over the keyword vocabulary $\mathcal{B}$ with categories $T_1, T_2, \ldots, T_N$ is*

$$\prod_{n=1}^{N} (2^{|T_n|} - 1) = (2^{|T_1|} - 1)(2^{|T_2|} - 1)\ldots(2^{|T_N|} - 1) \, .$$

2.2 Containing Relations between Sequences

Now, let us define some containing relations in $S_{\mathcal{B}}$.

First of all, sequence $a = A_1 A_2 \ldots A_n$ $(n > 0)$ is said to be *contained in* sequence $b = B_1 B_2 \ldots B_m$ $(m > 0)$ and denoted by $a \preceq b$, if there exist n integers $1 \leq i_1 < i_2 < \cdots < i_n \leq m$ such that

$A_1 \subseteq B_{i_1}, A_2 \subseteq B_{i_2}, \ldots, A_n \subseteq B_{i_n}$. In this case, we say that a is a *subsequence* of b and that b is a *super* sequence or an *extension* of a. Obviously, we have $|a| \leq |b|$.

Usually, sequence $a = A_1 A_2 \ldots A_n$ $(n > 0)$ is said to be usually *contained in* sequence $b = B_1 B_2 \ldots B_m$ $(m > 0)$, if there exist n integers $1 \leq i_1 < i_2 <$

$\cdots < i_n \leq m$ such that $A_1 = B_{i_1}, A_2 = B_{i_2}, \ldots, A_n = B_{i_n}$. In this case, we say that a is a usual *sub*-sequence of b and that b is a usual *super* sequence or *extension* of a.

In particular, the strong containing relation in $\mathcal{S}_\mathcal{B}$ means that sequence $A_1A_2 \ldots A_n$ is said to be strongly *contained in* sequence $B_1B_2 \ldots B_m$ and denoted by $A_1A_2 \ldots A_n \leq B_1B_2 \ldots B_m$, if there exist n *contiguous* integers $1 \leq i < i+1 < \cdots < i+n-1 \leq m$ such that

$A_1 = B_i, A_2 = B_{i+1}, \ldots, A_n = B_{i+n-1}$. In this case, we say that a is a strong *sub*-sequence of b and that b is a strong *super* sequence or *extension* of a.

For DNA sequences, we only consider the strong containing relation [10].

Given a sequence s, the set of sub-sequences of s is said to be the *language from* the sequence, denoted by $\mathcal{L}(s)$, which is equal to $\{v|v \preceq s, |s| > 0\}$. A sub-sequence of s is said to be a *pattern in* the sequence.

Theorem 2. *The size of the language from a sequence s, i.e., the number $|\mathcal{L}(s)|$ of patterns has an upper bound $\frac{2^{|\mathcal{B}|}-1}{2^{|\mathcal{B}|}-2}[(2^{|\mathcal{B}|}-1)^{|s|}-1]$.*

Corollary 3. *The size of the language from a DNA sequence u, i.e., the number of DNA (strong) sub-sequences has an upper bound $\frac{4}{3}(4^{|u|}-1)$, i.e., $|\mathcal{L}(u)| \leq \frac{4}{3}(4^{|u|}-1)$.*

Let $\mathcal{U}$ be a group (set) of sequences, $\mathcal{U} = \{s_1, s_2, \ldots, s_k, \ldots, s_K\}, K = |\mathcal{U}|$. Denote $L = \max(|s_1|, |s_2|, \ldots, |s_k|, \ldots, |s_K|)$.

The union $\mathcal{L}(\mathcal{U}) = \cup_{k=1}^{K} \mathcal{L}(s_k) = \cup_{k=1}^{K}\{v|v \preceq s_k, |v| > 0\}$ is said to be the *language from the group*, denoted by $\mathcal{L}(\mathcal{U})$.

Theorem 3. *Let $\mathcal{U}$ be a group of sequences and $\mathcal{U} = \{s_1, s_2, \ldots, s_k, \ldots, s_K\}$, K be the size of the group $K = |\mathcal{U}|$, and $L(\mathcal{U})$ be the maximum length of sequences in the group $L(\mathcal{U}) = \max(|s_1|, |s_2|, \ldots, |s_k|, \ldots, |s_K|)$. Then, the size $|\mathcal{L}(\mathcal{U})|$ of the language from the group, i.e., the number of patterns from the group has an upper bound $\frac{2^{|\mathcal{B}|}-1}{2^{|\mathcal{B}|}-2}[(2^{|\mathcal{B}|}-1)^{|L(\mathcal{U})|}-1]$.*

Corollary 4. *Let $\mathcal{U}$ be a group of DNA sequences, $\mathcal{U} = \{u_1, u_2, \ldots, u_K\}$, K be the size of the group $K = |\mathcal{U}|$, and $L(\mathcal{U})$ be the maximum length of sequences in the group $L(\mathcal{U}) = \max(|u_1|, |u_2|, \ldots, |u_k|, \ldots, |u_K|)$. Then, the size of the language from the group, i.e., the number of DNA patterns (strong sub-sequences) has an upper bound $\frac{4}{3}(4^{L(\mathcal{U})}-1)$, i.e., $|\mathcal{L}(\mathcal{U})| \leq \frac{4}{3}(4^{L(\mathcal{U})}-1)$.*

3 Frequent Patterns Contained in a Sequence Group

3.1 The Occurring Notation for Patterns

A particular sequence can be contained (or can "co-occur") in many sequences of a group as their common sub-sequence, i.e., common pattern.

First of all, for a sequence $w \in \mathcal{L}(\mathcal{U})$ in the language from group $\mathcal{U}$, we need to know how many sequences in $\mathcal{U}$ containing w. The number, denoted by t and so $t \leq |\mathcal{U}| = K$, of such sequences is called its *support/occurrence number* and it is said to be a *t-support/occurrence* pattern. Of course, a *t*-support/occurrence pattern is a t'-support/occurrence pattern whenever $t' \leq t$, and we prefer the number is a maximal one.

Furthermore, for a sequence $w \in \mathcal{L}(\mathcal{U})$ in the language from group $\mathcal{U}$, we need to know what sequences in $\mathcal{U}$ containing w. For a sequence $w \in \mathcal{L}(\mathcal{U})$ in the language from group $\mathcal{U}$, the sub-group (subset) of sequences in $\mathcal{U}$ containing w is $\{u \in \mathcal{U} | u \succeq w\}$, denoted by $w^{\mathcal{U}}$. Sub-group $w^{\mathcal{U}}$ consists of sequences in group $\mathcal{U}$ in which sequence w is contained, and is called the *support/occurrence group* of sequence w. So w is a $|w^{\mathcal{U}}|$-support/occurrence pattern. We also call w a $|w^{\mathcal{U}}|/|\mathcal{U}|$-*frequent* pattern.

For a given group $\mathcal{U}$ of sequences and a given pattern w, it is useful to know its support/occurrence group $w^{\mathcal{U}}$ in $\mathcal{U}$. When a pattern w is given, we also want to indicate its its support/occurrence group $w^{\mathcal{U}}$ simultaneously. Therefore, a particular notation is necessarily introduced for patterns. We call this notation the *occurring notation* for patterns. In this notation, pattern w is written as $w\{w^{\mathcal{U}}\}$ with its support/occurrence group $w^{\mathcal{U}}$ added and detailed to indicate the group when $w^{\mathcal{U}} \subset \mathcal{U}$. Notice that, in this notation, a *naked* pattern w means that its support/occurrence group is the whole group $\mathcal{U}$, i.e., $w^{\mathcal{U}} = \mathcal{U}$. In this case, w is abbreviated from $w\mathcal{U}$.

Theorem 4. *Let $\mathcal{U}$ be a group of sequences over $\mathcal{B}$. For two patterns $w, v \in \mathcal{L}(\mathcal{U})$, if $v \preceq w$ then $v^{\mathcal{U}} \supseteq w^{\mathcal{U}}$. That is, subsequence has super support/occurrence group, and super sequence has support/occurrence subgroup.*

Corollary 5. *Let $\mathcal{U}$ be a group of DNA sequences. For two patterns $w, v \in \mathcal{L}(\mathcal{U})$, if $v \leq w$ then $v^{\mathcal{U}} \supseteq w^{\mathcal{U}}$.*

Now, let us introduce the *concatenation* and *cut* operations on sequences. Given two sequences x, y, the *concatenation* $w = xy$ is the sequence which formed by x and then by y. We say w is a left *concatenation* of y by x or a right *concatenation* of x by y if $w = xy$; conversely, we say x is a right *cut* of w by y or y is a left *cut* of w by x.

Obviously, we have the following.

Theorem 5. *A concatenation is an extension; a cut is a subsequence of a concatenation.*

In fact, let $w = xy$ be a concatenation of x and y, $|w| = |x| + |y|; |x|, |y| > 0$. Then w is an extension of x and y, and x and y are subsequences of w.

Corollary 6. *Let $\mathcal{U}$ be a group of sequences over $\mathcal{B}$. For two patterns $w, v \in \mathcal{L}(\mathcal{U})$, if w is a concatenation of v or v is a cut of w then $v^{\mathcal{U}} \supseteq w^{\mathcal{U}}$. That is, a cut has super support/occurrence group, and a concatenation has support/occurrence subgroup.*

Corollary 7. *Let $\mathcal{U}$ be a group of DNA sequences. For two patterns $w, v \in \mathcal{L}(\mathcal{U})$, if w is a concatenation of v or v is a cut of w then $v^{\mathcal{U}} \supseteq w^{\mathcal{U}}$.*

3.2 Checking the Containing Relations

To check whether a sequence $s \in \mathcal{U}$ contains sequence w or not, we need to check the containing relation between two subsets A and B of $\mathcal{B}$, i.e., to compare two subsets. Therefore, we have the following algorithm.

Algorithm R for checking the containing relation between two sequences.

Let $w = A_1 A_2 \ldots A_n$ and $s = B_1 B_2 \ldots B_m$ be two sequences over $\mathcal{B}$. Algorithm R checks if $w \preceq s$, i.e., if there exist n integers $1 \leq i_1 < i_2 < \cdots < i_n \leq m$ such that $A_1 \subseteq B_{i_1}, A_2 \subseteq B_{i_2}, \ldots, A_n \subseteq B_{i_n}$.

Step 1. [Check A_1] Compare A_1 with B_i for $i = 1, 2, \ldots, m$. If $m < n$ or there is no B_i such that $A_1 \subseteq B_i$ then w is not a sub-sequence of s.

Step 2. [Check A_2] Let B_{i_1} be the first subset in s such that $A_1 \subseteq B_{i_1}$. Compare A_2 with B_i for $i = i_1 + 1, i_1 + 2, \ldots, m$. If $m - i_1 < n - 1$ or there is no such B_i that $A_2 \subseteq B_i$ then w is not a sub-sequence of s.

Step 3. [Check A_3] Let B_{i_2} be the first such subset in s that $A_2 \subseteq B_{i_2}$. Compare A_3 with B_i for $i = i_2 + 1, i_2 + 2, \ldots, m$. If $m - i_2 < n - 2$ or there is no such B_i that $A_3 \subseteq B_i$ then w is not a sub-sequence of s; and so on.

Step n. [Check A_n] Let $B_{i_{n-1}}$ be the first such subset in s that $A_{n-1} \subseteq B_{i_{n-1}}$. Compare A_n with B_i for $i = i_{n-1} + 1, i_{n-1} + 2, \ldots, m$. If $m = i_{n-1}$ or there is no such B_i that $A_n \subseteq B_i$ then w is not a sub-sequence of s. Otherwise, let B_{i_n} be the first such subset in s that $A_n \subseteq B_{i_n}$, and $w \preceq s$, i.e., there exist n integers $1 \leq i_1 < i_2 < \cdots < i_n \leq m$ such that $A_1 \subseteq B_{i_1}, A_2 \subseteq B_{i_2}, \ldots, A_n \subseteq B_{i_n}$.

Theorem 6. *The time complexity for checking whether $w \preceq s$ or not for two sequences w and s over $\mathcal{B}$ is $\frac{1}{2}|w|(2|s| - |w| + 1) = O(|w||s|)$.*

Corollary 8. *Let $\mathcal{U}$ be a group of sequences, $\mathcal{U} = \{s_1, s_2, \ldots, s_k, \ldots, s_K\}$, K be the size of the group $K = |\mathcal{U}|$, and $L(\mathcal{U})$ be the maximum length of sequences in the group $L(\mathcal{U}) = \max(|s_1|, |s_2|, \ldots, |s_k|, \ldots, |s_K|)$. Then, the time complexity for computing the support/occurrence group $w^{\mathcal{U}} = \{u \in \mathcal{U} | u \succeq w\}$ of a given sequence $w \in \mathcal{L}(\mathcal{U})$ is $O(|w|KL)$.*

Notice, however, a particular case for DNA application. To check whether a DNA sequence $s \in \mathcal{U}$ (strongly) contains the DNA sequence w or not, we need to check the equality relation between two nucleotides, i.e., to compare A, C, G, T to each other. It is interesting to see the difference between their algorithms for checking the respective containing relation.

Algorithm R' for checking the (strongly) containing relation between two DNA sequences.

Let $w = A_1A_2 \ldots A_n$ and $s = B_1B_2 \ldots B_m$ be two DNA sequences. Algorithm R' checks if $w \preceq s$, i.e., if there exist a integer $1 \leq i_1 \leq m$ such that $A_1 = B_{i_1}, A_2 = B_{i_1+1}, \ldots, A_n = B_{i_1+n-1}$.

Step 1. [Check from B_1] $(m \geq n)$ If $A_1 = B_1, A_2 = B_2, \ldots, A_n = B_n$ then w is a DNA sub-sequence of s. Otherwise, if $m = n$ then w is not a DNA sub-sequence of s; if $m > n$ go to step 2.

Step 2. [Check from B_2] $(m - 1 \geq n)$ If $A_1 = B_2, A_2 = B_3, \ldots, A_n = B_{n+1}$ then w is a DNA sub-sequence of s. Otherwise, if $m - 1 = n$ then w is not a DNA sub-sequence of s; if $m - 1 > n$ go to next step.

Step 3. [Check from B_3] $(m - 2 \geq n)$ If $A_1 = B_3, A_2 = B_4, \ldots, A_n = B_{n+2}$ then w is a DNA sub-sequence of s. Otherwise, if $m - 2 = n$ then w is not a DNA sub-sequence of s; if $m - 2 > n$ go to next step.

Step 4. [Check from B_4] $(m - 3 \geq n)$ If $A_1 = B_4, A_2 = B_5, \ldots, A_n = B_{n+3}$ then w is a DNA sub-sequence of s. Otherwise, if $m - 3 = n$ then w is not a DNA sub-sequence of s; if $m - 3 > n$ go to next step; and so on.

Step $m - n + 1$. [Check from B_{m-n+1}] If $A_1 = B_{m-n+1}, A_2 = B_{m-n+2}, \ldots,$ $A_n = B_m$ then w is a DNA sub-sequence of s. Otherwise, w is not a DNA sub-sequence of s.

Theorem 7. *The time complexity for checking whether $w \leq s$ or not for two DNA sequences w and s is $|w|(|s| - |w| + 1) = O(|w||s|)$.*

Corollary 9. *Let $\mathcal{U}$ be a group of DNA sequences, $\mathcal{U} = \{s_1, s_2, \ldots, s_K\}$, K be the size of the group $K = |\mathcal{U}|$, and $L(\mathcal{U})$ be the maximum length of sequences in the group $L(\mathcal{U}) = \max(|s_1|, |s_2|, \ldots, |s_k|, \ldots, |s_K|)$. Then, the time complexity for computing the support/occurrence group $w^{\mathcal{U}} = \{u \in \mathcal{U} | u \geq w\}$ of a given DNA sequence $w \in L(\mathcal{U})$ is $O(|w|KL)$.*

3.3 Patterns with Minimum Frequency

Generally, given a threshold $\tau > 0$, sequence $w \in L(\mathcal{U})$ is called a *τ-frequent pattern* if $|w^{\mathcal{U}}|/|\mathcal{U}| \geq \tau$. Here τ is called the *minimum support rate* or *minimum frequency*, and w is said to be a pattern with minimum support (rate) τ or [2, 11].

Notice that $0 \leq |w^{\mathcal{U}}|/|\mathcal{U}| \leq 1$. Thus, a threshold τ should take a value satisfying $0 < \tau \leq 1$. Usually, τ is given by a percentage.

In this paper, the frequency of a pattern is defined as the support/occurrence rate of the pattern in the group of sequences. For emphasis, we say again that pattern w is $|w^{\mathcal{U}}|$-support/occurrence or $|w^{\mathcal{U}}|/|\mathcal{U}|$-frequent.

The set of *τ-frequent patterns* with length l is denoted by $M(\mathcal{U}, \tau, l) = \{w \mid w \in L(\mathcal{U}), |w^{\mathcal{U}}|/|\mathcal{U}| \geq \tau, |w| = l\}$, where $l \leq L$.

Denote $M(\mathcal{U}, \tau) = \cup_{l=1}^{L} M(\mathcal{U}, \tau, l)$.

Theorem 8. *The set of k-support patterns is decreasing as k increasing from 1 to $|\mathcal{U}|$.*

4 Theorems for Finding Patterns

1-length patterns are called *units*. A 1-length pattern $w \in (2^{\mathcal{B}} - \{\emptyset\})$ is in $M(\mathcal{U}, \tau, 1)$ if and only if $|w^{\mathcal{U}}|/|\mathcal{U}| \geq \tau$.

Theorem 9. *For the customer sequences,* $|M(\mathcal{U}, \tau, l)| \leq (2^{|\mathcal{B}|} - 1)^l$ *for* $1 \leq l \leq L(\mathcal{U})$, *and* $|M(\mathcal{U}, \tau, l)| = 0$ *for* $l > L(\mathcal{U})$.

Corollary 10. *In document sequences with categories* $T_1, T_2, \ldots, T_N$,

$|M(\mathcal{U}, \tau, N)| \leq \prod_{n=1}^{N}(2^{|T_n|} - 1) = (2^{|T_1|} - 1)(2^{|T_2|} - 1)\ldots(2^{|T_N|} - 1)$, *where* $M(\mathcal{U}, \tau, N) = M(\mathcal{U}, \tau, N; \mathcal{B}), \mathcal{B} = T_1 \cup T_2 \cup \ldots \cup T_N; |M(\mathcal{U}, \tau, l)| = 0$ *for* $l > N$;

For $1 \leq l \leq N$, *given* l *categories* $T_{i_1}, T_{i_2}, \ldots, T_{i_l}$ *with* $1 \leq i_1 < i_2 < \cdots < i_l \leq N$, *we have* $|M(\mathcal{U}, \tau, l; T_{i_1} \cup T_{i_2} \cup \ldots \cup T_{i_l})| \leq \prod_{n=1}^{l}(2^{|T_{i_n}|} - 1)$, *and*

$$|M(\mathcal{U}, \tau, l)| \leq \sum_{(i_1, i_2, \ldots, i_l): 1 \leq i_1 < i_2 < \ldots < i_l \leq N} (2^{|T_{i_1}|} - 1)(2^{|T_{i_2}|} - 1)\ldots(2^{|T_{i_l}|} - 1) .$$

Corollary 11. *For DNA sequences,* $|M(\mathcal{U}, \tau, l)| \leq 4^l$ *for* $1 \leq l \leq L(\mathcal{U})$, *and* $|M(\mathcal{U}, \tau, l)| = 0$ *for* $l > L(\mathcal{U})$.

Theorem 10. *Every subsequence in a* τ-*pattern is also a* τ-*pattern.*

Corollary 12. *Each sequence in group* $\mathcal{U}$ *is a* $1/|\mathcal{U}|$-*pattern.*

Corollary 13. *Every subsequence of each sequence in group* $\mathcal{U}$ *is a* $1/|\mathcal{U}|$-*pattern.*

Corollary 14. *A cut in a* τ-*pattern is still a* τ-*pattern.*

Corollary 15. *For DNA sequences, a cut in a* τ-*pattern is still a* τ-*pattern.*

Corollary 16. *Every* τ-*pattern with length* $l > 1$ *either left cut or right cut by a 1-length pattern is still a* τ-*pattern.*

Corollary 17. *For DNA sequences, every* τ-*pattern with length* $l > 1$ *either left cut or right cut by a Nucleotide is still a* τ-*pattern.*

Theorem 11. *The concatenation of two* τ-*patterns is also a* τ-*pattern if its frequency is not decreased.*

Corollary 18. *The concatenation of a* τ-*pattern either left or right concatenating by a 1-length sequence is also a* τ-*pattern if its frequency is not decreased.*

Corollary 19. *For DNA sequences, the concatenation of a* τ-*pattern either left or right concatenating by a Nucleotide is also a* τ-*pattern if its frequency is not decreased.*

Corollaries 14 and 18 give us a way to construct τ-pattern $w \in M(\mathcal{U}, \tau, |w|)$ by left or right concatenating a 1-shorter τ-pattern $x \in M(\mathcal{U}, \tau, |w| - 1)$ and a 1-length τ-pattern or a Nucleotide $z \in M(\mathcal{U}, \tau, 1)$ by computing $|w^{\mathcal{U}}|/|\mathcal{U}| \geq \tau$. In this way we can establish the following theorem and corollaries and subsequently suggest the corresponding algorithm immediately.

Theorem 12. *1.* $M(\mathcal{U}, k/|\mathcal{U}|, 1) = \{z \in (2^{\mathcal{B}} - \{\emptyset\}) \mid |z^{\mathcal{U}}| \geq k\}$.
2. In $M(\mathcal{U}, k/|\mathcal{U}|, l+1)$, every pattern w can be expressed as either left or right concatenation $w = zx$ or $w = yz$ of such a unit z that $z \in M(\mathcal{U}, k/|\mathcal{U}|, 1)$ and $x, y \in M(\mathcal{U}, k/|\mathcal{U}|, l)$.

Conversely, given $z \in M(\mathcal{U}, k/|\mathcal{U}|, 1)$ and $x, y \in M(\mathcal{U}, k/|\mathcal{U}|, l)$, their concatenation $w = zx$ or $w = yz$ is in $M(\mathcal{U}, k/|\mathcal{U}|, l + 1)$ if its frequency is not decreased.

Corollary 20. *In customer sequences, for $1 < l \leq L(\mathcal{U})$, we have $|M(\mathcal{U}, \tau, l)| \leq (2^{|\mathcal{B}|} - 1)|M(\mathcal{U}, \tau, l - 1)|$.*

Corollary 21. *In document sequences with categories $T_1, T_2, \ldots, T_N$, we have $|M(\mathcal{U}, \tau, l)| = 0$ for $l > N$. For $1 < l \leq N$, given l categories $T_{i_1}, T_{i_2}, \ldots, T_{i_l}$ with $1 \leq i_1 < i_2 < \cdots < i_l \leq N$, we have*

$$|M(\mathcal{U}, \tau, l; T_{i_1} \cup T_{i_2} \cup \cdots \cup T_{i_l})|$$
$$\leq (2^{|T_{i_l}|} - 1)|M(\mathcal{U}, \tau, l - 1; T_{i_1} \cup T_{i_2} \cup \cdots \cup T_{i_{l-1}})|.$$

Corollary 22. *In DNA sequences, for $1 < l \leq L(\mathcal{U})$, we have $|M(\mathcal{U}, \tau, l)| \leq 4|M(\mathcal{U}, \tau, l - 1)|$.*

Algorithm Appending *for constructing $\tau = k/|\mathcal{U}|$-frequent patterns $M(\mathcal{U}, k/|\mathcal{U}|)$*
begin
    1. Find all 1-length patterns in $M(\mathcal{U}, k/|\mathcal{U}|, 1)$ from $z \in (2^{\mathcal{B}} - \{\emptyset\})$ by checking its occurring/support group $|z^{\mathcal{U}}| \geq k$.
    2. Find $M(\mathcal{U}, k/|\mathcal{U}|, l + 1)$ from $M(\mathcal{U}, k/|\mathcal{U}|, l)$ as follows.
    begin
    For all patterns in $M(\mathcal{U}, k/|\mathcal{U}|, l)$ of length l keep concatenating either left or right with 1-length patterns z in $M(\mathcal{U}, k/|\mathcal{U}|, 1)$
    For each resultant pattern w of length $l+1$ compute its support/occurrence group $w^{\mathcal{U}}$
    Add w to $M(\mathcal{U}, k/|\mathcal{U}|, l + 1)$ if $|w^{\mathcal{U}}| \geq k$
    end
end

Corollary 23. *Let $\mathcal{U}$ be a group of sequences over $\mathcal{B}$, $\mathcal{U} = \{s_1, s_2, \ldots, s_K\}$, K be the size of the group $K = |\mathcal{U}|$, and $L(\mathcal{U})$ be the maximum length of sequences in the group $L(\mathcal{U}) = \max(|s_1|, |s_2|, \ldots, |s_k|, \ldots, |s_K|)$. Then, the time complexity for computing $M(\mathcal{U}, \tau, l)$ for an l such that $1 < l \leq L(\mathcal{U})$ is $O(l(2^{|\mathcal{B}|} - 1)^l KL)$.*

Corollary 24. *The time complexity for appending $M(\mathcal{U}, \tau)$ of sequences over $\mathcal{B}$ is $O(L(2^{|\mathcal{B}|} - 1)^L KL)$, where $K = |\mathcal{U}|$, $L = \max_{u \in \mathcal{U}}(|u|)$.*

Corollary 25. *Let $\mathcal{U}$ be a group of DNA sequences, $\mathcal{U} = \{s_1, s_2, \ldots, s_K\}$, K be the size of the group $K = |\mathcal{U}|$, and $L(\mathcal{U})$ be the maximum length of sequences in the group $L(\mathcal{U}) = \max(|s_1|, |s_2|, \ldots, |s_k|, \ldots, |s_K|)$. Then, the time complexity for computing $M(\mathcal{U}, \tau, l)$ for an l such that $1 < l \leq L(\mathcal{U})$ is $O(l4^l KL)$.*

Corollary 26. *The time complexity for appending $M(\mathcal{U}, \tau)$ of DNA sequences is $O(L4^L KL)$, where $K = |\mathcal{U}|$, $L = \max_{u \in \mathcal{U}}(|u|)$.*

Algorithm Appending. for constructing $\tau = k/|\mathcal{U}|$-frequent patterns $M(\mathcal{U}, k/|\mathcal{U}|)$ in document sequences

begin
    1. Find all 1-length patterns in $M(\mathcal{U}, k/|\mathcal{U}|, 1)$ from $z \in (2^{T_i} - \{\emptyset\})$ by checking its occurring/support group $|z^{\mathcal{U}}| \geq k$ for $i = 1, 2, \ldots, N$.
    2. Find $M(\mathcal{U}, k/|\mathcal{U}|, l + 1)$ from $M(\mathcal{U}, k/|\mathcal{U}|, l)$ as follows.
    begin
    For all patterns in $M(\mathcal{U}, k/|\mathcal{U}|, l)$ of length l keep concatenating right with 1-length patterns z in $M(\mathcal{U}, k/|\mathcal{U}|, 1)$
    For each resultant pattern w of length $l+1$ compute its support/occurrence group $w^{\mathcal{U}}$
    Add w to $M(\mathcal{U}, k/|\mathcal{U}|, l + 1)$ if $|w^{\mathcal{U}}| \geq k$
    end
end

Corollary 27. *Let $\mathcal{U}$ be a group of document sequences over $T_1 \cup T_2 \cup \cdots \cup T_N$ and $\mathcal{U} = \{s_1, s_2, \ldots, s_k, \ldots, s_K\}$, K be the size of the group $K = |\mathcal{U}|$. Then, the time complexity for computing $M(\mathcal{U}, \tau, l)$ for an l such that $1 < l \leq N$ is*

$$O\left(l \sum_{(i_1, i_2, \ldots, i_l): 1 \leq i_1 < i_2 < \cdots < i_l \leq N} (2^{|T_{i_1}|} - 1)(2^{|T_{i_2}|} - 1) \cdots (2^{|T_{i_l}|} - 1)KN\right).$$

Corollary 28. *The time complexity for appending $M(\mathcal{U}, \tau)$ of document sequences over $T_1 \cup T_2 \cup \cdots \cup T_N$ is $O(N(2^{|T_1|} - 1)(2^{|T_2|} - 1) \cdots (2^{|T_N|} - 1)KN)$, where $K = |\mathcal{U}|$.*

5 Algorithm Appending for Finding Patterns

Our algorithm needs the following function:
    SaveFreqPattern(w, $M(\mathcal{U}, \tau, |w|)$) is a function for saving the τ-frequent pattern w in the occurring notation $w\{w^{\mathcal{U}}\}$ into $M(\mathcal{U}, \tau, |w|)$.

5.1 Customer Databases

First of all, we write algorithms for customer sequences as follows.

Algorithm 1 Append$M(\mathcal{U}, k/|\mathcal{U}|, 1)$ – for finding
$k/|\mathcal{U}|$-frequent patterns with length 1

```
for (every z ∈ (2^B − {∅})) {
    if (|z^U| ≥ k)
        SaveFreqPattern(z,M(U, 1/|U|, 1));
}
output M(U, τ, l);
```

Algorithm 2 Append$M(\mathcal{U}, k/|\mathcal{U}|, l)$ – for finding
$k/|\mathcal{U}|$-frequent patterns with length l
from $M(\mathcal{U}, k/|\mathcal{U}|, l-1)$ and $M(\mathcal{U}, k/|\mathcal{U}|, 1)$
for $l \geq 2$

```
String x, z, w; // w=x+z right concatenation by unit z
M(U, 1/|U|, l) = {}
for (every x ∈ M(U, k/|U|, l − 1)) {
    for (every z ∈ M(U, k/|U|, 1)) {
        w=x+z; // right concatenation of x by z
        if (|w^U| ≥ k) {
            SaveFreqPattern(w,M(U, k/|U|, |w|));
        }
    }
}
output {
    M(U, k/|U|, l);
}
```

Algorithm 3 Append$M(\mathcal{U}, k/|\mathcal{U}|)$ – for finding
all $k/|\mathcal{U}|$-frequent patterns

```
int l >= 2;
l = 2;
AppendM(U, k/|U|, 1);
for (j = 2; M(U, k/|U|, j) != {}; j + +) {
    AppendM(U, k/|U|, j);
    l + +;
}
Output {
    M(U, k/|U|)= M(U, k/|U|, 1) ∪ (∪_{j=2}^{l−1} M(U, k/|U|, j));
}
```

}

Theorem 13. *The time complexity for Algorithm 3 for finding all* $k/|\mathcal{U}|$*-frequent patterns in customer sequences is* $O((2^{|\mathcal{B}|}-1)^L KL^2)$*, where* $K = |\mathcal{U}|$*,* $L = \max_{u \in \mathcal{U}}(|u|)$*.*

5.2 DNA Databases

We can also propose algorithms for DNA sequences as follows.

Algorithm 4 Append$M(\mathcal{U}, k/|\mathcal{U}|, 1)$ – for finding
$k/|\mathcal{U}|$-frequent patterns with length 1

```
for (every character∈ {A, C, G, T}) {
    if (|character^U| ≥ k)
        SaveFreqPattern(character,M(U, 1/|U|, 1));
}
output M(U, τ, l);
```

Algorithm 5 is the same as **Algorithm 2** Append$M(\mathcal{U}, k/|\mathcal{U}|, l)$ – for finding
$k/|\mathcal{U}|$-frequent patterns with length l
from $M(\mathcal{U}, k/|\mathcal{U}|, l-1)$ and $M(\mathcal{U}, k/|\mathcal{U}|, 1)$
for $l \geq 2$

Algorithm 6 is the same as **Algorithm 3** Append$M(\mathcal{U}, k/|\mathcal{U}|)$ – for finding
all $k/|\mathcal{U}|$-frequent patterns

Theorem 14. *The time complexity for Algorithm 6 for finding all* $k/|\mathcal{U}|$*-frequent patterns in DNA sequences is* $O(4^L KL^2)$*, where* $K = |\mathcal{U}|$*,* $L = \max_{u \in \mathcal{U}}(|u|)$*.*

5.3 Document Databases

Finally, we write algorithms for document sequences as follows.

Algorithm 7 Append$M(\mathcal{U}, k/|\mathcal{U}|, 1)$ – for finding
$k/|\mathcal{U}|$-frequent patterns with length 1

```
M(U, τ, l) = {};
for (i = 1; i! = N; i + +){
for (every z ∈ (2^{T_i} − {∅})) {
    if (|z^U| ≥ k)
        SaveFreqPattern(z,M(U, 1/|U|, 1));
    }
```

```
    }
    output M(𝒰, τ, l);
```

Algorithm 8 Append$M(\mathcal{U}, k/|\mathcal{U}|, l)$ – for finding
  $k/|\mathcal{U}|$-frequent patterns with length l
  from $M(\mathcal{U}, k/|\mathcal{U}|, l-1)$ and $M(\mathcal{U}, k/|\mathcal{U}|, 1)$
  for $l \geq 2$

```
    String x, z, w; // w=x+z right concatenation by unit z
    M(𝒰, 1/|𝒰|, l) = {};
    for (every x ∈ M(𝒰, k/|𝒰|, l − 1)) {
        for (every z ∈ M(𝒰, k/|𝒰|, 1)) {
            w=x+z; // right concatenation of x by z
            if (|w^𝒰| ≥ k) {
                SaveFreqPattern(w,M(𝒰, k/|𝒰|, l));
            }
        }
    }
    output {
        M(𝒰, k/|𝒰|, l);
    }
```

Algorithm 9 Append$M(\mathcal{U}, k/|\mathcal{U}|)$ – for finding
  all $k/|\mathcal{U}|$-frequent patterns

```
    int l >= 2;
    l = 2;
    AppendM(𝒰, k/|𝒰|, 1);
    for (j = 2; M(𝒰, k/|𝒰|, j) != {}; j + +) {
        AppendM(𝒰, k/|𝒰|, j);
        l + +;
    }
    Output {
        M(𝒰, k/|𝒰|) = M(𝒰, k/|𝒰|, 1) ∪ (∪_{j=2}^{l−1} M(𝒰, k/|𝒰|, j; ));
    }
```

Theorem 15. *The time complexity for Algorithm 3 for finding all $k/|\mathcal{U}|$-frequent patterns in document sequences is $O((2^{|T_1|} - 1)(2^{|T_2|} - 1) \cdots (2^{|T_N|} - 1)KN^2)$, where $K = |\mathcal{U}|$.*

6 Finding Higher Occurrence Patterns

In the case where there is nothing known at the beginning, Algorithm Appending is a possible way to construct k-support/occurrence patterns for a given k. However, its computational cost is rather high. Fortunately, there is an easier way to find higher occurrence patterns, when we know all k-support/occurrence patterns at the outset.

The outline of a method to find these patterns is the following:

Algorithm Checking for finding higher occurrence patterns
For each pattern w in $M(\mathcal{U}, (k-1)/|\mathcal{U}|, l)$ of length l check its support/occurrence group $w^{\mathcal{U}}$ add w to $M(\mathcal{U}, k/|\mathcal{U}|, l)$ if $|w^{\mathcal{U}}| \geq k$

Algorithm 10 CheckM$(\mathcal{U}, k/|\mathcal{U}|, l)$ – for finding
$k/|\mathcal{U}|$-frequent patterns with length l for a given $k > 1$ from $M(\mathcal{U}, (k-1)/|\mathcal{U}|, l)$ for $l \geq 1$

```
for (every w ∈ M(U, (k − 1)/|U|, l)) {
    if (|w^U| ≥ k) {
        SaveFreqMotif(w,M(U, k/|U|, l));
    }
}
output {
    M(U, k/|U|, l);
}
```

Algorithm 11 CheckM$(\mathcal{U}, k/|\mathcal{U}|)$ – for finding
all $k/|\mathcal{U}|$-frequent patterns for a given $k > 1$

```
int l >= 1;
l = 1;
for (j = 1; M(U, k/|U|, j) != {}; j + +) {
    CheckM(U, k/|U|, j);
    l + +;
}
Output {
    M(U, k/|U|)= ∪_{i=1}^{l−1}M(U, k/|U|, i);
}
```

Theorem 16. *The time complexity of Algorithm 5.2 for finding all $k/|\mathcal{U}|$-frequent patterns in customer sequences for a given $k > 1$ is $O((2^{|\mathcal{B}|} - 1)^L)$.*

Theorem 17. *The time complexity of Algorithm 5.2 for finding all $k/|\mathcal{U}|$-frequent patterns in DNA sequences for a given $k > 1$ is $O(4^L)$.*

Theorem 18. *The time complexity of Algorithm 5.2 for finding all $k/|\mathcal{U}|$-frequent patterns in document sequences for a given $k > 1$ is $O((2^{|T_1|} - 1)(2^{|T_2|} - 1) \cdots (2^{|T_N|} - 1))$.*

7 Maximal Patterns

For the group of sequences $\mathcal{U}$ and its containing relation, there are two kinds of maximum to be considered: the first is maximal sequences over the group $\mathcal{U}$, the second is maximal patterns over the set of τ-frequent patterns for a given τ.

Given a threshold τ, in the set of τ-frequent patterns, we say that a τ-frequent pattern w is *maximal* if it cannot be τ-frequently extended further, i.e., if there is no τ-frequent pattern w' other than w such that $w' \succ w$. A maximal τ-frequent pattern is also called a τ-*key*. The set of τ-*keys* is denoted by $\max M(\mathcal{U}, \tau) = \max\{w | w \in M(\mathcal{U}, \tau)\}$.

Our conjecture is that keys play as an important role as that of the most important attributes and keys in databases and rough set theory, keywords and terms in text mining and information retrieval, etc. Therefore, our research on the mining of patterns focuses on keys.

Theorem 19. *All sub-sequences in a τ-key are τ-patterns that can be extended to the τ-key, and every τ-pattern can be extended to a τ-key.*

Corollary 29. *Every sequence in $\mathcal{U}$ itself can be extended to a $1/|\mathcal{U}|$-key.*

Let the support group be set $\mathcal{U}$ of sequences $u_1, u_2, \ldots, u_K; K > 0$. A sequence u is called a *maximal sequence* in $\mathcal{U}$ if there is no other extension sequence of u; i.e., there is no other sequence u' in $\mathcal{U}$ such that $u' \succ u$. The set of maximal sequences in $\mathcal{U}$ is denoted by $\max(\mathcal{U}) = \max\{u | u \in \mathcal{U}\}$.

If u is a maximal sequence then $u^{\mathcal{U}} = \{u\}$, i.e., its support/occurrence group consists of itself alone. In fact, u occurs in sequence u, so $u \in u^{\mathcal{U}}$. Moreover, u cannot occurs in other sequence s so that $u \prec s$ since then s would be a further sequence extension of u to contradict the sequence maximum of u in $\mathcal{U}$.

Let us denote the set of maximal sequences in $\mathcal{U}$ by $\max(\mathcal{U}) = \{u_{i_1}\{u_{i_1}\}, u_{i_2}\{u_{i_2}\}, \ldots, u_{i_k}\{u_{i_k}\}\}$, where $u_i\{u_i\}$ is sequence u_i expressed in the occurring notation $u_i\{u_i^{\mathcal{U}}\}$ with $u_i^{\mathcal{U}} = \{u_i\}$.

We suggest the following method to find maximal sequences in $\mathcal{U}$.

Algorithm Comparing.
begin
    Compare each sequence u in $\mathcal{U}$
    with every sequence v in $\mathcal{U} - \{u\}$
    to see if $u \prec v$
    if not then u is maximal and put it into $\max(\mathcal{U})$.
end

Now, we write Algorithm Comparing more formally as follows. Our algorithm needs the following function:

SaveMaxSequences(u,max($\mathcal{U}$)) is the function for saving the maximal sequence $u\{u\}$ in the occurring notation into $\max M(\mathcal{U})$.

Algorithm 1 Comparemax($\mathcal{U}$) – for finding
   maximal sequences
   from comparing $u \in \mathcal{U}$ with $v \in \mathcal{U} - \{u\}$

```
max(U) = {}
for (every u ∈ U) {
for (every v ∈ U − {u}) {
   int counter = 0;
   if (u≺ v) {
      counter++;
   }
   if ( counter == 0 )
      SaveMaxSequences(u,max(U)) ;
}}
output {
   max(U);
}
```

Theorem 20. *The Comparing Algorithm 1 for finding $max(\mathcal{U})$ requires $|\mathcal{U}|(|\mathcal{U}|-1)$ comparisons.*

The purpose of finding maximal sequences is to find frequent keys. For $\tau = 1/|\mathcal{U}|$ its keys are very easy to find as shown by the following theorem. It turns out the set of $1/|\mathcal{U}|$-frequent keys is just the set of maximal sequences.

Theorem 21. *Let the support group be $\mathcal{U} = \{u_1, u_2, \ldots, u_K\}, K > 0$. Then a pattern w is a $1/|\mathcal{U}|$-key if and only if w is a maximal sequence in $\mathcal{U}$. That is, $maxM(\mathcal{U}, 1/|\mathcal{U}|) = max(\mathcal{U})$.*

It is remarkable that Theorem 3 shows that keys for $\tau = 1/|\mathcal{U}|$ are very easy to find, whereas finding the set of patterns in $M(\mathcal{U}, 1/|\mathcal{U}|)$ is very complicated since its size is the biggest over all τ.

Algorithm 1. Print $maxM(\mathcal{U}, 1/|\mathcal{U}|)$
   int $K > 0$
   input $\mathcal{U} = \{u_1, u_2, \ldots, u_K\}$
   output
   $maxM(\mathcal{U}) = \{u_{i_1}\{u_{i_1}\}, u_{i_2}\{u_{i_2}\}, \ldots, u_{i_k}\{u_{i_k}\}\}.$

Theorem 22. *The time complexity for computing $maxM(\mathcal{U}, 1/|\mathcal{U}|)$ or $Max(\mathcal{U})$ is $O(K(K - 1)L^2)$, where $K = |\mathcal{U}|$ and $L = \max\{|u| \mid u \in \mathcal{U}\}$.*

Notice that the complexity for computing $maxM(\mathcal{U}, 1/|\mathcal{U}|)$ is only $O(K(K-1)L^2)$ while that for computing $M(\mathcal{U}, 1/|\mathcal{U}|)$ is much greater, up to

1. *$O((2^{|\mathcal{B}|} - 1)^L KL^2)$ for customer sequences by Theorem 9;*
2. *$O(4^L KL^2)$ for DNA sequences by Theorem 10;*
3. *$O((2^{|T_1|} - 1)(2^{|T_2|} - 1) \cdots (2^{|T_N|} - 1)KN^2)$ for document sequences by Theorem 11.*

So, to find maximal patterns of $\tau = 1/|\mathcal{U}|$ is rather easy. We now only need to find maximal patterns, keys, of $\tau = k/|\mathcal{U}|$ for $k \geq 2$.

We suggest the following method to find maximal patterns $maxM(\mathcal{U}, k/|\mathcal{U}|)$ for $k \geq 2$ based on the fact that $M(\mathcal{U}, k/|\mathcal{U}|)$ is obtained in hand:

Algorithm Comparing.
begin
    Compare each sequence w in $M(\mathcal{U}, k/|\mathcal{U}|)$
    with every sequence v in $M(\mathcal{U}, k/|\mathcal{U}|) - \{w\}$
    to see if $w \prec v$
    if not then w is maximal and put it into $maxM(\mathcal{U}, k/|\mathcal{U}|)$
end

Now, we write Algorithm Comparing more formally as follows. Our algorithm needs the following function:
SaveMaxPatterns$(w, maxM(\mathcal{U}, k/|\mathcal{U}|))$ *is the function for saving the maximal pattern $w\{w^{\mathcal{U}}\}$ in the occurring notation into $maxM(\mathcal{U}, k/|\mathcal{U}|)$.*

Algorithm 6.2 Compare $maxM(\mathcal{U}, k/|\mathcal{U}|)$ – for finding

maximal sequences
    from comparing $w \in M(\mathcal{U}, k/|\mathcal{U}|)$ with $v \in M(\mathcal{U}, k/|\mathcal{U}|) - \{w\}$

```
maxM(U, k/|U|) = {}
for (every w ∈ M(U, k/|U|)) {
for (every v ∈ M(U, k/|U|) − {u}) {
   int counter = 0;
   if (w≺ v) {
      counter++;
   }
   if ( counter == 0 )
      SaveMaxPatterns(w,maxM(U, k/|U|)) ;
}}
output {
   maxM(U, k/|U|);
}
```

Theorem 23. *The Comparing Algorithm 2 for finding $maxM(\mathcal{U}, k/|\mathcal{U}|)$ requires $|M(\mathcal{U}, k/|\mathcal{U}|)|(|M(\mathcal{U}, k/|\mathcal{U}|)| - 1)$ comparisons.*

In the case where the containing relation is strong (contiguous), taking account of the length of patterns may be beneficial.

Denote the set of τ-keys with length l by $maxM(\mathcal{U}, \tau, l)$.

Then $maxM(\mathcal{U}, \tau) = \cup_{l=1}^{L} maxM(\mathcal{U}, \tau, l)$.

Theorem 24. *Suppose that $x \in M(\mathcal{U}, \tau, |x|)$.*

If $|(xz)^{\mathcal{U}}|/|\mathcal{U}| < \tau$ and $|(zx)^{\mathcal{U}}|/|\mathcal{U}| < \tau$ (both Right and Left Concatenations) for all $z \in M(\mathcal{U}, \tau, 1)$ then x is a τ-frequent maximal pattern, $x \in \max M(\mathcal{U}, \tau, |x|)$.

The outline of the method to find maximal patterns $maxM(\mathcal{U}, k/|\mathcal{U}|, l)$ for $k \geq 2$ is the following comparing based on the fact that $M(\mathcal{U}, k/|\mathcal{U}|, l)$ and $M(\mathcal{U}, k/|\mathcal{U}|, l+1)$ are obtained in hand:

Algorithm Comparing.
begin
  Compare each pattern x in $M(\mathcal{U}, k/|\mathcal{U}|, l)$ of length l
  with every pattern w in $M(\mathcal{U}, k/|\mathcal{U}|, l+1)$ of length $l+1$
  Two comparisons for each pair (x, w) are to be taken to see
  if $w = xz$ or $w = zx$ with a $z \in M(\mathcal{U}, k/|\mathcal{U}|, 1)$
  if not then x is maximal and put it into $\mathrm{Max}M(\mathcal{U}, k/|\mathcal{U}|, l)$.
end

Now, write Algorithm Comparing more formally as follows. Our algorithm needs the following function:

SaveMaxPattern(x,$maxM(\mathcal{U}, \tau, |x|)$) is the function for saving the maximal τ-frequent pattern x in the occurring notation into $maxM(\mathcal{U}, \tau, |x|)$.

Algorithm 3 Comparemax$M(\mathcal{U}, \tau, l-1)$ – for finding maximal τ-frequent patterns with length $l-1$ for a given τ from comparing $M(\mathcal{U}, \tau, l-1)$ with $M(\mathcal{U}, \tau, l)$ and from $M(\mathcal{U}, \tau, 1)$
 for $l \geq 2$

```
maxM(U, τ, l − 1) = {}
for (every x ∈ M(U, τ, l − 1)) {
for (every w ∈ M(U, τ, l)) {
    int counter = 0;
    for (every z ∈ M(U, τ, 1)) {
       if (w=x+z) {
          counter++;
       }
    }
    if ( counter == 0 )
    for (every z ∈ M(U, τ, 1)) {
       if (w=z+x)
          counter++;
    }
    if ( counter == 0 )
       SaveMaxPattern(x,maxM(U, τ, |x|)) ;
}}
output {
   maxM(U, τ, l − 1);
}
```

Theorem 25. *The Comparing Algorithm 6.3 for finding* $maxM(\mathcal{U}, \tau, l-1)$ *with* $\tau \geq 2/|\mathcal{U}|$ *requires* $|M(\mathcal{U}, \tau, l-1)| \times |M(\mathcal{U}, \tau, 1)| \times |M(\mathcal{U}, \tau, l)|$ *comparisons.*

Theorem 26. *The time complexity for finding* $maxM(\mathcal{U}, k/|\mathcal{U}|)$ *with* $k \geq 2$ *in DNA sequences is* $O(4^{2L+2})$.

8 Keys and Sequence Occurrence Tables

Now, using the tools established so far, keys can be found. Among these, only those with threshold $\tau = 2/|\mathcal{U}|, 3/|\mathcal{U}|, \ldots, (|\mathcal{U}| - 1)/|\mathcal{U}|$ are interesting and those with $\tau = 1/|\mathcal{U}|, 1$ are trivial or very numerous. Thus, we can focus on keys with $1/|\mathcal{U}| < \tau < 1$.

Let us call keys with $1/|\mathcal{U}| < \tau < 1$ *meaningful keys* or *attributes*.

Two keys u, w are said to be *identical occurring*, and denoted by $u/w(\{u^{\mathcal{U}}\})$, if their occurrence sets are equal: $u^{\mathcal{U}} = w^{\mathcal{U}}$. Identical occurring keys occur in the same subset of $\mathcal{U}$ and so play the same role in discovering knowledge. Thus, we can choose the first one of them as a representative.

Now, let us introduce the *key occurrence table* as follows, where *attributes* are key representatives and *objects* are sequences.

Example 4. Continue Examples 1.

From $maxM(\mathcal{U}, \tau)$ for $\tau = 2/5, 3/5, 4/5$, the corresponding key occurrence table is the following.

$A \backslash \mathcal{U}$	s_1	s_2	s_3	s_4	s_5
$\{30\}\{40, 70\}$	0	1	0	1	0
$\{30\}\{90\}$	1	0	0	1	0
$\{30\}$	1	1	1	1	0
$\{70\}$	0	1	1	1	0
$\{90\}$	1	0	0	1	1

Example 5. Continue Examples 2.

From $maxM(\mathcal{U}, \tau)$ for $\tau = 2/10, 3/10, \ldots, 9/10$ the corresponding key occurrence table is the following.

$A \backslash \mathcal{U}$	d_1, d_2	d_3	d_4, d_5	d_6	d_7	d_8, d_9	d_{10}
$\{Canada, Iran, USA\}\{crude, ship\}$	1	0	0	0	0	0	0
$\{Canada\}\{trade, acq\}/$	0	0	0	0	0	1	0
$\{USA\}\{earn\}$	0	1	0	1	0	0	1
$\{USA\}\{jobs, cpi\}$	0	0	1	0	0	0	0
$\{Canada, USA\}$	1	0	0	0	0	1	1
$\{USA\}\{earn\}$	0	1	0	1	0	0	1
$\{USA\}\{cpi\}$	0	0	1	1	0	0	0
$\{Canada, USA\}$	1	0	0	0	0	1	1
$\{Canada\}$	1	0	0	0	0	1	1
$\{USA\}$	0	0	0	0	0	1	1

where $\{Canada\}\{trade, acq\}/$ represents
$[\{Canada\}\{trade, acq\}]/[\{USA\}\{trade, acq\}]$.

Example 6. Continue Examples 3. From $\max M(\mathcal{U}, \tau)$ for $\tau = 2/4, 3/4$ the corresponding key occurrence table is the following.

$A \backslash \mathcal{U}$	u_1	u_2	u_3	u_4
$CC/$	1	1	0	1
$AGTC$	1	1	1	0
$ACGTAAAA$	1	0	1	1
$CACACAGT$	0	1	1	1
$ACC/$	0	1	0	1
$CCC/$	1	0	0	1
$CGC/$	0	1	1	0
$CACGTA$	1	0	1	0
$CGTAAAAGTCACAC$	1	1	0	0

where

$CC/$ represents $CC/CGTAAAAGT/GTCACAC$,
$ACC/$ represents $ACC/CGTC/CCGTA/GTCACACAGT$,
$CCC/$ represents $CCC/CGTAC/GTAGC/ACGTAAAAGT$,
$CGC/$ represents $CGC/AGTCG/GTCGA$.

The *sequence-key table*, which is usually called the *information table*, is the transpose of the key occurrence table.

9 Discovering Associations between Patterns

When we have identified patterns using the simple "appending" algorithm and others, the question arises: can we find *associations between patterns*?

First of all, we should point out that fully-frequent patterns for $\tau = 1$ mean there is only one class and no classification on the whole group of sequences $\mathcal{U}$. So in this case we do not need to use rough set theory to mine for association rules since the theory is based on classification. We have a theorem for fully-frequent patterns immediately.

Theorem 27. *All fully-frequent patterns co-occur in every sequence in the group.*

Theorem 28. *A $\frac{1}{|\mathcal{U}|}$-frequent pattern $w\{u_i\}$ is a representative of the sequence u_i, i.e., w occurs and only occurs in the sequence u_i.*

On the other hand, for $\frac{1}{|\mathcal{U}|} < \tau < 1$ we can apply rough set theory to discover knowledge effectively and efficiently as illustrated as follows.

First of all, focusing on the maximal frequent patterns with $\frac{1}{|\mathcal{U}|} < \tau < 1$, we obtain the information table.

Our method to find the maximal frequent patterns can be summarised as follows.

1. *Use* Appending *Algorithm to find* $M(\mathcal{U}, 2/|\mathcal{U}|)$ *with cost* $O((2^{|\mathcal{B}|} - 1)^L K L^2)$.
2. *Use* Checking *Algorithm to find* $M(\mathcal{U}, 3/|\mathcal{U}|), \ldots, M(\mathcal{U}, (|\mathcal{U}| - 1)/|\mathcal{U}|)$ *with cost* $O((2^{|\mathcal{B}|} - 1)^L)$.
3. *Use* Comparing *Algorithm to find* $\max M(\mathcal{U}, 2/|\mathcal{U}|), \ldots, \max M(\mathcal{U}, (|\mathcal{U}| - 1)/|\mathcal{U}|)$.

In an information table, we say that a sequence u *supports* a given attribute w if $w \preceq u$. We say that a sequence u *supports* a given set of attributes X if $w \preceq u$ for all $w \in X$.

For a given attribute w, the set of sequences supporting w in group $\mathcal{U}$ is $w^{\mathcal{U}}$. For a given attribute set X, the set of sequences supporting X in group $\mathcal{U}$ is $\cap_{w \in X} w^{\mathcal{U}}$, denoted by $X^{\mathcal{U}}$.

The *strength* of an attribute (pattern) w in the information table, denoted by $stg(w)$, is the number of sequences u supporting w: $stg(w) = |w^{\mathcal{U}}|$. The *strength* of a given set of attributes X in the information table, denoted by $stg(X)$, is the number of sequences u supporting X: $stg(X) = |\cap_{w \in X} w^{\mathcal{U}}| = |X^{\mathcal{U}}|$.

An *association* between X and Y is an expression of the form $X \rightleftharpoons Y$, where X and Y are attribute sets; the *strength* of the association $stg(X \rightleftharpoons Y)$ is the strength of $X \cup Y$, i.e., the number of sequences supporting $X \cup Y$

$$stg(X \rightleftharpoons Y) = stg(X \cup Y) = |(X \cup Y)^{\mathcal{U}}|,$$ and the *confidence* of the association $cfi(X \rightleftharpoons Y)$ is $\frac{stg(X \cup Y)}{\max(stg(X), stg(Y))} = \frac{|(X \cup Y)^{\mathcal{U}}|}{\max(|X^{\mathcal{U}}|, |Y^{\mathcal{U}}|)}$

An *association rule* is a rule of the form $X \rightarrow Y$, where X and Y are attribute sets; the *strength* of the association rule $stg(X \rightarrow Y)$ is

$$stg(X \rightarrow Y) = stg(X \cup Y) = |(X \cup Y)^{\mathcal{U}}|,$$ and the *confidence* of the rule $cfi(X \rightarrow Y)$ is $stg(X \cup Y)/stg(X) = \frac{|(X \cup Y)^{\mathcal{U}}|}{|X^{\mathcal{U}}|}$.

Now, using the knowledge discovery method presented in [3, 3], we can find the *Association Rules*.

Example 7. Continue Example 1. From

$a_1 = \{30\}\{40, 70\}$, $a_1^{\mathcal{U}} = \{s_2, s_4\}$;

$a_2 = \{70\}$, $a_2^{\mathcal{U}} = \{s_2, s_3, s_4\}$;

we find that $\{a_1, a_2\}^{\mathcal{U}} = \{s_2, s_4\} \cap \{s_2, s_3, s_4\} = \{s_2, s_4\}$. So

$stg(a_1 \rightleftharpoons a_2) = |\{a_1, a_2\}^{\mathcal{U}}| = |\{s_2, s_4\}| = 2$, and the *confidence* of the association $cfi(a_1 \rightleftharpoons a_2)$ is

$$\frac{|\{a_1, a_2\}^{\mathcal{U}}|}{\max(|a_1^{\mathcal{U}}|, |a_2^{\mathcal{U}}|)} = \frac{|\{s_2, s_4\}|}{\max(|\{s_2, s_4\}|, |\{s_2, s_3, s_4\}|)} = \frac{2}{\max(2, 3)} = \frac{2}{3}.$$

The *strength* of the association rule $stg(a_1 \rightarrow a_2)$ is $|\{a_1, a_2\}^{\mathcal{U}}| = |\{s_2, s_4\}| = 2$, and the *confidence* of the rule $cfi(a_1 \rightarrow a_2)$ is $\frac{|\{a_1, a_2\}^{\mathcal{U}}|}{|a_1^{\mathcal{U}}|} = \frac{|\{s_2, s_4\}|}{|\{s_2, s_4\}|} = \frac{2}{2} = 1$.

The *strength* of the association rule $stg(a_1 \leftarrow a_2)$ is $|\{a_1, a_2\}^{\mathcal{U}}| = |\{s_2, s_4\}| = 2$, and the *confidence* of the rule $cfi(a_1 \leftarrow a_2)$ is $\frac{|\{a_1, a_2\}^{\mathcal{U}}|}{|a_2^{\mathcal{U}}|} = \frac{|\{s_2, s_4\}|}{|\{s_2, s_3, s_4\}|} = \frac{2}{3}$.

Example 8. Continue Example 2. From

$a_1 = [\{Canada\}\{trade, acq\}]/[\{USA\}\{trade, acq\}]$, $a_1^{\mathcal{U}} = \{d_8, d_9\}$;

$a_2 = \{Canada, USA\}$, $a_2^{\mathcal{U}} = \{d_1, d_2, d_8, d_9, d_{10}\}$; we find that

$\{a_1, a_2\}^{\mathcal{U}} = \{d_1, d_2, d_8, d_9, d_{10}\} \cap \{d_8, d_9\} = \{d_8, d_9\}$. So

$stg(a_1 \rightleftharpoons a_2) = |\{a_1, a_2\}^{\mathcal{U}}| = |\{d_8, d_9\}| = 2$, and the *confidence* of the

association $cfi(a_1 \rightleftharpoons a_2)$ is

$\frac{|\{a_1,a_2\}^{\mathcal{U}}|}{\max(|a_1^{\mathcal{U}}|,|a_2^{\mathcal{U}}|)} = \frac{|\{d_8,d_9\}|}{\max(|\{d_1,d_2,d_8,d_9,d_{10}\}|,|\{d_8,d_9\}|)} = \frac{2}{\max(5,2)} = \frac{2}{5}$. The *strength*

of the association rule $stg(a_1 \rightarrow a_2)$ is $|\{a_1, a_2\}^{\mathcal{U}}| = |\{d_8, d_9\}| = 2$, and the

confidence of the rule $cfi(a_1 \rightarrow a_2)$ is $\frac{|\{a_1,a_2\}^{\mathcal{U}}|}{|a_1^{\mathcal{U}}|} = \frac{|\{d_8,d_9\}|}{|\{d_8,d_9\}|} = \frac{2}{2} = 1$. The

strength of the association rule $stg(a_1 \leftarrow a_2)$ is $|\{a_1, a_2\}^{\mathcal{U}}| = |\{d_8, d_9\}| = 2$,

and the *confidence* of the rule $cfi(a_1 \leftarrow a_2)$ is $\frac{|\{a_1,a_2\}^{\mathcal{U}}|}{|a_2^{\mathcal{U}}|} = \frac{|\{d_8,d_9\}|}{|\{d_1,d_2,d_8,d_9,d_{10}\}|} =$

$\frac{2}{5}$.

Example 9. Continue Example 3. From

$a_1 = CC/CGTAAAAGT/GTCACAC$, $a_1^{\mathcal{U}} = \{u_1, u_2, u_4\}$;

$a_2 = CGC/AGTCG/GTCGA$, $a_2^{\mathcal{U}} = \{u_2, u_3\}$; we find that $\{a_1, a_2\}^{\mathcal{U}} =$

$\{u_1, u_2, u_4\} \cap \{u_2, u_3\} = \{u_2\}$. So $stg(a_1 \rightleftharpoons a_2) = |\{a_1, a_2\}^{\mathcal{U}}| = |\{u_2\}| = 1$,

and the *confidence* of the association $cfi(a_1 \rightleftharpoons a_2)$ is

$\frac{|\{a_1,a_2\}^{\mathcal{U}}|}{\max(|a_1^{\mathcal{U}}|,|a_2^{\mathcal{U}}|)} = \frac{|\{u_2\}|}{\max(|\{u_1,u_2,u_4\}|,|\{u_2,u_3\}|)} = \frac{1}{\max(3,2)} = \frac{1}{3}$. The *strength* of

the association rule $stg(a_1 \rightarrow a_2)$ is $|\{a_1, a_2\}^{\mathcal{U}}| = |\{u_2\}| = 1$, and the *confi-*

dence of the rule $cfi(a_1 \rightarrow a_2)$ is $\frac{|\{a_1,a_2\}^{\mathcal{U}}|}{|a_1^{\mathcal{U}}|} = \frac{|\{u_2\}|}{|\{u_1,u_2,u_4\}|} = \frac{1}{3}$. The *strength*

of the association rule $stg(a_1 \leftarrow a_2)$ is $|\{a_1, a_2\}^{\mathcal{U}}| = |\{u_2\}| = 1$, and the

confidence of the rule $cfi(a_1 \leftarrow a_2)$ is $\frac{|\{a_1,a_2\}^{\mathcal{U}}|}{|a_2^{\mathcal{U}}|} = \frac{|\{u_2\}|}{|\{u_2,u_3\}|} = \frac{1}{2}$.

From

$a_3 = ACGTAAAA$, $a_3^{\mathcal{U}} = \{u_1, u_3, u_4\}$;

$a_4 = CGTAAAAGTCACAC$, $a_4^{\mathcal{U}} = \{u_1, u_2\}$; we find that $\{a_3, a_4\}^{\mathcal{U}} =$

$\{u_1, u_3, u_4\} \cap \{u_1, u_2\} = \{u_1\}$. So $stg(a_3 \rightleftharpoons a_4) = |\{a_3, a_4\}^{\mathcal{U}}| = |\{u_1\}| = 1$,

and the *confidence* of the association $cfi(a_3 \rightleftharpoons a_4)$ is

$\frac{|\{a_3,a_4\}^{\mathcal{U}}|}{\max(|a_3^{\mathcal{U}}|,|a_4^{\mathcal{U}}|)} = \frac{|\{u_1\}|}{\max(|\{u_1,u_3,u_4\}|,|\{u_1,u_2\}|)} = \frac{1}{\max(3,2)} = \frac{1}{3}$. The *strength* of

the association rule $stg(a_3 \rightarrow a_4)$ is $|\{a_3, a_4\}^{\mathcal{U}}| = |\{u_1\}| = 1$, and the *confi-*

dence of the rule $cfi(a_3 \rightarrow a_4)$ is $\frac{|\{a_3,a_4\}^{\mathcal{U}}|}{|a_3^{\mathcal{U}}|} = \frac{|\{u_1\}|}{|\{u_1,u_3,u_4\}|} = \frac{1}{3}$. The *strength*

of the association rule $stg(a_3 \leftarrow a_4)$ is $|\{a_3, a_4\}^{\mathcal{U}}| = |\{u_1\}| = 1$, and the

confidence of the rule $cfi(a_3 \leftarrow a_4)$ is $\frac{|\{a_3,a_4\}^{\mathcal{U}}|}{|a_4^{\mathcal{U}}|} = \frac{|\{u_1\}|}{|\{u_1,u_2\}|} = \frac{1}{2}$.

10 Summary and Future Work

We have given a general treatment for some kind of sequences such as customer
sequences, document sequences, and DNA sequences, etc.

We have presented algorithms based on theorems developed here to find
maximal frequent patterns in sequences. Associations between these patterns

are then found by applying a data mining technique based on rough set analysis.

Further work and applications to discover knowledge about patterns in sequences are currently in process.

References

1. Agrawal, R.; Srikant, R. 1994, Fast algorithms for mining association rules, in *Proceedings of the 20th International Conference on Very Large Database (VLDB), Santiago, Chile, September 1994; Fast algorithms for mining association rules in large databases, IBM Research Report RJ 9839, June 1994 (expanded version)*.
2. Agrawal, R.; Srikant, R. 1994–1995, Mining sequential patterns, in *Proceedings of the 11th International Conference on Data Engineering, Taipei, Taiwan, March 1995; IBM Research Report RJ 9910, October 1994 (expanded version)*.
3. Bell, D.A.; Guan, J.W. 1998, "Computational methods for rough classification and discovery", *Journal of the American Society for Information Science, Special Topic Issue on Data Mining*, Vol. 49(1998), No. 5, 403–414.
4. Bell, D.A.; Guan, J.W. 2003, "Data mining for motifs in DNA sequences", in G. Wang et al (ed.) *Proceedings of the 9th International Conference on Rough Sets, Fuzzy Sets, Data Mining and Granular Computing (RSFDGrC'2003)*, Chongqing, China, October 19–22, 2003.
5. Feldman, R.; Aumann, Y.; Amir, A.; Zilberstain, A.; Kloesgen, W. Ben-Yehuda, Y. 1997, Maximal association rules: a new tool for mining for keyword co-occurrences in document collection, in *Proceedings of the 3rd International Conference on Knowledge Discovery (KDD 1997)*, 167–170.
6. Feldman, R.; Aumann, Y.; Zilberstain, A.; Ben-Yehuda, Y. 1998, Trend graphs: visualizing the evolution of concept relationships in large document collection, in *Proceedings of the 2nd European Symposium on Knowledge Discovery in Databases, PKDD'98, Nantes, France, 23–26 September 1998; Lecture Notes in Artificial Intelligence 1510: Principles of Data Mining and Knowledge Discovery*, Jan M. Zytkow Mohamed Quafafou eds.; Springer, 38–46.
7. Feldman, R.; Fresko, M.; Kinar, Y.; Lindell, Y.; Liphstat, O.; Rajman, M.; Schler, Y.; Zamir, O. 1998, Text mining at the term level, in *Proceedings of the 2nd European Symposium on Knowledge Discovery in Databases, PKDD'98, Nantes, France, 23–26 September 1998; Lecture Notes in Artificial Intelligence 1510: Principles of Data Mining and Knowledge Discovery*, Jan M. Zytkow Mohamed Quafafou eds.; Springer, 65–73.
8. Frawley, W.J., Piatetsky-Shapiro, G., & Matheus, C.J. 1991, Knowledge discovery in databases: an overview. In G. Piatetsky-Shapiro, W.J. Frawley (eds). Knowledge Discovery in Databases (pp. 1–27). AAAI/MIT Press.
9. Guan, J.W.; Bell, D.A. 1998, "Rough computational methods for information systems", *Artificial Intelligence – An International Journal*, Vol. 105 (1998), 77–104.
10. Guan, J.W.; Bell, D.A.; Liu, D.Y. 2004, "Discovering motifs in DNA sequences", *Fundamenta Informaticae*, IOS Press, Vol. 59 (2004), 119–133.

11. Kiem, H.; Phuc, D. 2000, "Discovering motif based association rules in a set of DNA sequences", in W. Ziarko & Y. Yao (ed.) *Proceedings of the Second International Conference on Rough Sets and Current Trends in Computing (RSCTC'2000)*, Banff, Canada, October 16–19, 2000; 348–352. ISBN 0828-3494, ISBN 0-7731-0413-5.
12. Landau, D.; Feldman, R.; Aumann, Y.; Fresko, M.; Lindell, Y.; Liphstat, O.; Zamir, O. 1998, TextVis: an integrated visual environment for text mining, in *Proceedings of the 2nd European Symposium on Knowledge Discovery in Databases, PKDD'98, Nantes, France, 23–26 September 1998; Lecture Notes in Artificial Intelligence 1510: Principles of Data Mining and Knowledge Discovery*, Jan M. Zytkow Mohamed Quafafou eds.; Springer, 56–64.
13. Pawlak, Z. 1991, Rough sets: theoretical aspects of reasoning about data. Kluwer.
14. Srikant, R.; Agrawal, R. 1995–1996, Mining sequential patterns: generalizations and performance improvements, in *Proceedings of the Fifth International Conference on Extending Database Technology (EDBT), Avignon, France, March 1996; IBM Research Report RJ 9994, December 1995 (expanded version).*

Mining Association Rules with Rough Sets

D.A. Bell[1], J.W. Guan[1,2], and D.Y. Liu[2]

[1] School of Computer Science, The Queen's University of Belfast, Belfast BT7
1NN Northern Ireland, UK
[2] College of Computer Science and Technology, Jilin University, 130012,
Changchun, P.R. China

Abstract. We say that there is an association between two sets of items when the
sets are likely to occur together in transactions. In information retrieval, an association between two keyword sets means that they co-occur in a record or document.
In databases, an association is a rule latent in the databases whereby an attribute
set can be inferred from another.

Generally, the number of associations may be large, so we look for those that are
particularly strong. Maximal association rules were introduced by [3, 4], and there
is only one maximal association.

Rough set theory has been used successfully for data mining. By using this
theory, rules that are similar to maximal associations can be found. However, we
show that the rough set approach to discovering knowledge is much simpler than
the maximal association method.

Key words: Rough Set, Text Mining, Association Rule

1 Introduction

An association is said to exist between two sets of items when a transaction
containing one set is likely to also contain the other. One example is the analysis of supermarket basket data where associations like "28% of all customers
who buy cheese also buy wine" may be discovered or mined. An association
is a rule latent in and possibly mined from databases, by which one attribute
set can be inferred from another.

For example, consider the Reuters-21578 database (Lewis D. 1997, "The
Reuters-21578, Distribution 1.0") *http://www.research.att.com/ lewis/reuters
21578*. This is a database which contains 21578 news articles categorized by
descriptions of features such as *country names, topics, people names, organizations* and *stock exchanges*. These are sets of keywords indicating the subject
matter of the documents.

D.A. Bell et al.: *Mining Association Rules with Rough Sets*, Studies in Computational Intelligence (SCI) **5**, 163–161 (2005)
www.springerlink.com

In a database like this, the number of associations may be large. For example, from a record "Canada, Iran, USA, crude, ship" we may discover a number of associations such as

"Canada, Iran, USA" is associated with "crude, ship";
"Iran, USA" is associated with "crude, ship";
"Canada, USA" is associated with "crude, ship";
"Canada, Iran" is associated with "crude, ship";
"Canada, Iran, USA" is associated with "ship";
"Canada, Iran, USA" is associated with "crude"; etc.

Now standard association rules are based on the notion of *frequent sets of attributes* which appear in many documents. We are concerned here with maximal association rules, which are based on *frequent maximal sets of attributes*. These attribute sets appear maximally in many documents. Maximal association rules, are rules of the form $X \to Y$ with attribute sets X and Y. The regular association rule $X \to Y$ means that if X then Y (with some *confidence*). The maximal association rule $X \to Y$ means that if X maximally then Y maximally. Roughly speaking, an association is maximal if *nearly all* of the descriptions are used in distinguishing *nearly all* the decisions.

There is only one maximal association between attributes, such as "countries: Canada, Iran, USA" and "topics: crude, ship".

Rough set theory is known to be a useful tool for data mining. By using this theory, rules that are similar to maximal associations can be found. However, we show that the rough set approach to discovering knowledge is much simpler than the maximal association method.

This paper is organized as follows. Section 1 introduces information tables in rough set theory. Section 2 discusses the support documents of a keyword set. Sections 3–5 introduce partitioning the indexing vocabulary of terms, the definition of maximality, and then discuss maximal associations. Section 6 discusses negated attributes. Finally, we introduce the rough set approach to association rule mining.

2 Information Tables

Let D be a document collection containing documents $d_1, d_2, \ldots, d_N$. The vocabulary of collection D and document d are denoted by V^D and V^d, respectively.

We now illustrate the method for finding maximal associations suggested by [3, 4], and in later sections we show how the same results can be found, more simply, using rough analysis.

In order to illustrate the Feldman method, we rework some of his illustrations below:

Example 1. A document collection $D = \{d_1, d_2, \ldots, d_{10}\}$:

$V^{d_1} = V^{d_2} = \{$Canada, Iran, USA, crude, ship$\}$

$V^{d_3} = \{$USA, earn$\}$

$V^{d_4} = V^{d_5} = \{$USA, jobs, cpi$\}$

$V^{d_6} = \{$USA, earn, cpi$\}$

$V^{d_7} = \{$Canada, sugar, tea$\}$

$V^{d_8} = V^{d_9} = \{$Canada, USA, trade, acq$\}$

$V^{d_{10}} = \{$Canada, USA, earn$\}$

Example 2. There are 10 articles about "corn" which are also about USA and Canada and another 20 articles concerning "fish" and the countries USA, Canada and France. That is,

$$\{$USA, Canada, corn$\} = V^{d_i} \text{ for } i = 1, 2, \ldots, 10$$

$$\{$USA, Canada, France, fish$\} = V^{d_j} \text{ for } j = 11, 12, \ldots, 30 \ .$$

In rough set theory, an *information system* $\mathcal{I}$ is a system $\langle U, A \rangle$, where

(1) $U = \{u_1, u_2, \ldots, u_i, \ldots, u_{|U|}\}$ is a finite non-empty set, called the *universe* or *object space*; elements of U are called *objects, records* or *documents*;

(2) $A = \{a_1, a_2, \ldots, a_j, \ldots, a_{|A|}\}$ is also a finite non-empty set; elements of A are called *attributes, terms,* or *keywords*;

(3) for every $a \in A$ there is a mapping a from U into some space $a : U \rightarrow a(U)$, and $a(U) = \{a(u) \mid u \in U\}$ is called the *domain* of attribute a.

In the rough set approach we use the concept of an information system, and we use the same mechanism for our present analysis. An information system can be expressed intuitively in terms of an *information table* as follows

$U \backslash A$	a_1	a_2	...	a_j	...	a_J
u_1	$a_1(u_1)$	$a_2(u_1)$	...	$a_j(u_1)$	...	$a_J(u_1)$
u_2	$a_1(u_2)$	$a_2(u_2)$	...	$a_j(u_2)$	...	$a_J(u_2)$
...	...	...	...	...	...	...
u_i	$a_1(u_i)$	$a_2(u_i)$	...	$a_j(u_i)$	...	$a_J(u_i)$
...	...	...	...	...	...	...
u_I	$a_1(u_I)$	$a_2(u_I)$	...	$a_j(u_I)$	...	$a_J(u_I)$

In the information table, each document d corresponds to a set of attributes $V^d = \{a \in A | a(d) = 1\}$, which is the set of *keywords* occurring in d.

Example 3. The document collection in Example 1 can be represented as an information table as follows:

An example of an Information System

U\A	Canada	Iran	USA	acq	cpi	crude	earn	jobs	ship	sugar	tea	trade
d_1	1	1	1	0	0	1	0	0	1	0	0	0
d_2	1	1	1	0	0	1	0	0	1	0	0	0
d_3	0	0	1	0	0	0	1	0	0	0	0	0
d_4	0	0	1	0	1	0	0	1	0	0	0	0
d_5	0	0	1	0	1	0	0	1	0	0	0	0
d_6	0	0	1	0	1	0	1	0	0	0	0	0
d_7	1	0	0	0	0	0	0	0	0	1	1	0
d_8	1	0	1	1	0	0	0	0	0	0	0	1
d_9	1	0	1	1	0	0	0	0	0	0	0	1
d_{10}	1	0	1	0	0	0	1	0	0	0	0	0

That is, in the information table for example 1.1,

$$A = V^D = \{t_1, t_2, t_3, \ldots, t_{10}, t_{11}, t_{12}\} \, ; U = D = \{d_1, d_2, d_3, \ldots, d_{10}\} \, ,$$

where $t_1 = Canada, t_2 = Iran, t_3 = USA, t_4 = acq, t_5 = cpi, t_6 = crude, t_7 = earn, t_8 = jobs, t_9 = ship, t_{10} = sugar, t_{11} = tea, t_{12} = trade.$

3 The Support Documents of a Keyword Set

Let $V^D = \{t_1, t_2, \ldots, t_n\}$, where t are keywords. Let $X \subseteq V^D$ be a subset of the collection vocabulary. Denote $X^D = \{d | X \subseteq V^d\}$, i.e. the set of documents in which X occurs. That is, in an information table, we say that a document d *supports* a given set of attributes X if $X \subseteq V^d$. The *strength* of X in the information table, denoted by $stg(X)$, is the number of documents d supporting X:

$$stg(X) = |\{d | X \subseteq V^d\}| = |X^D| \, .$$

An *association* between X and Y is an expression of the form $X \rightleftharpoons Y$, where X and Y are attribute sets; the *strength* of the association $stg(X \rightleftharpoons Y)$ is the strength of $X \cup Y$, i.e., the number of documents supporting $X \cup Y$

$$stg(X \rightleftharpoons Y) = stg(X \cup Y) = |\{d | X \cup Y \subseteq V^d\}| = |(X \cup Y)^D| \, ,$$

and the *confidence* of the association $cfi(X \rightleftharpoons Y)$ is

$$\frac{stg(X \cup Y)}{\max(stg(X), stg(Y))} = \frac{|\{d | X \cup Y \subseteq V^d\}|}{\max(|\{d | X \subseteq V^d\}|, |\{d | Y \subseteq V^d\}|)} = \frac{|(X \cup Y)^D|}{\max(|X^D|, |Y^D|)}$$

An *association rule* is a rule of the form $X \rightarrow Y$, where X and Y are attribute sets; the *strength* of the association rule $stg(X \rightarrow Y)$ is

$$stg(X \rightarrow Y) = stg(X \cup Y) = |\{d | (X \cup Y)^d)\}| = |(X \cup Y)^D| \, ,$$

and the *confidence* of the rule $cfi(X \to Y)$ is

$$stg(X \cup Y)/stg(X) = \frac{|\{d|X \cup Y \subseteq V^d)\}|}{|\{d|X \subseteq V^d\}|} = \frac{|(X \cup Y)^D|}{|X^D|} .$$

Example 4. For Example 2, it is supposed that
$V^{d_i} = \{USA, Canada, corn\}$ for $i = 1, 2, \ldots, 10$;
$V^{d_j} = \{USA, Canada, France, fish\}$ for $j = 11, 12, \ldots, 30$.

It can be found that associations are
$\{USA, Canada\} \rightleftharpoons \{fish\}$ with confidence

$$\frac{|(\{USA, Canada\} \cup \{fish\})^D|}{\max(|\{USA, Canada\}^D|, |\{fish\}^D|)}$$

$$= \frac{|\{d_{11}, \ldots, d_{30}\}|}{\max(|\{d_1, \ldots, d_{10}, d_{11}, \ldots, d_{30}\}|, |\{d_{11}, \ldots, d_{30}\}|)}$$

$$= \frac{20}{\max(30, 20)} = 66\% ,$$

$\{USA, Canada, France\} \rightleftharpoons \{fish\}$ with confidence

$$\frac{|(\{USA, Canada, France\} \cup \{fish\})^D|}{\max(|\{USA, Canada, France\}^D|, |\{\{fish\}^D|)}$$

$$= \frac{|\{d_{11}, \ldots, d_{30}\}|}{\max(|\{d_{11}, \ldots, d_{30}\}|, |\{d_{11}, \ldots, d_{30}\}|)}$$

$$= \frac{20}{\max(20, 20)} = 100\% ,$$

$\{USA, Canada\} \rightleftharpoons \{corn\}$ with confidence

$$\frac{|(\{USA, Canada\} \cup \{corn\})^D|}{\max(|\{USA, Canada\}^D|, |\{corn\}^D|)}$$

$$= \frac{|\{d_1, \ldots, d_{10}\}|}{\max(|\{d_1, \ldots, d_{10}, d_{11}, \ldots, d_{30}\}|, |\{d_1, \ldots, d_{10}\}|)}$$

$$= \frac{10}{\max(30, 10)} = 33\% .$$

Feldman et al argue that this confidence is too low to indicate a strong connection between USA-Canada and "corn".

Also, association rules are
$\{USA, Canada\} \rightarrow \{fish\}$ with confidence

$$\frac{|(\{USA, Canada\} \cup \{fish\})^D|}{|\{USA, Canada\}^D|}$$

$$= \frac{|\{d_{11}, \ldots, d_{30}\}|}{|\{d_1, \ldots, d_{10}, d_{11}, \ldots, d_{30}\}|} = 20/30 = 66\% ,$$

{USA, Canada, France}→ {fish} with confidence

$$\frac{|(\{USA, Canada, France\} \cup \{fish\})^D|}{|\{USA, Canada, France\}^D|}$$

$$= \frac{|\{d_{11}, \ldots, d_{30}\}|}{|\{d_{11}, \ldots, d_{30}\}|} = 20/20 = 100\% \,,$$

{USA, Canada}→ {corn} with confidence

$$\frac{|(\{USA, Canada\} \cup \{corn\})^D|}{|\{USA, Canada\}^D|}$$

$$= \frac{|\{d_1, \ldots, d_{10}\}|}{|\{d_1, \ldots, d_{10}, d_{11}, \ldots, d_{30}\}|} = 10/30 = 33\% \,.$$

Again, this confidence is too low to represent the strong connection between USA-Canada and "corn".

Therefore, Feldman et al introduce the definition of maximality. As a maximal association rule,

{USA, Canada}→ {corn} has confidence $10/10 = 100\%$ since whenever USA-Canada appear *maximally* without any other country (in and only in documents $d_1 - d_{10}$), "corn" also appears *maximally* without any other goods (also in and only in documents $d_1 - d_{10}$).

There are two ways to define maximality as follows: one way is by excluding association with negation of terms, and the other is by partitioning the indexing vocabulary of terms.

4 Partitioning the Indexing Vocabulary of Terms

For the definition of maximality, an underlying taxonomy τ of attributes is used. Then the "interesting" correlations between attributes from different categories can be obtained.

Let $A = V^D$ be partitioned to classes $T_1, T_2, \ldots, T_k, \ldots, T_K$. We denote the partition by

$$A/\tau : T_1, T_2, \ldots, T_k, \ldots, T_K \,;$$

i.e.,

$$A/\tau = \{T_1, T_2, \ldots, T_k, \ldots, T_K\} \,,$$

where τ is the corresponding equivalence relation of the partition such that $a\tau b$ for $a, b \in A$ if and only if a and b are in the same class, and

$$T_k \in 2^A; T_k \neq \emptyset; T_i \cap T_j = \emptyset \text{ when } i \neq j; A = \cup_{k=1}^K T_k \,.$$

The partition is called a *taxonomy*. Each class is called a *category*.

Example 5. For Example 1, the vocabulary of terms A can be partitioned as follows:

$$A/\tau = \{T_1, T_2\} \, ,$$

where

$$T_1 = countries = \{t_1, t_2, t_3\} \, ,$$

$$T_2 = topics = \{t_4, t_5, t_6, t_7, t_8, t_9, t_{10}, t_{11}, t_{12}\} \, .$$

The taxonomy is {countries, topics}, where countries and topics are categories.

Example 6. For Example 2, the vocabulary of terms A can be partitioned as follows:

$$A/\tau = \{T_1, T_2\} \, ,$$

where

$$T_1 = countries = \{USA, Canada, France\} \, ,$$

$$T_2 = topics = \{corn, fish\} \, .$$

The taxonomy is {countries, topics}, where countries and topics are categories.

In practice, there are of course other possible taxonomies. The choice is application dependent.

5 The Definition of Maximality

From a given underlying taxonomy τ of attributes, the definition of maximality can be introduced, and the "interesting" correlations between attributes from different categories can be obtained.

For a given subset $X \subseteq A$ ($X \neq \emptyset$), we can find a unique decomposition

$$X = X_1 \cup X_2 \cup \cdots \cup X_k \cup \cdots \cup X_K$$

such that

$$X_1 \subseteq T_1 \, , X_2 \subseteq T_2, \dots \, , X_k \subseteq T_k, \dots \, , X_K \subseteq T_K \, ,$$

where T_k ($k = 1, 2, \dots, K$) are categories of τ. We call T_k the *corresponding category* of X_k.

5.1 Maximally Supported Sets of Terms for One Category

Given a category T_k, for an $X_k \subseteq T_k$ ($X_k \neq \emptyset$) and document d, we say that d *supports* X_k or X_k *occurs* in d if $X_k \subseteq V^d \cap T_k$. The *support/occurrence collection* of X_k in the document collection D, denoted by $(X_k)^D$, is the subcollection of documents supporting X_k:

$$(X_k)^D = \{d | X_k \subseteq V^d \cap T_k\} \, .$$

This attribute set X_k of terms can also be denoted by a *taxonomy pair* $\{X_k : T_k\}$ with strength $stg(X_k)$, the number of documents supporting X_k:

$$stg(X_k) = |\{d|X_k \subseteq V^d \cap T_k\}| = |(X_k)^D| ,$$

and X_k itself can be denoted as

$$\{X_k : T_k\}(X_k)^D$$

with its support collection more precisely when $(X_k)^D \subset D$. We call X_k as a *clause* co-occurring in $(X_k)^D$.

Now for an $X_k \subseteq T_k$ ($X_k \neq \emptyset$) and document d, we say that d *maximally supports* X_k if $X_k = V^d \cap T_k$. The *max-support collection* of X_k in the document collection D, denoted by $(X_k)^{D,\max}$, is the sub-collection of documents maximally supporting X_k:

$$(X_k)^{D,\max} = \{d|X_k = V^d \cap T_k\} .$$

This attribute set X_k of terms can also be denoted by a *taxonomy pair* $\{X_k : T_k\}$ with max-strength $msg(X_k)$, the number of documents maximally supporting X_k:

$$msg(X_k) = |\{d|X_k = V^d \cap T_k\}| = |(X_k)^{D,\max}| ,$$

and X_k itself can be denoted as

$$\{X_k : T_k\}_{\max}(X_k)^{D,\max}$$

with its max-support collection more precisely when $(X_k)^{D,\max} \subset D$. We call X_k as a *sentence* co-occurring in $(X_k)^{D,\max}$.

Example 7. Continue Example 1.
For $T_1 = countries$, the following 7 clauses and their co-occurrence collection can be found:

$\{\{Canada, Iran, USA\} : countries\}\{d_1, d_2\}$
$\{\{Canada, Iran\} : countries\}\{d_1, d_2\}$
$\{\{Iran, USA\} : countries\}\{d_1, d_2\}$
$\{\{Iran\} : countries\}\{d_1, d_2\}$
$\{\{USA\} : countries\}\{d_1, d_2, d_3, d_4, d_5, d_6, d_8, d_9, d_{10}\}$
$\{\{Canada\} : countries\}\{d_1, d_2, d_7, d_8, d_9, d_{10}\}$
$\{\{Canada, USA\} : countries\}\{d_1, d_2, d_8, d_9, d_{10}\}$

Also, the following 4 sentences and their co-occurrence collection can be found:

$\{\{USA\} : countries\}_{\max}\{d_3, d_4, d_5, d_6\}$
$\{\{Canada, Iran, USA\} : countries\}_{\max}\{d_1, d_2\}$
$\{\{Canada\} : countries\}_{\max}\{d_7\}$
$\{\{Canada, USA\} : countries\}_{\max}\{d_8, d_9, d_{10}\}$

For $T_2 = topics$, the following 14 clauses and their co-occurrence collection can be found:

$\{\{crude, ship\} : topics\}\{d_1, d_2\}$
$\{\{crude\} : topics\}\{d_1, d_2\}$
$\{\{ship\} : topics\}\{d_1, d_2\}$
$\{\{earn\} : topics\}\{d_3, d_6, d_{10}\}$
$\{\{jobs, cpi\} : topics\}\{d_4, d_5\}$
$\{\{jobs\} : topics\}\{d_4, d_5\}$
$\{\{cpi\} : topics\}\{d_4, d_5, d_6\}$
$\{\{earn, cpi\} : topics\}\{d_6\}$
$\{\{sugar, tea\} : topics\}\{d_7\}$
$\{\{sugar\} : topics\}\{d_7\}$
$\{\{tea\} : topics\}\{d_7\}$
$\{\{trade, acq\} : topics\}\{d_8, d_9\}$
$\{\{trade\} : topics\}\{d_8, d_9\}$
$\{\{acq\} : topics\}\{d_8, d_9\}$

Also, the following 6 sentences and their co-occurrence collection can be found:

$\{\{crude, ship\} : topics\}_{\max}\{d_1, d_2\}$
$\{\{earn\} : topics\}_{\max}\{d_3, d_{10}\}$
$\{\{jobs, cpi\} : topics\}_{\max}\{d_4, d_5\}$
$\{\{earn, cpi\} : topics\}_{\max}\{d_6\}$
$\{\{sugar, tea\} : topics\}_{\max}\{d_7\}$
$\{\{trade, acq\} : topics\}_{\max}\{d_8, d_9\}$

Example 8. Continue Example 2.

For $T_1 = countries$, the following 7 clauses and their co-occurrence collection can be found:

$\{\{Canada, France, USA\} : countries\}\{d_{11}\text{-}d_{30}\}$
$\{\{Canada, USA\} : countries\}\{d_1\text{-}d_{10}, d_{11}\text{-}d_{30}\}$ or simply
    $\{\{Canada, USA\} : countries\}$ since $\{d_1\text{-}d_{30}\} = D$
$\{\{France, USA\} : countries\}\{d_{11}\text{-}d_{30}\}$
$\{\{Canada, France\} : countries\}\{d_{11}\text{-}d_{30}\}$
$\{\{Canada\} : countries\}\{d_1\text{-}d_{10}, d_{11}\text{-}d_{30}\}$
$\{\{USA\} : countries\}\{d_1\text{-}d_{10}, d_{11}\text{-}d_{30}\}$
$\{\{France\} : countries\}\{d_{11}\text{-}d_{30}\}|$

Also, the following 2 sentences and their co-occurrence collection can be found:

$\{\{Canada, France, USA\} : countries\}_{\max}\{d_{11}\text{-}d_{30}\}$
$\{\{Canada, USA\} : countries\}_{\max}\{d_1\text{-}d_{10}\}$

For $T_2 = topics$, the following 2 clauses and their co-occurrence collection can be found:

$$\{\{corn\} : topics\}\{d_1\text{-}d_{10}\}$$
$$\{\{fish\} : topics\}\{d_{11}\text{-}d_{30}\}$$

Also, the following 2 sentences and their co-occurrence collection can be found:

$$\{\{corn\} : topics\}_{\max}\{d_1\text{-}d_{10}\}$$
$$\{\{fish\} : topics\}_{\max}\{d_{11}\text{-}d_{30}\} \ .$$

Now, let us give some definitions as follows.

Definition 1. *Given a document collection D and a taxonomy τ of attributes V^D, $V^D/\tau : \{T_1, T_2, \ldots, T_k, \ldots, T_K\}$, for a category T_k the group of sentences is*

$$\{V^d \cap T_k \neq \emptyset \mid d \in D\} \ ,$$

and each sentence has its co-occurrence collection as follows:

$$\{V^d \cap T_k : T_k\}_{\max}(V^d \cap T_k)^{D,\max} \ ,$$

where

$$(V^d \cap T_k)^{D,\max} = \{d' \in D | V^d \cap T_k = V^{d'} \cap T_k\} \ .$$

Note that here the group of sentences $\{V^d \cap T_k \neq \emptyset | d \in D\}$ means the element (sentence) of this group is a non-empty subset $V^d \cap T_k$ for some $d \in D$ of category T_k since there may be some documents d sharing/co-supporting the same $V^d \cap T_k$ (i.e., $V^d \cap T_k$ co-occurs in those documents). Thus, here

$$(V^d \cap T_k)^{D,\max} = \{d' \in D | V^d \cap T_k = V^{d'} \cap T_k\} \ .$$

is to express those documents d' in collection D which share/co-support the same sentence $V^d \cap T_k$.

Example 9. Continue Example 1.
For $T_1 = countries$, the group of sentences is

$$\{\{Canada, Iran, USA\} , \{USA\} , \{Canada\} , \{Canada, USA\}\} \ .$$

For $T_2 = topics$, the group of sentences is

$$\{\{crude, ship\}, \{earn\}, \{jobs, cpi\}, \{earn, cpi\}, \{sugar, tea\}, \{trade, acq\}\}.$$

Example 10. Continue Example 2. For $T_1 = countries$, the group of sentences is

$$\{\{Canada, France, USA\}, \{Canada, USA\}\}.$$

For $T_2 = topics$, the group of sentences is

$$\{\{corn\}, \{fish\}\}.$$

Definition 2. *Given a document collection D and a taxonomy τ of attributes V^D, $V^D/\tau : \{T_1, T_2, \ldots, T_k, \ldots, T_K\}$, for a category T_k let G be the group of sentences, $G = \{s_1, s_2, \ldots, s_l\}$. Then, for each sentence $s \in G$, every nonempty subset of sentence s is a clause.*

Thus, clauses may be too numerous and we would like pay more attention to sentences.

Example 11. Continue Example 1.
For $T_1 = countries$, the group of clauses is

$$\{\{Canada, Iran, USA\}, \{Canada, Iran\}, \{Iran, USA\} ,$$

$$\{Iran\}, \{USA\}, \{Canada\}, \{Canada, USA\}\} .$$

For $T_2 = topics$, the group of clauses is

$$\{\{crude, ship\}, \{crude\}, \{ship\}, \{earn\}, \{jobs, cpi\}, \{jobs\}, \{cpi\},$$

$$\{earn, cpi\}, \{sugar, tea\}, \{sugar\}, \{tea\}, \{trade, acq\}, \{trade\}, \{acq\}\}.$$

Example 12. Continue Example 2.
For $T_1 = countries$, the group of clauses is

$$\{\{Canada, France, USA\}, \{Canada, USA\}, \{France, USA\} ,$$

$$\{Canada, France\}, \{Canada\}, \{USA\}, \{France\}\} .$$

For $T_2 = topics$, the group of clauses is

$$\{\{corn\}, \{fish\}\} .$$

5.2 Maximally Supported Sets of Terms for a Taxonomy

Given a taxonomy τ, for a given $X = X_1 \cup X_2 \cup \cdots \cup X_k \cup \cdots \cup X_K$ ($X \neq \emptyset$; $X_k \subseteq T_k$ for $k = 1, 2, \ldots, K$; and so there is at least one k such that $X_k \neq \emptyset$) and document d, we say that d *supports* X if $X_k \subseteq V^d \cap T_k$ for all k. The *support/occurrence collection* of X in the document collection D, denoted by X^D, is the sub-collection of documents supporting X:

$$X^D = \{d | X_k \subseteq V^d \cap T_k; k = 1, 2, \ldots, K\} .$$

This attribute set X of terms can also be denoted by a vector of *taxonomy pairs* $\{X_k : T_k\}$ with strength $stg(X)$, the number of documents supporting X:

$$stg(X) = |\{d | X_k \subseteq V^d \cap T_k; k = 1, 2, \ldots, K\}| = |X^D| ,$$

and X itself can be denoted as a vector

$$(\{X_1 : T_1\}, \{X_2 : T_2\}, \ldots, \{X_k : T_k\}, \ldots, \{X_K : T_K\})X^D$$

in a Cartesian product space

$$2^{T_1} \times 2^{T_2} \times \cdots \times 2^{T_k} \times \cdots \times 2^{T_K}$$

with its support collection more precisely when $X^D \subset D$. We call X as a *clause* co-occurring in X^D.

Now for an $X = X_1 \cup X_2 \cup \cdots \cup X_k \cup \cdots \cup X_K$ (where $X_k \neq \emptyset$ and $X_k \subseteq T_k$ for $k = 1, 2, \ldots, K$) and document d, we say that d *maximally supports* X if $X_k = V^d \cap T_k$ for all k. The *max-support collection* of X in the document collection D, denoted by $X^{D,\max}$, is the sub-collection of documents maximally supporting X:

$$X^{D,\max} = \{d | X_k = V^d \cap T_k; k = 1, 2, \ldots, K\} \, .$$

This attribute set X of terms can also be denoted by a vector of *taxonomy pairs* $\{X_k : T_k\}$ with max-strength $msg(X)$, the number of documents maximally supporting X:

$$msg(X) = |\{d | X_k = V^d \cap T_k; k = 1, 2, \ldots, K\}| = |X^{D,\max}| \, ,$$

and X itself can be denoted as a vector

$$(\{X_1 : T_1\}, \{X_2 : T_2\}, \ldots, \{X_k : T_k\}, \ldots, \{X_K : T_K\})_{\max} X^{D,\max}$$

in a Cartesian product space

$$2^{T_1} \times 2^{T_2} \times \cdots \times 2^{T_k} \times \cdots \times 2^{T_K}$$

with its max-support collection more precisely when $X^{D,\max} \subset D$. We call X as a *sentence* co-occurring in $X^{D,\max}$.

Example 13. For Example 1, the following sentences can be found:

$(\{Canada, Iran, USA\} : countries, \{crude, ship\} : topics)_{\max}\{d_1, d_2\}$
$(\{USA\} : countries, \{earn\} : topics)_{\max}\{d_3\}$
$(\{USA\} : countries, \{jobs, cpi\} : topics)_{\max}\{d_4, d_5\}$
$(\{USA\} : countries, \{earn, cpi\} : topics)_{\max}\{d_6\}$
$(\{Canada\} : countries, \{sugar, tea\} : topics)_{\max}\{d_7\}$
$(\{Canada, USA\} : countries, \{trade, acq\} : topics)_{\max}\{d_8, d_9\}$
$(\{Canada, USA\} : countries, \{earn\} : topics)_{\max}\{d_{10}\}$

Also, clauses can be found, for example, from a sentence

$(\{Canada, Iran, USA\} : countries, \{crude, ship\} : topics)_{\max}\{d_1, d_2\}$

along with $2^5 - 2 = 30$ clauses as follows:

$(\{Iran, USA\} : countries, \{crude, ship\} : topics)\{d_1, d_2\}$
$(\{Canada, USA\} : countries, \{crude, ship\} : topics)\{d_1, d_2\}$
$(\{Canada, Iran\} : countries, \{crude, ship\} : topics)\{d_1, d_2\}$

$(\{Canada, Iran, USA\} : countries, \{ship\} : topics)\{d_1, d_2\}$
$(\{Canada, Iran, USA\} : countries, \{crude\} : topics)\{d_1, d_2\}$
$(\{USA\} : countries, \{crude, ship\} : topics)\{d_1, d_2\}$
$(\{Canada\} : countries, \{crude, ship\} : topics)\{d_1, d_2\}$
$(\{Canada, Iran\} : countries, \{ship\} : topics)\{d_1, d_2\}$
$\cdots$

$(\{Canada\} : countries, \{\} : topics)\{d_1, d_2\}$
$(\{Iran\} : countries, \{\} : topics)\{d_1, d_2\}$
$(\{USA\} : countries, \{\} : topics)\{d_1, d_2\}$
$(\{\} : countries, \{crude\} : topics)\{d_1, d_2\}$
$(\{\} : countries, \{ship\} : topics)\{d_1, d_2\}$

Example 14. For Example 2, the following sentences can be found:

$(\{Canada, France, USA\} : countries, \{fish\} : topics)_{\max}\{d_{11}\text{-}d_{30}\}$
$(\{Canada, USA\} : countries, \{corn\} : topics)_{\max}\{d_1\text{-}d_{10}\}$

Also, clauses can be found from a sentence

$(\{Canada, France, USA\} : countries, \{fish\} : topics)_{\max}\{d_{11}\text{-}d_{30}\}$

along with $2^4 - 2 = 14$ clauses as follows:

$(\{Canada, France\} : countries, \{fish\} : topics)\{d_{11}\text{-}d_{30}\}$
$(\{Canada, USA\} : countries, \{fish\} : topics)\{d_{11}\text{-}d_{30}\}$
$(\{France, USA\} : countries, \{fish\} : topics)\{d_{11}\text{-}d_{30}\}$
$\cdots$

$(\{Canada, France, USA\} : countries, \{\} : topics)\{d_{11}\text{-}d_{30}\}$
$(\{\} : countries, \{fish\} : topics)\{d_{11}\text{-}d_{30}\}$

Also, clauses can be found from a sentence

$(\{Canada, USA\} : countries, \{corn\} : topics)_{\max}\{d_1\text{-}d_{10}\}$

along with $2^3 - 2 = 6$ clauses as follows:

$(\{Canada\} : countries, \{corn\} : topics)\{d_1\text{-}d_{10}\}$
$(\{USA\} : countries, \{corn\} : topics)\{d_1\text{-}d_{10}\}$
$(\{\} : countries, \{corn\} : topics)\{d_1\text{-}d_{10}\}$
$(\{Canada, USA\} : countries, \{\} : topics)\{d_1\text{-}d_{10}\}$
$(\{Canada\} : countries, \{\} : topics)\{d_1\text{-}d_{10}\}$
$(\{USA\} : countries, \{\} : topics)\{d_1\text{-}d_{10}\}$

Now, let us give some definitions as follows.

Definition 3. *Given a document collection D and a taxonomy τ of attributes V^D, $V^D/\tau : \{T_1, T_2, \ldots, T_k, \ldots, T_K\}$, a sentence for the taxonomy is a K-dimensional vector $s = (s_1, s_2, \ldots, s_k, \ldots, s_K)$, where component s_k is a sentence for category T_k. That is,*

$$s = (\{X_1 : T_1\}, \{X_2 : T_2\}, \ldots, \{X_k : T_k\}, \ldots, \{X_K : T_K\})_{\max} \,,$$

where

$$s_k = \{X_k : T_k\}, X_k = V^d \cap T_k \neq \emptyset \ \ for \ some \ d \in D \,,$$

and each sentence s has its co-occurrence collection as follows:

$$(V^d \cap T_1 : T_1, V^d \cap T_2 : T_2, \ldots, V^d \cap T_k : T_k, \ldots, V^d \cap T_K : T_K)_{\max} X^{D,\max} \,,$$

where

$$X^{D,\max} = \{d' \in D | V^d \cap T_k = V^{d'} \cap T_k; k = 1, 2, \ldots, K\} \,.$$

Example 15. For Example 1, sentences for the taxonomy are

$(\{Canada, Iran, USA\} : countries, \{crude, ship\} : topics)_{\max}\{d_1, d_2\}$
$(\{USA\} : countries, \{earn\} : topics)_{\max}\{d_3\}$
$(\{USA\} : countries, \{jobs, cpi\} : topics)_{\max}\{d_4, d_5\}$
$(\{USA\} : countries, \{earn, cpi\} : topics)_{\max}\{d_6\}$
$(\{Canada\} : countries, \{sugar, tea\} : topics)_{\max}\{d_7\}$
$(\{Canada, USA\} : countries, \{trade, acq\} : topics)_{\max}\{d_8, d_9\}$
$(\{Canada, USA\} : countries, \{earn\} : topics)_{\max}\{d_{10}\}$

Example 16. For Example 2, sentences for the taxonomy are

$(\{Canada, France, USA\} : countries, \{fish\} : topics)_{\max}\{d_{11}\text{-}d_{30}\}$
$(\{Canada, USA\} : countries, \{corn\} : topics)_{\max}\{d_1\text{-}d_{10}\}.$

Definition 4. *Given a document collection D and a taxonomy τ of attributes V^D, $V^D/\tau : \{T_1, T_2, \ldots, T_k, \ldots, T_K\}$, let s be a sentence for the taxonomy, $s = (s_1, s_2, \ldots, s_k, \ldots, s_K)$. Suppose that*

$$c_1 \subseteq s_1, c_2 \subseteq s_2, \ldots, c_k \subseteq s_k, \ldots, c_K \subseteq s_K \,,$$

where $c_k \neq s_k$ for at least one k and $c_i \neq \emptyset$ for at least one k. Then, $c = (c_1, c_2, \ldots, c_k, \ldots, c_K)$ is a clause for the taxonomy.

We call this kind of clause c of sentence s a *sub-clause* of the sentence. Obviously, we have $c^D \subseteq s^{D,\max}$. That is, sub-clauses may have more support documents.

Thus, clauses are too numerous and we would like pay more attention to sentences.

Example 17. For Example 1, sentence

$(\{Canada, Iran, USA\} : countries, \{crude, ship\} : topics)_{\max}\{d_1, d_2\}$

has $2^5 - 2 = 30$ sub-clauses

$(\{Iran, USA\} : countries, \{crude, ship\} : topics)\{d_1, d_2\}$
$(\{Canada, USA\} : countries, \{crude, ship\} : topics)\{d_1, d_2\}$

$(\{Canada, Iran\} : countries, \{crude, ship\} : topics)\{d_1, d_2\}$
$(\{Canada, Iran, USA\} : countries, \{ship\} : topics) : topics)\{d_1, d_2\}$
$(\{Canada, Iran, USA\} : countries, \{crude\} : topics)\{d_1, d_2\}$
$(\{USA\} : countries, \{crude, ship\} : topics)\{d_1, d_2\}$
$(\{Canada\} : countries, \{crude, ship\} : topics)\{d_1, d_2\}$
$(\{Canada, Iran\} : countries, \{ship\} : topics)\{d_1, d_2\}$
$\ldots$

$(\{Canada\} : countries, \{\} : topics)\{d_1, d_2\}$
$(\{Iran\} : countries, \{\} : topics)\{d_1, d_2\}$
$(\{USA\} : countries, \{\} : topics)\{d_1, d_2\}$
$(\{\} : countries, \{crude\} : topics)\{d_1, d_2\}$
$(\{\} : countries, \{ship\} : topics)\{d_1, d_2\}$

Example 18. For Example 2, sentence

$(\{Canada, France, USA\} : countries, \{fish\} : topics)_{\max}\{d_{11}\text{-}d_{30}\}$

has $2^4 - 2 = 14$ sub-clauses

$(\{Canada, France\} : countries, \{fish\} : topics)\{d_{11}\text{-}d_{30}\}$
$(\{Canada, USA\} : countries, \{fish\} : topics)\{d_{11}\text{-}d_{30}\}$
$(\{France, USA\} : countries, \{fish\} : topics)\{d_{11}\text{-}d_{30}\}$
$\ldots$

$(\{Canada, France, USA\} : countries, \{\} : topics)\{d_{11}\text{-}d_{30}\}$
$(\{\} : countries, \{fish\} : topics)\{d_{11}\text{-}d_{30}\}$
Also, clauses can be found from a sentence
$(\{Canada, USA\} : countries, \{corn\} : topics)_{\max}\{d_1\text{-}d_{10}\}$

along with $2^3 - 2 = 6$ clauses as follows:

$(\{Canada\} : countries, \{corn\} : topics)\{d_1\text{-}d_{10}\}$
$(\{USA\} : countries, \{corn\} : topics)\{d_1\text{-}d_{10}\}$
$(\{\} : countries, \{corn\} : topics)\{d_1\text{-}d_{10}\}$
$(\{Canada, USA\} : countries, \{\} : topics)\{d_1\text{-}d_{10}\}$
$(\{Canada\} : countries, \{\} : topics)\{d_1\text{-}d_{10}\}$
$(\{USA\} : countries, \{\} : topics)\{d_1\text{-}d_{10}\}$

6 Maximal Associations and Maximal Association Rules

There are two concepts to be clarified. These are, maximal association and maximal association rule, and frequent set and maximal frequent set.

6.1 Maximal Associations

A *maximal association* between V and W is an expression of the form $V \rightleftharpoons W$, where V and W are sentences for some different categories. The *strength* of the association is $msg(V \cup W)$, and confidence of the association is $msg(V \cup W)/max(msg(V), msg(W))$.

Example 19. For Example 1, the following maximal associations can be found:

$\{Canada, Iran, USA\} : countries \rightleftharpoons \{crude, ship\} : topics$ with strength $msg = |\{d_1, d_2\}| = 2$ and confidence $2/max(2, 2) = 1 = 100\%$
$\{USA\} : countries \rightleftharpoons \{earn\} : topics$ with strength $msg = |\{d_3\}| = 1$ and confidence $1/max(4, 2) = 25\%$
$\{USA\} : countries \rightleftharpoons \{jobs, cpi\} : topics$ with strength $msg = |\{d_4, d_5\}| = 2$ and confidence $2/max(4, 2) = 50\%$
$\{USA\} : countries \rightleftharpoons \{earn, cpi\} : topics$ with strength $msg = |\{d_6\}| = 1$ and confidence $1/max(4, 1) = 25\%$
$\{Canada\} : countries \rightleftharpoons \{sugar, tea\} : topics$ with support $msg = |\{d_7\}| = 1$ and confidence $1/max(1, 1) = 100\%$
$\{Canada, USA\} : countries \rightleftharpoons \{trade, acq\} : topics$ with strength $msg = |\{d_8, d_9\}| = 2$ and confidence $2/max(3, 2) = 66\%$
$\{Canada, USA\} : countries \rightleftharpoons \{earn\} : topics$ with strength $msg = |\{d_{10}\}| = 1$ and confidence $1/max(3, 2) = 33\%$

Example 20. For Example 2, the following maximal associations can be found:

$\{Canada, France, USA\} : countries \rightleftharpoons \{fish\} : topics$ with strength $msg = |\{d_{11}\text{-}d_{30}\}| = 20$ and confidence $20/max(20, 20) = 1 = 100\%$
$\{Canada, USA\} : countries \rightleftharpoons \{corn\} : topics$ with strength $msg = |\{d_1\text{-}d_{10}\}| = 10$ and confidence $10/max(10, 10) = 1 = 100\%$

6.2 Maximal Association Rules

A *maximal association rule* is an expression of the form $V \rightarrow W$, where V and W are sentences for some different categories. The *strength* of the association rule is $msg(V \cup W)$, and confidence of the association is $msg(V \cup W)/msg(V)$.

Example 21. For Example 1, the following maximal association rules can be found:

$\{Canada, Iran, USA\} : countries \rightarrow \{crude, ship\} : topics$ with strength $msg = |\{d_1, d_2\}| = 2$ and confidence $2/2 = 1 = 100\%$
$\{Canada, Iran, USA\} : countries \leftarrow \{crude, ship\} : topics$ with strength $msg = |\{d_1, d_2\}| = 2$ and confidence $2/2 = 1 = 100\%$
$\{USA\} : countries \rightarrow \{earn\} : topics$ with strength $msg = |\{d_3\}| = 1$ and confidence $1/4 = 25\%$

$\{USA\}$: $countries \leftarrow \{earn\}$: $topics$ with strength $msg = |\{d_3\}| = 1$ and confidence $1/2 = 50\%$

$\{USA\}$: $countries \rightarrow \{jobs, cpi\}$: $topics$ with strength $msg = |\{d_4, d_5\}| = 2$ and confidence $2/4 = 50\%$

$\{USA\}$: $countries \leftarrow \{jobs, cpi\}$: $topics$ with strength $msg = |\{d_4, d_5\}| = 2$ and confidence $2/2 = 100\%$

$\{USA\}$: $countries \rightarrow \{earn, cpi\}$: $topics$ with strength $msg = |\{d_6\}| = 1$ and confidence $1/4 = 25\%$

$\{USA\}$: $countries \leftarrow \{earn, cpi\}$: $topics$ with strength $msg = |\{d_6\}| = 1$ and confidence $1/1 = 100\%$

$\{Canada\}$: $countries \rightarrow \{sugar, tea\}$: $topics$ with strength $msg = |\{d_7\}| = 1$ and confidence $1/1 = 100\%$

$\{Canada\}$: $countries \leftarrow \{sugar, tea\}$: $topics$ with strength $msg = |\{d_7\}| = 1$ and confidence $1/1 = 100\%$

$\{Canada, USA\}$: $countries \rightarrow \{trade, acq\}$: $topics$ with strength $msg = |\{d_8, d_9\}| = 2$ and confidence $2/3 = 66\%$

$\{Canada, USA\}$: $countries \leftarrow \{trade, acq\}$: $topics$ with strength $msg = |\{d_8, d_9\}| = 2$ and confidence $2/2 = 100\%$

$\{Canada, USA\}$: $countries \rightarrow \{earn\}$: $topics$ with strength $msg = |\{d_{10}\}| = 1$ and confidence $1/3 = 33\%$

$\{Canada, USA\}$: $countries \leftarrow \{earn\}$: $topics$ with strength $msg = |\{d_{10}\}| = 1$ and confidence $1/2 = 50\%$

Example 22. For Example 2, the following maximal association rules can be found:

$\{Canada, France, USA\}$: $countries \rightarrow \{fish\}$: $topics$ with strength $msg = |\{d_{11}\text{-}d_{30}\}| = 20$ and confidence $20/20 = 1 = 100\%$

$\{Canada, France, USA\}$: $countries \leftarrow \{fish\}$: $topics$ with strength $msg = |\{d_{11}\text{-}d_{30}\}| = 20$ and confidence $20/20 = 1 = 100\%$

$\{Canada, USA\}$: $countries \rightarrow \{corn\}$: $topics$ with strength $msg = |\{d_1\text{-}d_{10}\}| = 10$ and confidence $10/10 = 1 = 100\%$

$\{Canada, USA\}$: $countries \leftarrow \{corn\}$: $topics$ with strength $msg = |\{d_1\text{-}d_{10}\}| = 10$ and confidence $10/10 = 1 = 100\%$

6.3 Frequent Sets and Maximal Frequent Sets

We would like to discover associations where

(1) the strength is above some user-defined threshold called the *minimal strength* and denoted by α (a natural number), and

(2) the confidence is above another user-defined threshold called the *minimal confidence* and denoted by β (a decimal number).

Thus, an attribute set with strength $\geq \alpha$ is called a *frequent set*. An attribute set of taxonomy pairs with maximal strength $\geq \alpha$ is called a *maximal frequent set*.

7 Negated Attributes

Maximal frequent sets of attributes are useful for efficiently finding association rules that include negated attributes, called *excluding associations*.

Let S be a subset of the attribute set A, $S = \{a_1, a_2, \ldots, a_{|S|}\}$. We denote the set $\{\neg a_1, \neg a_2, \ldots, \neg a_{|S|}\}$ by $\neg S$.

An excluding association is a rule of the form $S_1 \vee \neg S_2 \rightarrow S_3 \vee \neg S_4$, where S_1, S_2, S_3, S_4 are sets of terms. The intuitive meaning of such an association is that whenever one sees the attributes of S_1 and *not* those of S_2 then one should also expect (with some confidence) to find the attributes of S_3 and *not* those of S_4. For example, $\{mining\} \rightarrow \{coal\}$ is a regular association. However, adding negation one may find the excluding association as follows: $\{mining, \neg coal\} \rightarrow \{data\}$.

In general, there can be numerous excluding associations, most of which are redundant and noninteresting. So the problem is how to generate the "interesting" excluding associations.

Anyway, we can use excluding associations to express maximal associations. For Example 2 (see Example 2.8), the following maximal association

$\{Canada, USA\} : countries \rightleftharpoons \{corn\} : topics$

with strength $msg = |\{d_1 - d_{10}\}| = 10$ and confidence $10/\max(10, 10) = 1 = 100\%$ can be expressed as an excluding association as follows:

$\{Canada, USA, \neg France\} \rightleftharpoons \{corn\}$
with strength $stg = |\{d_1 - d_{10}\}| = 10$ and confidence $10/\max(10, 10) = 1 = 100\%$

8 Rough Set Approach to Association Rule Mining

Using the rough set method [1, 7, 16], we can discover similar knowledge to maximal association rules in a much simpler way. For a given sentence, we use the theory to recognize its support/occurrence documents. Now, the information table can be designed for sentences as follows.

$D \backslash 2^{T_1}$	T_1	T_2	...	T_k	...	T_K
d_1	$T_1 \cap V^{d_1}$	$T_2 \cap V^{d_1}$	...	$T_k \cap V^{d_1}$	...	$T_K \cap V^{d_1}$
d_2	$T_1 \cap V^{d_2}$	$T_2 \cap V^{d_2}$	...	$T_k \cap V^{d_2}$	...	$T_K \cap V^{d_2}$
...		...	...	...	...	...
d_i	$T_1 \cap V^{d_i}$	$T_2 \cap V^{d_i}$	...	$T_k \cap V^{d_i}$	...	$T_K \cap V^{d_i}$
...		...	...	...	...	...
d_N	$T_1 \cap V^{d_N}$	$T_2 \cap V^{d_N}$	...	$T_k \cap V^{d_N}$	...	$T_K \cap V^{d_N}$

Example 23. The information table for Example 1 can be designed as follows:

The Information Table for Example 1

$D \backslash 2^{T_k}$	$T_1 = countries$	$T_2 = topics$
d_1	$\{Canada, Iran, USA\}$	$\{crude, ship\}$
d_2	$\{Canada, Iran, USA\}$	$\{crude, ship\}$
d_3	$\{USA\}$	$\{earn\}$
d_4	$\{USA\}$	$\{jobs, cpi\}$
d_5	$\{USA\}$	$\{jobs, cpi\}$
d_6	$\{USA\}$	$\{earn, cpi\}$
d_7	$\{Canada\}$	$\{sugar, tea\}$
d_8	$\{Canada, USA\}$	$\{trade, acq\}$
d_9	$\{Canada, USA\}$	$\{trade, acq\}$
d_{10}	$\{Canada, USA\}$	$\{earn\}$

Then it can be found that partitions
$D/countries$ is

$$\{Canada, Iran, USA\}D_{11}, \{USA\}D_{12}, \{Canada\}D_{13}, \{Canada,USA\}D_{14},$$

where $D_{11} = \{d_1, d_2\}$, $D_{12} = \{d_3 : d_6\}$, $D_{13} = \{d_7\}$, $D_{14} = \{d_8, d_9, d_{10}\}$;
$D/topics$ is

$$\{crude, ship\}D_{21}, \{earn\}D_{22}, \{jobs, cpi\}D_{23},$$

$$\{earn, cpi\}D_{24}, \{sugar, tea\}D_{25}, \{trade, acq\}D_{26},$$

where $D_{21} = \{d_1, d_2\}$, $D_{22} = \{d_3, d_{10}\}$, $D_{23} = \{d_4, d_5\}$, $D_{24} = \{d_6\}$, $D_{25} = \{d_7\}$, $D_{26} = \{d_8, d_9\}$; and

$D/countries \cap topics$ is
$(\{Canada, Iran, USA\} : countries, \{crude, ship\} : topics)\{d_1, d_2\}$
$(\{USA\} : countries, \{earn\} : topics)\{d_3\}$
$(\{USA\} : countries, \{jobs, cpi\} : topics)\{d_4, d_5\}$
$(\{USA\} : countries, \{earn, cpi\} : topics)\{d_6\}$
$(\{Canada\} : countries, \{sugar, tea\} : topics)\{d_7\}$
$(\{Canada, USA\} : countries, \{trade, acq\} : topics)\{d_8, d_9\}$
$(\{Canada, USA\} : countries, \{earn\} : topics)\{d_{10}\}$

That is, we find the following associations/co-occurrences:

$\{Canada, Iran, USA\} : countries \rightleftharpoons \{crude, ship\} : topics$ with strength $msg = |\{d_1, d_2\}| = 2$ and confidence $2/max(2, 2) = 1 = 100\%$
$\{USA\} : countries \rightleftharpoons \{earn\} : topics$ with strength $msg = |\{d_3\}| = 1$ and confidence $1/max(4, 2) = 25\%$
$\{USA\} : countries \rightleftharpoons \{jobs, cpi\} : topics$ with strength $msg = |\{d_4, d_5\}| = 2$ and confidence $2/max(4, 2) = 50\%$
$\{USA\} : countries \rightleftharpoons \{earn, cpi\} : topics$ with strength $msg = |\{d_6\}| = 1$ and confidence $1/max(4, 1) = 25\%$

$\{Canada\}$: $countries \rightleftharpoons \{sugar, tea\}$: $topics$ with support $msg =$ $|\{d_7\}| = 1$ and confidence $1/max(1,1) = 100\%$
$\{Canada, USA\}$: $countries \rightleftharpoons \{trade, acq\}$: $topics$ with strength $msg =$ $|\{d_8, d_9\}| = 2$ and confidence $2/max(3,2) = 66\%$
$\{Canada, USA\}$: $countries \rightleftharpoons \{earn\}$: $topics$ with strength $msg =$ $|\{d_{10}\}| = 1$ and confidence $1/max(3,2) = 33\%$.

Example 24. The information table for Example 2 can be designed as follows:

The Information Table for Example 2

$D \backslash 2^{T_k}$	$T_1 = countries$	$T_2 = topics$
$d_1 : d_{10}$	$\{Canada, USA\}$	$\{corn\}$
$d_{11} : d_{30}$	$\{Canada, France, USA\}$	$\{fish\}$

Then it can be found that partitions

$D/countries$ is $\{Canada, USA\}D_{11}, \{Canada, France, USA\}D_{12}$, where $D_{11} = \{d_1 : d_{10}\}, D_{12} = \{d_{11} : d_{30}\}$;
$D/topics$ is $\{corn\}D_{21}, \{fish\}D_{22}$,

where

$D_{21} = \{d_1 : d_{10}\}, D_{22} = \{d_{11} : d_{30}\}$; and
$D/countries \cap topics$ is
$(\{Canada, USA\} : countries, \{corn\} : topics)\{d_1 : d_{10}\}$
$(\{Canada, France, USA\} : countries, \{fish\} : topics)\{d_{11} : d_{30}\}$

That is, we find the following associations/co-occurrences:

$\{Canada, France, USA\}$: $countries \rightleftharpoons \{fish\}$: $topics$ with strength $msg = |\{d_{11}\text{-}d_{30}\}| = 20$ and confidence $20/\max(20,20) = 1 = 100\%$
$\{Canada, USA\}$: $countries \rightleftharpoons \{corn\}$: $topics$ with strength $msg =$ $|\{d_1\text{-}d_{10}\}| = 10$ and confidence $10/\max(10,10) = 1 = 100\%$

These associations are the same as the maximal associations discovered in Examples 7 and 8. However, these examples show that the rough set approach to discovering knowledge is much simpler than the maximal association method.

9 Conclusion

Association rules are based on the notion of *frequent sets of attributes* which appear in many documents. Maximal association rules are based on *frequent maximal sets of attributes* which appear maximally in many documents. The regular association rule $X \rightarrow Y$ means that if X then Y (with some *confidence*). References [3, 4] argue that the use of confidence in the standard way is not adequate for strong connections and therefore they introduce maximal associations.

In this paper we work through an example of Feldman's approach and show its usefulness. However we also show that by using rough set theory, rules that are similar to maximal associations can be found. We show, by working through the same example as before, that the rough set approach to discovering knowledge is much simpler than the maximal association method.

References

1. Bell, D.A.; Guan, J.W. 1998, "Computational methods for rough classification and discovery", *Journal of the American Society for Information Science, Special Topic Issue on Data Mining*, Vol. 49(1998), No. 5, 403–414.
2. Bell, D.A.; Guan, J.W. 2003, "Data mining for motifs in DNA sequences", in G. Wang et al. (ed.) *Proceedings of the 9th International Conference on Rough Sets, Fuzzy Sets, Data Mining and Granular Computing (RSFDGrC'2003)*, Chongqing, China, October 19–22, 2003.
3. Feldman, R.; Aumann, Y.; Amir, A.; Zilberstain, A.; Kloesgen, W. Ben-Yehuda, Y. 1997, Maximal association rules: a new tool for mining for keyword co-occurrences in document collection, in *Proceedings of the 3rd International Conference on Knowledge Discovery (KDD 1997)*, 167–170.
4. Feldman, R.; Aumann, Y.; Zilberstain, A.; Ben-Yehuda, Y. 1998, Trend graphs: visualizing the evolution of concept relationships in large document collection, in *Proceedings of the 2nd European Symposium on Knowledge Discovery in Databases, PKDD'98, Nantes, France, 23–26 September 1998; Lecture Notes in Artificial Intelligence 1510: Principles of Data Mining and Knowledge Discovery*, Jan M. Zytkow Mohamed Quafafou eds.; Springer, 38–46.
5. Feldman, R.; Fresko, M.; Kinar, Y.; Lindell, Y.; Liphstat, O.; Rajman, M.; Schler, Y.; Zamir, O. 1998, Text mining at the term level, in *Proceedings of the 2nd European Symposium on Knowledge Discovery in Databases, PKDD'98, Nantes, France, 23–26 September 1998; Lecture Notes in Artificial Intelligence 1510: Principles of Data Mining and Knowledge Discovery*, Jan M. Zytkow Mohamed Quafafou eds.; Springer, 65–73.
6. Frawley, W.J.; Piatetsky-Shapiro, G.; Matheus, C.J. 1991, Knowledge discovery in databases: an overview. In G. Piatetsky-Shapiro, W.J. Frawley (eds). *Knowledge Discovery in Databases* (pp. 1–27). AAAI/MIT Press.
7. Guan, J.W.; Bell, D.A. 1998, "Rough computational methods for information systems", *Artificial Intelligence – An International Journal*, Vol.105(1998), 77–104.
8. Guan, J.W.; Bell, D.A.; Liu, D.Y. 2003, "Rough classification and knowledge discovery", *Proceedings of The 2003 International Conference on Artificial Intelligence (IC-AI'03)*, Monte Carlo Resort & Casino, 3770 Las Vegas Blvd., South, Las Vegas, Nevada, USA, June 23–26, 2003.
9. Guan, J.W.; Bell, D.A.; Liu, D.Y. 2003a, "Equivalence relations and rough set theory", *Proceedings of JCIS 2003 (7th Joint Conference on Information Sciences)*, Cary, North Carolina, USA, September 26–30, 2003, Vol. 1, 151–154.

10. Guan, J.W.; Bell, D.A.; Liu, D.Y. 2003b, "Rough set approach to association rule mining", *Proceedings of The Third IEEE International Conference on Data Mining (ICDM'03)*, Melbourne, Florida, USA, November 19–22, 2003, 151–154.

11. Guan, J.W.; Bell, D.A.; Liu, D.Y. 2004, "Discovering motifs in DNA sequences", *Fundamenta Informaticae*, IOS Press, NO. XXIII, 1001–1014.

12. Guan, J.W.; Bell, D.A.; Liu, D.Y. 2004a, "Discovering maximal frequent patterns in sequence groups", *Proceedings of the Fourth International Conference on Rough Sets and Current Trends in Computing (RSCTC'2004)*, Uppsala, Sweden, June 1–5, 2004, 276–283.

13. Guan, J.W.; Bell, D.A.; Liu, D.Y. 2004b, "Discovering frequent patterns in sequence groups", in D. Ruan ed. *Applied Computational Intelligence*, World Scientific, Singapore, 203–208.

14. Kiem, H.; Phuc, D. 2000, "Discovering motif based association rules in a set of DNA sequences", in W. Ziarko & Y. Yao (ed.) *Proceedings of the Second International Conference on Rough Sets and Current Trends in Computing (RSCTC'2000)*, Banff, Canada, October 16–19, 2000; 348–352. ISBN 0828–3494, ISBN 0-7731-0413-5.

15. Landau, D.; Feldman, R.; Aumann, Y.; Fresko, M.; Lindell, Y.; Liphstat, O.; Zamir, O. 1998, TextVis: an integrated visual environment for text mining, in *Proceedings of the 2nd European Symposium on Knowledge Discovery in Databases, PKDD'98, Nantes, France, 23–26 September 1998; Lecture Notes in Artificial Intelligence 1510: Principles of Data Mining and Knowledge Discovery*, Jan M. Zytkow Mohamed Quafafou eds.; Springer, 56–64.

16. Pawlak, Z. 1991, *Rough sets: theoretical aspects of reasoning about data*. Kluwer.

17. Srikant, R.; Agrawal, R. 1995–1996, Mining sequential patterns: generalizations and performance improvements, in *Proceedings of the Fifth International Conference on Extending Database Technology (EDBT), Avignon, France, March 1996; IBM Research Report RJ 9994, December 1995 (expanded version)*.

The Evolution of the Concept of Fuzzy Measure

Luis Garmendia

Facultad de Informática, Dep. Sistemas Informáticos y Programación, Universidad Complutense of. Madrid, Spain
lgarmend@fdi.ucm.es
fdi.ucm.es/profesor/lgarmend

Abstract. Most information discovery processes need to understand the reasons of the success of the inference methods or the usability of the new information, which can sometimes be somehow explained with a few useful measures on the premises or on the chosen relations, logics and implications. This chapter presents a state of art on the latest concepts of measure, from the additive measures, to monotone fuzzy measures and the latest monotone measures in relation to a preorder that gives an ordering for a measurable characteristic. A few measures on fuzzy sets and fuzzy relations are proposed. Namely, the entropy measures on fuzzy sets, which are a measure of fuzziness in itself, and the specificity measures on fuzzy sets, which cam be understood as measures of utility of the information contained in a fuzzy set. Some measures on fuzzy relations to be considered are the conditionality measures, which are degrees of generalization of modus ponens when making fuzzy inference. Those measures on fuzzy sets and fuzzy relations provide interesting information on fuzzy inference and approximate reasoning processes.

Key words: fuzzy measure, approximate reasoning, conditionality, specificity

1 Introduction

The discovery of useful information is the essence of any data mining process. Decisions are not usually taken based on complete real world data, but most of the times they deal with uncertainty or lack of information. Therefore the real world reasoning is almost always approximate. However it is not only necessary to learn new information in any data mining process, but it is also important to understand why and how the information is discovered. Most data mining commercial products are black boxes that do not explain the reasons and methods that have been used to get new information. However the "why and how" the information is obtained can be as important as the information on its own. When approximate reasoning is done, measures on fuzzy sets and fuzzy relations can be proposed to provide a lot of information that helps to

Luis Garmendia: *The Evolution of the Concept of Fuzzy Measure*, Studies in Computational Intelligence (SCI) **5**, 185–184 (2005)
www.springerlink.com

understand the conclusions of fuzzy inference processes. Those measures can even help to make decisions that allow to use the most proper methods, logics, operators for connectives and implications, in every approximate reasoning environment.

The latest concepts of measures in approximate reasoning is discussed and a few measures on fuzzy sets and fuzzy relations are proposed to be used to understand why the reasoning is working and to make decisions about labels, connectives or implications, and so a few useful measures can help to have the best performance in approximate reasoning and decision making processes.

Before some measures on fuzzy sets and fuzzy relations are proposed, this chapter collects all the latest new concepts and definitions on measures, and shows a few graphics that make a clear picture on how those measures can be classified.

Some important measures on fuzzy sets are the entropy measures and specificity measures. The entropy measures give a degree of fuzziness of a fuzzy set, which can be computed by the premises or outputs of an inference to know an amount of uncertainty crispness in the process. Specificity measures of fuzzy sets give a degree of the utility of information contained in a fuzzy set.

Other important measures can be computed on fuzzy relations. For example, some methods to measure a degree of generalisation of the MODUS PONENS property in fuzzy inference processes are proposed.

2 The Concept of Measure

The concept of measure is one of the most important concepts in mathematics, as well as the concept of integral respect to a given measure. The classical measures are supposed to hold the additive property. Additivity can be very effective and convenient in some applications, but can also be somewhat inadequate in many reasoning environments of the real world as in approximate reasoning, fuzzy logic, artificial intelligence, game theory, decision making, psychology, economy, data mining, etc., that require the definition of non additive measures and a large amount of open problems. For example, the efficiency of a set of workers is being measured, the efficiency of the same people doing teamwork is not the addition of the efficiency of each individual working on their own.

The concept of fuzzy measure does not require additivity, but it requires monotonicity related to the inclusion of sets. The concept of fuzzy measure can also be generalised by new concepts of measure that pretend to measure a characteristic not really related with the inclusion of sets. However those new measures can show that "x has a higher degree of a particular quality than y" when x and y are ordered by a preorder (not necessarily the set inclusion preorder).

The term fuzzy integral uses the concept of fuzzy measure. There are some important fuzzy integrals, as *Choquet* integral in 1974, which does not require an additive measure (as Lebesgue integral does). Michio Sugeno gives other new integral in 1974 for fuzzy sets, and so does *David Schmeidler* in 1982 for decision theory.

2.1 Preliminaries

A measurable space is a couple $(X, \wp)$ where X is a set and $\wp$ is a σ-**algebra** or set of subsets of X such that:

1. $X \in \wp$.
2. Let A be a subset of X. If $A \in \wp$ then $A' \in \wp$.
3. If $A_n \in \wp$ then $\bigcup_{n=1}^{\infty} A_n \in \wp$.

For example, when X is the set of real numbers and $\wp$ is the σ-algebra that contains the open subsets of X, then $\wp$ is the well-known Borel σ-algebra.

Note

The classical concept of measure considers that $\wp \subseteq \{0, 1\}^X$, but this consideration can be extended to a set of fuzzy subsets $\Im$ of X, $\Im \subseteq [0, 1]^X$, satisfying the properties of measurable space $([0, 1]^X, \Im)$.

2.2 Definition of Additive Measure

Let $(X, \wp)$ be a measurable space. A function $m: \wp \to [0, \infty)$ is an σ-additive measure when the following properties are satisfied:

1. $m(\varnothing) = 0$.
2. If $A_n, n = 1, 2, \ldots$ is a set of disjoint subsets of $\wp$ then
3. $m(\bigcup_{n=1}^{\infty} A_n) = \sum_{n=1}^{\infty} m(A_n)$

The second property is called σ-**additivity**, and the **additive** property of a measurable space requires the σ-additivity in a finite set of subsets A_n.

A well-known example of σ-**additive** is the probabilistic space $(X, \wp, p)$ where the probability p is an additive measure such that $p(X) = 1$ and $p(A) = 1 - p(A')$ for all subsets $A \in \wp$.

Other known examples of σ-additive measure are the Lebesgue measures defined in 1900 that are an important base of the XX century mathematics. The Lebesgue measures generalise the concept of length of a segment, and verify that if $[c, d] \subset \bigcup_{i=1}^{n} [a_i, b_i)$ then $d - c \leq \sum_{i=1}^{n} (b_i - a_i)$. Other measures given by Lebesgue are the exterior Lebesgue measures and interior Lebesgue measures. A set A is Lebesgue measurable when both interior and exterior Lebesgue measures are the same. Some examples of Lebesgue measurable sets are the compact sets, the empty set and the real numbers set $\Re$.

2.3 Definition of Normal Measure

Let $(X, \wp)$ be a measurable space. A measure $m: \wp \to [0, 1]$ is a normal measure if there exists a minimal set A_0 and a maximal set A_m in $\wp$ such that:

1. $m(A_0) = 0$.
2. $m(A_m) = 1$.

For example, the measures of probability on a space $(X, \wp)$ are normal measures with $A_0 = \emptyset$ and $A_m = X$. The Lebesgue measures are not necessarily normal.

2.4 Definition of Sugeno Fuzzy Measure [17]

Let $\wp$ be an σ-algebra on a universe X. A **Sugeno fuzzy measure** is $g: \wp \to [0, 1]$ verifying:

1. $g(\emptyset) = 0, g(X) = 1$.
2. If $A, B \in \wp$ and $A \subseteq B$ then $g(A) \le g(B)$.
3. If $A_n \in \wp$ and $A_1 \subseteq A_2 \subseteq \ldots$ then $\lim_{n \to \infty} g(A_n) = g(\lim_{n \to \infty} A_n)$

Property 2 is called monotony and property 3 is called Sugeno's convergence.

The Sugeno measures are monotone but its main characteristic is that additivity is not needed.

Banon [1981] shows that several measures on finite algebras, as probability, credibility measures and plausibility measures are Sugeno measures. The possibility measures on possibility distributions introduced by Zadeh [1978] gives Sugeno measures.

2.5 Theory of Evidence

The theory of evidence is based on two dual non-additive measures: belief measures and plausibility measures.

Given a measurable space $(X, \wp)$, a belief measure is a function Bel: $\wp \to [0, 1]$ verifying the following properties:

1. $\text{Bel}(\emptyset) = 0$.
2. $\text{Bel}(X) = 1$.
3. $\text{Bel}(A \cup B) \ge \text{Bel}(A) + \text{Bel}(B)$.

Property 3 is called **superadditivity**. When X is infinite, the superior continuity of the function Bel is required. For every $A \in \wp$, $\text{Bel}(A)$ is interpreted as a belief degree for some element to be in the set A.

From the definition of belief measure, it can be proved that $\text{Bel}(A) + \text{Bel}(A') \le 1$.

Given a belief measure, its dual plausibility measure can be defined as $\text{Pl}(A) = 1 - \text{Cred}(A')$.

Given a measurable space $(X, \wp)$ a measure of plausibility is a function Pl: $\wp \to [0, 1]$ such that

1. $\text{Pl}(\emptyset) = 0$.
2. $\text{Pl}(X) = 1$.
3. $\text{Pl}(A \cup B) \leq \text{Pl}(A) + \text{Pl}(B)$.

Property 3 is called **subadditivity**.

When X is infinite, the inferior continuity of the function Pl is required.

It can be proved that $\text{Pl}(A) + \text{Pl}(A') \geq 1$.

The measures of credibility and plausibility are defined by a function m: $\wp \to [0, 1]$ such that $m(\emptyset) = 0$ and $\sum_{A \in \wp} m(A) = 1$ where m represents a proportion of the shown evidence that an element of X is in a subset A.

2.6 Theory of Possibility

The theory of possibility is a branch of theory of evidence where the plausibility measures verify that $\text{Pl}(A \cup B) = \max\{\text{Pl}(A), \text{Pl}(B)\}$. Such plausibility measures are called **possibility measures**. In the theory of possibility, the belief measures satisfy that $\text{Bel}(A \cap B) = \min\{\text{Bel}(A), \text{Bel}(B)\}$ and are called **necessity measures**.

Definition 1 (Zadeh; 1978, Higashi & Klir; 1983).

Let $(X, \wp)$ be a measurable space. A possibility measure is a function Π: $\wp \to [0, 1]$ that verifies the following properties:

1. $\Pi(\emptyset) = 0, \Pi(X) = 1$.
2. $A \subseteq B \Rightarrow \Pi(A) \leq \Pi(B)$
3. $\Pi(\bigcup_{i \in I} A_i) = \sup_{i \in I}\{\Pi(A_i)\}$ for a set of indexes I.

The possibility measures are sub additive normal measures.

Definition 2 (Zadeh; 1978, Higashi & Klir; 1983).

Let $(X, \wp)$ be a measurable space. A necessity measure is a function Nec: $\wp \to [0, 1]$ that verifies the following properties:

1. $\text{Nec}(\emptyset) = 0, \text{Nec}(X) = 1$.
2. $A \subseteq B \Rightarrow \text{Nec}(A) \leq \text{Nec}(B)$
3. $\text{Nec}(\bigcap_{i \in I} A_i) = \inf_{i \in I}\{\text{Nec}(A_i)\}$ for any set I.

Possibility measures are plausibility measures and necessity measures are belief measures, so:

1. $\Pi(A) + \Pi(A') \geq 1$.
2. $\text{Nec}(A) + \text{Nec}(A') \leq 1$.
3. $\text{Nec}(A) = 1 - \Pi(A')$.

4. $\max\{\Pi(A), \Pi(A')\} = 1$.
5. $\min\{\text{Nec}(A), \text{Nec}(A')\} = 0$.
6. $\text{Nec}(A) > 0 \Rightarrow \Pi(A) = 1$.
7. $\Pi(A) < 1 \Rightarrow \text{Nec}(A) = 0$.

The [27] theory of evidence stands that the probability of an element or a set is related to its complementary one. It includes concepts of "low probability" and "high probability", that are related to the measures of possibility and necessity in the sense that for any subset A, $\text{Nec}(A) \le P(A) \le \Pi(A)$.

The theory of possibility also stands on fuzzy sets, where $\wp$ is a family of fuzzy subsets in X.

A measure of possibility is not always a Sugeno fuzzy measure [22]. However a normal possibility distribution on a finite universe X is a Sugeno measure.

2.7 Definition of Fuzzy Measure [Nguyen & Walker, 1996, 183]

Let $(X, \wp)$ be a measurable space. A function $m\colon \wp \to [0, \infty)$ is a fuzzy measure (or monotone measure) if it verifies the following properties:

1. $m(\emptyset) = 0$.
2. If $A, B \in \wp$ and $A \subseteq B$ then $m(A) \le m(B)$.

Property 2 is called monotony.

For example, all σ-additive measures (as probability) are fuzzy measures. Some other fuzzy measures are the necessity measures, the possibility measures and the Sugeno measures.

The Fig. 1 represents a typology of the most important monotone and normal measures.

2.8 Definition of Fuzzy Sugeno λ-Measure

Sugeno [1974] introduces the concept of fuzzy λ-measure as a normal measure that is λ-additive. So the fuzzy λ-measures are fuzzy (monotone) measures.

Let $\lambda \in (-1, \infty)$ and let $(X, \wp)$ be a measurable space. A function $g_\lambda\colon \wp \to [0, 1]$ is a fuzzy λ-measure if for all disjoint subsets A, B in $\wp$, $g_\lambda(A \cup B) = g_\lambda(A) + g_\lambda(B) + \lambda g_\lambda(A)g_\lambda(B)$.

For example, if $\lambda = 0$ then the fuzzy λ-measure is an additive measure.

2.9 S-Decomposable Measures

Weber [1984] defined the S-decomposable measures providing a general concept of the fuzzy λ-measures and the possibility measures.

Let S be a t-conorm, and let $(X, \wp)$ be a measurable space. A S-decomposable measure is a function $m\colon \wp \to [0, 1]$ that verifies the following conditions.

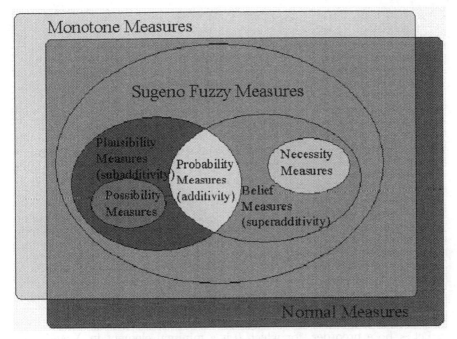

Fig. 1. Monotone and normal measures

1. $m(\emptyset) = 0$.
2. $m(X) = 1$.
3. For all disjoint subsets A and B in $\wp$, $m(A \cup B) = S(m(A), m(B))$.

The property 3 is called **S-additivity.**

For example, the probability measures are W*-decomposable measures, where $W*$ is the Łukasiewicz t-conorm. The W_λ^*-decomposable measures, where W_λ^* is the t-conorm $W_\lambda^*(x,y) = x + y + \lambda xy$ are fuzzy λ-measures.

Let m be a S-decomposable measure on $(X, \wp)$. If X is finite then given a subset A in $\wp$, $m(A) = S_{x \in A}\{m(\{x\})\}$.

The Fig. 2 shows the relation between Sugeno fuzzy measures and S-decomposable measures.

2.10 Fuzzy ≺-Measure (Fuzzy Preorder-Monotone Measure)

Trillas and Alsina [Trillas & Alsina; 1999] give a general definition of fuzzy measure. When a characteristic, namely – volume, weight, etc. – needs to be measured on the elements of a set X, a preorder relation that allows to stand that "x shows the characteristic less than y shows it" for all x and y in X is necessary to be set. That reflexive and transitive relation is denoted $x \prec y$.

A fuzzy ≺-measure is defined as follows:

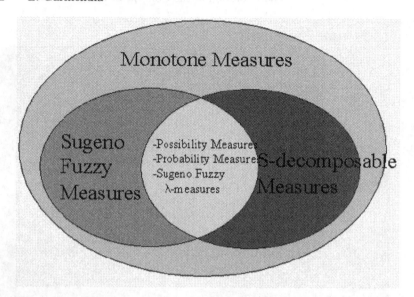

Fig. 2. Relation between S-decomposable measures and Sugeno measures

Let $\prec$ be a preorder, for which 0 is a minimal element in X and **1** is a maximal element in X. Then a fuzzy $\prec$-measure is a function $m: \wp(X) \to [0,1]$ that verifies the following conditions:

1. $m(0) = 0$
2. $m(1) = 1$
3. If $x \prec y$ then $m(x) \le m(y)$.

A good example of fuzzy $\prec$-measure on the set of natural numbers N is the Sarkovskii measure, which is defined as a measure of approximately even numbers, given by the following function m:

$$m(n) = \begin{cases} 1 & \text{if } n = 2^k \text{ for } k = 0,1,2,\dots \\ 0 & \text{if } n = 2^k + 1 \text{ for } k = 1,2,\dots \\ 1 - \frac{1}{2^k} & \text{if } n = 2^k(2p+1) \text{ for } k = 1,2,\dots p = 1,2,\dots \end{cases}$$

Then m is a fuzzy $\prec$-measures, not for the normal natural numbers order, but for the *Sarkovskii* order, for which the lowest number is 3, and the greatest number is 1. It is a well-known order used in dynamic systems and given defined as follows:

$$3 \prec 5 \prec 7 \prec \cdots \prec 2.3 \prec 2.5 \prec \cdots \prec 2^2.3 \prec 2^2.5 \prec \cdots \prec 2^3.3 \prec 2^3.5$$
$$\prec \cdots 2^3.3 \prec 2^3 \prec 2^2 \prec 2 \prec 1$$

Other fuzzy $\prec$-measure are all previous defined fuzzy measures, which are monotone measures with respect to the set inclusion preorder, that is now generalised in both classic set inclusion and fuzzy set inclusion cases.

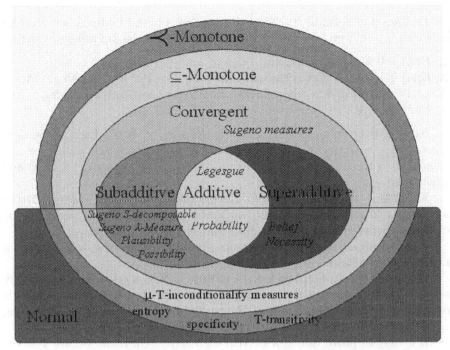

Fig. 3. Relation between different measures and a classification of some examples

The Sugeno [Sugeno; 1974] fuzzy measure concept is also generalised: if $\wp$ is a partial order lattice, then $x \prec y$ if and only if $x \wedge y = x$, and the three Sugeno properties are satisfied. If the lattice is ortocomplemented then there exists a dual function $m^*(x) = 1 - m(x')$ that at the same time is a fuzzy $\prec$-measure.

Then, the probability measure on a Boole algebra of probabilistic successes is also a fuzzy $\prec$-measure.

Let $\wp$ be the set of fuzzy subsets on a given set, the entropy measure introduced by De Luca and Termini [1972], and the possibility or necessity measures [Higashi & Klir; 1983] are also a fuzzy $\prec$-measures.

The Fig. 3 shows graphically the relation between different measures and a classification of many of the given examples.

3 Some Measures On Fuzzy Sets and Fuzzy Relations

3.1 Entropy or Measures of Fuzziness

Let X be a set and let $\wp\,(X)$ be the set of fuzzy sets on X. The measures of fuzziness or **entropies** give a degree of fuzziness for every fuzzy set in $\wp$.

Some entropy measures have influence from the *Shannon* probabilistic entropy, which is commonly used as measures of information.

De Luca and Termini [1972] consider the entropy E of a fuzzy set of $\wp(X)$ as a measure that gives a value in $[0, \infty]$ and satisfies the following properties:

1. $E(A) = 0$ if A is a crisp set.
2. $E(A)$ is maximal if A is the constant fuzzy set $A(x) = \frac{1}{2}$ for all $x \in X$.
3. $E(A) \geq E(B)$ if A is "more fuzzy" than B by the "sharpen" order.
4. $E(A) = E(A')$.

Note that the defined entropy measure of a fuzzy set is a fuzzy $\prec$-measure where the $\prec$ preorder is the $\leq_S$ sharpen order, in which $B \leq_S A$ if for any element x in the universe of discourse when $A(x) \leq \frac{1}{2}$ then $B(x) \leq A(x)$ and when $A(x) \geq \frac{1}{2}$ then $B(x) \geq A(x)$

Kaufmann [1975] proposes a fuzziness index as a normal distance. Other authors as Yager [1979] and [14] understand the entropy measures as the difference between a fuzzy set and its complementary fuzzy set.

A new proposal for entropy measure given by a t-norm, a t-conorm and a negation is as follows:

Let H: $\wp \to [0, 1]$, $\mathbf{H}(\mu) = \mathbf{k}\, \mathbf{S}_{x \in U}\{\mathbf{T}(\mu(\mathbf{x}), \mathbf{N}(\mu(\mathbf{x})))\}$, where μ is a fuzzy set on U, k is a constant that is used to normalise the measure and depends on the chosen continuous t-conorm S, the continuous t-norm T and the strong negation N.

It is easy to prove that if μ is crisp then $H(\mu) = 0$. If the t-norm T is the minimum or is in the family of the product t-norm, then $H(\mu) = 0$ if and only if μ is a crisp set. It is proved that $H(\mu) = H(\mu^c)$ by the symmetry property of the t-norms. When the t-norm T is the minimum or in the family of the product, then if $B \leq_S A$ then $H(B) \leq H(A)$.

So the new proposed entropy verifies the properties given by De Luca and Termini [1972] when T is the minimum or in the family of the product t-norm.

If the fuzzy set is on an infinite domain, the same result can be extended to the new expression

$$H(\mu) = k \int_{x \in E} \{T(\mu(x), N(\mu(x)))\}.dx \ .$$

3.2 Measures of Specificity

The specificity measures introduced by [53] are useful as measures of tranquillity when making a decision. Yager introduces the specificity-correctness tradeoff principle. The output information of expert systems and other knowledge-based systems should be both specific and correct to be useful. Yager suggests the use of specificity in default reasoning, in possibility-qualified statements and data mining processes, giving several possible manifestations of this measure. Kacprzyk [1990] describes its use in a system for inductive learning. Dubois and Prade [1999] introduce the minimal specificity principle and show the role of specificity in the theory of approximate reasoning. Reference [14]

introduce a closely related idea called non-specificity. The concept of granularity introduced by Zadeh [1978] is correlated with the concept of specificity.

Let X be a set with elements $\{x_i\}$ and let $[0, 1]^X$ be the class of fuzzy sets
of X. A measure of specificity Sp is a function Sp: $[0, 1]^X \rightarrow [0, 1]$ such that:

1. $Sp(\mu) = 1$ if and only if μ is a singleton ($\mu = \{x_1\}$).
2. $Sp(\varnothing) = 0$
3. If μ and η are normal fuzzy sets in X and $\mu \subset \eta$, then $Sp(\mu) \geq Sp(\eta)$.

A general expression [8] that can be used to build measures of specificity from
three t-norms and negations is an application Sp_T: $[0, 1]^X \rightarrow [0, 1]$ defined by
$\mathbf{Sp_T(\mu) = T_1(a_1, N(T_2^*}_{j=2,...,n}\{T_3(a_j, w_j)\}))$ where μ is a fuzzy set in a
finite set X, and a_i is the membership degree of the element x_i ($\mu(x_i) = a_i$),
the membership degrees $a_i \in [0, 1]$ are totally ordered with $a_1 \geq a_2 \geq \cdots \geq$
a_n, N is a negation, let T_1 and T_3 be any t-norms, T_2* a n-argument t-conorm
and $\{w_j\}$ is a weighting vector.

For example, when N is the negation $N(x) = 1 - x$, T_1 and T_2 are the
Tukasiewicz t-norm defined by $T_1(a, b) = \max\{0, a + b - 1\}$, so
$T_2 * (a_1, \ldots, a_n) = \min\{1, a_1 + \cdots + a_n\}$, and T_3 is the product, then the
previous expression gives [53] linear measure of specificity, defined as

$$Sp(\mu) = a_1 - \sum_{j=2}^{n} w_j \, a_j \, .$$

The measures of specificity are not monotone measures, because the measure
of specificity of a fuzzy set is lower when some membership degrees that are
not the highest degree are increased. However the measures of specificity of
fuzzy sets are fuzzy $\prec$-measures, where $\prec$ is a preorder that classifies the fuzzy
sets by the utility of the contained information, or by a given distance to a
singleton.

Let A be a fuzzy set on an infinite universe X and let A_α be its α-cut.

Reference [58] defines a **measure of specificity on a continuous do-
main** Sp as $Sp(A) = \int_0^{\alpha_{\max}} F(M(A_\alpha)) \, d\alpha$, where $\alpha_{\max}$ is the maximum
membership degree of A and F is a function F: $[0, 1] \rightarrow [0, 1]$ verifying that
$F(0) = 1, F(1) = 0$, and $F(x) \leq F(y) \leq 0$ when $x > y$.

For example, the expression $\int_0^{\alpha_{\max}} \text{Length}(A_{\alpha i}) d\alpha$ can be interpreted as
the area under the fuzzy set A. So, the measure of specificity under a fuzzy
set under an interval $[a, b]$ can be interpreted as

$$\alpha_{\max} - \frac{\text{area under } A}{b - a} \, .$$

The measures of specificity of fuzzy sets on infinite universes can be given by
an expression

$$MS(A) = T_1\left(\alpha_{\max}, N\left(\int_0^{\alpha_{\max}} T_2(M(A_\alpha), d\alpha)\right)\right)$$

where $\int_0^{\alpha_{max}}$ is a Choquet integral, T_1 and T_2 are t-norms and N is a negation.

Reference [54] introduced the concept of specificity of a fuzzy set under similarities using the [61] concept of similarity or Min-indistinguishability.

The α-cut of a similarity S is a classical equivalence relation denoted S_α. Let π_α be the set of equivalence classes of S for a given value α. Let μ_α/S be the set of equivalence classes of π_α defined in the following way: class $\pi_\alpha(i)$ belongs to μ_α/S if there exists an element x contained in $\pi_\alpha(i)$ and in the μ's α-cut (μ_α).

Reference [54] definition of measure of specificity of a fuzzy set μ under a similarity is the following:

$$S_p(\mu/S) = \int\limits_0^{\alpha_{max}} \frac{1}{Card(\mu_\alpha/S)} d\alpha \ .$$

3.3 Measures of μ-T-Unconditionality

The μ-T-conditionality property of fuzzy relations generalises the modus ponens property when making fuzzy inference. A fuzzy relation R: $E_1 \times E_2 \to [0,1]$ is μ-T-conditional if and only if $T(\mu(a), R(a,b)) \leq \mu(b)$ for all (a,b) in $E_1 \times E_2$.

Some ways to measure a degree of verification of this property are discussed, which are monotonous measures on the measurable space $(\Re, \Im, M)$, where $\Re$ is the set of fuzzy relations R: $E_1 \times E_2 \to [0,1]$, $\Im$ the set of measurable subsets of $\Re$ and M is a measure of μ-T-unconditionality. There are two ways to define those measures [9]. A first way computes a generalised distance between a fuzzy relation R and the greatest μ-T-conditional relation that is contained in R. The other way measures the difference between $T(\mu(a), R(a,b))$ and $\mu(b)$ in all points (a,b) in which R is not μ-T-conditional.

Let $J^T(x,y) = \text{Sup}\{z : T(x,z) \leq y\}$ be the residual operator of a t-norm T, let $J_\mu^T(a,b) = J^T(\mu(a), \mu(b))$, and let $T_R^\mu(a,b) = T(\mu(a), R(a,b))$. In [7] it is proved that $J^T(R(a,b), J_\mu^T(a,b)) = J^T(T_R^\mu(a,b), \mu_2(a,b))$ for all continuous t-norms, which shows that when a generalised distance defined from a residuated operator of the T-norm is used, both methods set the same measures of μ-T-unconditionality of fuzzy relations.

So, given any continuous t-norm T, for all (a,b) in $E_1 \times E_2$, the distance 1-J^T between a fuzzy relation R in the point (a,b) and its μ-T-conditionalized relation $J_\mu^T(a,b)$, is the same than the distance 1-J^T between $T_R^\mu(a,b)$ and $\mu(b)$.

Some operators are frequently used to make fuzzy inference. The fuzzy operators are fuzzy relations on the universe $E_1 \times E_2 = [0,1] \times [0,1]$.

The following examples show the evaluation of the measures for some implication operators. For all of them, the fuzzy set μ is taken as the identity (that is, as a function $\mu:[0,1] \to [0,1]$ such that $\mu(x) = x$).

The chart bellow gives the measures of the most used residual implication operators, S-implications, QM-implications and conjunctions for the t-norms minimum, product and Tkasiewicz:

Table 1. Measures of Id-T-unconditionality of some operators

Operator	$T = $ Min	$T = $ Prod	$T = W$
J^{Min}	0	0	0
J^{Prod}	$\frac{1}{3}$	0	0
$\text{Max}(1-x,y)$	$\frac{5}{24}$	$\frac{1}{30}$	0
$1-x+xy$	$\frac{1}{3}$	$\frac{3}{2}-2\ln 2$	0
$\text{Min}(x,y)$	0	0	0
$\text{Prod}(x,y)=xy$	$\frac{1}{3}$	0	0
$1(x,y)=1$	$\frac{1}{3}$	$\frac{1}{4}$	$\frac{1}{6}$

The measures of μ-T-unconditionality of fuzzy relations are monotone measures on the measurable space $(\Re, \Im, M)$ where $\Re$ is the set of fuzzy relations R: $E_1 \times E_2 \to [0,1]$, $\Im$ is the set of measurable subsets of $\Re$ and M is a measure of μ-T-unconditionality.

4 Conclusions

This paper presents an overview of the evolution of the concept of fuzzy measure. Some of these are relevant to understand the process of inference, even when these are neither additive nor monotone. Proposals for non-monotone measures on fuzzy sets (entropy and specificity) are presented. They can be considered to be measures of information on the premises or conclusions in approximate reasoning. Finally some results are discussed on a monotone measure on fuzzy relations to understand whether or with which logics is the modus ponens generalised when making fuzzy inference.

Acknowledgment

This research is partially supported by the Spanish MCyT project BFM2002-00281.

References

1. Denneberg D. (1994) Non-additive Measure and Integral, Kluwer Academic Publishers, Dordrecht.

2. Drewnowski L. (1978) On the continuity of certain non-additive set functions. Colloquium Math. 38 pp. 243–253.

3. Dubois D., Prade H. (1980) Fuzzy Sets and Systems. Theory and its Applications. Academic Press, New York.

4. Dubois D., Prade H. (1982) A class of fuzzy measures based on triangular norm. A general framework for the combinations of uncertain information. Int. J. Gen. Syst. 8, 1, pp. 43–61.

5. Dubois D., Prade H. (1987) The principle of minimum specificity as a basis for evidential reasoning. In: Uncertainty in Knowledge-based Systems, Bouchon, B. & Yager RR (Eds.). Springer-Verlag, Berlin, pp. 75–84.

6. Dubois D., Prade H. (1995) A note on measures of specificity for fuzzy sets. International Journal of General Systems 10, pp. 279–283.

7. Garmendia L., Campo C., Cubillo S., Salvador A. (1999) A Method to Make Some Fuzzy Relations T-Transitive. International Journal of Intelligence Systems. 14 (9), pp. 873–882.

8. Garmendia L., Yager R., Trillas E., Salvador A. (2003) On t-norms based specificity measures. Fuzzy Sets and Systems 133–2, pp. 237–248.

9. Garmendia L. Trillas E., Salvador A., Moraga C. (2005) On measuring μ-T-conditionality of fuzzy relations. Soft Computing. Springer-Verlag. Volume 9, Number 3, pp. 164–171.

10. Grabisch M., Murofushi T., Sugeno M. (1992) Fuzzy measure of fuzzy events defined by fuzzy integrals. Fuzzy Sets and Systems 50, pp. 293–313.

11. Grabisch M., Murofushi T., Sugeno M. (2000) Fuzzy Measures and Integrals Theory and Applications. Physica-Verlag.

12. Ha M., Wang X. (1997) Some notes on the regularity of fuzzy measures on metric spaces. Fuzzy Sets and Systems 87, pp. 385–387.

13. Halmos P.R. (1968) Measure Theory, Van Nostrand, Princeton, NJ (1962). Van Nostrand Reinhold, New York.

14. Higashi M., Klir G.J. (1983) Measures of uncertainty and information based on possibility distributions. International Journal of General Systems 9, pp. 3–58.

15. Jang L.C., Kwon J.S. (2000) On the representation of Choquet integrals of set-valued functions, and null sets, Fuzzy Sets and Systems, 112, pp. 233–239.

16. Jiang Q., Suzuki H. (1996) Fuzzy measures on metric spaces. Fuzzy Sets and Systems 83, pp. 99–106.

17. Murofushi T., Sugeno M. (1989) An interpretation of fuzzy measures and the Choquet integral as an integral with respect to a fuzzy measure. Fuzzy sets and Systems, pp. 201–227.

18. Murofushi T., Sugeno M. (1991) Fuzzy t-conorm integral with respect to fuzzy measures: generalization of Sugeno integral and Choquet integral, Fuzzy Sets and Systems 42, pp. 57–71.

19. Murofushi T., Sugeno M. (1991) A theory of fuzzy measures: representations, the Choquet integral and null sets. J. Math. Anal. Appl. 159, pp. 532–549.

20. Murofushi T., Sugeno M., Machida M. (1994) Non-monotonic fuzzy measures and the Choquet integral. Fuzzy Sets and Systems 64, pp. 73–86.

21. Pradera A., Trillas E., Cubillo S. (2000) On modus ponens generating functions. Internat. J. Uncertain. Fuzziness Knowledge Based Systems 8, 1, pp. 7–19.

22. Puri M.L., Ralescu D. (1982) A possibility measure is not a fuzzy measure (short communication). Fuzzy Sets and Systems, pp. 311–313.

23. Riera T., Trillas E. (1982) From measures of fuzziness to Booleanity control. Fuzzy information and decision processes, 3-16. North-Holland, Amsterdam-New York.
24. Schmeidler D. (1986) Integral representation without additivity. Proc. Amer. Math. Soc. 97, pp. 253–261.
25. Schweizer B., Sklar A. (1960) Statistical metric spaces. Pacific J. Math. 10, pp. 313–334.
26. Schweizer B., Sklar A. (1983) Probabilistic Metric Spaces. North-Holland. New York.
27. Shafer G.A. (1976) A Mathematical Theory of Evidence. Princeton.
28. Sugeno M. (1975) Theory of Fuzzy Integrals and its applications. Ph.D. Dissertation. Tokyo Institute of Technology.
29. Trillas E., Riera T. (1978) Entropies in finite fuzzy sets. Inform. Sci. 15, 2, pp. 159–168.
30. Trillas E., Sanchis C. (1979) On entropies of fuzzy sets deduced from metrics. Estadistica Española 82-83, pp. 17–25.
31. Trillas E., Cubillo S. (1992) Condicionales Exactos. Actas II Conferencia Estylf., pp. 1–15.
32. Trillas E. (1992) On exact conditionals. Stochastica 13, 1, pp. 137–143.
33. Trillas E. (1992) On Exact and Inexact Conditionals. Prc. IV International Conference On Information Processing and Management of Uncertainty in Knowledge-Based Systems, pp. 649–655.
34. Trillas E. (1992) On fuzzy conditionals generalising the material conditional. IPMU'92. Advanced methods in artificial intelligence, pp. 85–100.
35. Trillas E., Alsina C. (1993) Logic: going farther from Tarski? Fuzzy Sets and Systems, 53, pp. 1–13.
36. Trillas E., Cubillo S. (1994) On monotonic fuzzy conditionals. Journal of Applied Non-Classical Logics, 4, 2, pp. 201–214.
37. Trillas E., Cubillo S., Rodriguez A. (1994) On the identity of fuzzy material conditionals. Mathware Soft Computing 1, 3, pp. 309–314.
38. Wang Z., Klir G. (1992) Fuzzy Measure Theory, Plenum Press, New York.
39. Wang Z., Klir G.J. (1996) Fuzzy measures defined by fuzzy integral and their absolute continuity. J. Math. Anal. Appl. 203, pp. 150–165.
40. Wu C., Wang S., Ma M. (1993) Generalized fuzzy integrals: Part I. Fundamental concepts. Fuzzy Sets and Systems 57, pp. 219–226.
41. Wu C., Ha M. (1994) On the regularity of the fuzzy measure on metric fuzzy measure spaces. Fuzzy Sets and Systems 66, pp. 373–379.
42. Xuechang L. (1992) Entropy, distance measure and similarity measure of fuzzy sets and their relations. Fuzzy Sets and Systems 52, pp. 305–318.
43. Yager R.R. (1979) On the measure of fuzziness and negation, part I: membership in the unit interval. International Journal of General Systems 5, pp 221–229.
44. Yager R.R. (1982) Measuring tranquillity and anxiety in decision making: An application of fuzzy sets. International Journal of General Systems 8, pp. 139–146.
45. Yager R.R. (1983) Entropy and specificity in a mathematical theory of evidence. International Journal of General Systems 9, pp. 249–260.
46. Yager R.R. (1984) Approximate reasoning as a basis for rule-based expert systems. IEEE Trans. Systems Man Cybernet. 14 4, pp. 636–643.

47. Yager R.R. (1985) Measures of specificity for possibility distributions. In Proceedings of the IEEE Workshop on Languages for Automation: Cognitive Aspects in Information Processing, Palma de Mallorca, Spain, pp. 209–214.

48. Yager R.R. (1985) Reasoning with uncertainty for expert systems. Proceedings of the Ninth International Joint Conference on Artificial intelligence. Los Angeles, pp. 1295–1297.

49. Yager R.R. (1986) Toward a general theory of reasoning with uncertainty part I: nonspecificity and fuzziness. International Journal of Intelligent Systems 1, pp. 45–67.

50. Yager R.R. (1986) The entailment principle for Dempstes-Shafer granules. International Journal of Intelligent Systems 1, pp. 247–262.

51. Yager R.R., Ovchinnikov S, Tong R, Nguyen H (1987) Fuzzy sets and applications. Selected papers by L. A. Zadeh. John Wiley & Sons, New York.

52. Yager R.R. (1988) On ordered weighted averaging aggregation operators in multicriteria decision making. IEEE Trans. Systems, Man, Cybernet. 18, pp. 183–190.

53. Yager R.R. (1990) Ordinal measures of specificity. International Journal of General Systems 17, pp. 57–72.

54. Yager R.R. (1991) Measures of specificity of possibility distributions. Proceedings of the Tenth NAFIPS meeting, U. Of Missouri, Columbia, MO, pp. 240–241.

55. Yager R.R. (1991) Similarity based measures of specificity. International Journal of General Systems 19, pp. 91–106.

56. Yager R.R. (1992) Default knowledge and measures of specificity. Information Sciences 61, pp. 1–44.

57. Yager R.R. (1992) On the specificity of a possibility distribution. Fuzzy Sets and Systems 50, pp. 279–292.

58. Yager R.R. (1998) On measures of specificity, Computational Intelligence: Soft Computing and Fuzzy-Neuro Integration with Applications, edited by Kaynak, O., Zadeh L. A., Turksen, B., and Rudas, I. J., Springer-Verlag, Berlin, pp. 94–113.

59. Yao J., Chang S. (2000) Fuzzy measure based on decomposition theory, Fuzzy Sets and Systems, 112, pp. 187–205.

60. Zadeh L.A. (1965) Fuzzy sets. Information and Control 8, pp. 338–353.

61. Zadeh L.A. (1971) Similarity relations and fuzzy orderings, Inform. Sci. 3, pp. 177–200.

Economic and Management Applications

Part II

Economic and Management Application

Association Rule Based Specialization
in ER Models

Martine De Cock[1], Chris Cornelis[1], Ming Ren[2], Guoqing Chen[2], and
Etienne E. Kerre[1]

[1] Ghent University, Fuzziness and Uncertainty Modelling Research Unit,
   Krijgslaan 281 (S9), 9000 Gent, Belgium
   {martine.decock|chris.cornelis|etienne.kerre}@ugent.be
[2] Tsinghua University, School of Economics and Management, Beijing 100084,
   China
   {renm|chengq}@em.tsinghua.edu.cn

Abstract. Association rules (ARs) emerged in the domain of market basket analysis and provide a convenient and effective way to identify and represent certain dependencies between attributes in a database. In this paper, we demonstrate that they also act as an appropriate aid in the construction and enrichment of entity-relationship (ER) models, structuring tools that provide high-level descriptions of data. In particular, we present different conceptual ideas for semi-automated specialization of ER models based on AR mining.

Key words: entity-relationship model, association rules, fuzzy sets, database design

1 Introduction

The entity-relationship (ER) model is a conceptual model that describes real world phenomena in terms of entities, relationships between those entities, and attributes of both of them. In an ER model for a grocery store for instance we typically encounter entity classes such as *product* and *customer*, having attributes such as *price* and *freshness date* (for the product) and *age* and *sex* (for the customer). *Purchase* is an example of a relationship class between these two entity classes, while *quantity* and *time* are examples of attributes of the purchase relationship class.

The ER model is a powerful means for business and data modelling that helps to identify essential elements of the domain of interest in a conceptual and integrated manner. Initially introduced in [10], the methodology itself has evolved considerably and has become widely accepted as a standard design tool for relational databases [11]. During the past decades, basic ER concepts have been extended in various ways, resulting in enhanced ER models. In this

Martine De Cock et al.: *Association Rule Based Specialization in ER Models*, Studies in Computational Intelligence (SCI) **5**, 203–200 (2005)
www.springerlink.com

paper we will focus on specialization: the process of defining subclasses for a given entity or relationship class. Referring to the grocery store example, for the entity class *product* we might define the subclasses[1]: *product with price in* [0, 20[, *product with price in* [20, 100[and *product with price in* [100, +∞[. *Product* is then called a superclass of these subclasses.

One of the advantages of ER models is that they are easy to understand with a minimum of training; hence they are very suitable to communicate and discuss the database design with the end user (e.g. the shop owner). Also in this respect it is more convenient to denote the subclasses of attribute values by linguistic terms in the ER model. Indeed a linguistic expression such as *cheap product* corresponds better to the shop owner's daily use of language than *product with price in* [0, 20[. Now regarding the implementation of the ER model, one might argue that it is against intuition to call a product of 19.90 EUR cheap and one of 20 EUR not. The transition between being cheap and not being cheap is not abrupt but gradual. Hence it makes more sense to model linguistic terms such as *cheap, medium* and *expensive* by fuzzy sets [30], characterized by membership functions that associate with every price a number between 0 and 1 indicating the degree to which this price can be called *cheap, medium* and *expensive* respectively. So-called *fuzzy* ER models have been proposed from different perspectives [4, 20, 31].

Traditionally, ER models are built upon the knowledge of business managers and database designers. However, as the real world phenomena represented by the ER model change, and our understanding of the world improves, the need arises for an extension or enrichment of the original ER model. In this paper we propose to use assocation rule mining as a tool for the semi-automatic construction and enrichment of ER models, more in particular for the specialization process. We treat the specialization of entities and relations separately (so called E-specialization and R-specialization).

Our main aim where E-specialization is concerned, is to identify subclasses of entities denoted by linguistic expressions that are common in the domain of interest. For instance typical subclasses of the entity class *product* in an interior decoration store are *bathroom accessories, beds and mattresses, storage systems, chairs, cookware, for the pets* etc. Each of those in turn can be divided into subclasses; for instance for *cookware* this might be *kitchen storage and containers, kitchen utensils and accessories, knives and chopping board, pots, pans and ovenware*. Our aim is to construct such a specialization relation between linguistic expressions (further on called "terms") automatically. To this end, we start from a collection of documents containing text that is highly related to the phenomena represented by the ER model. From this collection, we generate a document-term table. In this table we mine for association rules between terms.

If available at all, a text collection related to the domain of interest is usually already at our disposal when domain experts start constructing the

---

[1] [a, b[denotes the real interval that contains a but not b.

ER model. As such, the automated E-specialization described above is in the first place intended as a means of support for the domain expert, summarizing the knowledge of all those that have contributed somehow to writing the documents in the collection. An ER model is usually intended as a design for a relational database. Once the model is built and a significant amount of data has been gathered in the database, we can in turn use this data to discover new knowledge and enrich our model. Examples of such knowledge are relations that were unknown at the time when the model was built (e.g. due to the limitation of the scope of expertise) or that have previously been unimportant and/or uninteresting but now become significant and worth considering. The main focus here is R-specialization, but as will become clear later on, this process also involves a second kind of E-specialization.

In Sect. 2 we recall basic notions about (fuzzy) association rule mining. Drawing upon earlier techniques for the generation of fuzzy thesauri (see [14] for an overview), in Sect. 3 we explain how the discovery of association rules in an automatically generated document-term table provides an outline for user-defined E-specialization at the stage of the initial database design. We also briefly refer to other work done in the framework of association rule mining that can be applied to attribute-based E-specialization when a significant amount of data has already been gathered in the database. In Sect. 4 we explain how these results, together with association rule mining in the data tables corresponding to the ER model, can be used as a stepping stone for R-specialization (see also [8]).

2 Association Rules

The idea of association rule (AR) mining already dates back to Hájek et al. (see e.g. [17, 18, 19]). Its application to market basket analysis gained high popularity soon after its re-introduction by Agrawal et al. [1] at the beginning of the 1990's. The straightforwardness of the underlying ideas as well as the increasing availability of transaction data from shops certainly helped to this end.

In the context of AR mining, data is represented as a table. The rows correspond to objects (e.g. transactions, patients,. . .) while the columns correspond to attributes (e.g. items bought in a transaction, symptoms,. . .). Ones and zeros in the data matrix denote whether or not the object has a specific attribute (whether or not *cheese* was purchased in the *5th transaction*, whether or not patient *John* has *fever*,. . .). In this way, we can think of an object as a set of attributes. The purpose of AR mining is to detect rules of the form

$$A \to B$$

in the data, indicating that an object containing the attribute A is likely to contain B as well, for example *bread* $\to$ *milk*.

Rules containing more than one attribute in the antecedent and/or the consequent have been studied as well. However mining algorithms tend to generate too many rules to be digested by users, and using this more general rule format can easily lead to a rule explosion problem. There is a trend to focus on simple association rules, i.e. those containing only one attribute in the consequent, and use them as building blocks to construct more general rules if required [7, 9]. For our purposes in this paper it is sufficient to consider only the simplest possible association rules, namely those with one attribute in antecedent and consequent.

Association rules can be rated by a number of quality measures, among which *support* and *confidence* stand out as the two essential ones. Support measures the statistical significance of a candidate rule $A \to B$ as the fraction of objects in which both A and B occur. Confidence assesses the strength of a rule as the fraction of objects containing A that contain B as well. The basic problem is then to generate all association rules $A \to B$ that have support and confidence greater than user-specified thresholds.

Suppose we have a non-empty data table D containing records described by their values for the binary attributes in $\mathcal{A} = \{A_1, \ldots, A_m\}$. In the framework of market basket analysis, the attributes correspond to items which customers may purchase, while the records represent customer transactions (market baskets). If the value of record x for attribute A_i $(i = 1, \ldots, m)$ is 1, we say that x contains A_i, otherwise x does not contain A_i. For each attribute A of $\mathcal{A}$ (also called an *item*), by D_A we denote the set of records that contain the attribute A.

Support. The support of an association rule $A \to B$ is usually defined as

$$\text{supp}(A \to B) = \frac{|D_A \cap D_B|}{|D|} \tag{1}$$

i.e. the number of records containing both A and B, scaled to a value between 0 and 1. The idea behind the definition of support is to measure the statistical significance by counting *positive examples*, i.e. transactions that explicitly support the hypothesis expressed by the association rule. It is worth noting that the positive examples of $A \to B$ are also those of the rule $B \to A$, i.e. support is a symmetric measure. Hence, as can be expected, it reveals only part of the global picture [13]. This is why we also need the confidence measure, to assess the strength of a rule.

Confidence. Traditionally, if a rule $A \to B$ generates a support exceeding a user-specified threshold, it is meaningful to compute its confidence, i.e. the proportion of correct applications of the rule:

$$\text{conf}(A \to B) = \frac{|D_A \cap D_B|}{|D_A|} \tag{2}$$

Note that $|D_A|$ will not be 0 if we assume that the confidence is computed only when the support exceeds a certain threshold (which should be greater than 0 to be meaningful).

Many efficient algorithms, like Apriori [25], exist for mining ARs whose support and confidence values exceed thresholds fixed by the user.

Quantitative Association Rules. In most real life applications, databases contain many other attribute values besides 0 and 1. Very common for instance are quantitative attributes such as *age* or *quantity*, taking values from a partially ordered, numerical scale, often a subset of the real numbers. One way of dealing with a quantitative attribute like *age* is to replace it by a few other attributes that form a crisp partition of the range of the original one, such as $\langle age, [0, 30[\rangle$, $\langle age, [30, 65[\rangle$ and $\langle age, [65, +\infty[\rangle$. Now we can consider these new attributes as binary ones that have value 1 if the *cost* attribute equals a value within their range, and 0 otherwise (see Table 1 for an example). In this way, the problem is reduced to the mining procedure described above. The generated rules are called quantitative association rules [27].

Table 1. (a) Original attribute values.

	age
x_1	10
x_2	32
x_3	60
x_4	28
x_5	69

Table 1. (b) Attribute values for quantitative association rule mining

	$\langle age, [0, 30[\rangle$	$\langle age, [30, 65[\rangle$	$\langle age, [65, +\infty[\rangle$
x_1	1	0	0
x_2	0	1	0
x_3	0	1	0
x_4	1	0	0
x_5	0	0	1

Fuzzy Association Rules. When a database contains values between 0 and 1, it is also possible to extend the classical mining algorithm using fuzzy set theoretical operations to obtain fuzzy association rules (see e.g. [6, 13, 15, 21, 22, 29]). The associated interpretation is that a record x can have, or contain, an attribute A to a certain extent $D_A(x)$. The $D \to [0,1]$ mapping D_A is the fuzzy set of records that contain A. Support and confidence can be generalized by means of fuzzy cardinality and a t-norm[2] T, i.e. for items A and B,

$$\text{supp}(A \to B) = \frac{\sum_{x \in D} T(D_A(x), D_B(x))}{|D|} \tag{3}$$

$$\text{conf}(A \to B) = \frac{\sum_{x \in D} T(D_A(x), D_B(x))}{\sum_{x \in D} D_A(x)} \tag{4}$$

3 E-specialization

Recall that we assume to have a text collection available that is highly related to our domain of interest. One way to obtain this is to launch one or several WWW queries containing keywords related to the phenomena under consideration, to focus the search engine on a set of relevant documents. The returned document collection (which typically consists of the top-ranked pages for the query) then serves as a context, containing more or less related terms, for this particular domain of interest.

Our aim is to discover a hierarchical relationship between entities corresponding to terms occurring in the documents. The first step in the construction process is to transform the document collection into a document-term table, in which the objects (i.e. the rows) correspond to documents, and the attributes (i.e. the columns) correspond to terms featuring in the documents. ER models are concerned with semantics of real world phenomena; we treat the documents accordingly by preprocessing them. Filter words, i.e. frequently occurring, insignificant words such as *and, it, is,* etc. (also commonly called stop words) can be removed from the documents using so-called stop word lists that are freely available on the WWW[3]. In most cases, morphological variants of words have similar semantic interpretations and can be considered as equivalent for our purposes. Therefore a so-called stemming algorithm, or stemmer, such as Porter's [23], can be applied to reduce a word to its stem or root form. For instance the words *juice* and *juicy* are treated as one and the same term.

The most straightforward way to build a document-term table is by marking occurrences: a 0 indicates that a term does not occur in a document, a

---

[2] Recall that a t-norm is an increasing commutative, associative $[0,1]^2 \to [0,1]$ mapping T that satisfies $T(1, x) = x$ for all x in $[0,1]$. The intersection $D_A \cap_T D_B$ of two fuzzy sets D_A and D_B in D by means of T is defined by, for x in D,

$$(D_A \cap_T D_B)(x) = T(D_A(x), D_B(x))$$

Often used t-norms are the minimum and the product.

[3] A list of filter words for English that is used in the SMART system at Cornell University can be found at ftp://ftp.cs.cornell.edu/pub/smart/english.stop

1 means that the term appears at least once in the document. To this table we can apply a traditional association rule mining algorithm. Let t_1 and t_2 be two terms. A high confidence value for candidate association rule $t_1 \rightarrow t_2$ indicates that when t_1 occurs in a document, t_2 is likely to occur as well. A high confidence value for $t_1 \rightarrow t_2$ combined with a low confidence for $t_2 \rightarrow t_1$ strenghtens our belief that term t_1 is narrower or more specific than t_2. Hence we can build a hierarchical "is narrower than" relationship among terms based on their co-occurrences in documents. When the degree of confidence is interpreted as a degree of specialization (a degree of "being narrower than"), we obtain a fuzzy relation among terms. Also note that high confidence values for both $t_1 \rightarrow t_2$ and $t_2 \rightarrow t_1$ strenghtens our belief that t_1 and t_2 are *similar* terms, but this is not the focus of our interest here.

A refinement to the scheme above is to take into account the number of times a term appears in a document, assuming that words that appear more frequently are more important and/or more relevant for the document. In information retrieval this is known as a term frequency (TF) model. This effort only makes sense when the documents are significantly long enough; if we for instance only consider short summaries of webpages returned by a search engine (like Google's "snippets") there is no need for this refinement.

An important family of term weighting schemes combines term frequencies (which are relative to each document) with an absolute measure of term importance called inverse document frequency (IDF), giving rise to the following formula for the importance of term t for document d (see e.g. [2]):

$$\text{TF}_{t,d} \cdot \log \left(\frac{N}{\text{DF}_t} \right)$$

$\text{TF}_{t,d}$ is the number of occurrences of term t in document d; DF_t is the number of documents in which term t occurs and N is the total number of documents. Terms that are globally rare receive a higher weight, whereas a term that appears in all documents has weight 0 for every document. In our application however we assume that all documents are relevant for the domain of interest; hence it is likely that many of them contain the most important terms for our domain, and we cannot penalize or exclude them from the document-term table for this reason. Hence the TF-IDF measure is unsuitable for our purposes.

The weighting schemes discussed so far are purely based on the number of occurrences and co-occurrences of the terms in the documents. However (web) documents typically contain more information besides the occurring words. Layout characteristics such as font size and type are valuable indications about the importance of terms for documents; e.g. a word occurring in a large bold-faced font will typically be of high importance. Similar information can be drawn from the position of a term in the document (search engines that only take into account e.g. the first 20 lines rely on this) and the use of XML-tags. In [24] a fuzzy IF-THEN rule base that takes these kind of features into account is presented.

In all of these refined weighting schemes, the resulting document-term table will no longer be binary but (possibly after normalization) contains numbers between 0 and 1 as its attribute values. Hence we can use fuzzy AR mining to generate a narrower-than relation between the terms.

As an aside, we mention that automatic construction of such a relation for terms also comes in handy for the mining of generalized association rules [26]. In this domain such a term-term relation is usually called a taxonomy, represented as a directed acyclic graph whose leaf nodes all correspond to attributes belonging to $\mathcal{A}$ (basic items) while the interior nodes represent higher-level concepts. The aim of generalized association rule mining is to discover rules involving those higher level concepts, such as *outerwear* → *shoes*, even if rules that contain the more specific basic items such as *jackets* → *shoes* and *clothes* → *shoes* do not hold. Fuzzy generalized association rule mining has been studied as well [5].

Great advantages of an automated process for the construction of a draft for a taxonomy include objectivity, high speed and low cost. Indeed, such a process does not have to be hindered by background knowledge or points of view of a few experts only. Moreover, as information structures representing associations such as synonymy, specification, and generalization between linguistic terms seem to pop up in many domains that require the semantical representation of language (such as information retrieval, and natural language processing techniques like machine translation) and under a variety of different names (apart from taxonomies one also encounters thesauri and ontologies), this application is steadily gaining in importance.

Let us return to ER models. Formally, an ER model can be denoted as a triple $(\mathcal{E}, \mathcal{R}, \mathcal{A})$ where $\mathcal{E}$ is the set of entity classes, $\mathcal{R}$ is the set of relationship classes between those entities, and $\mathcal{A}$ is a mapping that maps every element of $\mathcal{E} \cup \mathcal{R}$ onto its set of attributes. For example $\mathcal{A}(product)$ is the set of attributes of the entity class *product*, typically containing elements such as *price*, *freshness date* and *category*. A subclass E of an entity class F arises by limiting the range of an attribute of F from its original domain X to a value of X, a subset of X or even a fuzzy set in X (i.e. a $X \rightarrow [0,1]$ mapping) [4]. F is then also called a superclass of E, denoted by

$$E \subseteq F$$

The process of defining such subclasses is called E-specialization.

The automatically generated narrower-than relation between terms can assist the business expert and the database designer to identify E-specializations that are of practical interest in the domain under consideration. An example of an E-specialization is graphically depicted in Fig. 1.

This kind of specialization can already be carried out at the stage of the initial database design, i.e. before any data is gathered. An ER model however usually serves as a design for a relational database, and once a sufficient amount of data is collected, AR mining in the tables of this database can be

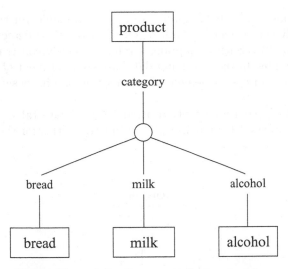

Fig. 1. E-specialization on attribute `category`

used to provide feedback on the initial ER model. The ideas presented in the remainder of the paper are based on this feedback principle.

In Sect. 2 we mentioned that quantitative attributes such as *age* or *quantity* are often encountered in real-life applications. In the framework of quantitative AR mining, the range of such an attribute is typically partitioned in intervals such as $\langle age,[0,30[\rangle$, $\langle age,[30,65[\rangle$ and $\langle age,[65,+\infty[\rangle$. We can even go a step further and use linguistic terms represented by fuzzy sets to this end, such as $\langle age,young\rangle$, $\langle age,middle\text{-}aged\rangle$ and $\langle age,old\rangle$. Table 2 illustrates this for the original attribute values of Table 1. It is clear that fuzzy association rule mining can be used on a table with attributes like the newly constructed ones.

Table 2. Attribute values for fuzzy association rule mining

	⟨age, **young**⟩	⟨age, **middle-aged**⟩	⟨age, **old**⟩
x_1	1.0	0.0	0.0
x_2	0.4	0.6	0.0
x_3	0.0	0.8	0.2
x_4	0.6	0.4	0.0
x_5	0.0	0.3	0.7

Note that partitioning of the attribute range should not be done arbitrarily: besides relying on expert knowledge we can use clustering techniques (see e.g. [28]) as well as partitioning algorithms developed specifically for AR mining (see e.g. [16]).

Now suppose that the mapping from the available relational database scheme to ER components is known (either a priori or detected by reverse engineering [3, 12]). Such a mapping s relates a newly constructed attribute $\langle A, v \rangle$ from a table in the relational database to an element Q of $\mathcal{E} \cup \mathcal{R}$ such that $A \in \mathcal{A}(Q)$ and v is an element of, a subset of, or a fuzzy set in the domain of A.

In this sense, partitioning of an attribute in a data table corresponds to specialization in the ER model. Figure 2 illustrates this for the attribute *age* of the entity *customer*.

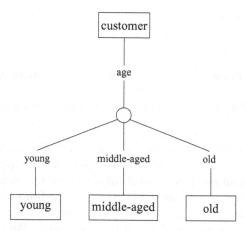

Fig. 2. E-specialization on attribute **age**

4 R-specialization

Relationships are more complicated than entities, and so is the corresponding R-specialization. There are basically two forms of R-specialization. The first one is analogous to E-specialization based upon the fact that relationships, like entities, have attributes, and a subclass R of a relationship S arises by limiting the allowed values of an attribute of S. Again S is called a superclass of R, denoted by $R \sqsubseteq S$. In Fig. 3 a subclass of the *purchase* relationship is defined based on the attribute *quantity*.

In addition, a more general setting of R-specialization can be made not only based on an attribute of a relationship R, but also based on attributes of R's participating entity classes. Examples of this are *"middle-aged customers purchase milk"* or even more sophisticated *"middle-aged customers purchase milk in the evening"*. The first example involves E-specialization of the entities *customer* and *product* on their attributes *age* and *category* respectively, while the second example additionally involves the above-discussed form of

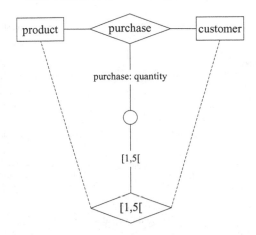

Fig. 3. R-specialization on attribute `quantity`

R-specialization of the relationship *purchase* on the attribute *time*. In both cases a subclass of the *purchase* relationship arises.

We propose to use AR mining in the relational database to generate subclasses of relationships. The semantics reflected by the discovered association rules may either have been unknown at the time when the ER model was built (e.g. due to the limitation of the scope of expertise), or have previously been unimportant/uninteresting but now become significant and worth considering. Thus, incorporating the new knowledge into the model is necessary and enables us to gain a more integrated and informative view of the domain. Assume we have an association rule

$$\langle A, v \rangle \rightarrow \langle B, w \rangle$$

discovered in a table of a relational database. Let s be the above mentioned mapping from the relational scheme to ER components, then $s(\langle A, v \rangle)$ and $s(\langle B, w \rangle)$ are either entities or relationships.

If they are both entities, then $\langle A, v \rangle$ and $\langle B, w \rangle$ by themselves will give rise to E-specializations of two entities (say) E and F. If a relationship R between E and F already exists, the association rule may lead to an R-specialization of R. An example corresponding to the association rule

$$\langle \text{age, middle-aged} \rangle \rightarrow \langle \text{category, milk} \rangle$$

is shown in Fig. 4. However, there are cases where the discovered rules may reflect new relationships among entity class which are not represented in the original model in an either general or specific fashion. A particular new relationship of interest is that among subclasses of the same class, i.e. with

$$s(\langle A, v \rangle) = s(\langle A, w \rangle)$$

An example of such a case is shown in Fig. 5. The corresponding association rule is

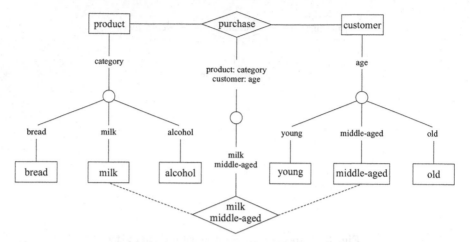

Fig. 4. R-specialization on attributes of entities

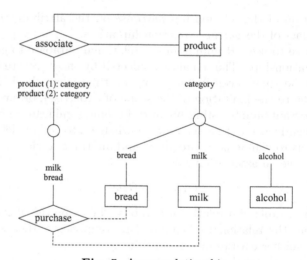

Fig. 5. A new relationship

$$\langle \text{category, bread} \rangle \rightarrow \langle \text{category, milk} \rangle$$

Yet another case occurs when $\langle A, v \rangle$ and $\langle B, w \rangle$ are both mapped onto the relationship class R, giving rise to a direct extension of the case of Fig. 3 as illustrated in Fig. 6. The corresponding association rule is

$$\langle \text{time, night} \rangle \rightarrow \langle \text{quantity}, [1,5[\rangle$$

Finally it can happen that one of $\langle A, v \rangle$ and $\langle B, w \rangle$ is mapped onto an entity class and the other onto a relationship class, say $s(\langle A, v \rangle) = E$ and $s(\langle B, w \rangle) = R$. Such an association rule may lead to an R-specialization based on two attributes: one is an attribute of a relationship class R, and the other is an

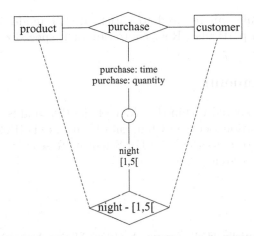

Fig. 6. R-specialization on attributes of a relationship

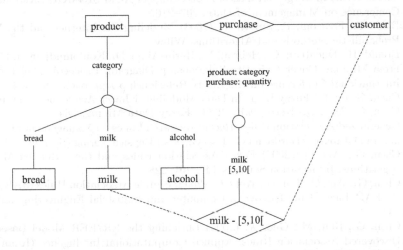

Fig. 7. R-specialization on attributes of an entity and a relationship

attribute of R's participating entity class E. An example corresponding to the association rule

$$\langle \text{category, milk} \rangle \rightarrow \langle \text{quantity, } [5,10[\rangle$$

is depicted in Fig. 7.

5 Conclusion

In this paper we have gathered different conceptual ideas for semi-automated specialization of ER models based on AR mining. By bridging the paradigms of AR and ER, a framework of semantical correspondence is obtained, where an association rule can be used to enrich an ER model by introducing a specialization (in particular, R-specialization or E-specialization).

Acknowledgements

This work was prepared within the scope of the Bilateral Scientific and Technological Cooperation Between China and Flanders (174B0201). Martine De Cock would like to thank the Fund For Scientific Research – Flanders (FWO) for funding her research.

References

1. Agrawal, R., Imielinski, T., Swami, A. (1993) Mining Association Rules between Sets of Items in Large Databases. Proceedings of ACM SIGMOD International Conference on Management of Data, 207–216.
2. Baldi, P., Frasconi, P., Smyth, P. (2003) Modeling the Internet and the Web. Probabilistic Methods and Algorithms. Wiley.
3. Briand, H., Ducateau, C., Hebrail, Y., Hérin-Aime, D., Kouloumdjian, J. (1987) From Minimal Cover to Entity-Relationship Diagram. Proceedings of the 7th International Conference on the Entity Relationship Approach, 287–304.
4. Chen, G. (1998) Fuzzy Logic in Data Modeling. Kluwer Academic, Boston.
5. Chen, G., Wei, Q., Kerre, E.E. (2000) Fuzzy Data Mining: Discovery of Fuzzy Generalized Association Rules. Recent Research Issues on Management of Fuzziness in Databases (Bordogna G., Pasi G., eds.) Physica-Verlag (Springer), 45–66.
6. Chen, G., Wei, Q. (2002) Fuzzy Association Rules and the Extended Mining Algorithms. Information Sciences 147: 201–228.
7. Chen, G., Wei, Q., Liu, D., Wets, G. (2002) Simple Association Rules (SAR) and the SAR–based Rule Discovery. Computer and Industrial Engineering 43:721–733.
8. Chen, G., Ren, M., Guo, X. (2004) Enriching the ER/EER Model Based on Discovered Association Rules. Applied Computational Intelligence (Ruan, D., D'hondt, P., De Cock, M., Nachtegael, M., Kerre, E.E., eds.), World Scientific, 179–184.
9. Chen, G., Yan, P., Kerre, E.E. (2004) Computationally Efficient Mining for Fuzzy Implication–based Association Rules in Quantitative Databases. International Journal of General Systems 33 (23):163–182.
10. Chen, P.P. (1976) The Entity-Relationship Model – Toward a Unified View of Data. ACM Transactions on Database Systems 1(2): 9–36.
11. Chen, P.P. (2002) Entity-Relationship Modeling: Historical Events, Future Trends, and Lessons Learned. Software Pioneers: Contributions to Software Engineering (Broy M. and Denert, E., eds.), Springer-Verlag, 100–114.
12. Davis, K.H., Arora, A.K. (1987) Converting a Relational Database Model into an Entity-Relational Model. Proceedings of the 7th International Conference on the Entity Relationship Approach, 271–285.
13. De Cock, M., Cornelis, C., Kerre, E.E. (2005) Elicitation of Fuzzy Association Rules from Positive and Negative Examples. Fuzzy Sets and Systems 149(1): 73–85.

14. De Cock, M., Guadarrama, S., Nikravesh, M. (2005) Fuzzy Thesauri for and from the WWW. Soft Computing for Information Processing and Analysis (Nikravesh, M., Zadeh, L.A., Kacprzyk, J., eds.), Studies in Fuzziness and Soft Computing 164, Springer-Verlag.
15. Delgado, M., Marín, N., Sanchéz, D., Vila, M.A. (2003) Fuzzy Association Rules: General Model and Applications. IEEE Transactions on Fuzzy Systems 11(2): 214–225.
16. Gyenesei, A., Teuhola, J. (2004) Multidimensional fuzzy partitioning of attribute ranges for mining quantitative data. International Journal of Intelligent Systems 19(11): 1111–1126.
17. Hájek, P., Havel I., Chytil M. (1966) The GUHA Method of Automatic Hypotheses Determination. Computing, 1: 293–308.
18. Hájek, P., Havránek, T. (1977) On Generation of Inductive Hypotheses. International Journal of Man-Machine Studies 9: 415–438.
19. Hájek, P., Havránek, T. (1978) Mechanizing Hypothesis Formation: Mathematical Foundations for a General Theory. Springer-Verlag, http://www.cs.cas.cz/~hajek/guhabook/.
20. Kerre, E.E., Chen, G.C. (1995) An Overview of Fuzzy Data Models. Fuzziness in Database Management Systems (Bosc, P., Kacprzyk, J., eds.), Studies in Fuzziness and Soft Computing 5, Physica-Verlag, 23–41.
21. Kuok, C.M., Fu, A., Wong, M.H. (1998) Mining Fuzzy Association Rules in Databases. ACM SIGMOD Record, 27(1): 41–46.
22. Lee, J.H., Hyung, L.K. (1997) An Extension of Association Rules using Fuzzy Sets. Proceedings of the 7th World Conference of the International Fuzzy Systems Association (IFSA 1997), 399–402.
23. Porter, M. (1980) An Algorithm for Suffix Stripping. Program 14(3):130–137.
24. Ribeiro, A., Fresno, V., Garcia-Alegre, M.C., and Guinea, D. (2003) A Fuzzy System for the Web Page Representation. Intelligent Exploration of the Web (P. S. Szczepaniak, J. Segovia, J. Kacprzyk, L.A. Zadeh, eds.), Studies in Fuzziness and Soft Computing 111, Springer-Verlag, 19–37.
25. Srikant, R., Agrawal, R. (1994) Fast Algorithms for Mining Association Rules. Proceedings of the Proceedings of 20th International Conference on Very Large Data Bases, 487–499.
26. Srikant, R., Agrawal, R. (1995) Mining Generalized Association Rules. Proceedings of the 21st International Conference on Very Large Databases, 407–419.
27. Srikant, R., Agrawal, R. (1996) Mining Quantitative Association Rules in Large Relational Tables. Proceedings of the ACM SIGMOD Conference on Management of Data, v.25 n.t2, 1–12.
28. Verlinde, H., De Cock, M., Boute R. (2005) Fuzzy Versus Quantitative Association Rules: A Fair Data Driven Comparison Submitted.
29. Yan, P., Chen, G., Cornelis, C., De Cock, M., Kerre, E.E. (2004) Mining Positive and Negative Fuzzy Association Rules. Lecture Notes in Computer Science (Subseries LNAI) 3213 (M.G. Negoita, R.J. Howlett, L.C. Jain, eds.), Springer-Verlag, 2004, 270–276.
30. Zadeh, L.A. (1975) The Concept of a Linguistic Variable and its Application to Approximate Reasoning I, II, III. Information Sciences 8:199–249, 301–357, 9:43–80.
31. Zvieli, A., Chen, P.P. (1985) Entity-Relationship Modeling and Fuzzy Databases. Proceedings of the 2nd International Conference on Data Engineering, 320–327.

Discovering the Factors Affecting the Location Selection of FDI in China*

Li Zhang**, Yujie Zhu, Ying Liu, Nan Zhou, and Guoqing Chen

School of Economics and Management, Tsinghua University, Beijing, 100084, China
zhangl34@em.tsinghua.edu.cn

Abstract. Since the late 1970s, Foreign Direct Investment (FDI) has played an important role in the economic development of China. However, the growth of FDI in China has an increasingly unbalanced development between the eastern and western provinces. In this chapter, the temporal association rule (TAR) mining method is applied in discovering the factors affecting the location selection of FDI. In the light of the data set at the provincial level in the period of 1984–2001, our analysis reveals that the factors, including incoming and saving, residential consumption, GDP, infrastructure, population, education and cumulative FDI amount, have significant impacts on attracting FDI, which is consistent with existing studies. But some of other factors such as means of transportation, the loan policy, and the degree of modernization of the eastern rural area have seldom been discussed, and may represent certain special and unique characteristics of China. Such findings may be helpful for the local governments to take proper measures for attracting more FDI.

Key words: Temporal Association Rule, Foreign Direct Investment, location selection of FDI

1 Introduction

By the end of December 2003, China had ushered in a total of US$501.5 billion in actual FDI, making it the second largest destination for FDI in the world, after the United States. Since the late 1970s, FDI has played an important role in the economic development of China. However, the unbalanced distribution of FDI across provinces within China has widened the gap in economic

*   Partly supported by the National Natural Science Foundation of China (79925001/70231010), and the Bilateral Scientific & Technological Cooperation between China and the Flanders.
**   Corresponding author.

development between the eastern, central and western provinces by accelerating the development of the east provinces. The "By Region" column of Table 1 shows that from 1985 to 1998, the eastern region received a lion's share of the total FDI amount, more than 85 percent, while the central and the western regions together only received less than 15 percent [17]. The reasons why the eastern region has always been the key region of the direct investment of foreigners refer to excellent commercial geography, history, some open and preferential policies, etc.

Table 1. Distribution of FDI across region/provinces within China

	Distribution of FDI (%)					
	By Region			By Province		
Year	Eastern	Central	Western	Guangdong	Fujian	Others
1985	90.18	5.56	4.26	46	11	44
1986	87.95	7.28	4.76	50	6	44
1987	88.58	6.02	5.40	36	4	59
1988	87.00	5.94	7.06	43	5	52
1989	92.16	3.84	3.99	38	11	51
1990	93.87	3.87	2.26	46	9	45
1991	92.46	4.48	3.06	44	11	45
1992	91.30	6.82	1.89	34	13	53
1993	87.38	8.88	3.74	28	11	62
1994	87.83	7.85	4.31	28	11	60
1995	87.71	9.21	3.08	28	11	62
1996	88.04	9.52	2.45	28	10	62
1997	86.90	10.56	2.54	27	9	64
1998	88.04	9.86	2.10	30	10	61

Source: DRI CEIC Database and various issues of the Statistical Yearbook of China

The increasing gap has created some social and political problems and will deter the durative development of national economy. In order to decrease the differences among the regions, China's central government has adopted a series of measures including the policies encouraging FDI in the central and western regions. However, the situation of the Central and Western shows that the policy-makers need more information about how the region attracts more FDI to guide the investment inflow.

The studies on FDI have grown rapidly in resent years. There have been quite a few research efforts on FDI so far. The pioneer industrial organization theory was put forward to explain FDI by Hymer [12]. The well-known "OLI" theory [13] presents that the host country must possess three advantages including Ownership, Location and Internalization (OLI), which explain the who, where and how of FDI respectively, if it tries to attract more FDI. The international trade theory was used to explain the allocation aspects of FDI in [2] and [11]. In our study, we focus on the location perspective, which is often used to explain why a multinational corporation would choose to invest in a particular host country. It can also be used to explain why foreign investors would choose to invest in a specific location within a particular host country [17].

Existing studies have identified some factors that may attract FDI in one country. The studies on FDI in USA [1, 5, 7, 10, 15] accounted for most of the effort compared with that in other countries. The factors found include market potential, labor cost, transportation network, state expenditures, population size, and so on. In recent years, more and more attention has been paid to FDI in China. Chen [4] found such factors as market potential, labor cost, infrastructure (railway) and R&D. Borensztein et al. [6] identified the labor quality to have impact on attracting FDI. Zhang et al. [26] showed that GDP, GDP per capita, cumulative FDI amount, and transportation density were important factors affecting the location selection of FDI in China. Sun [24] identified the preferential policy, industrial structure, degree of openness, market and education as the determinants of FDI across the provinces within China. Based upon these findings, 7 potentially important factors associated with FDI distribution are summarized in Table 2.

In this chapter, we will investigate the factors that may affect the way regions across China attract FDI. First of all, we will examine whether the findings from existing studies are also applicable for the location selection of FDI in China. Secondly, we will explore whether there are factors that are only applicable in certain regions, which may reflect the national characteristics. Such a study is deemed meaningful for policy-makers.

Table 2. Summary of factors

Factor	Index
market potential	It means that the underlying market and the actual market scale, including the indexes such as GDP, GDP per capita, resident consumption, and incoming and saving
Infrastructure	The level of infrastructure is mainly represented by transportation and communication facilities
labor cost	wage
labor quality	education
Population	population size, population growth
degree of openness	It is often measured with the rate of Import/GDP
cumulative FDI amount	–
preferential policy	–
Finance	state expenditures, taxes

Instead of using the conventional approaches that apply empirical analysis or statistics methods based on models of selected variables/factors, our study attempts at employing a novel approach that is data-driven to identify the variables/factors that are associated. This approach, namely temporal association rule (TAR) mining, considers all possible combinations of factors in terms of their interconnections based on certain association measures, and is regarded particularly useful in the case where the data volume is huge primarily with a large number of variables/factors. It is worth indicating that this approach may be used in a supplementary manner to conventional approaches in that TAR identified factors, some of which may be new and interesting, could be further described in terms of their analytical relationships.

The chapter is organized as follows. Section 2 explains how the data is prepared. The TAR method is discussed in Sect. 3. In Sect. 4 and Sect. 5 the mining process with respective data and the mining results are discussed. Notably, for the sake of convenience, the terms variable, factor and index are used interchangeably in the text (otherwise indicated where necessary).

2 Data Preparation

The data set is available at the website of China Economic Information Network (www.cei.gov.cn) developed by China Economic Information Network Data Co. Ltd. whose holding company is State Information Center. The data provided by this website were collected from some authorities including the National Planning Commission, National Bureau of Statistics of China, and State Information Center. The data set over the years 1984–2001 for each of 31 provinces consists of ten sections in terms of the attributes, i.e., finance, foreign trade, GDP (including the primary industry, second industry, tertiary industry and architecture industry), investment in fixed assets, culture & education and sanitation, population, employment and labor cost, people's life, price index and the order of economic indexes for each province. Furthermore, some detail indexes are included in each section. The total number of the indexes is 372. Using the same data set, [26] examined the location selection of FDI in China by building a regression model based on certain selected factors.

Although the data source appears to be authoritative, the data may not be ready directly for use due to existence of certain noises, missing values, inconsistency and/or redundancy. Thus, data cleaning is necessary. Furthermore, data normalization and discretization are carried out in order to obtain appropriate formats for mining purposes.

2.1 Data Cleaning

The data cleaning process attempts to fill out missing values, smooth out noise, remove the redundancy, and correct inconsistencies in the data. First, for each of 31 provincial tables, each of 372 attributes and each of 18 records

Table 3. Steps and results of data cleaning

Step	Object	Operation	Before Operation	After Operation
Step 1	province	Removal of incomplete data at the provincial level	31	19
Step 2	attribute	Elimination of redundant attributes and the attributes with too many missing values	372	More than 200
Step 3	Record (year)	Filling out the missing value	–	–

are examined. Then, correction measures are taken on the deficient data. The steps and result of data cleaning are shown in Table 3.

Step 1 Removal of incomplete data at the provincial level. The whole data set includes 31 data tables. Note that Chongqing is excluded from the list because it was established only in 1997. Further, in some western provinces, such as Tibet, Xinjiang, Qinghai and Guizhou, the receipt of FDI was negligible and they are excluded from the list. Finally, the following 19 provinces are included: namely Anhui, Beijing, Fujian, Gansu, Guangdong, Hainan, Hebei, Heilongjiang, Hubei, Hunan, Jiangsu, Jiangxi, Jilin, NingXia, Shandong, Shannxi, Shanghai, Tianjin, and Yuan'nan. Some of these provinces are located along the Chinese coast, and others are situated in the Center or West of China.

Step 2 Elimination of the attributes with too many missing values and of repeated and redundant attributes. Firstly, the attributes whose values are unbearably incomplete should be deleted. On the other hand, in fact, some indexes are repeated because of multiple data sources. In addition, some other indexes are redundant, for instance, "the number of doctors" have a linear relation with "the number of doctors per ten thousands". In the above-mentioned cases, the repeated or redundant attributes will be eliminated.

Step 3 Filling out the missing value in the records. The missing values are values that actually exist but have gone astray. Clearly where possible they should be looked for, but this may either be impossible or just too expensive and time-consuming. One of the solutions to the problem is to find a suitable value as the replacement of the missing value. Since most missing values are between 1999 and 2001 while economy developed steadily in China during the three years, it is reasonable that the missing values are estimated with the previous and next values of the missing value. The formulation of the filled-in value used is $a_n = (a_{n-1} + a_{n+1})/2$, where a_n, a_{n-1}, a_{n+1} represent the attribute values of the nth, $(n-1)$th and $(n+1)$th year, respectively. For

Table 4. TV cover rate, with missing value for 1995 and 2001

Year	92	93	94	95	96	97	99	00	01
Value	70	77	79	**80**	81	84	87	89	**91**

example, in Table 4, value of 1995 is $a_{95} = (a_{94} + a_{96})/2 = 80$, and the value of 2001 is $a_{01} = 2 * a_{00} - a_{99} = 91$.

2.2 Data Normalization and Discretization

With data normalization, attribute values are scaled so as to fall within a small specified range, such as -1.0 to 1.0. In the chapter, the values with initially large ranges are normalized to the increase percentage as:

$$IP_n = \frac{a_n - a_{n-1}}{a_{n-1}} \times 100\% \tag{1}$$

where IP_n is the increase percentage of the nth year, a_n and a_{n-1} represents the attribute values of the nth, and $(n-1)$th year, respectively. Another issue that should be considered in data preparation is discretization. For most mining methods, discretization of numeric attributes is helpful and sometimes necessary. For our data, it is rational that the data be divided into several intervals using equidistance subsection method. That is, with max(a) and min(a) representing the maximum and minimum values of an attribute, the interval [min(a), max(a)] is equally divided into some sub-intervals. And then these sub-intervals are replaced by some discrete numbers denoting different categories. As shown in Fig. 1, the attribute values are classified into four categories: $[\min(a), \min(a) + c]$ $[\min(a) + c \min(a) + 2c]$ $[\min(a) + 2c \min(a) + 3c]$ and $[\min(a) + 3c \max(a)]$ where c stands for the length of each sub-interval and is calculated as $c = (\max(a) - \min(a))/4$. These four sub-intervals are labeled categories $C1, C2, C3$ and $C4$.

3 The Method

Data mining includes several kinds of technologies such as association rule analysis, classification, clustering, sequential pattern etc. In the chapter, we focus on association rule mining since it has been applied in many fields and

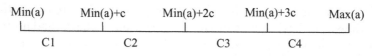

Fig. 1. Four categories

considered an important method for discovering associations among data [23]. Let $I = \{I_1, I_2, \ldots, I_m\}$ be a set of binary items, and D be a database of transactions. Each transaction t is represented as a binary vector, with $t[k] = 1$ if the item I_k occurred, and $t[k] = 0$ otherwise. Let X be a set of some items in I. We say that a transaction t satisfies X if for all items I_k in X, $t[k] = 1$. An association rule is an expression $X \Rightarrow Y$ meaning that if X occurs, then Y occurs at the same time, where X and Y are sets of items, $X \subset I, Y \subset I$, and $X \cap Y = \emptyset$.

Table 5. Number of beds in sanitary organization of a province during 1987 to 2001

Year	The Number of Beds	The Increase Percentage	Categorical Value
1987	10.11		
1988	10.45	0.034	4
1989	10.65	0.019	3
1990	10.78	0.012	2
1991	11.13	0.032	4
1992	11.45	0.029	4
1993	11.67	0.019	3
1994	11.94	0.023	3
1995	11.98	0.003	1
1996	12.08	0.008	1
1997	12.31	0.019	3
1998	12.35	0.003	1
1999	12.29	0.004	1
2000	12.39	0.008	1
2001	12.52	0.01	1

During recent years, many scholars have improved the conventional mining method and corresponding algorithm, namely Apriori algorithm [19], mainly in two directions: algorithmic efficiency [3, 16, 18, 23] and model extensions [18, 20]. However, most of these research efforts do not take into account the time dimension. On the other hand, time series data are common in many domains, such as business and economics in which temporal relationships are very important. In the research on discovering the factors affecting the location selection of FDI, some factors may not affect FDI immediately but in a period of time (e.g., years). Therefore, the approach to discovery of temporal association rules will be used.

Some researchers have extended the classical association rules mining with temporal semantics. However, most of temporal mining is not applied on the data but on the rules extracted from the data at various time points [14]. Reference [8] proposed an approach to extending the notion of a typical $X \Rightarrow Y$ to be a rule of the form $X \overset{T}{\Rightarrow} Y$, stating that if X occurs then Y will occur

Table 6. The reconstructed data set of one province (min_support = 0.3, min-confidence = 0.6)

year	Attri.1	Attri.2	......	Attri.N	FDI	FDI-1	FDI-2	FDI-3
1984	1	1		1	1	1	1	2
1985	1	1		1	1	1	2	3
1986	1	2		1	1	2	3	......
1987	1	2		1	2	3	......	2
1988	2	1		2	3	......	2	1
......	......	......	......	......	......	2	1	3
1999	3	2		2	3	3	4	
2000	4	3		3	3	4		
2001	4	4		4	4			

within time T. They first form subsequences by sliding a window through the time series, and then cluster these subsequences by using a suitable measure of time series similarity. The discretized version of the time series is obtained by taking the cluster identifiers corresponding to the subsequence. Once the time-series is obtained, simple rule finding method is used to obtain rules from sequence. Chen & Yu extended the traditional association rules to temporal association rules with a delay time T [17], which is of the form

$$X \overset{T}{\Rightarrow} Y (T \geq 0) .$$

When $T = 0$, $X \overset{T}{\Rightarrow} Y$ degenerates to traditional association rule $X \Rightarrow Y$. When $T > 0$, it is a temporal association rule showing that if X occurs, Y will occur in the next T units of time. Concretely, after reconstructing the data set (as the shadow area shown in Table 6 of Sect. 4, in the case of $T = 1$), the number of the records in the reconstructed data set is $|D| - T$. Thus, the support degree (Dsupport) and confidence degree (Dconfidence) can be represented as follows [17].

$$\text{Dsupport}(X \overset{T}{\Rightarrow} Y) = \|X \underset{T}{\cup} Y\| / (|D| - T)(T > 0)$$

where $\|X \underset{T}{\cup} Y\|$ is the number of occurrences in that X is followed by Y in delay time T.

$$\text{Dconfidence}(X \overset{T}{\Rightarrow} Y) = \|X \underset{T}{\cup} Y\| / \|X\|'(T > 0)$$

where $\|X\|'$ is the number of occurrences of X in the reconstructed data set. Furthermore if we have $X \overset{T}{\Rightarrow} Y$ for any T value in an interval $[T_1, T_2]$, then we may use an extended form

$$X \overset{[T_1, T_2]}{\Rightarrow} Y (T_1 \leq T_2 \ T_1 \ T_2 > 0)$$

to reflect the association that occurs during the interval.

4 Mining Temporal Association Rules

4.1 Rules at The Provincial Level

For the purpose of discovering the factors affecting the location selection of FDI in a delay-time, a new data set is needed. Let us concentrate on the individual province, as shown in Table 6, where data for FDI-i ($i = 1, 2, 3$) are the FDI values with a delay of i years. Here FDI-i is regarded as a new attribute. Hence, with the reconstructed dataset and regarding each new tuple as a transaction, an association rule has a specific form:

$$\text{the trend of } X \overset{T}{\Rightarrow} \text{the trend of FDI } (T = 0, 1, 2, 3)$$

where the consequent of the rule is the definite item FDI, and X is a subset of the attribute set {Attri.1, Attri.2, ..., Attri.N, FDI-1, FDI-2, FDI-3}.

Using the Apriori algorithm on the reconstructed dataset, the mining process has discovered a good number of rules when $T = 2$ or $T = 3$, as listed in Table 7. Such results indicate that there exist some temporal relationships between the factors and FDI. For instance, the rule FDI - 1 ↑ $\overset{1}{\Rightarrow}$ FDI ↑ represents that the receipt of FDI in the previous year may have a positive effect on attracting more FDI in this year. Here and thenafter, symbol ↑ is used to stand for trend of increase.

For example, rules Education ↑ $\overset{2}{\Rightarrow}$ FDI ↑ and Education ↑ $\overset{3}{\Rightarrow}$ FDI ↑ mean that the improvement of education has a prominently positive effect on the inflow of FDI in two or three years.

4.2 Rules at The Regional Level

In the above stage, TARs are obtained at the provincial level. This subsection discusses the rules discovered, which are based upon the rules extracted from the provincial data.

Let R_i be the rule set of the ith province, so the rule set of all provinces is expressed as follows:

$$R = \overset{19}{\underset{i=1}{\cup}} R_i$$

The support count of a rule at the regional level is defined as the number of the provinces in the region in which the rule is applicable. Obviously, the support degree of the rule is

$$\text{Region\_Dsup} = \frac{|R_x|}{|\text{Region}|} \times 100\%$$

Table 7. The number of the rules when $T = 0, 1, 2, 3$

No.	Province	The Number of the Rules with Delay Time T				
		$T = 0$	$T = 1$	$T = 2$	$T = 3$	Total
1	Anhui	4	5	22	17	48
2	Beijing	1	0	0	0	1
3	Fujian	20	12	43	62	137
4	Gansu	2	18	36	13	69
5	Guangdong	5	26	10	27	68
6	Hainan	7	7	20	15	49
7	Hebei	1	26	39	36	102
8	Heilongjiang	3	9	30	18	60
9	Hubei	6	27	49	31	113
10	Hunan	1	9	20	21	51
11	Jiangsu	5	19	1	10	35
12	Jiangxi	1	3	9	28	41
13	Jilin	1	8	24	33	66
14	NingXia	1	10	43	30	84
15	Shandong	4	11	30	22	67
16	Shannxi	9	12	34	30	85
17	Shanghai	5	21	34	35	95
18	Tianjin	1	21	24	29	75
19	Yuannan	0	2	0	57	29
20	Total	77	246	468	514	1305

Where |Region| represents the number of provinces belonging to the region/country, $|R_x|$ represents in the region/country the support count of the rule Rx. If Region_Dsup of one rule at the regional level is not less than the support threshold, we say the rule is applicable in the region/country and can be considered as one of the regional characteristics.

The mining result shows that no rule exists in the whole country, which may be due to the significant economic difference among provinces. Mining the rules in a particular region, in which the economic levels of the provinces are similar, may be interesting and meaningful. In China, the economy of the Eastern, the Central and the Western is at developed, developing and underdeveloped level, respectively. Furthermore, China's western development strategy is one of the most important projects at present. The research on the difference between the Eastern and the Western has attracted remarkable attention.

Let R be divided into three disconnect subsets

$$R_e = \bigcup_{i=1}^{11} R_i \quad R_m = \bigcup_{i=1}^{4} R_i \quad R_w = \bigcup_{i=1}^{4} R_i$$

where R_e is the rule set of the eleven eastern provinces, including Beijing, Tianjin, Hebei, Shandong, Shanghai, Jiangsu, Fujian, Hainan, Guangdong,

Heilongjiang, and Jilin; R_m is the rule set of the four central provinces, including Anhui, Hunan, Hubei, and Jiangxi; and R_w contains four provinces in the Western: Shannxi, Gansu, Ningxia, and Yunnan. If the support degree Region_ Dsup is equal to or larger than the threshold (e.g., min_sup = 0.6), the rule is deemed applicable in the region.

5 Results and Discussions

As shown in Table 2, the factors identified by existing studies are macroeconomic indexes, such as labor cost, labor quality, market potential and infrastructure. Our study considers these factors, along with some microcosmic variables, such as the number of the students in college, and the TV cover rate. In addition, these variables are classified into the following categories in Eastern, Middle, and Western area; income and saving, resident consumption, finance, infrastructure, education, population, GDP, investment in fixed asserts, income from tourism, and FDI, as summarized in the Appendix.

The results of our study reveal that some factors, such as income and saving, resident consumption, infrastructure, population, education and GDP, etc., had significant effects on the location selection of FDI in each region to different degrees, which are generally consistent with those of existing studies. Furthermore, by breadthwise summarizing and comparing these economic indexes, some of other factors such as means of transportation, the currency policy, and the degree of modernization of the eastern rural area have seldom been discussed previously, and may represent certain special and unique characteristics of China and also confirm the existence of gap between the Eastern and the Western. These factors and the corresponding rules are summarized in Table 8. Except for the last row, each index listed in the third column is the antecedent of the rule, which has a prominently positive effect on the location selection of FDI in the following three years. The last column is the region in which the rule can be applicable. For instance, in the Eastern, "Percentage of investment in fixed assets, all entities" has a positive effect on attracting FDI, which can be denoted by

$$\text{Percentage of investment in fixed assets, all entities} \uparrow \overset{[1,3]}{\Rightarrow} \text{FDI} \uparrow$$

Result 1 rules about investment in fixed asserts. The investment in fixed asserts refers to the economic activities of constructing and purchasing fixed asserts, which mainly calculates the actual investment in fixed asserts in one region and can reflect the capital scale. The investment in fixed asserts is a synthetical factor, involving the investment in transportation and communication facilities, farmland and water conservancy, power facilities and houses. All of these are indeed the most concerned part for foreign investors. With more investment in fixed asserts, the regions have a better investment environment and could attract more FDI. At present, most western provinces

Table 8. Factors and rules that could represent certain special characteristics of China

Result No.	Factors	Indexes (X) Rule: $X \uparrow \overset{[1,3]}{\Rightarrow}$ FDI $\uparrow$	Region
1	Investment in fixed asserts	Percentage of investment in fixed assets, all entities	Eastern
2	means of transportation	Added value index of transportation, storage, and postal and telecommunication services	Eastern
		Tons of cargo carried through railway	Central Western
		Tons of freight carried on waterways	Centra
		Tons of cargo carried through highways	Western
3	education	Enrollment at secondary schools	China
4	currency policy	Total amount of deposits in state-owned commercial banks	
		Total amount of loans from state-owned commercial banks	Central Western
		Total amount of loans to commercial enterprises from state-owned commercial banks	
		Total amount of loans to industrial enterprises from state-owned commercial banks	Western
5	the degree of modernization of the eastern rural area	Transaction amount in bazaar trade Total retail sales on consumer products in rural areas Total mechanical power owned in rural areas	Eastern
		Total retail sales on consumer products of collectively-owned businesses and others	Western
		Total retail sales of consumer products of individually-owned businesses	Central
6	self-reinforcing effect of FDI	$FDI \uparrow \overset{1}{\Rightarrow}$ FDI $\uparrow$	Eastern

have begun to seize the opportunities of implementing the west development strategy to strengthen the investment in fixed assets.

Result 2 rules about means of transportation. In each region, the factor infrastructure is always significant. But, note that the factor is materialized by tons of cargo carried through railway and highways for the Western, tons of cargo carried through railway and tons of freight carried on waterways for the Central, and added value index of transportation, storage, and postal and telecommunication services for the Eastern respectively. These rules are consistent with the fact that the road and railway are the main means of transportation in the western region while in the Central the water carriage is very often because all the four chosen provinces situating by the Yangtze River and Yellow River. Many means of transportation are used in the Eastern where possesses excellent geographical environment so that the synthetical factor "Added value index of transportation, storage, and postal and telecommunication services" emerged.

Result 3 rules about education. Education is a publicly accepted factor that could affect the location selection of FDI in the research of other countries, as well as of China. In the Appendix, we can find that the representation of Education only involves the indexes relating to secondary schools. The reason is that the investments of foreign companies are mainly clustered in the manufacturing and processing industries that do not require a high labor quality. However, FDI in China is coming into a new stage with China's entry into WTO. As shown in the appendix, in the Eastern and Central, the tertiary industry has begun affecting FDI. It is expected that Education will be more important in the future and foreign companies will employ more people that are better educated.

Result 4 rules about the currency policy. These rules show that governments play an important role in improving the investment environment of the Central and Western to attract FDI. In undeveloped regions, the Central and Western have to improve the investment environments by largely relying on the loan of national banks, while the Eastern as a developed region needs fewer support of this kind from governments.

Result 5 rules about the degree of modernization of the eastern rural area. Transaction amount in bazaar trade and Total retail sales on consumer products in rural areas in the Eastern shows that the east rural area has a relative large market potential and development from town to city. While both Total retail sales on consumer products of collectively-owned businesses and others in the Western, and Total retail sales of consumer products of individually-owned businesses in the Central could impact on the location selection of FDI. The above rules reflect the consumption market structure across regions in China, and meanwhile, such structure affects the distribution of FDI.

Result 6 rules about self-reinforcing effect of FDI. This is a public held rule in exiting studies. However, it is worth noticing that the rule holds only in the Eastern. Observing the data values of FDI of each province, we find that the volume of FDI is very small in the western provinces and they can hardly produce the "scale effect". As the Central and the Western get more developed, and with more FDI attracted, it is expected that the self-reinforcing effect will emerge.

6 Conclusions

Different from most of existing studies, this chapter has used a novel approach, namely temporal association rules mining, to discovering the factors that affect attracting FDI in China. The results have revealed factors that are consistent with those obtained in existing studies, along with certain other factors that may only pertain to the context of China.

Appendix

Income & Saving	
Eastern	Balance of saving deposit of town resident in year-end
	Average annual per capita net income of rural families
Middle	Balance of saving deposit of town resident in year-end
Western	Balance of saving deposit of town resident in year-end
	Average annual per capita disposable income
	Index of government purchasing prices for agriculture products

Resident Consumption	
Eastern	Consumption level of total residents
	The final consumption
	Average annual per capita consumptive expenditure of town residents
	Total retail sales on consumer products in rural areas
	Transaction amount in bazaar trade
Middle	Resident consumption
	Total retail sales of consumer products of individually-owned businesses
Western	Consumption level in total residents
	Total retail sales on consumer products of collectively-owned businesses and others
	Average annual per capital expenditure on food consumption of rural households

Finance

Eastern	Tax revenue of local government, all items
Middle	Tax revenue of local government, all items
	Local Financial Expenditure
	Local financial assistance to agricultural and non-profit units
	Total amount of deposits in state-owned commercial banks
	Total amount of loans from state-owned commercial banks
Western	Local financial expenditures
	Local financial assistance to agricultural and non-profit units
	Total amount of deposits in state-owned commercial banks
	Total amount of loans from state-owned commercial banks
	Total amount of loans to commercial / industrial enterprises from state-owned commercial banks

Infrastructure

Eastern	Added value index of transportation, storage, and postal and telecommunication services
	Total transactions of postal and telecommunication services
	Total number of passenger cars owned
	Total number of telephone connections owned
	Medical facility
	Electronic power consumption in rural areas
	Total mechanical power owned in rural areas
Middle	Total transactions of postal and telecommunication services
	Total number of passenger cars owned
	Tons of cargo carried through Railway
	Tons of freight carried on waterways
	Medical facility
Western	Total number of passenger cars owned
	Tons of cargo carried through railway
	Tons of cargo carried through highways
	Total number of passenger carried

Education

Eastern	Enrollment and full-time teachers at secondary schools (excluding professional schools)
	Full-time classroom teachers at secondary schools
	Number of art communities
	Number of journals published
Middle	Enrollment at secondary schools (excluding professional schools)
	Enrollment at secondary schools
	Full-time classroom teachers at secondary schools
	Number of journals published
	School attendance rate
Western	Enrollment at secondary schools
	Number of books published

Population		
Eastern	Rural population	
	Urban population	
	Death rate	
Middle	Year-end Permanent Population	
Western	Death rate	

Income from tourism	
Middle	Foreign exchange income from international tourism

Investment in fixed assets	
Eastern	Percentage of investment in fixed assets, all entities
	Investment in fixed assets in state-owned enterprises

GDP	
Eastern	Primary Industry:
	Gross output index of Agriculture, Forestry, Animal Husbandry, Fishery
	Gross output index of agriculture
	Gross output index of forestry
	Gross output index of fishery
	Second Industry:
	Gross output of medium-sized industrial enterprises
	Gross output of tobacco, iron, and cement products
	Tertiary Industry:
	Added value index of wholesale, retail trade and food services
Middle	Primary Industry:
	Gross output index of Agriculture, Forestry, Animal Husbandry, Fishery
	Second Industry:
	Gross output of industry
	Gross output of yarn, salt, and steel
	Tertiary Industry:
	Added value index of wholesale, retail trade and food services
Western	Primary Industry:
	Gross output index of Agriculture, Forestry, Animal Husbandry, Fishery
	Second Industry:
	Gross output of large industrial enterprises
	Gross output of industry
	Gross output of salt, coal, and sugar
	Electronic power generation

FDI	
Eastern	The self-reinforcing of FDI

References

1. Bagchi-Sen, S., Wheeler, J. (1989), A spatial and temporal model of foreign direct investment in the United States. Economic Geography 65 (2), 113–129.
2. Bhagwati, J.N., Srinavasan, T.N. (1983), Lectures in International Trade. MIT Press, Cambridge, MA.
3. Brin S., Motwani R., Ullman J., Tsur S. (1997), Dynamic Itemset Counting and Implication Rules for Market Basket Data. *In Proc. Of the 1997 ACM-SIGMOD Int'l conf on the Management of Data*, 255–264.
4. Chen, Chien-Hsun (1996), Regional determinants of foreign direct investment in Mainland China. Journal of Economic Studies 23:18–30.
5. Coughlin, C.C., Terza, J.V., Arromdee, V. (1991), State characteristics and the location of foreign direct investment within the United States. Review of Economics and Statistics 73 (4), 675–683.
6. Borensztein, E., De Gregorio, J., Lee. J.-W. (1997), How does foreign direct investment affect economic growth? Journal of International Economics 45, 115–135.
7. Friedman, J., Gerlowski, D.A., Silberman, J. (1996), Foreign direct investment: the factors affecting the location of foreign branch plants in the United States. Global Finance Journal 7 (2), 209–222.
8. Gautam Das, King-Ip Lin, Heikki Mannila, Gopal Renganathan, Padhraic Smyth (1998), Rule discovery from time series. In *Proceedings of the international conference on KDD and Data Mining (KDD-98)*.
9. Gautam Das, King-Ip Lin, Heikki Mannila, Gopal Renganathan, Padhraic Smyth Graham, E., Krugman, P. (1991), Foreign Direct Investment in the United States, 2nd ed. Institute for International Economics, Washington, DC.
10. Grossman, G.M., Helpman, E. (1991), Innovation and Growth in the Global Economy. MIT Press, Cambridge, MA.
11. Hymer, S. (1960), The International Operations of National Firms: A Study of Direct Foreign Investments. MIT Press, Cambridge, MA.
12. Dunning, J. (1977), Trade, Location of Economic Activity and the Multinational Enterprise: A Search for an Eclectic Approach, in B. Ohlin, P.O. Hesselborn and P.M. Wijkman (eds.), *The International Allocation of Economic Activity*, Macmillan, London, pp. 395–418.
13. John F. Roddick, Myra Spiliopoulou (2002), A survey of temporal knowledge discovery paradigms and methods. IEEE transactions on knowledge and data engineering, vol. 14, No. 4, July/August.
14. Lipsey, R. (1993), Foreign direct investment in the United States: changes over three decades. In: Froot, Kenneth (Ed.), Foreign Direct Investment. University of Chicago Press, Chicago, pp. 113–170.
15. Park J.S., Chen M., Yu P.S. (1995), An Effective Hash Based Algorithm For Mining Association Rules. *In Proc. of SIGMOD Intl. Conf. on Management of Data*, San Jose, CA, May 1995. 175–186.
16. Qian Sun, Wilson Tong, Qiao Yu (2002), Determination of foreign direct investment across China, *Journal of International Money and Finance*, 21 (2002) 79–11.
17. R. Agrawal, R. Srikant (1994), Fast Algorithms for Mining Association Rules, *Proceedings of VLDB Conference*, Santiago, Chile, Sept. 1994. Expanded version available as IBM Research Report RJ9839, June.

18. R. Agrawal, T. Imielinski, A. Swami (1993), Mining association rules between sets of items in large database. In *Proc. of the ACM SIGMOD Conference on Management of Data*, pp. 207–216, Washington, D.C., May.
19. R. Srikant, R. Agrawal (1995), Mining Generalized Association Rules, *Proceedings of the 21st VLDB Conference* Zurich, Swizerland.
20. R. Srikant, R. Agrawal (1996), Mining Quantitative Association Rules in Large Relational Tables, *SIGMOD'96 6/96* Montreal, Canada.
21. Rastogi R., Shim K. (1998). Mining Optimized Association Rules with Categorical and Numerical Attributes, *In Proc. of the 14th Int'l Conf. on Data Engineering*, 503–512.
22. Savasere, E. Omiecinski, S. Navathe (1995), An Efficient Algorithm for Mining Association Rules in Large Databases, *Proceedings of the VLDB Conference*, Zurich, Switzerland, September.
23. Sun J. (2002), the Analysis of the factors of FDI location selection in China, China Economic Quarterly, Vol. 1, No. 3, Apr.
24. Yu. W., Chen G. (2002), Mining delayed association rules based on temporal data, *computer application and research*, November.
25. Zhang L., Gong Y.C. (2002), Determinants of foreign direct investment across provinces in China. The research report of Tianjin University (social science version), Vol. 4, Issue 2: 117–121, June.
26. Zhu Y.J., Empirical Analysis the factor affecting regional selection of FDI in China, International Economic Cooperation, 2003. 11.

Penalty-Reward Analysis with Uninorms: A Study of Customer (Dis)Satisfaction

Koen Vanhoof[1], Pieter Pauwels[2], József Dombi[3], Tom Brijs[1], and Geert Wets[1]

[1] Dept. of Economics, Limburgs Universitair Centrum, Belgium, Universitaire Campus, 3590 Diepenbeek, Belgium
[2] Dept. of Marketing, Maastricht University, The Netherlands
[3] Dept. of Applied Informatics, University of Szeged, Hungary
koen.vanhoof@luc.ac.be

Abstract. In customer (dis)satisfaction research, analytic methods are needed to capture the complex relationship between overall (dis)satisfaction with a product or service and the underlying (perceived) performance on the product's or service's attributes. Eventually, the method should allow to identify the attributes that need improvement and that most significantly enhance the business relationship with the customer. This paper presents an analytic design based on uninorms, which is able to capture the nature of the relationship between attribute-level (dis)satisfaction and overall (dis)satisfaction in the context of different attributes. In contrast to alternative statistical approaches, ours allows for full reinforcement and compensation in the satisfaction model without a priori defining the formal role of each attribute. Impact curves visualize the relationships between attribute-level (dis)satisfaction and overall satisfaction. Penalty-reward analysis on the basis of uninorms is illustrated on a satisfaction study of an energy supply firm. The analysis confirms the three-factor structure of (dis)satisfaction. The interpretation of the impact curves allow managers optimizing their attribute scores in order to maximize customer (dis)satisfaction.

1 Introduction

Over the last forty years, consumer (dis)satisfaction has taken a prominent position in the marketing research literature (e.g., [5, 6, 9, 10, 25, 26, 29, 45]). This attention is justified since consumer (dis)satisfaction (directly or indirectly) impacts upon repurchase intention [35] , consumer retention [1, 24], on firm performance [2], and eventually on shareholder value [3]. Consumer (dis)satisfaction is a summarizing response that results from a consumer's post-consumption cognitive and affective evaluation of a product or service performance given pre-purchase expectations [5, 27, 29, 36].

Koen Vanhoof et al.: *Penalty-Reward Analysis with Uninorms: A Study of Customer (Dis)Satisfaction*, Studies in Computational Intelligence (SCI) **5**, 237–236 (2005)
www.springerlink.com   © Springer-Verlag Berlin Heidelberg 2005

The debate on the nature of the relationship between consumer (dis)satisfaction and its antecedents is ongoing. Initially, the effects of the antecedents and in particular of attribute-level performance on consumer (dis)satisfaction were assumed linear, full-compensatory and symmetric [25, 31, 32]. Only recently, marketing scholars have questioned this double assumption on the basis of economic and psychological theory e.g., [4, 5]. Moreover, performance-importance analysis [23], regression analysis [25] and critical incident analysis (e.g., [33]) among others have empirically confirmed non-linearity, non-compensation and asymmetry in the relationship between attribute-level performance and overall (dis)satisfaction. However, these model-based analyses require that specific rules on the relationship between attributes and the overall (dis)satisfaction are postulated a priori. [22] and [40], among others, have called for the use of artificial neural networks to study customer (dis)satisfaction.

In this chapter we present an analytic design that is able to capture the relationship between consumer (dis)satisfaction and its different attributes. Overall (dis)satisfaction can have multiple causes and certain combinations may interact to increase the probability of (dis)satisfaction. While we presume non-linearity and non-compensatory effects in these relationships, we may not be able to postulate a priori which attribute plays which rule and how the attributes interact to affect the overall (dis)satisfaction score. Therefore, we need an empirical way to summarize the importance of each attribute while taking into account the existence of multiple causes of (dis)satisfaction.

In the remainder of this chapter, we firstly develop and adopt uninorms to summarize the satisfaction evaluation process. Thereby, we acknowledge and integrate particular characteristics of the consumer's information-processing process such as reinforcement (i.e., high (low) attribute-level satisfaction results in even higher (lower) overall satisfaction) and compensation (i.e., at the attribute level high performance scores compensate low scores) without defining a priori how each attribute will behave in the customer's information-processes process. Secondly, we use this model to calculate impact figures of attribute scores on the overall (dis)satisfaction score as well as to categorize the attributes through penalty-reward analysis [7]. In a case study, we show the impact curves and the corresponding classification. We round up this article with a discussion of academic and managerial consequences.

2 Heuristics in (Dis)Satisfaction Theory

It is widely accepted that customers generate a multi-attribute-based response on their satisfaction with a certain product or service. For instance, with respect to his car, a customer may be satisfied with the engine's performance, the car's reliability, image, etc. yet unsatisfied with the car's fuel consumption, its luxury level, etc. Next, a consumer is expected to aggregate the attribute-level responses of (dis)satisfaction to the product or service level: an overall reflection of (dis)satisfaction. This aggregation process is presumed to be a heuristics-based decision-making process. Basically, a consumer processes information to come to a decision on whether and to what extend s/he is (dis)satisfied with a product or service. In this respect, two well-described phenomena stand out: "anchoring and adjustment" and "reinforcement".

Anchoring is a classic decision-making heuristic [37, 38]. While forming an overall satisfaction response, a consumer assesses the attribute-level satisfaction scores against an individual product-level anchor or norm. This norm emerges of previous experiences with this or a comparable product or service. Therefore, it is highly complementary to the "experience-based norm" as put forward in the context of satisfaction research [42]. During the assessment, a consumer basically evaluates whether an attribute-level satisfaction response is below, above or equal to his/here individual norm.

Typical for an experience-based norm is that it captures both ideal and realistic expectations [8]. A major consequence of this is that this norm allows for a "zone of tolerance" which is defined as the range between realistic expectations (based on and adapted after previous experience) and ideal expectations [46]. Within this zone of tolerance, the effect of a discrepancy between the attribute-level satisfaction response and the norm is not reflected in the overall satisfaction score. Both inertia [20] and assimilation [30] explain the size of this zone of tolerance. As both ideal expectations and the experience with the product or service are consumer-specific, so are the zone of tolerance and the overall product-level norm.

A second heuristic is reinforcement. Mainly due to contrast effects [6, 11, 28] people increasingly exaggerate evaluations when they fall short of or exceed expectations. As a consequence, average attribute-level satisfaction scores that all fall below (exceed) the product-level norm are expected to aggregate to an overall product-level satisfaction score that is lower (higher) than the weighted average of the attribute-level scores.

Anchoring and reinforcement define a typology of satisfaction aggregation (Table 1).

Table 1. Evaluation Patterns

Situation	s_1	s_2	s_3	s	Phenomenon
A	5	3	6	4	Compensation
B	5	6	5	7	Upward reinforcement
C	3	3	4	2	Downward reinforcement
D_1	5	7	6	6	Individual i
D_2	5	7	6	5	Individual j

In a linear, full-compensatory perspective, one expects situation A in Table 1. The overall satisfaction S is a weighted average of the attribute-level satisfaction responses s_1 to s_3. Attribute-level responses of satisfaction compensate each other. Although compensation behavior does occur in real life [31], it is expected that it does not suffice to capture the relationship between attribute-level and overall satisfaction. Situations B and C represent reinforcement [44]. A collection of high scores reinforce each other to give a resulting score more affirmative then any of the individual score alone. A collection of low scores can reinforce each other to give a resulting score more "disfirmative" then any of the individual scores. Situation B is called upward reinforcement and C is called downward reinforcement. Finally, situations D_1 and D_2 in Table 1 are identical with respect to the attribute-level satisfaction responses. Yet, the overall satisfaction is different. We see two potential reasons: the attribute weights are different for each consumer and/or D_1 and D_2 differ with respect the individual norm. D_1 and D_2 differ to the anchoring effect.

Given this typology, an evaluation procedure is needed that can cover full reinforcement, compensation and personalization in one framework, without a priori identifying how each attribute behaves in the model. More specifically, we present a method that acknowledges: (1) the fact that attribute-level scores may have different weights, (2) the impact of a consumer-specific product-level norm as a basis for disconfirmation (anchoring), (3) the existence and neutral effect of an individual zone of tolerance, and (4) the possibility of positive/negative reinforcement.

3 Aggregation Theory

3.1 Uninorms

Aggregation operators serve as a tool for combining various scores into one numerical value. The theoretical properties of more than 90 different families of operators have been studied and the reader is referred to [14, 47] for an overview. The family of uniform aggregation operators (uninorms) has two interesting characteristics that are of use in this research. Uninorms incorporate full reinforcement and compensating behavior.

Formally, a uninorm is a function $U : [0,1] \times [0,1] \rightarrow [0,1]$ having the following properties:

$$-U(x,y) = U(y,x)$$ Commutativity
$$-U(x,y) \leq U(u,y), if\ x \leq u$$ Monotonicity
$$-U(x,U(y,z)) = U(U(x,y),z)$$ Associativity
$$-\exists\ !e \in [0,1]\ \forall x \in [0,1] : U(x,e) = x$$ e is the neutral element

Applied to our satisfaction problem, a uninorm has the following characteristics. Being commutative, the overall satisfaction is independent of the order of the attribute-level satisfaction responses to be aggregated. The uninorm is monotone in that the overall satisfaction will not decrease if one attribute-level satisfaction response increases. Associativity enables the extension of a two argument function to an n-argument function. Finally a uninorm has a neutral element which is a specific attribute-level satisfaction response that does not influence the overall satisfaction. This neutral value – or anchor – is unique for each respondent. Attribute-level satisfaction responses higher than the neutral value can be seen as positive scores, attribute-level satisfaction responses lower can be seen as a negative score. Reference [43] showed that uninorms show full reinforcement behavior, if the neutral element e is different from zero or one. Reference [12] showed that uninorms incorporate a compensating behavior between positive and negative scores $(\min(x, y) = U(x, y) = \max(x, y))$. In other words, uninorms show a compensating behavior between positive and negative scores.

An important class of uninorms are the representable uninorms with generator function g holding the following characteristics [19]:

$$g : [0, 1] \rightarrow [0, 1] \text{ which define the uninorm via}$$
$$U(x, y) = g^{-1}(g(x) + g(y)) \qquad (1)$$

Examples of representable uninorms and their plots are provided in [19]. This generator function g is continuous, strictly monotonously increasing and has a neutral value for which $g(e) = 0, g(0) = -\infty$ and $g(1) = \infty$. The inverse generator function has a non-linear sigmoid shape and has properties conform to the properties of the relationship between consumer (dis)satisfaction and its antecedents [4, 40].

With every representable uninorm we can also define a corresponding negation function $n : [0, 1] \rightarrow [0, 1] : n(x) = g^{-1}(1 - g(x))$, for which the following properties hold:

1. $n(0) = 1$ and $n(1) = 0$
2. $n(n(x)) = x$
3. $n(e) = e$
4. $U(x, e) = x$
5. $U(x, n(x)) = e$

When we can learn the generator function from the empirical data, we can construct the corresponding uninorm, the corresponding negation function and use property three to calculate the neutral value. Reference [39] showed that the neutral value can be considered as a proxy for customer expectations, which on their turn fulfill the anchor function in the customer's decision-making heuristic [8]. In sum, aggregation theory, and more in particular representable uninorms construct an evaluation function that captures human evaluation acknowledging both compensation and reinforcement behavior.

3.2 Fitting the Data with Uninorms

Reference [13] showed that if $g(x)$ is the generator function of the uninorm operator then the function displaced by $\alpha : g(x + \alpha) = g_\alpha(x)$ also possesses the properties of the generator function. The neutral value "e" varies, which allows the formation of uninorm operators with different neutral values from one generator function. Examples of these generator functions can be found in [13]. We apply the generator function g_α of Dombi's aggregative operator [13]. By extending ((1) to n arguments and rewriting (1) as

$$g_\alpha(x_1) + \ldots + g_\alpha(x_n) - g_\alpha(U(x_1, \ldots, x_n)) = 0$$

and filling in $x_1, \ldots x_n$ with $s_1 .. s_n$ the measured satisfaction levels of the attributes and replacing $(U(x_1, \ldots, x_n))$ with s the measured overall satisfaction we obtain

$$g_\alpha(x_1) + \ldots + g_\alpha(x_n) - g_\alpha(u(x_1, .., x_n)) = g_\alpha(s_1) + \ldots + g_\alpha(s_n) - g_\alpha(s) = 0 \quad (2)$$

From (2) we can learn the value of α and with this value we can calculate the neutral value e for every correspondent. Examples of cases with three attribute satisfaction levels s_i and an overall satisfaction level s are given in Table 2 with the corresponding calculated neutral value as output from (2).

Table 2. Calculated Neutral Values

Situation	s_1	s_2	s_3	s	Calculated Neutral Value(e)
A	5	3	6	4	5.0
B	5	6	5	7	4.4
C	3	3	4	2	4.1
D_1	5	7	6	6	6.0
D_2	5	7	6	5	6.5

From Table 2, we can read the impact of the neutral value as reference point for the aggregation of $s_1 - s_3$ to the overall degree of satisfaction s. For D_1 and D_2, the difference in s is only explainable through the neutral value, which can be considered as a calculated expectation level [39].

Next, we calculate the individual learned evaluation function or individual uninorm U_i, which has the following properties:

1. it is associative
2. it is continuous, except for points $(0, 1)$ and $(1, 0)$
3. it is strictly monotonously in $(0, 1) \, x \, (0, 1)$
4. it is commutative.
5. $U_i(x, n(x)) = e_i, x \neq 0, x \neq 1$
6. $U_i(x, y, e_i) = U_i(x, y)$
7. if $x, y \leq v$, then $\min(x, y) \geq U_i(x, y)$
8. if $x, y \geq v$, then $U_i(x, y) \geq \max(x, y)$
9. if $x \leq v \leq y$, then $\min(x, y) \leq U_i(x, y) \leq \max(x, y)$

The combination of properties 2, 3 and 6 enables us to calculate specific contributions.

We define the impact of an attribute value as its mean contribution to overall (dis)satisfaction. This contribution of value x of attribute j to overall satisfaction of customer i can be defined by the following difference:

$$\text{Contrib}\,(x, j, i) = U_i(x_1, \ldots, x_{j-1}, x, x_{j+1}, \ldots, x_n)$$
$$- U_i((x_1, \ldots, x_{j-1}, e_i, x_{j+1}, \ldots, x_n) \tag{3}$$

with U_i the uninorm of customer i
     e_i the neutral value of the uninorm
     $1..n$ the number of satisfaction attributes

As such, the effect on overall satisfaction of replacing the attribute score by the neutral score is calculated. This effect can be positive or negative. As a result, we can determine the importance of each attribute value while taking into account the existence of the other attributes. This is important because the attributes show a high intercorrelation. This indicates that we do not know which attribute is the true cause of (dis)satisfaction and that we do not set specific roles per attribute a priori. Linear statistical models tend to choose one of the highly correlated variables at the expense of the others. Since we are trying to make specific recommendations a priori attribute selection is counterproductive. Unless we have theoretical reasons to prefer one attribute over another as the cause of (dis)satisfaction, it is better to keep all attributes in the system, even when they show high intercorrelation.

All satisfaction patterns of Table 1 are easily understood when we consider the value of the neutral element as a personal, product-level expectation value or as an anchor in the decision-making heuristic. Next table extends the previous table with the contribution figures.

From Table 3, we can conclude the following. Situation A represents compensation as the scores are higher and lower than expected. The contributions are positive and negative or zero when the score equals the expectation level. Situation B represents upward reinforcement. The scores of the attributes are higher than expected, the contributions are positive and the global score is the highest value. Situation C is the opposite of B and represents downward reinforcement. The scores of the attributes are lower than expected, the

Table 3. Comparing the attribute scores with the neutral value

Situation	s_1	Contrib(s_1)	s_2	Contrib(s_2)	s_3	Contrib(s_3)	s	e_i
A	5	+0.0	3	-2.0	6	+1.0	4	5.0
B	5	+0.5	6	+1.4	5	+0.5	7	4.4
C	3	-0.9	3	-0.9	4	-0.1	2	4.1
D_1	5	-0.1	7	+0.1	6	+0.0	6	6.0
D_2	5	-1.5	7	+0.6	6	-0.6	5	6.5

contributions are negative and the global score is the lowest value Finally, D illustrates how a difference in the initial expectations levels with similar attribute scores results in a different global score.

3.3 Calculating Impact Figures

Next, we want to assess the dynamic effect of each attribute (i.e., the degree of (dis)satisfaction of each attribute) on the overall degree of (dis)satisfaction. The impact figure for every attribute value a is the contribution of that attribute averaged over cases with attribute value a, or:

$$\text{Impact}(x = \text{``}a\text{''}, j) = \sum \text{contr}(x, j, i) / \#(i | x = \text{``}a\text{''}, j) \qquad (4)$$

We can draw the impact values on the Y-axis and the attribute value on the X-axis and obtain an impact curve. For classification, we adopt the three-factor theory of satisfaction which builds upon [18] quality model. The three-factor theory distinguishes between basic factors, excitement factors and performance factors (see Fig. 1). Basic or utility-preserving factors do not generate customer satisfaction if fulfilled. If the customer's expectations on these basic factors are meet or even exceeded, they merely avoid dissatisfaction. The customer regards these as prerequisites or qualifiers. In contrast, excitement factors increase customer satisfaction if delivered but do not cause dissatisfaction if they are not delivered. Excitement factor may surprise the customer and generate "delight". Performance factors lead to satisfaction if performance is high and to dissatisfaction if performance is low [21].

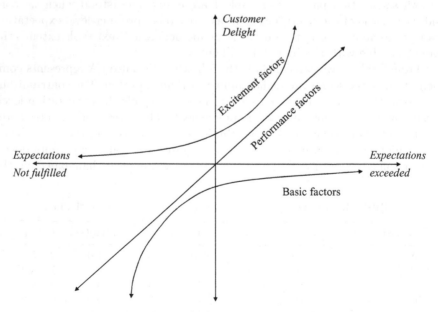

Fig. 1. Three-factor theory [18, 21]

4 Case Study: Illustrating the Three-Factor Theory

In this section, we present a case study to illustrate the potential of uninorms to capture the relationship between attribute-level satisfaction and overall satisfaction, and more in particular the existence of basic, excitement, and performance factors. We rely upon a recent customer satisfaction study of an utility market and service provider. The database provides the satisfaction scores of 1310 randomly chosen users, who have self-reported their (dis)satisfaction with 17 service attributes as well as their overall satisfaction with the service. Overall satisfaction was measured on a 10-point performance scale. Attributes are measured on a 5-point scale.

4.1 Results

Following the analytic approach discussed earlier in this chapter, we created impact curves for each of the attribute scores across respondents. Next, we grouped the attributes in the three aforementioned categories. Figure 2 shows the impact curves of the attributes waiting time, accessibility, quality of solution, convenience of solution, clear and correct invoice, quality of employee.

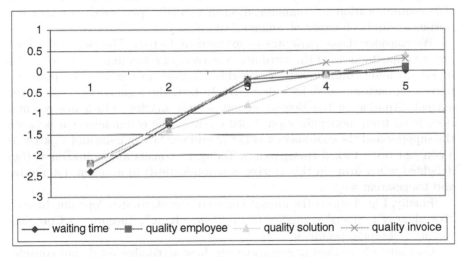

Fig. 2. Impact Curves of Basic Factors

The attributes illustrated in Fig. 2 are considered basic factors. Customers report clear expectations on each of these factors. Figure 2 illustrates that a "high quality of solution", "high quality of employees", "correct and clear invoices", "accessible firm with short waiting times" are qualifiers. Failing to perform at least as expected will result in a significant decrease in the overall satisfaction score. Even when the firm reaches maximal scores on each of these factors, the impact on the overall satisfaction is 0.5 at best.

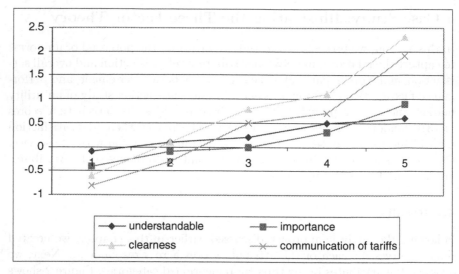

Fig. 3. Impact curves of excitement factors

Figure 3 illustrates the impact curves of the attributes "understand-able communication", "communication of tariffs", "perceived customer importance", and "clearness of communication".

We consider these attributes as excitement factors. The two upper lines are the impact curves of the attribute "clearness of communication" (triangles in Fig. 3) and "communication of tariffs" (crosses in Fig. 3). The potential impact on customer delight is dramatic. In fact, the explanation rests in the current situation of the Belgian energy supply market, which has recently been freed from monopolization. Until recently, the communication between the supplier and the customers was not (could not be) an issue and – partly as a consequence – lacked transparency. Today, customers are (would) now (be) delighted when firms in this market communicate(d) in an open, interactive and transparent way.

Finally, Fig. 4 shows the impact curves of the attributes "opening hours", "friendliness", "flyers", "quality of problem solving", "accuracy of problem solving", "keeping promises".

Customers have clear expectations on these attributes yet do not consider performance on these attributes as obvious. Poor as well high performance significantly impacts upon the overall satisfaction score. Therefore, we consider these attributes performance factors.

4.2 Validation

When confronted with the results of this study, the company's quality experts agreed that the methodology, presented in this paper, was able to add

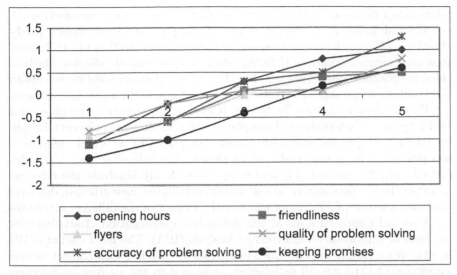

Fig. 4. Impact curves of performance factors

information to their present way of analyzing data. The results were in congruence with previous findings and therefore have face validity.

The results confirm recent empirical work on consumer (dis)satisfaction (e.g., [4, 25]. More specifically, a difference was made between utility-preserving attributes or qualifiers (e.g., bacterial hygiene in restaurants) and utility-enhancing attributes or success factors (e.g., social skills of waiters). In line with the marketing strategy [15, 16] and the service marketing literature (e.g., [17]), negative asymmetry (i.e., negative performance has more impact than positive performance) was reported for utility-preserving attributes and positive asymmetry (i.e., positive performance has more impact than negative performance) for utility-enhancing attributes [21, 25]. This negative and positive symmetry is clearly demonstrated by the impact curves of the basic and excitement factors and is also present in the impact curves of the performance factors.

5 Discussion and Conclusion

In this section, we discuss the potential of uninorms to capture customer (dis)satisfaction in contrast to more traditional methods. Secondly, we shortly point at relevant managerial consequences from this approach.

5.1 Methodological Power of Uninorms

To date and to the best our knowledge four alternative methods have been adopted in marketing to assess the asymmetric and non-linear character of the relationship between attribute-level satisfaction and overall satisfaction: (1) the Critical Incident Technique (CIT), (2) Importance-Performance Analysis (IPA), (3) logarithmic regression with dummies (LRD), and (4) Penalty-Reward Contrast Analysis (PRCA). Hereafter, we shortly discuss/illustrate these approaches and assess whether they offer a (better) solution to capture the typology as presented in Table 1 and/or to assess the three-factor theory.

The Critical Incidence Technique (CIT) is a qualitative technique that builds upon content analysis of complaints and compliments on events. Despite its limitations (as discussed for instance in [22], CIT is widely used as it has proven to be a powerful tool to identify the role of service attributes. Indeed, [17], Stauss and [34], and many more clearly illustrate the relevance of a three-factor perspective on satisfaction attributes. Nevertheless, due to its qualitative nature, CIT cannot measure the impact of the different attributes.

A second approach aims at the descriptive classification of attributes on the basis of Importance-Performance Analysis (IPA). The basic output of IPA is a two-dimensional grid in which attributes are plotted with respect to their importance to the overall satisfaction score and to the current performance level of this attribute. While basically assuming linearity and symmetry, IPA can account for asymmetry in the effect of attributes. For instance, instead of one, [21] assigned two regression coefficients to each attribute – one to measure the impact when performance is low, the other one when performance is high. This way, IPA can distinguish between utility-preserving and utility-enhancing attributes. At best, however, this approach assumes asymmetry (not non-linearity) in the relationship more than it tests it. Moreover, [22] question the theoretical underpinning of IPA. Nevertheless, this approach has proven especially useful for managerial recommendations on investment levels per attribute.

A third approach builds upon logistic regression with dummies. Exemplary in this respect is [25] who have extensively explored the measurement and analysis of the non-linear and asymmetric relationship between attribute-level and overall satisfaction. The authors compare absolute impact scores of positive versus negative attributes to confirm significant asymmetry in the relationship between attribute-level and overall satisfaction. Negative scores have more effect than positive scores. Their results on a priori postulated non-linearity and more in particular on diminishing returns, are mixed. Non-linear functions (both polynomial and natural logarithmic) confirm the relevance of diminishing returns in the positive yet not in the negative domain. Moreover, an integrative non-linear solution did not outperform a linear model. Reference [25] findings are significant, yet their methodological approach does not allow capturing the characteristics embedded in the aforementioned typology of satisfaction aggregation (Table 1) for four reasons. First, [25] focus on the direct relationship between attribute-level satisfaction/performance and overall satisfaction, thereby abstracting from the disconfirmation logic and foregoing the indirect effect of an individual product-level norm (Types D_1 and D_2). Second, the logarithmic model does not allow for full compensatory decision-making (Type A). Third, both polynomial and logarithmic models basically capture a non-compensatory conjunctive relationship and forego possible disjunctive effects. Finally, all rules and attribute roles are postulated a priori which is second-best to no a priori defined roles as in a neural network approach [40]. More advanced non-linear functions have been specified (an overview in [31]). Some of these do account for disjunctive and well as for conjunctive relationships with weighted profile scatters per attribute. None of them, however, does allow for the modeling of all five types of Table 1. The main reason is that they cannot integrate the simultaneous effect of the direct relationship between attribute-level satisfaction and the indirect effect of an individual (i.e., not constant) norm against which these attribute-level scores are assessed, thereby allowing for an effect-neutral zone of tolerance.

Finally, Penalty-Reward Contrast Analysis (PRCA) has been proven more robust and powerful than IPA [22]. In contrast to IPA, PRCA allows for the independent measurement of performance and importance of attributes [7]. While the underlying idea of PRCA is valid, it focuses exclusively on asymmetry and does not allow capturing the typology defined in Table 1. Moreover, [22] empirically confirmed that PRCA can only explain a small amount of variance. This is due to the necessary use of dichotomized independent variables.

All four alternative approaches have their merits and contribute significantly to the understanding of the impact of attribute-level satisfaction on overall satisfaction. Nevertheless, we argue that an evaluation operator approach, such as ours, outperforms alternative techniques since (1) it is able to measure the weights of attribute-level scores, (2) it allows acknowledges a case-specific (i.e., individual) product-level norm as a basis for disconfirmation (anchoring), (3) it allows for a neutral effect of an individual zone of tolerance, and (4) it acknowledges the possibility of positive/negative reinforcement simultaneously.

5.2 Managerial Consequences

The potential of the uninorms to unravel the decision-making heuristics of customers when "deciding" on a certain level of (dis)satisfaction seems to outperform that of more traditional statistical models. This is due to the power of uninorms to allow for compensatory and non-compensatory effects as well as to the fact that researchers should not a priori postulate the roles the different attributes will take. Although the evaluation operator (uninorm) approach is/appears analytically complex, it may yield parsimonious results, even at the level of the individual respondent. This chapter illustrates and confirms earlier advice that managers should classify attributes as basic, excitement or performance factor and then identify the "optimal" performance level for each attribute. The goal should be to optimize, not to maximize attribute-level performance at a level where the payoff in terms of overall customers (dis)satisfaction is maximized. This optimal level can be determined by analyzing the different impact curves offered by our method.

References

1. Anderson, Eugene W. (1994) Cross-Category Variation in Customer Satisfaction and Retention. *Marketing Letters* 5 (January): 19–30.
2. Anderson, Eugene W., C. Fornell and D.R. Lehmann (1994) Customer Satisfaction, Market Share and Profitability: Findings from Sweden. *Journal of Marketing* 58 (July): 63–66.
3. Anderson, Eugene W., C. Fornell and S.K. Mazvancheryl (2004) Customer Satisfaction and Shareholder Value. *Journal of Marketing* 68 (October): 172–185.
4. Anderson, Eugene W. and V. Mittal (2000) Strengthening the Satisfaction-Profit Chain *Journal of Service Research* 3(2): 107–120.
5. Anderson, Eugene W. and M.W. Sullivan (1993) The Antecedents and Consequences of Customer Satisfaction for Firms. *Marketing Science* 12(2): 125–143.
6. Anderson, Rolph E. (1973) Consumer Dissatisfaction: The Effect of Disconfirmed Expectancy on Perceived Product Performance. *Journal of Marketing Research* 10 (February): 38–44.
7. Brandt, D. Randall (1987) A Procedure for Identifying Value-enhancing Service Components Using Customer Satisfaction Survey Data. In: Carol Surprenant (Ed.) *Add Value to Your Service*. American Marketing Association, Chicago, pp. 61–65.
8. Cadotte, Ernest R., R.B. Woodruff, and R.L. Jenkins (1987) Expectations and Norms in Models of Consumer Satisfaction. *Journal of Marketing Research* 24 (August): 305–314.
9. Cardozo, Richard N. (1965) An Experimental Study of Customer Effort, Expectations, and Satisfaction. *Journal of Marketing Research* 2 (August), 244–249.
10. Churchill, Gilbert A. Jr. and C. Surprenant (1982) An Investigation into the Determinants of Consumer Satisfaction. *Journal of Marketing Research* 19: 491–504.
11. Dawes, Robin M., D. Singer, and F. Lemons (1972) An Experimental Analysis of the Contrast Effect and Its Implications for Intergroup Communication and the Indirect Assessment of Attitude. *Journal of Personality and Social Psychology* 21 (March): 281–295.
12. De Baets, Bernard and J. Fodor (1997) On the structure of uninorms and their residual implicators, proceedings of the Eighteenth Seminar on Fuzzy set theory.: Linz Austria, 1997.
13. Dombi, József (1982) Basic Concepts for a Theory of Evaluation: The Aggregative Operator. *European Journal of Operational Research* 10: 282–293.
14. Dubois, D. and H. Prade. 1985. A Review of Fuzzy Set Aggregation Connectives. *Information Sciences* 36: 85–121.
15. Grunert, Klaus G. and C. Ellegaard (1993) The Concept of Key Success Factors: Theory and Method. In: M.J. Baker (Ed.) *Perspectives on Marketing Management* Vol.3. John Wiley and Sons.
16. Grunert, Klaus G. and L. Hildebrandt (2004) Success Factors, Competitive Advantage and Competitive Development. *Journal of Business Research* 57: 459–461.
17. Johnston, Robert (1995) The Determinants of Service Quality: Satisfiers and Dissatisfiers. *International Journal of Service Industry Management* 6(5): 53–71.
18. Kano, N. (1984) Attractive Quality and Must-be Quality. *Hinshitsu: The Journal of the Japanse Society for Quality Control* (April): 147–156.

19. Klement, E.P., R. Mesiar and E. Pap (2000). *Triangular Norms*. Kluwer: Dordrecht, The Netherlands.
20. Liljander, Veronica and T. Strandvik (1993) Estimating Zones of Tolerance in Perceived Service Quality and Perceived Service Value. *International Journal of Service Industry Management* 4(2): 6–28.
21. Matzler, Kurt E., F. Bailom, H.H. Hinterhuber, B. Renzl and J. Pichler (2004) The Asymmetric Relationship Between Attribute-level Performance and Overall Customer Satisfaction: A Reconsideration of the Importance-Performance Analysis. *Industrial Marketing Management* 33: 271–277.
22. Matzler, Kurt, E. Sauerwein (2002) The Factor Structure of Customer Satisfaction. An Empirical Test of the Importance Grid and the Penalty-Reward-Contrast Analysis. *International Journal of Service Industry Management* 13(4): 314–332.
23. Matzler, Kurt, E. Sauerwein, and K.A. Heischmidt (2003) Importance-Performance Analysis Revisited: The Role of the Factor Structure of Customer Satisfaction. *The Service Industries Journal* 23(2): 112–129.
24. Mittal, Vikas and W. Kamakura (2001) Satisfaction, Repurchase Intention, and Repurchase Behavior: Investigating the Moderating Effect of Customer Characteristics. *Journal of Marketing Research* 38 (February), 131-142.
25. Mittal, Vikas, W.T. Ross Jr., and P.M. Baldasare (1998) The Asymmetric Impact of Negative and Positive Attribute-Level Performance on Overall Satisfaction and Repurchase Intentions. *Journal of Marketing* 62 (January), 33–47.
26. Oliver, Richard L. (1980) A Cognitive Model of the Antecedents and Consequences of Satisfaction Decisions. *Journal of Marketing Research* 17 (November): 460–469.
27. Oliver, Richard L. (1993) Cognitive, Affective and Attribute Bases of the Satisfaction Response. *Journal of Consumer Research* 20: 418–430.
28. Oliver, Richard L. and W.S. DeSarbo (1988) Response Determinants in Satisfaction Judgments. *Journal of Consumer Research* 14 (March): 495–507.
29. Phillips, Diane M. and H. Baumgartner (2002) The Role of Consumption Emotions in the Satisfaction Response. *Journal of Consumer Psychology* 12(3): 243–252.
30. Pieters, Rik, K. Koelemeijer and H. Roest (1995) Assimilation Processes in Service Satisfaction Formation. *International Journal of Service Industry Management* 6(3): 17–33.
31. Sethi, Vikram and R.C. King (1999) Nonlinear and Noncompensatory Models in User Information Satisfaction Measurement. *Information Systems Research* 10(1): 87–96.
32. Spreng, Richard A., S.B. McKensey, and R.W. Olshavsky (1996) A Reexamination of the Determinants of Consumer Satisfaction. *Journal of Marketing* 60 (July): 15–32.
33. Stauss, Bernd (1993) Using the Critical Incident Technique in Measuring and Managing Service Quality. In: Eberhard E. Scheuing and William F. Christopher (Eds) *The Service Quality Handbook*. New York: American Management Association, 408–427.
34. Stauss, Bernd and B. Hentschel (1992) Attribute-based versus Incident-based Measurement of Service Quality: Results of an Empirical Study in the German Car Service Industry. In: Paul Kunst and Jos Lemmink (Eds.) *Quality Management in Services*. Van Gorcum, Assen, pp. 59–78.

35. Szymanski, David M. and D.H. Henard (2001) Consumer Satisfaction: A Meta-Analysis of the Empirical Evidence. *Journal of the Academy of Marketing Science* 29(1): 16–35.
36. Tse, David K. and P.C. Wilton (1988) Models of Consumer Satisfaction Formation: An Extension. *Journal of Marketing Research* 25 (May): 204–212.
37. Tversky, Amos and D. Kahneman (1974) Judgement under Uncertainty: Heuristics and Biases. *Science* 185: 1124–1131.
38. van Bruggen, Gerrit H., A. Smidts and B. Wierenga (1998) Improving Decision Making by Means of a Marketing Decision Support System. *Management Science* 44(5): 645–658.
39. Vanhoof, Koen, T. Brijs and G. Wets (2003) An Indirect Measurement for Customer Expectation. In: B. De Baets and J. Fodor (eds.): *Princ. of Fuzzy Preference Modelling and Decision Making*, Academia Press, 109–122.
40. West, Patricia, M., P.L. Brockett and L.L. Golden (1997) A Comparative Analysis of Neural Networks and Statistical Methods for Predicting Consumer Choice. *Marketing Science* 16(4): 370–391.
41. Westbrook, Robert A. (1987) Product/Consumption-Based Affective Responses and Postpurchase Processes. *Journal of Marketing Research* 24 (August): 258–270.
42. Woodruff, Robert B., E.R. Cadotte and R.L. Jenkins (1983) Modeling Consumer Satisfaction Processes Using Experience-based Norms. *Journal of Marketing Research* 20 (August): 296–304.
43. Yager, Ronald. and A. Rybalov (1998) Full Reinforcement Operators in Aggregation Techniques. *IEEE Transactions on Systems, Man and Cybernetics* 28: 757–769.
44. Yager, Ronald and A. Rybalov (1996) Uninorm Aggregation Operators. *Fuzzy Sets and Systems* 80: 111–120.
45. Yi, Youjae (1991) A Critical Review of Consumer Satisfaction. In: Valarie A. Zeithaml (Ed.) *Review of Marketing* Chicago: American Marketing Association, pp. 68–123.
46. Zeithaml, Valerie, L.A. Berry and A. Parasuraman (1993) The Nature and Determinants of Customer Expectations of Service. *Journal of the Academy of Marketing Science* 2(1): 1–12.
47. Zimmerman, Hans Jurgen (1996). *Fuzzy Set Theory and its Applications*. Kluwer: Boston.

Using an Adapted Classification Based on Associations Algorithm in an Activity-Based Transportation System

Davy Janssens, Geert Wets, Tom Brijs, and Koen Vanhoof

Limburgs Universitair Centrum, Transportation Research Institute, Universitaire Campus, gebouw D, B-3590 Diepenbeek, Belgium
{davy.janssens; geert.wets; tom.brijs; koen.vanhoof}@luc.ac.be

Abstract. A lot of research has been carried out in the past by using association rules to build more accurate classifiers. The idea behind these integrated approaches is to focus on a limited subset of association rules. However, these integration approaches have not been tested yet within the context of transportation research. The aim of this chapter is therefore to evaluate the performance of an adapted well-known associative classification algorithm on the datasets that are used in the Albatross transportation modelling system. The presented work is an extension of previous research efforts in the sense that it now becomes possible to use the adapted CBA system for multi-class problems. Experiments showed that the original CBA system achieved the best average performance for the three classifiers under evaluation. While the adapted CBA still generated better average results than CHAID, the performance with respect to original CBA was slightly worse.

1 Introduction

Over the last decade, activity-based transportation models have received increased attention for modelling travel demand. The most important characteristic in these models is that travel demand is derived from the activities that individuals and households need or wish to perform. The main advantage is that travel has no longer an isolated existence in these models, but is perceived as a way to perform activities and to realize particular goals in life.

One of the most advanced operational activity-based transportation models is the **Albatross** system (A Learning Based Transportation Oriented Simulation System), developed by [1] for the Dutch Ministry of Transport. **Albatross** is a multi-agent rule-based system that predicts which activities are conducted where, when, for how long, with whom and the transport mode involved. It uses decision rules to predict each of those facets (where, when, etc.) and to support scheduling decisions. These decision rules were originally derived by means of a decision tree induction algorithm that is based on a

Davy Janssens et al.: *Using an Adapted Classification Based on Associations Algorithm in an Activity-Based Transportation System*, Studies in Computational Intelligence (SCI) **5**, 253–252 (2005)

chi-squared test for determining the subsequent splits in the tree (hereafter referred to as CHAID; [10]). Since the choice of the learning algorithm is one of the most important factors to determine the overall performance of a transportation model, several studies have been conducted in the past to measure the performance of different algorithms. For instance, in comparative studies by [18] evidence was found that different kinds of decision tree induction algorithms achieve comparable results; while improvements were found with respect to the reduction in complexity of the decision model in [14]. A previous study by [8] suggested that Bayesian networks achieved better performance than decision trees and that they are better suited to capture the complexity of the underlying decision-making process. While the studies by [18] and [14] mainly focus on classification based (supervised) learning, the work by [8] is more an association based learning (unsupervised) approach. However, in recent years, extensive research has been carried out to integrate supervised and unsupervised learning (also referred to as associative classification). Despite the solid performance of these integration approaches in previous research efforts, the comprehensiveness and complexity of dealing with the often-large number of association rules have lead to difficulties and (accuracy versus generality) trade-off questions, which are part of a lot of research that is currently going on. In addition to this, it is important to state that these integration approaches have not yet been tested within the context of transportation research. The aim of this chapter is therefore to evaluate the performance of a well-known associative classification algorithm, as well as an adaptation of this algorithm within the context of the `Albatross` transportation model.

As mentioned above, association and classification rules are the two main learning algorithms in associative classification. The study of association rules is focused on using exhaustive search to find all rules in data that satisfy user-specified minimum support and minimum confidence criteria. On the other hand, classification rules aim to discover a small set of rules to form an accurate classifier. Given a set of cases with class labels as a training set, the aim of classification is to build a model (called classifier) to predict future data objects for which the class label is unknown. Now, by focusing on a limited subset of association rules, i.e. those rules where the consequent of the rule is restricted to the classification class attribute, it is possible to build classifiers. Several publications (for instance [4, 5, 17] and others) have shown that associative classification in general generates at least equal accuracy than state-of-the-art classification algorithms such as C4.5 [15]. The reasons for the good performance are obvious. Association rules will search globally for all rules that satisfy minimum support and minimum confidence norms. They will therefore contain the full set of rules, which may incorporate important information. The richness of the rules gives this technique the potential of reflecting the true classification structure in the data [17]. However, the comprehensiveness and complexity of dealing with the often-large number of association rules have lead to difficulties and (accuracy versus generality) trade-off questions. Contributions to tackle a number of these difficulties can be found in [5, 11] and in [17]. Reference [13] proposed an improvement of their original CBA (classification based on associations)-system [12] in [13] to cope with the weaknesses in the system. In spite of the fact that the presented adaptations of CBA are valuable, some important issues still remain unsolved. Some unsolved issues were already tested and implemented with success within the context of the 16 UCI Repository of Machine Learning binary class databases [8]. In this work, it was found that the adapted CBA algorithm has the lowest average error rate in comparison with the four other classifiers under comparison (i.e. original CBA system, C4.5., C4.5. with discretized values and Naïve Bayes). The aim of this chapter is to extend this adaptation towards multi-class problems and to test the performance of the adapted CBA algorithm within the context of the datasets that are used in Albatross.

The remainder of this chapter is organized as follows. First, the conceptual framework underlying the Albatross-system is briefly discussed in order to provide some background information with respect to this transportation model. Section 3 explains the original CBA-algorithm [12]. Weaknesses and adaptations of the original system are suggested in Sect. 4. Section 5 then describes the design of the experiments that were carried out to validate the new approach and gives an overview of the data that were used. A discussion of the empirical results is also given in this section. Finally, conclusions and recommendations for further research are presented in Sect. 6.

2 The Albatross System

The Albatross system [1] is a computational process model that relies on a set of decision rules to predict activity-travel patterns. Rules are typically extracted from activity diary data. The activity scheduling agent of Albatross is the core of the system which controls the scheduling processes in terms of a sequence of steps. These steps are based on an assumed sequential execution of decision tables to predict activity-travel patterns (see Fig. 1). The first step in the figure involves for each person decisions about which activities to select, with whom the activity is conducted and the duration of the activity. The order in which activities.

The assignment of a scheduling position to each selected activity is the result of the next two steps. After a start time interval is selected for an activity, trip-chaining decisions determine for each activity whether the activity has to be connected with a previous and/or next activity. Those trip chaining decisions are not only important for timing activities but also for organizing trips into tours. The next steps involve the choice of transport mode for work (referred to as mode1), the choice of transport mode for other purposes (referred to as mode2) and the choice of location. Possible interactions between different facets of the transportation model, such as for instance between mode and location choices are taken into account by using location information as

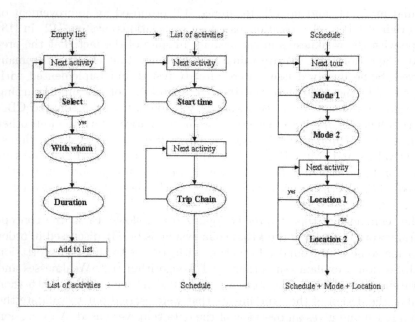

Fig. 1. Albatross' scheduling engine [1]

conditions of mode selection rules. As mentioned before, each decision in the Albatross system (see oval boxes of Fig. 1) is extracted from activity travel diary data using the CHAID-based [10] algorithm.

3 Classification Based on Associations

Before elaborating on the improvements that can be made to CBA, an overview of the original algorithm will be provided. First, a definition of association rules is given. Hereafter, class association rules (CARs) are introduced.

3.1 Association Rules

Let $I = \{i_1, i_2, \ldots, i_k\}$ be a set of literals, called items. Let D be a set of transactions, where each transaction T is a set of items such that $T \subseteq P$. We say that a transaction T contains X, a set of items in I, if $X \subseteq T$. An association rule is an implication of the form $X \Rightarrow Y$, where $X \subset I, Y \subset I$ and $X \cap Y = \emptyset$. The rule $X \Rightarrow Y$ holds in the transaction set D with confidence c if $c\%$ of transactions in D that contain X also contain Y. The rule $X \Rightarrow Y$ has support s in the transaction set D if $s\%$ of transactions in D contain $X \cup Y$. Given a set of transactions D, the problem of mining association rules is to generate all association rules that have support and confidence greater than a user-specified minimum support (minsup) and minimum confidence (minconf) [2].

3.2 Class Association Rules

To make association rules suitable for the classification task, the CBA method focuses on a special subset of association rules, i.e. those rules with a consequent limited to class label values only; or so-called class association rules (CARs). Thus, only rules of the form $A => c_i$ where c_i is a possible class, need to be generated by the CBA-algorithm. Therefore, the Apriori algorithm which is widely used for generating association rules, was modified to build the CARs. Details about these modifications can be found in [12].

To reduce the number of rules generated, the algorithm performs two types of pruning. The first type is the pessimistic error rate used in [15]. The second type of pruning is known as database coverage pruning [11]. Building a classifier in CBA is therefore also largely based on this coverage pruning method, which is applied after all the CARs have been generated. The original algorithm, which is used in CBA, is shown in Fig. 2. Before the pruning, the algorithm will first rank all the CARs and sort them in the descending order. As it will be shown in the next section, this rank will be subject to one of the modifications that were implemented. The ranking is as follows: given two rules r_i and $r_j, r_i > r_j$ (or r_i is said having higher rank than r_j), if (1) conf (r_i) > conf (r_j); or (2) conf (r_i) = conf (r_j), but sup (r_i) > sup (r_j); or (3) conf (r_i) = conf (r_j) and sup (r_i) = sup (r_j), but r_i is generated before r_j.

If at least one case among all the cases covered by the rule is classified correctly by the rule, the rule is inserted into the classifier by following this sorted descending sequence order and all the cases it covers are removed from the database. The rule insertion stops when either all of the rules are used

```
R=sort (R);
for each rule r ∈ R in sequence do
        temp = ø;
        for each case d ∈ D do
        if d satisfies the conditions of r then store d.id in temp and mark r if it
        correctly classifies d;
        end
        if r is marked then
            insert r at the end of C;
            delete all the cases with the ids in temp from D;
            selecting a default class for the current C;
            compute the total number of errors of C;
        end
    end
end
```

Fig. 2. Building a classifier in CBA [12]

or no cases are left in the database. The majority class among all cases left in the database is selected as the default class. The default class is used in case when there are no covering rules. Then, the algorithm computes the total number of errors, which is the sum of the number of errors that have been made by the selected rules in the current classifier and the number of errors made by the default class in the training data.

After this process, the first rule that has the least number of errors is identified as the cutoff rule. All the rules after this rule are not included in the final classifier since they will only produce more errors [12].

4 Identifying Weaknesses and Proposing Adaptations to CBA

4.1 Limits of Conditional Probability (Confidence)

A profound examination of the algorithm identified a potential weakness in the way the rules are sorted. Since rules are inserted in the classifier following the sorted confidence order, this will determine to a large extent the accuracy of our final classifier. Confidence is a good measure for the quality of (class) association rules but it also suffers from certain weaknesses. The aim of this section is to elaborate on them.

The first weakness is that the conditional probability of a rule $X => Y$ is invariable when the size of $s(Y)$ or D varies. The subset of the cases which are covered by the consequent of the rule is given by $s(Y)$, while D is the total number of observations in the dataset. The confidence property is also insensitive to cardinal dilatation (i.e. the size of the subsets increases in the same proportion). Figure 3 graphically displays the problem. It can be seen from the figure that the three cases have the same confidence (indicated by the intersections between the ovals). Nevertheless, the rule $X => Y$ is more likely to happen when the size of $s(Y)$ increases or when the size of D decreases; and furthermore, the implication will be more meaningful when the size of all the sets grows in the same proportion. Figure 3 shows that it is not surprising

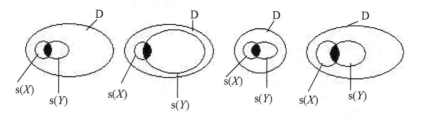

(a) Reference case   (b) s(Y) increases   (c) D decreases   (d) Cardinal dilatation

Fig. 3. Three cases with constant conditional probability

that, when $s(Y)$ is close to the size of D, the observations which are covered by the antecedent X of the rule, are also included in $s(Y)$. In Fig. 3 it is evident that when the size of $s(X)$ and the size of $s(Y)$ are close to that of D, $s(X)$ and $s(Y)$ share many objects. And also in the final case (Fig. 3), a more important sample will allow to be more confident in the statistical implication $X => Y$.

The second drawback of the use of conditional probability is that when for a particular class, the minsup parameter is set to 1% or even lower, it might very well happen that some rules have a high confidence parameter but on the other hand they might be confirmed by a very limited number of instances, and that those rules stem from noise only. This is why it is always dangerous to look for implications with small support even though these rules might look very "interesting". This danger seems to exist all the more in CBA because the application which implements the algorithm even offers a possibility to include rules with high confidence that do not satisfy the minimum support threshold in the final classifier. As a result, choosing the most confident rules may not always be the best selection criterion. Therefore, a suitable measure which takes both drawbacks that were identified above into account, was used to perform the sorting. The next section elaborates on this.

4.2 Intensity of Implication

Intensity of implication, introduced by [6] measures the distance to random choices of small, even non statistically significant, subsets. In other words, it measures the statistical surprise of having so few examples on a rule as compared with a random draw [7]. Consider a database D, where $|D|$ is the total number of observations in the database, and an association rule $X \Rightarrow Y$. Now, let U and V be two sets randomly chosen from D with the same cardinality as X and Y respectively, i.e., $s(X) = s(U)$ and $s(Y) = s(V)$, and let $\neg Y$ mean "not Y" as shown in Fig. 4.

Let $s(U \wedge \neg V)$ be the random variable that measures the expected number of random negative examples under the assumption that U and V are independent, and $s(X \wedge \neg Y)$ the number of negative examples observed on the rule. Now, if $s(X \wedge \neg Y)$ is unusually small compared with $s(U \wedge \neg V)$, the one we would expect at random, then we say that the rule $X \Rightarrow Y$ has a strong statistical implication. In other words, the intensity of implication for a rule $X \Rightarrow Y$ is stronger, if the quantity $P[s(U \wedge \neg V) \leq s(X \wedge \neg Y)]$ is smaller. Intensity of implication is then defined as $1 - P[s(U \wedge \neg V) \leq s(X \wedge \neg Y)]$. Since the random variable $s(U \wedge \neg V)$ follows the hypergeometric law, which means $\Pr[s(U \wedge \neg V) = k] = \Pr[\text{of } s(U) \text{ examples selected at random, exactly } k \text{ are not in } V] = \dfrac{C_{s(\neg V)}^{k} \times C_{s(V)}^{s(U)-k}}{C_{|D|}^{s(U)}}$ and taking into account that $s(U) = s(X)$ and that $s(V) = s(Y)$, the intensity of implication can be written as:

$$1 - \sum_{k=\max(0,s(X)-s(Y))}^{s(X\wedge\neg Y)} \frac{C^k_{s(\neg Y)} \times C^{s(X)-k}_{s(Y)}}{C^{s(X)}_{|\mathcal{D}|}} \tag{1}$$

This formula for intensity of implication is suitable as long as the number of cases in the database, i.e. $|D|$, is reasonably small. Otherwise, the combination numbers in the above formula explode quite soon. Therefore, [16] came up with an approximation of this formula for big datasets. They argue that if $s(U \wedge \neg V)$ is small, which is often the case in rule discovery, then Poisson approximations can be applied. In that case, the above formula for intensity of implication reduces to a much simpler version that is easier to compute, i.e.:

$$1 - \sum_{k=\max(0,s(X)-s(Y))}^{s(X\wedge\neg Y)} \frac{C^k_{s(\neg Y)} \times C^{s(X)-k}_{s(Y)}}{C^{s(X)}_{|\mathcal{D}|}}$$

$$\approx 1 - \sum_{k=0}^{s(X\wedge\neg Y)} \frac{\lambda^k}{k!} e^{-\lambda} \tag{2}$$

with

$$\lambda \equiv \frac{s(X) \times (|\mathcal{D}| - s(Y))}{|\mathcal{D}|}$$

where $|D|$ is the number of cases, $s(X)$ is the number of cases covered by the antecedent and $s(Y)$ is the number of cases covered by the consequent of the rule. The coefficient $s(X \wedge Y)$ represents the number of cases that are covered by the antecedent and the consequent of the rule, while $s(X \wedge \neg Y)$ stands for the number of cases which are covered by the antecedent but not by the consequent of the rule. Since confidence and support are standard measures for determining the quality of association rules, it would be suitable if those could be incorporated in (2). This procedure is quite straightforward.

Rewriting $s(X \wedge \neg Y)$ gives:

$$s(X \wedge \neg Y) = s(X) - s(X \wedge Y)$$

$$= s(X \wedge Y) \times \left(\frac{s(X)}{s(X \wedge Y)} - 1 \right)$$

$$= \frac{s(X \wedge Y)}{|D|} \times |D| \times \left(\frac{1}{\frac{s(X\wedge Y)}{s(X)}} - 1 \right)$$

$$= \text{support} \times \text{cases} \times \left(\frac{1}{\text{confidence}} - 1 \right)$$

Rewriting λ gives:

$$
\lambda = \frac{s(X) \times (|D| - s(Y))}{|D|}
$$

$$
= \frac{\frac{s(X \wedge Y)}{|D|}}{\frac{s(X \wedge Y)}{s(X)}} \times (|D| - s(Y))
$$

$$
= \frac{\text{support}}{\text{confidence}} \times (\text{cases-abssupcons})
$$

Substituting both derivations in (2) gives

$$
1 - \sum_{k=0}^{\text{support} \times \text{cases} \times \left(\frac{1}{\text{confidence}} - 1\right)} \frac{\left(\left(\frac{\text{support}}{\text{confidence}}\right) \times (\text{cases-abssupcons})\right)^{k}}{k!}
$$

$$
\times e^{-\left(\left(\frac{\text{support}}{\text{confidence}}\right) \times (\text{cases-abssupcons})\right)} \tag{3}
$$

By means of formula (3), we are now ready to adapt the CBA algorithm. This is done by using intensity of implication as the primary criteria when doing the sorting work in the first rule of Fig. 2. Rule r_i has a higher rank than rule r_j if it has a larger value of intensity of implication. When two rules have the same values of intensity of implication, they are ranked according to the sorting mechanism of the original CBA, which is mentioned in Sect. 4.1. Reference [7] claim that the relevance of the discovered association rules can be significantly improved by using intensity of implication. The following section examines whether the same conclusion can be found when the adaptation is evaluated within the context of the multi-class datasets that are used for the Albatross modelling system.

5 Empirical Section

5.1 Description of the Data

The activity diary data used in this study were collected in 1997 in the municipalities of Hendrik-Ido-Ambacht and Zwijndrecht in the Netherlands (South Rotterdam region) to develop the Albatross system [1]. The data involve a full activity diary, implying that both in-home and out-of-home activities were reported. The sample covered all seven days of the week, but individual respondents were requested to complete the diaries for two designated consecutive days. Respondents were asked, for each successive activity, to provide information about the nature of the activity, the day, start and end time, the location where the activity took place, the transport mode, the travel time, accompanying individuals and whether the activity was planned or not. A pre-coded scheme was used for activity reporting.

There are some general variables that are used for each choice facet of the Albatross model (i.e. each oval box). These include (among others) household and person characteristics that might be relevant for the segmentation of the sample. Each dimension also has its own extensive list of more specific variables, which are not described here in detail.

5.2 Empirical Results

In this section, the CBA algorithm is evaluated on the datasets that are used within the context of the Albatross modelling system. To be able to test the validity of the presented models on a holdout sample, only a subset of the cases is used to build the models (i.e., "training set"). The decline in goodness-of-fit between this "training" set and the validation set ("test set") is taken as an indicator of the degree of overfitting. The purpose of the test set is also to evaluate the predictive ability of the three techniques for a new set of cases. For each decision step, we used a random sample of 75% of the cases to build and optimise the models. The other subset of 25% of the cases were presented as "unseen" data to the models; this part of the data was used as the test set. The accuracy percentages that indicate the predictive performance of the three models on the training and test sets are presented in Table 1. The nine datasets that are used in Albatross are multiple class datasets. The number of classes has been indicated between brackets in Table 1.

Table 1. Accuracy Results

Dataset	CHAID		CBA		Adapted CBA	
	Train (%)	Test (%)	Train (%)	Test (%)	Train (%)	Test (%)
Duration (3)	41.30	38.80	44.71	39.21	40.67	40.88
Location1 (7)	57.50	58.90	66.34	62.72	64.54	68.13
Location2 (6)	35.40	32.60	52.58	41.14	26.78	26.29
Mode for work (4)	64.80	66.70	83.57	73.68	74.71	76.84
Mode other (4)	52.80	49.50	66.53	60.88	54.92	54.82
Selection (2)	72.40	71.60	79.56	78.74	79.07	79.19
Start time (6)	39.80	35.40	34.48	33.68	33.33	33.06
Trip chain (4)	83.30	80.90	83.86	80.40	82.72	81.95
With whom (3)	50.90	48.40	61.14	56.20	54.65	48.07
Average	**55.36**	**53.64**	**63.64**	**58.52**	**56.82**	**56.58**

In order to get a more comprehensive evaluation, results are also compared with the original CHAID algorithm that is used in Albatross besides the obvious comparison with original CBA. When the average results of the three algorithms are compared, it is clear that the CBA classifier produces better results than CHAID. Only with respect to the "start time", and "trip chain"-datasets, the CBA algorithm performs somewhat worse. Unfortunately, while the adapted CBA algorithm does better than the CHAID decision tree approach, the average (original) CBA accuracy could not be improved. However, in this case, it remains important to notice that the adapted CBA algorithm contains less rules and that also the degree of overfitting is significantly lower. While a significant improvement could be seen on the "location 1"-dataset, the performance on the "location 2"-dataset is significantly worse with respect to the original CBA algorithm. It can be seen from Table 1 that the "location 2"-dataset deteriorates to a large extent the general average performance. The reason for this is not clear and is likely attributable towards the nature of the dataset.

Generally speaking, adapted CBA was able to achieve an improvement in performance on the test set for 5 out of the 9 datasets. As mentioned before, in previous work [8], it was shown on 16 binary UCI datasets that the performance of adapted CBA achieved an important improvement with respect to the other classification algorithms under evaluation (CBA, C4.5 and Naïve Bayes). For this reason, the reason for the disagreement in results is not immediately obvious. One plausible assumption is that intensity of implication performs worse when dealing with multi-class problems. In order to evaluate this assumption, 10 additional experiments were carried out on multi-class UCI data. The datasets that were used in these additional tests are Annealing, Automobile, Glass Identification, Iris Plant, Led Display, Lymphography, Vehicle, Waveform, Wine recognition and Zoo [3]. The average error rate over the 10 datasets was 18.05% for adapted CBA, while it was 15.54% for original CBA. This finding is of the same magnitude as the result that was found in Table 1. Based on this additional external validation, it is fair to say that intensity of implication in its current form performs somewhat worse when dealing with multi-class problems. The fact that Albatross is dealing with real-life (and not artificial) data does not seem to be a differentiator in results.

Finally, it is important to stress that the accuracy versus generality trade-off remarks that were already made in the introductory section of this chapter, needs to be kept in mind in the interpretation of the results. Indeed, if accuracy is the main objective, the original CBA algorithm (with huge number of rules) is likely the best option to choose, while CHAID or adapted CBA will be more favourable choices when the complexity of the classifier is an issue.

6 Conclusions

The research presented in this chapter focused on the integration of supervised and unsupervised learning. In doing so, a modified version of the CBA algorithm, which can be used to build classifiers based on association rules, has been proposed. The modified CBA algorithm was tested within the context of the datasets that are used in Albatross, which is a multi-agent rule-based system that predicts which activities are conducted where, when, for how long, with whom and the transport mode involved. For this reason, CBA needed to be extended for application on multi-class datasets. CBA was adapted by coupling it with intensity of implication, a measure to calculate the distance to random choices of small, even non statistically significant, subsets.

Experiments proved that CBA achieved better average performance than the CHAID decision tree algorithm that is used in Albatross. While adapted CBA still generated better average results than CHAID, the performance with respect to original CBA was slightly worse. However, adapted CBA generated less decision rules than original CBA and also the degree of overfitting was significantly lower. Also, the average result was mainly deteriorated by a significantly worse performance of adapted CBA on one particular dataset. In spite of both arguments, the result was somewhat surprising in comparison with previous good performances of the adapted algorithm (on binary class problems).

For explanatory purposes, tests on 10 additional multi-class datasets were therefore carried out. Due to the fact that a similar finding was found on these datasets, the authors conclude that the implementation of intensity of implication in its current form performs somewhat worse on multi-class problems.

Finally, it needs to be stated that in its current form, it is still possible that rules with a relatively low confidence occasionally may have a high intensity of implication. Therefore, the evaluation of a more advanced sorting mechanism needs to be the subject of additional and future research.

References

1. Arentze, T.A. and H.J.P. Timmermans (2000), Albatross: A Learning-Based Transportation Oriented Simulation System. European Institute of Retailing and Services Studies. Eindhoven, The Netherlands.
2. Agrawal, R., T. Imielinski and A. Swami (1993), Mining Association Rules between Sets of Items in Large Databases, in: Proc. of the ACM SIGMOD Conference on Management of Data, Washington D.C., pp. 207–216.
3. Blake C.L. and Merz, C.J. (1998), UCI Repository of machine learning databases[http://www.ics.uci.edu/~mlearn/MLRepository.html].Irvine,CA: University of California, Department of Information and Computer Science.
4. Chen, J., H.Y. Liu and G.Q. Chen (1999), Mining insightful classification rules directly and efficiently IEEE/SMC'99.
5. Dong, G., Zhang, X., Wong, L., & Li, J. (1999), CAEP: Classification by aggregating emerging patterns, in Proc. of the Second International Conference on Discovery Science, Tokyo, Japan, pp. 30–42.
6. Gras, R. and A. Lahrer (1993), L'implication statistique: une nouvelle méthode d'analyse des données, Mathématiques, Informatique et Sciences Humaines no. 120.
7. Guillaume, S., F. Guillet and J. Philippé (1998), Improving the discovery of association rules with intensity of implication, in Principles of Data Mining

and Knowledge Discovery, vol. 1510 of Lecture Notes in Artificial Intelligence, pp. 318–327.

8. Janssens, D., Wets, G., Brijs, T., Vanhoof, K., Timmermans, H.J.P., Arentze, T.A. (2004), Improving the Performance of a Multi-Agent Rule-Based Model for Activity Pattern Decisions Using Bayesian Networks Forthcoming in Journal of the Transportation Research board, also in: Electronic conference proceedings of the 83rd Annual Meeting of the Transportation Research Board (CD-ROM)

9. Janssens, D., Lan, Y., Wets G., Chen G. and T. Brijs (2004), Empirically validating an adapted classification based on associations algorithm on UCI data, Applied Computational Intelligence (D. Ruan, P. D'hondt, M. De Cock, M. Nachtegael, E. E. Kerre, eds.), World Scientific, ISBN 981-238-873-7.

10. Kass, G.V. (1980), An Exploratory Technique for Investigating Large Quantities of Categorical Data, Applied Statistics, Vol. 29, 119–127.

11. Li, W., J. Han and J. Pei (2001), CMAR: Accurate and Efficient Classification Based on Multiple Class-Association Rules, in ICDM 2001, San Jose, California, pp. 369–376.

12. Liu, B., W. Hsu and Y. Ma (1998), Integrating Classification and Association Rule Mining, in Proc. of the Fourth International Conference on Knowledge Discovery and Data Mining (KDD-98), New York, pp. 80–86.

13. Liu, B., Y. Ma and C. Wong (2001), Classification using Association Rules: Weaknesses and Enhancements, To appear in Vipin Kumar, et al. (eds.), Data mining for scientific and engineering applications

14. Moons, E., G. Wets, M. Aerts, T.A. Arentze and H.J.P. Timmermans, (2004) The Impact of Irrelevant Attributes on the Accuracy of Classifier Systems in Generating Activity Schedules. Conditionally accepted for publication in Environment and Planning A.

15. Quinlan J.R., (1993), C4.5: Programs for Machine Learning, Morgan Kaufmann, Los Altos.

16. Suzuki, E. and Y. Kodratoff (1998), Discovery of surprising exception rules based on intensity of implication, in PKDD'98, pp. 10–18.

17. Wang, K., S. Zhou and Y. He (2000), Growing decision tree on support-less association rules, in Proc. of the Sixth ACM SIGKDD International Conference on Knowledge Discovery and Data Mining (KDD'00),Boston, pp. 265–269.

18. Wets, G., K. Vanhoof, T.A. Arentze and H.J.P. Timmermans (2000), Identifying Decision Structures Underlying Activity Patterns: An Exploration of Data Mining Algorithms. Transportation Research Record: Journal of the Transportation Research Board, No. 1718, TRB, National Research Council, Washington, D.C., 1–9.

Evolutionary Induction of Descriptive Rules in a Market Problem

M.J. del Jesus[1], P. González[1], F. Herrera[2], and M. Mesonero[3]

[1] Department of Computer Science, University of Jaén, 23071 – Jaén, Spain
mjjesus@ujaen.es, pglez@ujaen.es
[2] Department of Computer Science and Artificial Intelligence, University of
Granada., 18071 – Granada, Spain
herrera@decsai.ugr.es
[3] Department of Department of Organisation and Marketing, University of
Mondragón, 20560 – Oñati (Gipuzkoa), Spain
mmesoner@eteo.muni.es

Abstract. Nowadays, face to face contact with the client continues to be fundamental to the development of marketing acts. Trade fairs are, in this sense, a basic
instrument in company marketing policies, especially in Industrial Marketing. Due
to the elevated investment in term of both time and money it is necessary the automatic extraction of relevant and interesting information which helps to improve fair
planning policies.

In this paper, we analyse this problem and the kind of knowledge the user is
interested in. We study the use of Soft Computing methodologies, specifically Fuzzy
Logic and Genetic Algorithms, in the design of the Data Mining algorithms most
proper to this problem, descriptive induction algorithms for subgroup discovery.
Then we present an evolutionary model for the descriptive induction of fuzzy or
crisp rules which describe subgroups. The proposal includes a GA in an iterative
model which extracts rules while examples are left uncovered and the rules obtained
surpass an specified confidence level.

Key words: Data Mining algorithms, Fuzzy Logic, Genetic Algorithms, descriptive
induction, Subgroup Discovery, Market problems

1 Introduction

In the field of business an interesting problem is the study of the influence
that the planning variables of a trade fair have over the successful achievement
of its objectives. Faced with a real problem of this type a data mining algorithm should extract relevant interesting information concerning each of the
effectiveness groups by which the stands have been arranged. The information
obtained must be open to interpretation so as to be useful for the policies of
trade fair planning.

M.J. del Jesus et al.: *Evolutionary Induction of Descriptive Rules in a Market Problem*, Studies
in Computational Intelligence (SCI) **5**, 267–265 (2005)
www.springerlink.com   © Springer-Verlag Berlin Heidelberg 2005

In this chapter this problem is approached with an algorithm based on Soft Computing methodologies, specifically using an evolutionary data mining algorithm which allows for the extraction of a set of descriptive fuzzy rules.

The term Soft Computing applies to a methodology whose main characteristic is the tolerance to inaccuracy and uncertainty which gives it the capacity to adapt and allows us to solve problems comprehensively in a changeable environment and at low cost. The basic underlying principle in Soft Computing is the hybridization of techniques in the development of computing methods which give us an acceptable solution at low cost through the search for an approximate solution to a problem which has been set precisely or imprecisely. Within Soft Computing methodologies such as Fuzzy Logic, Neural Networks and Evolutionary Computing are included [54].

In the process of Knowledge Discovery in Databases and specifically in the process of Data Mining there are different tasks which can be treated and solved as search and optimisation problems. Evolutionary algorithms imitate the principles of natural evolution in carrying out global search and optimisation processes and are applicable to the development of data mining algorithms [26].

In Data Mining one of the objectives to be considered, besides predictive accuracy and interest, is the user's potential comprehension of the results. In this respect, Fuzzy Logic represents a tool for knowledge representation which allows us to adjust for uncertainty and inaccuracy simply and easily interpretable for the user.

This chapter is arranged in the following way: In Sect. 2, the market problem and the kind of knowledge the user is interested in are dealt with. In Sect. 3 the type of data mining algorithm to be applied to the problem is defined, descriptive rule induction for subgroup discovery. In addition some proposals carried out in the area of subgroup discovery will be analysed and the rule quality measures that are most frequently used will be described. In Sect. 4 one of the methodologies used in our proposal, Genetic Algorithms, are described and some proposals in specialised literature for rule induction with Genetic Algorithms will be analysed. The second soft methodology used in our method, the Fuzzy Logic and its use in Data Mining algorithms is explained in Sect. 5. The evolutionary approach to obtain descriptive fuzzy rules is described in Sect. 6, and in Sect. 7 the experimentation carried out and the analysis of results are explained. Finally, the conclusions and further research are outlined.

2 The Extraction of Useful Information on Trade Fairs

This study deals with a market problem studied in the Department of Organisation and Marketing of the University of Mondragón, Spain: the extraction of useful information on trade fairs [45].

In today's increasingly virtual world, face to face contact with the client continues to be fundamental to the development of marketing acts. Trade fairs are, in this sense, a basic instrument in company marketing policies, especially those which are run in industrial surroundings, or Industrial Marketing. AFE (Spanish Trade Fairs Association)'s data shows that to be so: during 2002, 419 events took place in Spain, in which participated 114.646 exhibitors, with an attendance of 15.739.986 visitors.

Businesses consider trade fairs to be an instrument which facilitates the attainment of commercial objectives such as contact with current clients, the securing of new clients, the taking of orders, and the improvement of the company image amongst others [32]. One of the main inconveniences in this type of trade fair is the elevated investment which they imply in terms of both time and money. This investment sometimes coincides with a lack of planning which emphasises the impression that trade fairs are no more than an "expense" which a business must accept for various reasons such as tradition, client demands, and not giving the impression that things are going badly amongst other factors [48]. Therefore convenient, is the automatic extraction of information about the relevant variables which permit the attainment of unknown data, which partly determines the efficiency of the stands of a trade fair.

Trade Fair planning models indicate the convenience of dividing participation into three moments in time: before, during and after the fair [32, 48].

- Pre-Fair. The exhibitor should reflect, amongst other things, on the following aspects:
  - The trade fair which s/he will attend
  - Objectives set
  - Budget
  - Visitors present at the fair
  - Stand design
  - Company personnel to attend
  - Pre-advertising campaign
- During the fair. The exhibiting company should analyse the way in which contact will be made with visitors and the communicative acts to be developed at the stand itself, amongst other things.
- Post-fair. The close of the fair is the moment to harvest the fruit of the effort made during the fair. To this end, it is necessary to evaluate participation, follow up on contacts made and measure the results achieved.

In the Machinery and Tools biennial held in Bilbao in March 2002, information was collected on all these aspects. To be precise, 104 variables of 228 exhibitors were analysed. Of these variables, 7 are continuous and the rest are categorical features, result of an expert discretization. Additionally, for each exhibitor, based on various marketing criteria, the stand's global efficiency was rated as high, medium or low, in terms of the level of achievement of objectives set for the trade fair.

For this real problem, the data mining algorithm should extract information of interest about each of the three efficiency groups of the stands. The extracted information should be interpretable so that it can be used the policies of fair planning. This makes the application of a data mining algorithm appropriate for the extraction of simple rules. The rules generated will determine the influence which the different fair planning variables have over the results obtained by the exhibitor, therefore allowing fair planning policies to be improved.

3 Descriptive Rule Induction Processes

Knowledge discovery through a process of descriptive rule induction is included within the area of Knowledge Discovery in Databases (KDD) which is defined as the non trivial process of identifying valid, original, potentially useful patterns which have comprehensible data [21]. KDD is a wide ranging process which covers distinct stages: the comprehension of the problem, the comprehension of the data, pre-processing (or preparation) of the data, Data Mining and post-processing (assessment and interpretation of the models). The data mining stage is responsible for automatic knowledge discovery of a high level and from information obtained from real data [22].

Data Mining can discover knowledge using different representation models and techniques from two different perspectives:

- *Predictive induction*, whose objective is the discovery of knowledge for classification or prediction [13, 14, 46]. Under this focus, models are generated with the objective of achieving a high degree of prediction from the data base. In this area, attention can be drawn to supervised learning processes of classification rules.

- *Descriptive induction*, whose fundamental objective is the discovery of interesting knowledge from the data. In this area, attention can be drawn to the discovery of association rules following an unsupervised learning model [1], subgroup discovery [36, 58], and other approaches to non classificatory induction.

Considering the characteristics of the problem to be solved, the obtention of simple rules which provide conclusive information about the efficiency of the stands in trade fairs, the most suitable approach is descriptive induction.

A subdivision of descriptive induction algorithms which has recently received a lot of attention from researchers is subgroup discovery which, given a set of data and having a property of interest to the user, attempts to locate subgroups which are statistically "most interesting". The discovered subgroups must be as large as possible and have the most unusual statistical (distributional) characteristics with respect to the property of interest. Subgroup discovery methods take relations between independent (explaining) variables and a dependent (target) variable into account. The concept was initially formulated by Klosgen in his rule learning algorithm EXPLORA [36], and by Wrobel in the algorithm MIDOS [58, 59].

In the majority of proposals in the specialised literature, subgroup discovery is described as a task half way between predictive induction and descriptive induction. It could be said that subgroup discovery aims to generate models based on rules whose objective is descriptive, using a predictive perspective in the majority of cases in order to obtain them [44]. It is a form of descriptive induction which induces rules from labelled training instances with the objective of categorising subgroups of the population which fulfil the given interest property. This suggests the use of classic learning techniques in order to solve the task. But the adaptation of this type of tasks should not lose sight of the final aim of subgroup discovery, which is to find individual rules or interest patterns, simply represented, which allows their effective use by potential users of the information.

Subgroup discovery is a descriptive induction mechanism that focuses not necessarily on finding complete relations between the target and the explaining variables. For some problems, partial relations, i.e., subgroups with "interesting" characteristics, with a significant deviation from the total population, are sufficient. This means that the rules generated do not have to fulfil high support criteria, which is necessary for other data mining approaches, e.g. the induction of association rules. Neither do they have to verify common demands for algorithms for obtaining classification systems, for example that the rules have to have a given degree of accuracy or that the final set of rules is as different as possible in order to cover all the areas of the search space. In some subgroup discovery problems, it is interesting to obtain overlapped rules, as these can show information on a single area of examples from different perspectives.

In this section we will describe some subgroup discovery algorithms presented in the KDD literature and the quality measures used frequently in this kind of inductive algorithms.

3.1 Approaches for Subgroup Discovery

The first algorithms developed for subgroup discovery are EXPLORA [36] and MIDOS algorithms [58, 59]. Both use a rule-extraction model based on decision trees, in order to obtain the best subgroups among the population. In order to evaluate the subgroups evaluation measurements are defined which determine the interest of an expression through a combination of unusualness [42] and size. MIDOS tackles, within this same approach, the problem of discovery in multi-relational databases. Another system which uses a multi-relational hypothesis space is the SubgroupMiner system [37].

In addition to these first proposals, different methods have been developed which obtain descriptions of subgroups represented in different ways and using different quality measurements:

- The SD algorithm [28] induces rules guided by expert knowledge. In this proposal, instead of defining an optimal measurement for the search for and the automatic selection of subgroups as do the EXPLORA and MIDOS algorithms, the objective is to help the expert to carry out flexible and effective searches over a wide range of optimal solutions. In consequence, the final decision over which subgroups will be selected is left to the expert.
- The CN2-SD algorithm [43] induces subgroups in the form of rules using as quality measurement the relationship between true positive rate and false positive rate. This algorithm is applied in [44] to a marketing problem, specifically to a relational database which stores information about interviews with potential clients. Based on the clients' responses to whether they recognise, use and appreciate given brands, the objective of the descriptive induction algorithm is to discover which brands can potentially be more used by clients, and to direct a specific marketing campaign towards these clients.
- In data analysis with high uncertainty it is useful to present a subgroup of the population by listing its support factors, instead of using a subgroup discovery approach which obtains descriptions of subgroups in the form of rules. In [11], this approach is applied to a marketing problem in which the objective is to identify the characteristics of clients who do not recognise and/or use a given brand of non-alcoholic drink.

What the described proposals have in common is the adaptation of classification rule extraction models for the subgroup discovery task. Currently, interest is starting to be shown in the development of subgroup discovery approaches which modify association rule extraction algorithms [41].

3.2 Rule Quality Measurements

Any rule induction algorithm uses a measurement of rule quality or of the set of rules which guides the data mining process. In addition, other quality measurements are calculated which evaluate the result of the induction process as a whole. In both cases, we can distinguish between objective and subjective quality measurements [52]. In order to solve subgroup discovery tasks, both objective and subjective measurements must be considered and the more appropriate quality criteria will depend on the application. Obviously, for automatic rule induction only objective quality criteria can be applied. However, to evaluate the quality of descriptions of induced subgroups and their use in decision making, subjective criteria are more important, although also more difficult to evaluate. In this section we will centre exclusively on objective criteria, which are those which are included in data mining algorithms.

Among the objective quality measurements, some of those most used in the literature are:

- *Size* of the rule or set of rules.
- *Predictive accuracy*, defined as the percentage of correctly classified instances.
- *Support* or relative frequency of correctly classified covered examples.
- *Significance*, called *"evidence"* in [36], which indicates how significant a subgroup is compared to the null hypothesis of statistical independence [13].
- *Unusualness*, defined as the weighted relative accuracy of a rule [42]. The average relative accuracy can be described as the balance between the coverage of a rule (a measure of generality, computed as the relative frequency of all the examples covered by the rule) and its accuracy gain: the higher a rule's unusualness, the more relevant is it. Unusualness measures the novelty in the distribution of the subgroups, considering its coverage.
- *Interest*, which can be determined objectively or subjectively, and which can be found defined in various ways in the literature. In our proposal (see Sect. 6) we apply an interest measurement based on information gain.
- *Accuracy gain* is the difference between the accuracy of a rule and the default accuracy [44].

The measurements for the evaluation of each individual rule can be complemented with variations which calculate the average of the set of subgroups obtained. This allows the comparison of different subgroup discovery algorithms [41].

The appropriateness of the quality measurements used in a problem will determine to a great extent the effectiveness of the mechanism of descriptive rule induction used for the discovery of subgroups. This means that the study of appropriate measurements of the quality of rules is an aspect which is currently being researched.

4 Genetic Algorithms in Rule Induction Processes

In any knowledge extraction process in databases, and specifically in a data mining process there are different tasks or problems which can be approached and solved as optimisation and search problems. Evolutionary Algorithms [6] imitate the principles of natural evolution in order to form procedures for searching and overall optimisation and can be used as data mining algorithms, as pre- or post-processing algorithms or as tools for the optimisation of the parameters of other algorithms [26]. Among the different types of Evolutionary Algorithms, currently the most widely used in the field of Data Mining are Genetic Algorithms and Genetic Programming. In this chapter we will centre on Genetic Algorithms and their application to descriptive rule induction.

In this section we will briefly describe Genetic Algorithms and we will analyse their use for rule induction in the specialised literature.

4.1 Genetic Algorithms

Genetic Algorithms (GAs) are optimisation and search algorithms inspired in natural evolution processes and initially defined by Holland [30, 34].

Stated simply, they work as follows: the system starts with an initial population of individuals who encode, through a form of genetic representation, candidate solutions for the proposed problem. This population of individuals (called chromosomes) evolves in time through a process of competition and controlled variation. Each chromosome in the population is associated with a fitness function in order to determine which chromosomes will be selected to form part of the new population in the competition process. The new population will be created using genetic operators of crossover and mutation. This evolutionary cycle continues until a given stop condition is reached: a given maximum number of evaluations of individuals is calculated, the population reaches a given maximum number of generations, a solution with a given fitness function value is reached, or the population does not generate new individuals for a given number of generations, among other possible stop conditions.

Bäck et al. in [6] gives a complete description of Genetic Algorithms as well as other examples of Evolutionary Algorithms.

The application of a GA to solve a problem must determine:

- A genetic representation (called genotype) of the solutions of the problem (called phenotype).
- A way to create an initial population of solutions.
- An evaluation function (called fitness function) which provides a quality value to each chromosome.
- Operators which modify the genetic composition of the descendants during reproduction.
- Values for the parameters used (population size, probabilities of application of genetic operators, etc.).

4.2 Rule Induction Using Genetic Algorithms

GAs have several advantages as a rule induction method:

- They tend to cope well with attribute interaction because they usually evaluate a rule as a whole via fitness function, rather than evaluating the impact of adding/removing one condition to/from a rule.
- They have the ability to scour a search space thoroughly and the ability to allow arbitrary fitness functions in the search [19]. The fitness function can contain different criteria such as the ability to penalise overlap among rules or rule sets with too many rules or a problem-specific quality measure, etc.
- In addition, the genetic search performs implicit backtracking in its search of the rule space, thereby allowing it to find complex interactions that other non-backtracking searches would miss.

- An additional advantage over other conventional rule-learning algorithms is that the search is carried out among a set of competing candidate rules or rule sets.

However, this is not to say that GAs are inherently superior to rule induction algorithms as no rule discovery algorithm is superior in all cases [20, 47].

In KDD literature different GA proposals have been presented with predictive or descriptive aims. As we indicated in the previous section, rule induction algorithms for subgroup discovery (the aim of which is fundamentally descriptive) share characteristics with algorithms which guide the induction process using predictive quality measurements. In this section we will describe some of the principle GA proposals for rule induction, no matter what is their final aim.

In the design of any rule induction GA, the following aspects, among others, must be determined:

1. A genetic representation of the solutions of the problem (rule or set of rules).
2. A fitness function that provides a quality value for each rule or set of rules.
3. Operators that modify the genetic composition of the descendant during reproduction.

The first aspect is perhaps the most determining of the rest of the characteristics of any proposal. The GAs follows two approaches in order to encode rules within a population of individuals:

- The *"Chromosome = Rule"* approach, in which each individual codifies a single rule.
- The *"Chromosome = Set of rules"*, also called the *Pittsburgh* approach, in which each individual represents a set of rules. GABIL [17] and GIL [35] are examples of classification GAs which use this representation model.

In turn, within the *"Chromosome = Rule"* approach, there are three generic proposals:

- The *Michigan* approach in which each individual codifies a single rule but the final solution will be the final population. In this case, it is necessary to evaluate the behaviour of the whole set of rules and what the individual rules add to it. ZCS [54] and XCS [55] algorithms are examples of GAs with this type of representation.
- The *IRL* (Iterative Rule Learning) approach, in which each chromosome represents a rule, but the GA solution is the best individual obtained and the global solution is formed by the best individuals obtained when the algorithm is run multiple times. In [15] and [31] two proposals with this model are described.
- The *cooperative-competitive* approach, in which the complete population or a subset of it codifies the rule base. COGIN [33] and REGAL [29] are examples of GAs with this type of representation.

The choice of representation model depends, among other aspects, on the task to be carried out by the data mining algorithm and therefore on the type of rule to be discovered.

The most noteworthy systems based on GAs in the specialised literature, grouped according to their approach, are described below.

The "*Chromosome = Rule*" Approach

- GA-NUGGETS. Freitas in [25] presents a GA designed to discover descriptive rules. The algorithm uses a special codification which depends on the problem, in which each individual represents the antecedent of a rule. The consequent is calculated in the evaluation of the individual such that the predictive capacity of the rule and the interest of the chosen consequent are maximised. The authors propose in [49] a modification of this algorithm, in which the fitness function consists of two parts: the first part measures the degree of interest of the rule (computed by an information-theoretical measurement, the information gain, and refers to the antecedent and to the consequent of the rule) and the second part, its predictive accuracy.
- GA-PVMINER is a parallel GA developed by Araujo, Lopes and Freitas [2] that uses Parallel Virtual Machine to discover descriptive rules for dependence modelling. In GA-PVMINER the fitness function is based on the J-measure proposed by Smyth and Goodman [53] which measures the degree of interest of a rule. The algorithm uses a crossover operator based on that used in GA-MINER, two types of mutation and a removal operator which acts in the consequent part.
- Fidelis et al. in [22] proposes a GA for predictive rule induction with a codification model which is very appropriate for the treatment of quantitative variables. The antecedent of a rule is represented in each individual and the consequent is fixed in each run of the GA.
- GLOWER, developed by Dhar et al. [19]. The design of this GA was motivated by financial prediction problems, and incorporates ideas from tree induction and rule learning.
- Carvalho and Freitas present in [8, 9] a hybrid decision tree/genetic algorithm approach for a predictive rule induction process, which follows the following model: examples belonging to large disjuncts are classified by rules produced by a decision-tree algorithm while examples belonging to small disjuncts are classified by rules produced by a GA specifically designed for this task. In the GA each individual represents a small-disjunct rule identified by a decision tree leaf node. The genome of an individual consists of the conditions in the antecedent part of the rule and the consequent part of the rule, which specifies the predicted class, is not represented in the genome. Rather, it is fixed for a given GA run, so that all individuals have the same rule consequent during the whole of that run, as occurs in [22].

The antecedent of the rule, with variable length, is represented in a fixed-length genome equal to $n - k$, where n is the total number of variables and k the number of ancestor nodes of the decision tree leaf node identifying the small disjunct in question. The GA uses a new operator specially designed to improve the comprehensibility of the rules. The authors propose in [10] some modifications to the original GA. The most significant change is the use of a sequential niching method [7] to foster population diversity and avoid the GA convergence to a single rule.

- ARMGA. Yan, Zhang y Zhang propose in [60] a GA with codification of the consequent for descriptive induction model for association rules without the need for previous specification by the user of the minimum levels of confidence and support.

The "*Chromosome = Set of Rules*" Approach

- GA-RULE. Pei, Goodman y Punch in [50] propose a GA to induce predictive rules in disjunctive normal form. In the genotype both the antecedent and the consequent of the rules are encoded.
- dAR is a GA developed by Au y Chan [4] the objective of which is the predictive induction of association rules. For this, each individual represents a set of rules only through their antecedents, since the rule consequents are determined in the fitness function computation, in the same way as in GA-NUGGETS. The authors describe in [5] DEML, a GA for predictive rule induction with applications for churn prediction. The algorithm uses a codification scheme similar to that used in dAR but the mutation operator includes a hill-climbing search procedure to search for improvements even when premature convergence occurs.
- GA-MINER, developed by Flockart and Radcliffe [23]. The GA-MINER system incorporates domain information to form a hybrid genetic algorithm using explicit representation of rules. The algorithm, apart from using the conventional genetic operators, experiments with adding two new operators, the addition and the subtraction operators and with different expressions for the fitness function. This algorithm is designed for running GAs on large scale parallel data mining.
- EDRL-MD (Evolutionary Decision Rule Learner with Multivariate Discretization). In this GA, developed by Kwedlo and Kretowski [38], a single chromosome encodes a set of rules in a variable length genome. The consequent is not coded, it remains fixed for each iteration of the GA (as in [8, 9, 10, 22]). As well as the standard genetic operators it includes four specific for rule induction: one operator which modifies a chromosome in order to include an additional positive example, another which modifies the chromosome in order to eliminate a negative example, a rule elimination operator and a rule inclusion operator. This predictive induction GA is modified in [39] so that the algorithm minimises misclassification cost rather than the number of classification errors.

Wang et al. present in [55] an algorithm to generate predictive rules which incorporates both approaches. The proposed algorithm consists of two phases: knowledge integration, which integrates rules from different sources using a GA with *Pittsburgh* approach; and knowledge refinement which improves the rule set performance using a GA with the *"Chromosome = Rule"* approach.

5 Fuzzy Logic in Rule Induction Processes

Fuzzy Logic allows us to model inaccurate and quantitative knowledge, as well as to handle uncertainty and deal naturally to a reasonable extent with human reasoning. Ever since it was proposed in 1965 by [61], it has been applied to many areas of research, fundamentally because of its proximity to human reasoning and because it provides an effective way of capturing the approximate and inexact nature of the real world.

As we know, the principle objective of any process of Data Mining is the identification of interesting patterns and their description in a concise and significant manner. The use of Fuzzy Logic in Data Mining is sensible because fuzzy models represent a description of the data directed towards the user through a set of qualitative models which establish significant and useful relationships between variables. Fuzzy sets allow us to establish flexible limits between the different levels of meaning, without ignoring or overemphasising the elements closest to the edges, in the same way as human perception does. In any process of knowledge extraction there is a component of human interaction and fuzzy sets allow us to represent knowledge in a linguistic form, to easily incorporate previous knowledge and to provide interpretable solutions.

In rule induction processes, Fuzzy Logic is included in such a way that the models extracted are fuzzy rules. In the most interpretable type of fuzzy rules, linguistic fuzzy rules, and therefore the most appropriate for Data Mining, the continuous variables are defined as linguistic variables; that is, variables which take as possible values linguistic labels, the semantics of which are represented by an associated fuzzy set [62]. The use of Fuzzy Logic in rule induction processes with quantitative variables eases the interpretability of the knowledge which is finally extracted, the incorporation of qualitative knowledge of the problem, the treatment of lost values and classes with limits which are not well defined, and the processing of noise in variables which are the result of real measurements [3, 62].

One of the fundamental aspects when working with fuzzy rules is the definition of membership functions associated with the fuzzy sets used. In Kouk's algorithm [40] for the extraction of fuzzy rules the expert needs to give the algorithm the continuous variables and their corresponding membership functions. In this case, the quality of the results obtained by the algorithm depends on the suitability of the fuzzy sets. For many applications it is very difficult to know from the outset which fuzzy sets will be the most appropriate, and so algorithms such as that in [27] have been described, which determine the fuzzy sets and generate the membership functions through clustering techniques.

However, in order to increase the interpretability of the results obtained in some proposals such as [3], knowledge of the problem is introduced in the initial definition of the fuzzy sets, such that the rules obtained are based on these fuzzy sets. Our proposal is centred on this approach.

The interpretability of fuzzy rules can be increased by the use of linguistic hedges. Chen and Wei in [12] propose an algorithm for the extraction of association fuzzy rules with linguistic hedges. These operators allow the knowledge extracted to be more comprehensible and closer to the user, and also enrich the semantics of the rules; the rules are therefore more detailed.

6 An Evolutionary Approach to Obtain Descriptive Fuzzy Rules

In the evolutionary model of extraction of fuzzy rules for subgroup discovery which we present, two components can be distinguished:

- An iterative model of extraction of fuzzy rules for the description of attributes of subgroups supported by different areas (not necessarily apart) of the instance space. This model includes the hybrid GA described below.
- A hybrid genetic algorithm for the extraction of one fuzzy rule that is simple, interpretable, and has an adequate level of support and confidence.

Both elements will be described in the next subsections.

6.1 Iterative Model for the Extraction of Descriptive Fuzzy Rules

The objective of the model for the extraction of descriptive fuzzy rules is to obtain a set of rules which give information on the majority of available examples for each value of the target variable.

The proposal follows the IRL (Iterative Rule Learning) approach: it includes a hybrid GA which generates a rule in an iterative plan. The iterative model allows new rules to be obtained while the generated rules reach a minimum level of confidence and give information on areas of search space in which examples which are not described by the rules generated by the previous iterations, remain.

The complete algorithm diagram is as follows:

```
START
  RuleSet ← ∅
  REPEAT
    Execute the GA obtaining rule R
    Local Search (R)
    RuleSet ← RuleSet + R
    Modify the set of examples
  WHILE confidence(R) ≥ minimum confidence and
      R represents new examples
END
```

The repetition mechanism promotes the generating of different rules (in the sense that they give information on different groups of examples). This is achieved by penalizing – once a rule is obtained – the set of examples represented by the same in order to generate future rules. It is important to point out that this penalization does not impede the extraction of concealed rules. In subgroup discovery algorithms, the possibility of extracting information on described examples is not eliminated since redundant descriptions of subgroups can show the properties of groups from a different perspective.

As can be seen in the extraction model diagram, in each iteration the confidence of the obtained rule must be higher than a previously specified minimum value. In the algorithms of descriptive induction, one of the fundamental problems, and partly significant to the quality of the obtained results, is the specification of minimum support (minimum confidence) required of the rules to be extracted. This value depends greatly on the problem to be solved and its solution is a problem which is still not completely resolved. Zhang et al. describes in [63] a method based on fuzzy logic for the setting of the minimum confidence level.

6.2 Genetic Algorithm for the Extraction of a Descriptive Fuzzy Rule

In order to obtain the best fuzzy rule, a hybrid GA which, following the evolutionary obtainment of the fuzzy rule, applies a stage of post-processing, a hill-climbing process is used.

The elements of the GA will be described below.

Chromosome Representation

The objective of the GA is to discover rules whose consequent is formed by a target variable which has been defined previously. The rules generated will be fuzzy or crisp, according to whether the variables involved are continuous or categorical, and they are coded according to the *"Chromosome = Rule"* approach.

Only the antecedent is represented in the chromosome and all the individuals in the population are associated with the same value of the target feature. This form of categorizing the target feature, used in algorithms for classification rule extraction [35] means that the evolutionary algorithm must be run many times in order to discover the rules of the different classes. As we commented in Sect. 4.2, in the specialized literature other proposals have been considered, such as codifying within the individual's genome [17] (in this way consequent evolution is also possible) or the deterministic selection for each rule of the value of the target feature [27, 49]. In problems such as the one we present, the focus we have adopted is suitable because it is necessary to describe all the values of the target feature, and the two alternatives mentioned above do not ensure information extraction relating to all the classes.

Some of the variables of the problem are continuous variables which are treated as linguistic variables with linguistic labels. The fuzzy sets corresponding to the linguistic labels are defined by a uniform fuzzy partition with triangular membership functions, as shown in Fig. 1.

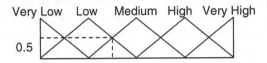

Fig. 1. Example of fuzzy partition for a continuous variable

All the information relating to a rule is contained in a fixed-length chromosome for which we use an integer representation model (the i-th position indicates the value adopted by the i-th variable). The set of possible values for the categorical features is that indicated by the problem plus an additional value which, when it is used in a chromosome, indicates that the corresponding variable does not take part in the rule. For continuous variables the set of values is the set of linguistic terms determined heuristically or with expert information, plus the value indicating the absence of the variable. In Fig. 2 we show an example of the rule and the chromosome which codifies it.

Zone	Employees	Sector		Bar
2	7	2	...	1

Fig. 2. Whole encoding model of a rule

In this example, the variable "Employees" does not influence the rule because the possible values are from 1 to 6, and in the corresponding gene the value is 7, which indicates the absence of this variable in the rule.

Fitness Function

In this process of rule discovery the objective is to obtain rules with high predictive capacity, and which are understandable and interesting. This objective can be achieved by using a weighted lineal combination of these three measurements [49]. In our proposal, the fitness function reflects these three quality factors by using a weighted lineal combination thus:

$$\text{fitness}(c) = \frac{\omega_1 \cdot \text{Support}(c) + \omega_2 \cdot \text{Interest}(c) + \omega_3 \cdot \text{Confidence}(c)}{\omega_1 + \omega_2 + \omega_3}$$

- *Confidence.* This measurement determines the accuracy of the rule, in that it reflects the degree to which the examples within the zone of the space marked by the antecedent verify the information indicated in the consequent of the rule. In order to calculate this factor we use an adaptation of Quinlan's accuracy expression [51] in order to generate fuzzy classification rules [16]: the sum of the degree of membership of example of this class to the zone determined by the antecedent, divided the sum of the degree of membership of all the examples (irrespective of their class) to the same zone. In order to calculate these membership degrees, we use triangular membership functions and the minimum t-norm. In the case of non-fuzzy rules, the degrees of membership correspond to the classic sets, 0 or 1.

- *Support.* This is the measurement of the degree of coverage that the rule offers to examples of that class. It is calculated as the quotient between number of new examples in the class covered by the rule and the number of examples that are not covered. This way of measuring support is sensible, when using the GA within an iterative process, in order to obtain different rules each time the GA is run. From the second iteration rules which cover examples belonging to zones delimited by previously obtained rules are penalised, because the support factor only considers examples which have not been described by already-obtained rules. No distance function is used as differences are penalised on a phenotypical level.

- The *interest* of a rule can be determined objectively (based on data) or subjectively (based on the user). In the specialised literature there are proposals of both forms, depending on the specific problem to which a data mining algorithm is applied, and neither of the two approaches has a significant advantage over the other. However, it seems evident that in practice both approaches should be used: the objective criteria as a filter to select potentially interesting rules and the subjective criteria in order that the final user can determine which rules are really interesting [24].

We follow this focus in our proposal, and in the GA the degree of interest is assessed objectively. To this end we use the interest criteria provided by Noda et al. in a dependence modelling process in [49] where the level of interest of a rule is determined by two terms, one referring to the antecedent and the other to the consequent. In our proposal the consequent is prefixed and only the antecedent is used for the interest calculation, based on an information measurement given by the following expression:

$$\text{Interest} = 1 - \left(\frac{\sum_{i=1}^{n} \text{Gain}(A_i)}{n \cdot \log_2 (|\text{dom}(G_k)|)} \right)$$

Where Gain is information gain, n is the number of variables which appear in the antecedent of the rule and $|\text{dom}(Gk)|$ is the cardinality of the objective variable (the number of values possible for the variable which is considered as a class). The denominator term is introduced to standardise the global value.

As discussed in [24] variables with high information gain are suitable for predicting a class when they are considered individually. However, from the point of view of the interest of a rule, it is understood that the user already knows which are the most predictive variables for a specific application domain, and therefore the rules which contain these variables are less interesting, as they are less surprising and carry less information. Therefore, it is understood that the antecedent of a rule is more interesting if it contains attributes with a small quantity of information.

The overall objective of the evaluation function is to direct the search towards rules which maximise accuracy, minimising the number of negative and not-covered examples.

Reproduction Model and Genetic Operators

A steady-state reproduction model [6] is used: the original population is only modified through the substitution of the worst individuals by individuals resulting from crossover and mutation.

The genetic operators used are a multi-point crossover operator and a random mutation operator which is biased such that half the mutations carried out have the effect of eliminating the corresponding variable, in order to increase the generality of the rules.

Post-Processing Phase of the Genetic Algorithm: Local Search Algorithm

The post-processing phase follows the algorithm bellow:

```
START
   Best_ Rule ← R; Best_support ← support(R); Better ← True
   REPEAT WHILE Better
      Better ← False
      FOR (i=1 to gene_number)
         R'i ← R without considering variable i
         IF (support (R'i) >= support (R))
            Better ← True
            IF (support (R'i) > Best_support)
               Best_support ← support (R'i)
               Best_Rule ← R'i
      END FOR
      IF (Better AND support(Best_Rule) >= min_supp)
         Return Best_Rule
      ELSE
         Return R
   END WHILE
END
```

It improves the rule obtained by a hill-climbing process, modifies the rule while increasing the degree of support. To accomplish this, in each iteration a variable is determined such that when it is eliminated, the support of the resulting rule is increased; in this way more general rules are obtained. Finally, the optimised rule will substitute the original only if it overcomes minimum confidence. The diagram is as follows:

7 Experimentation

The experimentation is carried out with the market dataset obtained in the Machinery and Tools biennial held in Bilbao in March 2002 (see Sect. 2). From total set of 104 variables, marketing experts have made a selection of variables which reduces the total set to a subset of 18 variables (in Table 1, a brief definition of these variables is described). The evolutionary rule induction algorithm has been applied to this set of variables.

Table 1. Variables description

Name	Description
Efficiency	Target variable: Global efficiency for the stands stated by *high*, *medium* or *low*
Zone	Geographic zone of the company
Sector	Sector to which the exhibitor belongs
Fair utility	Utility provided by the fairs
Annual fair number	Number of fairs participating annually as exhibitor
Written objectives	Existence of objectives for the BIEMH in writing
Previous promotion	Accomplishment of previous promotion to the fair
Promotion listings	Listings of clients to inform of the presence in the fair
Monitoring promotion	Monitoring of the previous promotion campaign
Operation after fair	Importance granted to the operations conducted after the fair
Client Contacts	Importance granted to the number of contacts made with present clients
Quality	Importance granted to the quality of the contacts made
Visitors number importance	Importance granted to the number of visitors in stand
Telephone calls	Accomplishment of telephone calls
Machinery demo type	Type of demonstrations of machinery
New features	Presence of new features in stand
Stewardesses	Presence of stewardesses
Stand at entrance	Stand in pavilion entrance
Near of stairs	Stand near the stairs

Parameters of the experimentation:

- The proposed algorithm is run five times for each one of the target variable values (global efficiency *high*, *medium* or *low*).
- Number of chromosomes in the GA: 100.
- Maximum number of evaluations of individuals in each GA run: 5000.
- Fitness function weights. Support: 0.4; confidence: 0.3; interest: 0.3.
- Minimum confidence value: 60.

In Table 2 the best results obtained are described. Here, for each value of the target variable the confidence, support and interest corresponding to each rule induced are shown (by means of three real numbers belonging to [0,100]). In Tables 3, 4 and 5 the rule expressions are described.

Table 2. Quality measurements of the rules extracted

Class	Rule	Support	Confidence	Interest
	1	10,526	100,000	61,282
	2	13,158	100,000	60,663
	3	18,421	100,000	58,341
1	4	7,895	100,000	58,248
	5	7,895	100,000	59,971
	6	5,263	100,000	57,806
	7	5,263	100,000	53,024
	1	10,811	100,000	59,112
	2	10,135	100,000	55,906
	3	6,081	100,000	58,062
	4	3,378	100,000	61,805
	5	6,081	100,000	59,567
	6	3,378	100,000	57,870
2	7	4,730	100,000	59,923
	8	3,378	100,000	60,617
	9	2,027	100,000	60,929
	10	3,378	100,000	59,232
	11	95,946	64,840	62,340
	12	0,676	100,000	60,977
	1	4,762	100,000	62,110
	2	9,524	100,000	59,904
3	3	11,905	100,000	59,045
	4	4,762	100,000	59,845
	5	7,143	100,000	60,580

We can observe that the algorithm induces set of rules with a high confidence (higher than the minimum confidence value) and interest level, around 60 in most cases. This high level of interest, according to the definition of the interest measurement used (see Sect. 6.2.2.) indicates that the variables

Table 3. Rules for *medium* efficiency

---

1   IF Zone = *North* and Fairs utility = *Low* AND Visitors number
     importance = *Medium* AND Stand at pavilion entrance = *Yes*
     THEN Efficiency = *Medium*

2   IF Zone = *North* AND Importance of contacts quality = *High* AND
     Telephone calls = *Yes* AND New features = "Catalogue"
     THEN Efficiency = *Medium*

3   IF Sector = *Rest* AND Importance of operations after the fair = *Medium*
     AND New features = *Product improvement*
     THEN Efficiency = *Medium*

4   IF Sector = *Starting+Deformation* AND Number of annual fairs = *More
     than 11*
     THEN Efficiency = *Medium*

5   IF Previous promotion = *Yes* AND Visitors number importance = *Low*
     AND Stand at pavilion entrance = *Yes*
     THEN Efficiency = *Medium*

6   IF Sector = *Rest* AND Importance of operations after the fair = *Low* AND
     Visitors number importance = *High*
     THEN Efficiency = *Medium*

7   IF Zone = *North* AND Sector = *Starting+Deformation* AND Fairs utility =
     Low AND Previous promotion = *Yes* AND Importance of contacts
     quality = *Medium*
     THEN Efficiency = *Medium*

8   IF Importance of contacts quality = *Medium* AND Stewardesses = *Yes*
     THEN Efficiency = *Medium*

9   IF Previous promotion = *No* AND Importance of contacts quality = *High*
     AND Stand at pavilion entrance = *Yes*
     THEN Efficiency = *Medium*

10 IF Sector = *Rest* AND Importance of operations after the fair = *Low* AND
     Importance of contacts quality = *Medium*
     THEN Efficiency = *Medium*

11 IF Number of annual fairs = *Less than 11*
     THEN Efficiency = *Medium*

12 IF Number of annual fairs = *More than 11* AND Importance of contacts
     quality = *Medium*
     THEN Efficiency = *Medium*

---

which intervene in the general rules are variables with low information gain
value, more surprising to the user and they carry more information. We must
note that variables with high information gain are suitable for predicting a
class when they are considered individually. However, from the point of view
of the interest of a rule, it is understood that the user already knows which
are the most predictive variables for a specific application domain, and therefore the rules which contain these variables are less interesting, as they are
less surprising and carry less information. Therefore, it is understood that the

Table 4. Rules for *low* efficiency

---

1 IF Sector = *Starting+Deformation* AND Written objectives = *No* AND
   Previous promotion = *No*
   THEN Efficiency = *Low*

2 IF Written objectives = *No* AND Importance of present clients contacts =
   Low AND Importance of contacts quality = *High* AND Stand at pavilion
   entrance = *No* AND Near of stairs = *No*
   THEN Efficiency = *Low*

3 IF Zone = *North* AND Sector = *Starting+Deformation* AND Written
   objectives = *No* AND Telephone calls = *Yes* AND New features =
   Product improvement AND Stand at pavilion entrance = *No*
   THEN Efficiency = *Low*

4 IF Importance of present clients contacts = *Low* AND Importance of
   contacts quality = *Low*
   THEN Efficiency = *Low*

5 IF Zone = *East* AND Written objectives = *No* AND Existence of promotion
   listings = *No* AND Importance of operations after the fair = *High* AND
   Stand at pavilion entrance = *No* AND Near of stairs = *No*
   THEN Efficiency = *Low*

6 IF Zone = *North* AND Fairs utility = *Low* AND Importance of present
   clients contacts = *Medium* AND New features = *Product improvement*
   THEN Efficiency = *Low*

7 IF Sector = *Starting+Deformation* AND Promotion campaign monitoring =
   No AND Importance of present clients contacts = *High* AND Machinery
   demonstrations type = *Sporadic operation* AND Stewardesses = *Yes*
   THEN Efficiency = *Low*

---

antecedent of a rule is more interesting if it contains attributes with a small
quantity of information, as the rule induces here.

The rule support, except for some rules, is low. The model induces, for
this problem, specific rules which represent a small number of examples. The
market problem used in this work is a difficult real problem in which inductive
algorithms tend to obtain small disjuncts, more common in datasets than
one might think at first glance. However, the small disjunct problem, as we
described in Sect. 3, is not a determining factor in the induction process
for subgroup discovery. This is because partial relations, i.e., subgroups with
interesting characteristics, with a significant deviation from the rest of the
dataset, are sufficient.

The knowledge discovered for each one of the target variable values is
understandable by the user due to the use of Fuzzy Logic, and the small
number of rules and conditions in the rule antecedents.

Table 5. Rules for *high* efficiency

---

1   IF Written objectives = *Yes* AND Stewardesses = *No* AND Stand at
    pavilion entrance = *Yes* AND Near of stairs = *Yes*
    THEN Efficiency = *High*
2   IF Sector = *Rest* AND Number of annual fairs = *More than 11* AND New
    features = *Authentic newness*
    THEN Efficiency = *High*
3   IF Zone = *East* AND Sector = *Rest* AND Fairs utility = *High* AND
    Importance of contacts quality = *High* AND New features = *Authentic
    newness*
    THEN Efficiency = *High*
4   IF Zone = *East* AND Sector = *Rest* AND Number of annual fairs = *Less
    than 11* AND Existence of promotion listings = *Yes* AND Importance of
    operations after the fair = *High* AND Importance of contacts quality =
    Medium AND Stand at pavilion entrance = *No*
    THEN Efficiency = *High*
5   IF Fairs utility = *High* AND Written objectives = *Yes* AND New features =
    Authentic newness AND Stand at pavilion entrance = *No* AND Near of
    stairs = *No*
    THEN Efficiency = *High*

---

8 Concluding Remarks

The area of Soft Computing provides a set of tools which, independently or
together, are being successfully used in knowledge extraction tasks.

Fuzzy Logic allows the user to incorporate directly linguistic knowledge
into the data mining process, to mix this knowledge with non-linguistic information and to treat appropriately incomplete data or data with noise. But
perhaps one of the characteristics which is most important for the use of
fuzzy logic in this type of algorithm is its ability to represent knowledge in a
linguistic form which is directly interpretable, through fuzzy rules.

Genetic Algorithms carry out a global search which is independent of the
domain. This makes them a strong tool which can be applied to various stages
of the knowledge extraction process.

In this chapter we describe an evolutionary model for the descriptive induction of fuzzy or crisp rules which describe subgroups. The proposal includes
a GA in an iterative model which extracts rules when some examples are left
uncovered, and the rules obtained surpass a given confidence level which is
specified by the user.

We have applied this proposal to a real knowledge extraction problem in
trade fairs. The experiment carried out has determined a simple set of rules
which use few variables and therefore has a simple structure. The information
extracted is comprehensible for and usable by the final user.

In future studies, we will examine the use of a more flexible structure
for the rule and the study of an appropriate interest measurement for this

structure. Moreover, we are working in the development of a multiobjective version of this model [18].

Acknowledgments

This work was supported by the Spanish Ministry of Science and Technology and by the European Fund. FEDER under Projects TIC-04036-C05-01 and TIC-04036-C05-04.

References

1. Agrawal R, Mannila H, Srikant R, Toivonen H, Verkamo I (1996) Fast discovery of association rules. In: Fayyad UM, Piatetsky-Shapiro G, Smyth P, Uthurusamy R (eds) Advances in Knowledge Discovery and Data Mining. AAAI Press, pp. 307–328
2. Araujo DLA, Lopes HS, Freitas AA (1999) A parallel genetic algorithm for rule discovery in large databases. In: Proceedings IEEE Conference on Systems, Man and Cybernetics, vol III, pp. 940–945
3. Au WH, Chan KCC (1998) An effective algorithm for discovering fuzzy rules in relational databases. In: Proceedings of IEEE International Conference on Fuzzy Systems (Fuzz IEEE'98), pp. 1314–1319
4. Au WH, Chan KCC (2002) An evolutionary approach for discovering changing patterns in historical data. In: Proceedings of 2002 SPIE 4730, Data Mining and Knowledge Discovery: Theory, Tools and Technology IV, pp. 398–409
5. Au WH, Chan KCC, Yao X (2003) A novel evolutionary data mining algorithm with applications to churn prediction. IEEE Transactions on Evolutionary Computation 7 (6): 532–545
6. Bäck T, Fogel D, Michalewicz, Z (1997) Handbook of Evolutionary Computation, Oxford University Press, Oxford
7. Beasly D, Bull DR, Martin RR (1993) A sequential niche technique for multimodal function optimization. Evolutionary Computation 1: 101–125
8. Carvalho DR, Freitas AA (2000) A genetic algorithm-based solution for the problem of small disjuncts. In: Zighed DA, Komorowski J, Zytkow J (eds) Principles of Data Mining and Knowledge Discovery (Proc. 4th European Conf. PKDD-2000, Lyon, France). Springer-Verlag, pp. 345–352
9. Carvalho DR, Freitas AA (2002) A genetic algorithm for discovering small-disjunct rules in data mining. Applied Soft Computing 2: 75–88
10. Carvalho DR, Freitas AA (2002) A genetic algorithm with sequential niching for discovering small-disjunct rules. In: Langdon WB, Cantu-Paz E, et al. (eds) Proceedings Genetic and Evolutionary Computation Conference (GECCO-2002). Morgan Kaufmann, New York San Francisco, pp. 1035–1042
11. Cestnik B, Lavrac N, Zelezny F, Gamberger D, Todorovski L, Kline M (2002) Data mining for decision support in marketing: A case study in targeting a marketing campaign. In: Proceedings of the ECML/PKDD-2002 Workshop on Integration and Collaboration Aspects of Data Mining, Decision Support and Meta-Learning, pp. 25–34

12. Chen G, Wei Q (2002) Fuzzy association rules and the extended mining algorithms. Information Sciences 147: 201–228
13. Clark P, Niblett T (1989) The CN2 induction algorithm. Machine Learning, 3(4): 261–283
14. Cohen W (1995) Fast effective rule induction. In Proceedings of the Twelfth International Conference on Machine Learning. Morgan Kaufmann, pp. 115–123
15. Cordón O, Herrera F (1997) A three-stage evolutionary process for learning descriptive and approximative fuzzy logic controller knowledge bases from examples. International Journal of Approximate Reasoning 17 (4): 369–407
16. Cordón O, del Jesus MJ, Herrera F (1998) Genetic Learning of Fuzzy Rule-based Classification Systems Co-operating with Fuzzy Reasoning Methods. International Journal of Intelligent Systems 13 (10/11): 1025–1053
17. De Jong KA, Spears WM, Gordon DF (1993) Using genetic algorithms for concept learning. Machine Learning 13 (2–3): 161–188
18. Deb K (2001) Multi-Objective Optimization using Evolutionary Algorithms. John Wiley & Sons
19. Dhar V, Chou D, Provost F (2000) Discovering interesting patterns for investment decision making with Glower-a Genetic Learner Overlaid With Entropy Reduction. Data Mining and Knowledge Discovery 4: 251–280
20. Domingos P (1998) Occam's two razors: the sharp and the blunt. In: Proceedings of the Fourth International Conference on Knowledge Discovery and Data Mining (KDD'98), pp. 37–43
21. Fayyad U, Piatesky-Shapiro G, Smyth P (1996) From data mining from knowledge discovery in databases. In: Fayyad UM, Piatetsky-Shapiro G, Smyth P, Uthurusamy R (eds): Advances in Knowledge Discovery & Data Mining. AAAI/MIT, pp. 1–34
22. Fidelis MV, Lopes HS, Freitas AA (2000) Discovering compresible classification rules with a genetic algorithm. In: Proceedings on Evolutionary Computation (CEC'2000), pp. 805–810
23. Flockhart IW, Radcliffe NJ (1995) GA-MINER: Parallel data mining with hierarchical genetic algorithms (Final Report by the University of Edimburgh, UK, EPCC-AIKMS-GA-Miner-Report 1.0)
24. Freitas AA (1999) On Rule Interestingness Measures. Knowledge-Based Systems 12: 309–315
25. Freitas AA (1999) A genetic algorithm for generalized rule induction. In: Roy R, Furuhashi T, Chawdhry PK (eds) Advances in Soft Computing – Engineering Design and Manufacturing (Proc. WSC3, 3rd on-line world conf., hosted on the internet), pp. 340–353
26. Freitas AA (2002) Data Mining and Knowledge Discovery with Evolutionary Algorithms, Springer Verlag
27. Fu AW, Wong MH, Sze SC, Wong WC, Wong WL, Yu WK (1998) Finding fuzzy sets for the mining of fuzzy association rules for numerical attributes. In: Proceedings of the First International Symposium on Intelligent Data Engineering and Learning (IDEAL'98), pp. 263–268
28. Gamberger D, Lavrac N (2002) Expert guided subgroup discovery: Methodology and application. Journal of Artificial Intelligence Research, 17: 501–527
29. Giordana A, Neri F (1995) Search-intensive concept induction. J Evolutionary Computation, 3 (4): 375–416

30. Goldberg DE (1989) Genetic algorithms in search, optimization and machine learning. Addison-Wesley
31. González A, Pérez R (1998) Completeness and consistency condicions for learning fuzzy rules. Fuzzy Sets and Systems 96: 37–51
32. Gopalakrishna S, Lilien GL, Williams JD, Sequeira IK (1995) Do trade shows pay off. Journal of Marketing 59: 75–83
33. Greene DP, Smith SF (1993) Competition-based induction of decision models from examples. Machine Learning 3: 229–257
34. Holland JH (1975) Adaptation in natural and artificial systems. University of Michigan Press
35. Janikow CZ (1993) A knowledge-intensive genetic algorithm for supervised learning. Machine Learning 13: 189–228
36. Klösgen W (1996) Explora: a multipattern and multistrategy discovery assistant. In: Fayyad V, Piatetsky-Shapiro G, Smyth P, Uthurusamy R. (eds) Advances in Knowledge Discovery and Data Mining. MIT Press pp. 249–271
37. Klösgen W (2002) Handbook of data mining and knowledge discovery. Oxford University Press
38. Kwedlo W, Kretowski M (1999) An evolutionary algorithm using multivariate discretization for decision rule induction. In: Proceedings of the 3rd European Conference on Principles of Data Mining and Knowledge Discovery (PKDD'99), pp. 392–397
39. Kwedlo W, Kretowski M (2001) An evolutionary algorithm for cost-sensitive decision rule learning. Lecture Notes in Artificial Intelligence 2167: 288–299
40. Kuok C, Fu A, Wong M (1998) Mining fuzzy association rules in databases. ACM SIGMOD Record, 27: 41–46
41. Lavrac N, Cestnik B, Gamberger D, Flach P (2004) Decision support through subgroup discovery: three case studies and the lessons learned. J Machine Learning 57 (1–2): 115–143
42. Lavrac N, Flach P, Zupan B (1999) Rule evaluation measures: A unifying view. In: Proceedings of the Nineth International Workshop on Inductive Logic Programming. Springer, pp. 74–185
43. Lavrac N, Flach P, Kavsek B, Todorovski L (2002) Adapting classification rule induction to subgroup discovery. In: Proceedings of the Second IEEE International Conference on Data Mining. IEEE Computer Society, pp. 266–273
44. Lavrac N, Kavsec B, Flach P, Todorovski L (2004) Subgroup discovery with CN2-SD. Machine Learning Research 5: 153–188
45. Mesonero M (2004) Toward an effective trade show planning model based on genetic algorithms. Ph.D. thesis (in Spanish), Mondragon Unibertsitatea.
46. Michalski RS, Mozetic I, Hong J, Lavrac N (1986) The multi-purpose incremental learning system AQ15 and its testing application on three medical domains. In: Proceedings of the Fifth National Conference on Artificial Intelligence. Morgan Kaufmann, pp. 1041–1045
47. Michie D, Spiegelhalter DJ, Taylor CC (1994) Machine learning, neural and estatistical classification. Ellis Horwood
48. Millar S (2003) How to get the most of the trade shows. NTC Publishing Group
49. Noda E, Freitas AA, Lopes HS (1999) Discovering Interesting Prediction Rules with a Genetic Algorithm. In Proceedings of the Congress on Evolutionary Computation 2, pp. 1322–1329

50. Pei M, Goodman ED, Punch III WF (1997) Pattern discovery from data using genetic algorithms. In: Proceedings of the First Pacific-Asia Conference on Knowledge Discovery and Data Mining. KDD: Techniques and applications, pp. 264–276
51. Quinlan JR (1987) Generating production rules Machine Learning. Morgan Kaufmann
52. Silberschatz A, Tuzhilin A (1995) On subjective measures of interestingness in knowledge discovery. In: Proceedigns of the First International Conference on Knowledge Discovery and Data Mining, pp. 275–281
53. Smyth P, Goodman RM (1991) Rule induction using information theory. In: Piatetsky-Shapiro G, Frawley J (eds.) Knowledge Discovery in Databases. MIT Press, Cambridge, pp. 159–176
54. Tettamanzi A, Tomassini M (2001) Soft Computing. Integrating Evolutionary, Neural and Fuzzy Systems. Springer
55. Wang CH, Hong TP, Tseng SS (1998) A hybrid genetic knowledge-integration strategy. In: Proceedings of the 1998 IEEE International Conference on Evolutionary Computation, pp. 587–591
56. Wilson SW, Goldberg DE (1994) ZCS: a zeroth order classifier system. Evolutionary Computation 2: 1–18
57. Wilson SW (1995) Classifier system based on accuracy. Evolutionary Computation 3 (2): 149–175
58. Wrobel S (1997) An algorithm for multi-relational discovery of subgroups. In: Proceeding of the First European Conference on Principles of Data Mining and Knowledge Discovery, pp. 78–87
59. Wrobel S (2001) Inductive logic programming for knowledge discovery in databases. In: Dzeroski S, Nada Lavrac N (eds) Relational Data Mining, Springer: 74–101
60. Yan X, Zhang C, Zhang S (2003) A database-independent approach of mining association rules with genetic algorithms. In: Proceedings of IDEAL 2003, pp. 882–886
61. Zadeh LA (1965) Fuzzy Sets. Inform. Control. 9: 338–352
62. Zadeh LA (1975) The concept of a linguistic variable and its applications to approximate reasoning, Parts I, II, III. Information Sciences 8–9: 199–249, 301–357, 43–80
63. Zhang S, Lu J, Zhang C (2004) A fuzzy logic based method to acquire user threshold of minimum-support for mining association rules. Information Sciences 164: 1–16

Personalized Multi-Stage Decision Support in Reverse Logistics Management

Jie Lu and Guangquan Zhang

Faculty of Information Technology, University of Technology, Sydney, PO. Box 123, Broadway, NSW 2007, Australia
{jielu,zhangg}@it.uts.edu.au

Abstract. Reverse logistics has gained increasing importance as a profitable and sustainable business strategy. As a reverse logistics chain has strong internal and external linkages, the management of a reverse logistics chain becomes an area of organizational competitive advantage, in particular, with the growth of e-commerce applications. To effectively manage a reverse logistics chain always involves a decision optimization issue in which uncertain information, individual situation, multiple criteria and dynamic environment all need to be considered. This paper addresses the need of supporting reverse logistics managers in selecting an optimal alternative for goods return under their business objectives. Through analyzing the characteristics of reverse logistics chain, this paper proposes a personalized multi-stage decision-support model for reverse logistics management. It then presents a personalized fuzzy multi-criteria decision-making approach to assist managers to lead and control the reverse logistics within an uncertain and dynamic system.

1 Introduction

During the last decade many companies have realized that the opportunity to improve operations lies largely with procurement, distribution and logistics–the supply chain. As companies are increasing their levels of outsourcing, buying goods or services instead of producing or providing them by themselves, they are therefore spending increasing amounts on supply related activities. Logistics is one of the key elements of supply chain management [14, 17]. It refers to decide the best way of the movement of goods within a facility [12]. Logistics has become a hot competitive advantage as companies struggle to get the right stuff to the right place at the right time.

There are two logistics channels in a supply chain system of a company. Forward logistics channel concerns the movement of goods from source to the point of consumption. A backward movement can be happened to return goods to suppliers called reverse logistics [2, 10]. Forward logistics usually brings profit to all operational departments involved, while reverse logistics usually cannot. Some companies even perceive goods return as failure of their

Jie Lu and Guangquan Zhang: *Personalized Multi-Stage Decision Support in Reverse Logistics Management*, Studies in Computational Intelligence (SCI) **5**, 293–292 (2005)
www.springerlink.com   © Springer-Verlag Berlin Heidelberg 2005

operations. However, the high rate of goods return from online purchases, the increasing environmental regulations and standards, and the growing consumer awareness of recycling have brought a need to rethink the significance of reserve logistics [15]. Some reports have shown that companies trying to hide from the significance of reverse logistics miss tremendous profit making opportunities [5, 6]. The reason is that companies can use reverse logistics as an opportunity for maintaining customer support, building good customer relationship and reach the ultimate business objective of profitability [14]. Moreover, many companies have discovered that effective management for a reverse logistics chain such as the reductions in inventory carrying costs, transportation costs and waste disposal costs can be also substantial with the supply chain program [13]. Companies like IBM, HP have tailored reverse logistics to their industry with it [7].

To effectively manage a reverse logistics chain involves finding the best way of movement of goods by evaluating a number of alternatives of goods return disposals under a set of business objectives. In the evaluation, a set of criteria are as constraints, such as buyer's demand, vendors' quota flexibility, repairer's capacity, purchase and repair values of the returned items, and time [3, 11]. In principle, this is a multi-criteria decision-making problem. However, there are several issues to result in a normal multi-criteria decision-making approach that cannot effectively support the decision-making in such reverse logistics management:

1. Multi-stage and dynamic: A reverse logistics chain involves a series of stages (operational functions). All the stages involved in the chain are interrelated in a way that a decision made at one stage affects the performance of next stages. That is, the decision objective(s) and alternatives at each stage (except the first one) are dynamically affected by the decision(s) made in previous stages/functions. A normal multi-criteria decision-making approach is not able to handle the multi-stage dynamic decision feature.
2. Personalization: Managers at different service stations of a reverse logistics chain making decisions are based on different evaluation criteria and different alternatives. For example, the alternatives to deal with a goods return in a collection station are totally different from one in a redistribute station. Managers at different stations need a personalized decision support, while the normal multi-criteria approach could hardly support such "personalized" decision-making of reverse logistics managers.
3. Uncertainty and imprecision: In practice, reverse logistics managers often imprecisely know the values of related constraints and evaluation criteria in selecting an optimal alternative. For example, they can only estimate inventory carrying costs and transportation costs of a particular set of goods to be returned. Also, the evaluation for any alternative of a goods return, logistics managers need assigning values for a number of selection criteria descriptors according to his/her specialized experience. These values assigned are often in linguistic terms, such as "high reusability", "low

reusability" for a set of goods to be returned. Obviously, the normal multi-criteria decision-making approach is not efficient to solve these problems in which uncertain information and imprecise linguistic expressions are involved.

This study aims to propose a decision-making approach which extends a normal multi-criteria approach to effectively handle the three issues: multi-stage, personalization, and uncertainty in reverse logistics management. This paper is organized as follows. Section 2 analyses the main operational functions in a reverse logistics chain and summarizes the characteristics of decision making in selecting the best way to handle goods return. A personalized multi-stage decision support model for reverse logistics management is established. Based on this established model, Sect. 3 proposes a personalized fuzzy multi-criteria decision-making approach which takes the form of optimizing procedures to provide an optimal way for logistics managers through evaluating related alternatives at any stage of a reverse logistic chain. A case-study example illustrates the power and details of the proposed approach in Sect. 4. Finally, a conclusion and future research plan are given in Sect. 5.

2 A Reverse Logistics Decision Support Model

This section firstly analyses the composition of a reverse logistics chain and the characteristics of goods return decision-making. It then presents a personalized multi-stage reverse logistics decision support model.

2.1 Reverse Logistics Chain

It is easy to think of logistics as managing the flow of products from the point of the view of the raw material acquisition to end customers. But the life of a product, from a logistics viewpoint, does not end with delivery to the end customer [1]. For many companies there is a reverse logistics chain that must be managed as well. Products may become obsolete, damaged or non-functioning and therefore need to be returned to their source points for repair or disposition. This procedure forms a reverse logistics chain. The reverse logistics chain may utilize all or some stages of the forward logistics chain or require a separate design, and terminates with the final disposition of a product. As a fairly new concept, a company's supply chain consists of both forward logistics and reverse logistics. The European working group on reverse logistics puts forward the following definition of reverse logistics including the goal and the process involved: "the process of planning, implementing and controlling flows of raw materials, in process inventory, and finished goods, from a manufacturing, distribution or use point to a point of recovery or point of proper disposal."

A reverse logistics chain involves a series of stages, each concerns a kind of activities associated with the management of goods (can be products, materials or components) return, with different facilities. These stages/facilities are interrelated in a way that a decision made at previous stage affects the decision making in the following stages. In general, the stages of a reverse logistics chain typically includes collection, combined testing/sorting/inspection/separation process, reprocessing/repairing or direct recovery and redistribution/resale/ reusing or disposal which can be also happened with other operational functions such as testing [2, 16]. As shown in Fig. 1, Supply, Manufacture, Distribution and Consumer form a flow of forward logistics. A reverse logistics flow has a backward movement from 'Consumer" to "Supply." Stage "Collection" refers to all activities rendering goods to be returned available and physically moving them to some point where a further treatment is taken care of. Testing (or inspection) determines whether collected goods are in fact reusable or how much work needs to be paid in order to make it usable. Sorting (or separation) decides what to do with each or a set of collected goods, including reprocessing and disposal. Thus, testing and sorting will result in splitting the flow of collected goods according to distinct treatment options. Reprocessing means the actual transformation of returned goods into usable products again. The transformation may take different forms including recycling, reconditioning, and remanufacturing. Disposal could be an option at this stage as well. Redistribution refers to directing reusable products to a potential reuse market and to physically moving them to future end customers. Therefore, the reverse logistics can simply be just reselling a product, or can be accompanies by a series of processes, as shown in Fig. 1, from collection to reuse or disposal [2, 16].

The important degrees of these operational functions are different in a goods return. Some functions may play more important roles than others for a particular goods return. The degree of importance of each operational

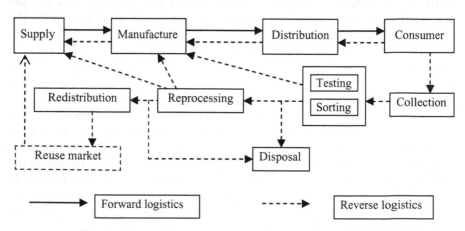

Fig. 1. Forward logistics chain and reverse logistics chain

function is also variable for different goods returns. This variance is mainly dependent on the business objective of the reverse logistics management. For example, if the business objective of a company's reverse logistics management is to provide customer services in warranties, then the function of "collection" may play a more important role in the reverse logistics chain than the reprocessing for the disassembly of products. If the business objectives is more environmentally related such as "reclaiming parts", the function of "sorting" may be more important. For a particular reverse logistics flow, some operation functions may not appear. For example, "reuse market" will not appear for many kinds of goods returned.

2.2 Characteristics of Goods Return Decision Making

There are several kinds of actors involved in reverse logistics activities in practice. They are independent intermediaries, specific recovery companies, reverse logistics service providers, municipalities taking care of waste collection, and public-private foundations created to take of goods recovery. The aims of different kinds of actors in a reverse logistics chain are different. For example, a manufacture may do recycling in order to prevent jobbers reselling its products at a lower price, while a collector may collect used products in order to establish a long-term customer relationship. These actors can also be logically differentiated into returners, receivers, collectors, processors and sales persons based on the features of their roles in a reverse logistics chain [2, 9]. The most important type of actors is "returner" as any stage can be a returner, including customers, in the whole reverse logistics chain, hence suppliers, manufactures, wholesalers and retailers.

Returners, at any operational stage of a reverse logistics chain, always need to decide how to best move current returned goods such as to return it to a factory for repairing or disposal it locally. Returners at different stages or at the same stage but with different goods returns may have different alternatives and different selection criteria to find the best way from these alternatives. For example, at the stage of "collection", the decision is mainly about planning and scheduling of recovery operations, and the transportation and the warehousing of returns have to be dealt with. At the stage of "sorting", returners need to determine whether or not to do recovery and which type of recovery if do. The recovery options are thus taken into account and judged. The decisions for a goods return at a previous stage will become constraints given for and impact directly on the decision activities of its following stages. For example, when one product is identified to be not usable any other decisions on storage, treatment, transportation for reusing process are not considerable except transportation for disposing processed wastes. Therefore, every decision has to bear the impact on the decisions at its previous stages.

Table 1. Example of relationships among returners' types, their business objectives and alternatives in a reverse logistics chain

Returner Types	Business Objectives (O)	Alternatives (A)
Collector	Maximizing customer relationship Minimizing customer service cost in warranties	Replacement Local storage Customer postal
Tester/Sorter	Minimizing total operational cost Maximizing customer relationship Maximizing satisfying environmental regulation	Recycling Remanufacturing Reuse Disposal
Processor	Minimizing total operational cost Maximizing customer services in warranties of repair	Local remanufacturing Recycling Disposal
Redistributor	Maximizing business profit Maximizing reclaiming parts Minimizing time	Resale Disposal Storage

The following characteristics have been seen through the above analysis:

1. reverse logistics management involves decision making at multiple stages;
2. decisions made at different stages are based on different alternatives and selection criteria;
3. at each stage, returners' business objectives, related alternatives and evaluation criteria are dynamic changed. The change is caused by both the features of returned goods and the actions of previous functions of the reverse logistics chain. The analysis reminds a personalized multi-stage decision support model to help the selection of the best way to handle a goods return in a reverse logistics chain.

In order to build the model, two sets of relationships have to be discussed. One is the dependence relationship between business objectives and alternatives, and the other is between business objectives and selection criteria.

Based on Fig. 1, returners can be classified into four basic types: collector, tester/sorter, processor, and redistributor, as shown in Table 1. The four types of returners are at four main functional stages of a reverse logistics chain respectively. For each type of returners, possible business objectives are shown in the column two of Table 1. Once a returner's business objectives for a particular goods return are determined, a set of alternatives can be identified. For example, two business objectives of a collector are to maximize customer relationship and to minimize customer services cost in warranties. Related alternatives are thus recycling, reconditioning and disposal as shown in the column three of Table 1. However, different companies may set up different

Table 2. Example of relationships among business objectives, selection criteria and related items in a reverse logistics chain

Objectives (O)	Selection Criteria (C)	Related Items
Minimizing total operational cost	Cost	Collection cost, storage cost, treatment cost, transportation cost for reusing processed wastes, transportation cost for disposing processed wastes, repair cost
Minimizing customer services in warranties	Time	Collecting time, treatment time, and transportation time
Maximizing customer relationship	Customer satisfaction	Product life stages (Introduction; Growth; Maturity; Decline) Time Usability
Maximizing business profit	Benefit Cost	Reusability Resale income Repair cost Transportation cost Redistribute cost

business objectives and related different alternatives for each type of returners. Related data can be obtained through data mining and other methods.

To evaluate these alternatives, a number of selection criteria are set up. Each criterion is described by one or more related items which are strongly dependent on the corresponded business objectives. For example, when a company's business objective for a goods return is to minimize customer services in warranties, time including collect time, treatment time and transportation time, is the only assessment item for selection of a solution from related alternatives. Table 2 lists the possible business objectives, related selection criteria and involved assessment items. Same as Table 1, different companies may set up different criteria for the same business objective.

2.3 A Personalized Multi-stage Decision Support Model

Figure 2 shows the proposed personalized multi-stage decision support model. This model describes a whole decision-making process of a returner at any stage of a reverse logistics chain. In the model, when a returner's type is known, its business objectives can be identified based on the relationships shown in Table 1. After business objectives are determined, the returner is allowed to indicate a weight for each objective based on individual experience and knowledge. Related alternatives are then determined based on the relationships shown in Table 1 as well. As the alternatives of a goods return decision are totally related to its business objectives, when an objective's

Decision support system   Returner

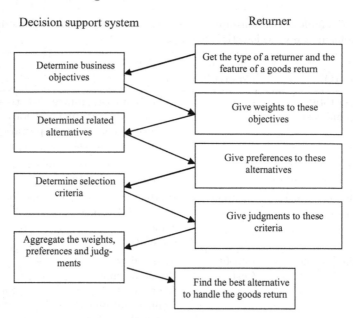

Fig. 2. A personalized multi-stage decision support model of reverse logistics management

weight is very low, its related alternatives and selection criteria will not be considered. To evaluate these alternatives, a set of selection criteria is determined based on information shown in Table 2. The types of returners and their preferences for business objectives may result in different sets of alternatives. Obviously, this decision process involves multiple layers of relationships: from the type of a returner to determining its business objectives, and then alternatives and finally selection criteria. This process has a personalized feature as each individual logistic manager may have a set of individual alternatives and individual preferences for assessing these alternatives with a particular set of goods return.

Uncertainty and imprecision are involved in the model. In practice, returners often describe and measure the degree of weights and their preferences in linguistic terms, such as "preferable" and "not really", "high" or "low" since a numerical evaluation is sometimes unacceptable. These linguistic terms are obviously with uncertainties [8]. Each criterion may involve a number of related selection items, estimation of these items' values is needed and these estimated values are often with imprecision. For example, when minimizing the total operational cost is the business objective of a goods return at an operational stage, five major time-varying cost items may need to be estimated and measured: collection cost, storage cost, treatment cost, transportation cost for reusing processed wastes, and transportation cost for disposing processed wastes [10]. All these estimations and measures often involve imprecise values.

The uncertainty and imprecision features will affect on the processing of a decision evaluation. When several layers of a goods return decision evaluation are synthesized into an aggregated result, that is, the weights of business objectives will be combined with the preferences of related criteria to selection alternatives, the uncertainty and imprecision features will be integrated into the final outcome, an optimal plan, for the particular goods to be returned. Therefore, the uncertainty issue has to be in the proposed decision-making approach.

3 A Personalized Fuzzy Multi-Criteria Decision-Making Approach for Reverse Logistics Management

As uncertainty is incorporated in the personalized multi-stage goods return decision process, the proposed decision-making approach must take into account the presentation and processing of imprecise information, and deal with its personalization and multi-stage issues at the same time. This section gives a personalized fuzzy multi-criteria decision-making approach to handle the three features for reverse logistics management problems.

3.1 Preliminaries of Fuzzy Sets

This section briefly reviews some basic definitions and properties of fuzzy sets from [18, 21, 22, 24]. These definitions and notations will be used throughout the paper until otherwise stated.

Let $F*(R)$ be the set of all finite fuzzy numbers on R. By the decomposition theorem of fuzzy set, we have

$$\tilde{a} = \bigcup_{\lambda \in (0,1]} \lambda [a_\lambda^L, a_\lambda^R] , \qquad (1)$$

for every $\tilde{a} \in F(R)$.

Definition 1. *If $\tilde{a}$ is a fuzzy number and $a_\lambda^L > 0$ for any $\lambda \in (0, 1]$, then $\tilde{a}$ is called a positive fuzzy number. Let $F_+^*(R)$ be the set of all finite positive fuzzy numbers on R.*

Definition 2. *For any $\tilde{a}$, $\tilde{b} \in F_+^*(R)$ and $0 < \lambda \in R$, the sum, scalar product and product of two fuzzy numbers $\tilde{a} + \tilde{b}, \lambda \tilde{a}$ and $\tilde{a} \times \tilde{b}$ are defined by the membership functions*

$$\mu_{\tilde{a}+\tilde{b}}(t) = \sup_{t=u+v} \min \left\{ \mu_{\tilde{a}}(u), \mu_{\tilde{b}}(v) \right\} , \qquad (2)$$

$$\mu_{\lambda \tilde{a}}(t) = \max\{0, \sup_{t=\lambda u} \mu_{\tilde{a}}(u)\} , \qquad (3)$$

$$\mu_{\tilde{a} \times \tilde{b}}(t) = \sup_{t=u \times v} \min \left\{ \mu_{\tilde{a}}(u), \mu_{\tilde{b}}(v) \right\} . \qquad (4)$$

where we set $\sup\{\phi\} = -\infty$.

Theorem 1. *For any* $\tilde{a}$, $\tilde{b} \in F_+^*(R)$ *and* $0 < \alpha \in R$,

$$\tilde{a} + \tilde{b} = \bigcup_{\lambda \in (0,1]} \lambda \left[a_\lambda^L + b_\lambda^L, \ a_\lambda^R + b_\lambda^R\right],$$

$$\alpha \tilde{a} = \bigcup_{\lambda \in (0,1]} \lambda \left[\alpha a_\lambda^L, \ \alpha a_\lambda^R\right],$$

$$\tilde{a} \times \tilde{b} = \bigcup_{\lambda \in (0,1]} \lambda \left[a_\lambda^L \times b_\lambda^L, \ a_\lambda^R \times b_\lambda^R\right].$$

Definition 3. *For any* $\tilde{a} \in F_+^*(R)$ *and* $0 < \alpha \in Q_+$ *(Q_+ is a set of all positive rational numbers), the positive fuzzy number* $\tilde{a}$ *power of* λ *is defined by the membership function*

$$\mu_{\tilde{a}^\alpha}(t) = \sup_{t = u^\alpha} \min \left\{\mu_{\tilde{a}(u)}\right\} \tag{5}$$

where we set $\sup\{\phi\} = -\infty$.

Theorem 2. *For any* $\tilde{a} \in F_+^*(R)$ **and** $0 < \alpha \in Q_+$,

$$\tilde{a}^\alpha = \bigcup_{\lambda \in (0,1]} \lambda \left[(a_\lambda^L)^\alpha, \ (a_\lambda^R)^\alpha\right].$$

Definition 4. *Let* $\tilde{a}$ *and* $\tilde{b}$ *be two fuzzy numbers. Then* $\tilde{a} = \tilde{b}$ *if* $a_\lambda^L = b_\lambda^L$ *and* $a_\lambda^R = b_\lambda^R$ *for any* $\lambda \in (0, 1]$.

Definition 5. *If* $\tilde{a}$ *is a fuzzy number and* $0 < a_\lambda^L \leq a_\lambda^R \leq 1$, *for any* $\lambda \in (0, 1]$, *then* $\tilde{a}$ *is called a normalized positive fuzzy number.*

Definition 6. *A linguistic variable is a variable whose values are linguistic terms.*

Definition 7. *Let* $\tilde{a}$, $\tilde{b} \in F^*(R)$, *then the quasi-distance function of* $\tilde{a}$ *and* $\tilde{b}$ *is defined as*

$$d(\tilde{a}, \tilde{b}) = \left(\int_0^1 \frac{1}{2} \left[\left(a_\lambda^L - b_\lambda^L\right)^2 + \left(a_\lambda^R - b_\lambda^R\right)^2\right] d\lambda\right)^{\frac{1}{2}} \tag{6}$$

Definition 8. *Let* $\tilde{a}$, $\tilde{b} \in F^*(R)$, *then fuzzy number* $\tilde{a}$ *is closer to fuzzy number* $\tilde{b}$ *as* $d(\tilde{a}, \tilde{b})$ *approaches 0.*

Proposition 1. *If both* $\tilde{a}$ *and* $\tilde{b}$ *are real numbers, then the quasi-distance measurement* $d(\tilde{a}, \tilde{b})$ *is identical to the Euclidean distance.*

Proposition 2. *Let* $\tilde{a}$, $\tilde{b} \in F^*(R)$ *(1). If they are identical, then* $d(\tilde{a}, \tilde{b}) = 0$.
   2) If $\tilde{a}$ *is a real number or* $\tilde{b}$ *is a real number and* $d(\tilde{a}, \tilde{b}) = 0$, *then* $\tilde{a} = \tilde{b}$.

Proposition 3. *Let* $\tilde{a}$, $\tilde{b}$, $\tilde{c} \in F^*(R)$, *then* $\tilde{b}$ *is closer to* $\tilde{a}$ *than* $\tilde{c}$ *if and only if* $d(\tilde{b}, \tilde{a}) < d(\tilde{c}, \tilde{a})$.

Proposition 4. *Let* $\tilde{a}$, $\tilde{b} \in F^*(R)$. *If* $d(\tilde{a}, 0) < d(\tilde{b}, 0)$, *then* $\tilde{a}$ *is closer to 0 than* $\tilde{b}$.

3.2 Process of Personalized Fuzzy Multi-Criteria Decision-Making Approach for Reverse Logistics Management

Based on the model proposed in Sect. 2.3, we integrate the normal multi-criteria decision-making approach [19] and fuzzy number techniques [20] into our proposed personalized multi-stage decision model to accommodate the requirement of goods return decision-making in a reverse logistics chain, called the personalized fuzzy multi-criteria decision-making (PFMCDM) approach.

In this approach, we use any form of fuzzy numbers, called general fuzzy numbers, to handle linguistic terms and other uncertain values. The proposed approach is designed to include nine steps as follows:

Step 1. Setting up weights for business objectives and each objective's related evaluation criteria.

When a returner's type is identified, a set of business objectives $O = \{O_1, O_2, \ldots, O_n\}$are determined based on the information shown in Table 1. Let $WO = \{WO_1, WO_2, \ldots, WO_n\}$ be the weights of these objectives, $OW_i \in \{Absolutely\ not\ important,\ Strongly\ not\ important,\ Weakly\ not\ important,\ Medium\ important,\ Weakly\ more\ important,\ Strongly\ more\ important,\ Absolutely\ more\ important\}$ and are described by general fuzzy numbers $a_1, a_2, \ldots a_n$. For an objective O_i, let $C_i = \{C_{i1}, C_{i2}, \ldots, C_{it_i}\}, i = 1, 2, \ldots, n$, be a set of the selected criteria corresponding to the objective. Let $WC_i = \{WC_{i1}, WC_{i2}, \ldots, WC_{it_i}\}, i = 1, 2, \ldots, n$, be the weights for the set of criteria, where $WC_{ij} \in \{Absolutely\ not\ important,\ Strongly\ not\ important,\ Weakly\ not\ important,\ Medium\ important,\ Weakly\ more\ important,\ Strongly\ more\ important,\ Absolutely\ more\ important\}$ and are described by general fuzzy numbers $c_1, c_2, \ldots c_t$. Both WO and WC_{ij} are given by reverse logistics decision makers.

Step 2. Finalizing the objectives and selection criteria by following rules

The objective (and its selection criteria) can be ignored when

(1) it has a very low weight;
(2) the degree of its weight is much less than others; or
(3) its related criteria is a subset of another selected objective's one.

Step 3. Setting up the relevance degree of each criterion on each alternative

Let $A = \{A_1, A_2, \ldots, A_m\}$ be a set of alternatives for a goods return decision, $AC_i^k = \{AC_{i1}^k, AC_{i2}^k, \ldots, AC_{it_i}^k\}$ be the relevance degree of C_i on alternatives $A_k, i = 1, 2, \ldots, n, k = 1, 2, \ldots, m$, provided by returners, where $AC_{ij}^k \in \{Very\ low,\ Low,\ Medium\ low,\ Medium,\ Medium\ high,\ High,\ Very\ high\}$ and are described by general fuzzy numbers $b_1, b_2, \ldots, b_m$.

Step 4. Weight normalization

The weights for criteria are normalized based on $WC_i = \{WC_{i1}, WC_{i2}, \ldots, WC_{it_i}\}, i = 1, 2, \ldots, n, WC_{ij} \in \{a_j, j = 1, 2, \ldots 7\}$ and denoted as

$$WC_{ij}^* = \frac{WC_{ij}}{\sum_{j=1}^{t_i} WC_{ij0}^R}, \quad \text{for } i = 1, 2, \ldots, n, \ j = 1, 2, \ldots, t_i.$$

Step 5. Relevance degree calculation

To calculate the relevance degree OA_i^k of O_i on the alternatives $A_k, i = 1, 2, \ldots, n, k = 1, 2, \ldots, m$, by using $OA_i^k = WC_i^* \times AC_i^k = \sum_{j=1}^{t_i} WC_{ij}^* \times AC_{ij}^k, i = 1, 2, \ldots, n, k = 1, 2, \ldots, m$.

Step 6. Relevance degree normalization

The relevance degree OA_i^k of O_i on the alternatives $A_k, i = 1, 2, \ldots, n, k = 1, 2, \ldots, m$, are normalized based on $OA^k = \{OA_1^k, OA_2^k, \ldots, OA_n^k\}, k = 1, 2, \ldots, m$,

$$\overline{OA}_i^k = \frac{OA_i^k}{\sum_{i=1}^n OA_{i0}^{kR}}, \quad \text{for } i = 1, 2, \ldots, n, k = 1, 2, \ldots, m .$$

Step 7. Objective relevance degree calculation

Calculating the relevance degree S_k of O on alternatives $A_k, k = 1, 2, \ldots, m$, by using $S_k = \overline{OA}^k \times WO = \sum_{j=1}^{t_i} \overline{OA}_i^k \times WO_i, k = 1, 2, \ldots, m$.

Step 8. The results $S_k, k = 1, 2, \ldots, m$ are normalized to be positive fuzzy numbers, and their ranges belong to the closed interval $[0, 1]$. We define fuzzy positive-ideal alternative (FPIS, S^*) and fuzzy negative-ideal alternative (FNIS, S^-) as:

$$S^* = 1 \quad \text{and} \quad S^- = 0.$$

The distance between each S_k and S^* is called a positive distance, and the distance between S_k and S^- is called a negative distance. The two kinds of distances are calculated respectively by

$$d_k^* = d(S_k, S^*) \quad \text{and} \quad d_k^- = d(S_k, S^-), \quad k = 1, 2, \ldots, m, \text{ where}$$

$$d(\tilde{a}, \tilde{b}) = \left(\int_0^1 \frac{1}{2} \left[\left(a_\lambda^L - b_\lambda^L \right)^2 + \left(a_\lambda^R - b_\lambda^R \right)^2 \right] d\lambda \right)^{\frac{1}{2}}$$

is the distance measure between any two fuzzy numbers $\tilde{a}, \tilde{b}$.

Step 9. A closeness coefficient is defined to determine the ranking order of alternatives once the d_k^* and d_k^- of each alternative A_k ($k = 1, 2, \ldots, m$) are obtained. The closeness coefficient of each alternative is calculated as:

$$D_k = \frac{1}{2}(d_k^* + (1 - d_k^-)), \quad k = 1, 2, \ldots, m .$$

The alternative A_k that corresponds to the largest D_k, is the best suitable alternative for the particular goods return decision problem.

4 A Case-Study Example

This section gives an example to illustrate how to use the proposed approach to support goods return decision-making in reverse logistics management practice.

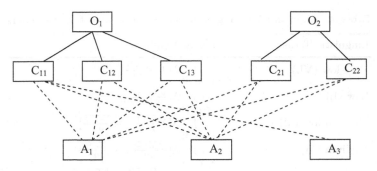

Fig. 3. An example of the interrelation among objectives, criteria and alternatives

A returner at the stage of collection for a reverse logistics chain needs to make a decision for a particular goods return. The returner has currently two objectives $O = \{O_1, O_2\}$ and three alternatives $A = \{A_1, A_2, A_3\}$ for the goods return. The first objective can be evaluated by three criteria (C_{11}, C_{12}, C_{13}), and the second one can be evaluated by two criteria (C_{21}, C_{22}). The relationships among these business objectives, alternatives and evaluation criteria are shown in Fig. 3. By using the proposed approach, a solution from the alternatives which can maximally reach these business objectives will be selected.

As all linguistic terms provided by returners can be described by any kind of fuzzy numbers in the proposed approach, we assume that these linguistic terms are described by fuzzy numbers as shown in Table 3 and Table 4 respectively.

Table 3. An example of linguistic terms and related fuzzy numbers

Linguistic Terms	Fuzzy Numbers
Absolutely not important (ANI)	$\bigcup\limits_{\lambda \in [0,1]} \lambda[0, \frac{\sqrt{1-\lambda}}{10}]$
Strongly not important (SNI)	$\bigcup\limits_{\lambda \in [0,1]} \lambda[\frac{\sqrt{\lambda}}{10}, \frac{\sqrt{9-8\lambda}}{10}]$
Weakly not important (WNI)	$\bigcup\limits_{\lambda \in [0,1]} \lambda[\frac{\sqrt{8\lambda+1}}{10}, \frac{\sqrt{25-16\lambda}}{10}]$
Medium important (MI)	$\bigcup\limits_{\lambda \in [0,1]} \lambda[\frac{\sqrt{16\lambda+9}}{10}, \frac{\sqrt{49-24\lambda}}{10}]$
Weakly more important (WI)	$\bigcup\limits_{\lambda \in [0,1]} \lambda[\frac{\sqrt{24\lambda+25}}{10}, \frac{\sqrt{81-32\lambda}}{10}]$
Strongly more important (SI)	$\bigcup\limits_{\lambda \in [0,1]} \lambda[\frac{\sqrt{32\lambda+49}}{10}, \frac{\sqrt{100-19\lambda}}{10}]$
Absolutely more important (AI)	$\bigcup\limits_{\lambda \in [0,1]} \lambda[\frac{\sqrt{19\lambda+81}}{10}, 1]$

Table 4. An example of linguistic terms and related fuzzy numbers

Linguistic Terms	Fuzzy Numbers
Very low (VL)	$\bigcup\limits_{\lambda \in [0,\,1]} \lambda[0, \frac{\sqrt{1-\lambda}}{10}]$
Low (L)	$\bigcup\limits_{\lambda \in [0,\,1]} \lambda[\frac{\sqrt{\lambda}}{10}, \frac{\sqrt{9-8\lambda}}{10}]$
Medium low (ML)	$\bigcup\limits_{\lambda \in [0,\,1]} \lambda[\frac{\sqrt{8\lambda+1}}{10}, \frac{\sqrt{25-16\lambda}}{10}]$
Medium (M)	$\bigcup\limits_{\lambda \in [0,\,1]} \lambda[\frac{\sqrt{16\lambda+9}}{10}, \frac{\sqrt{49-24\lambda}}{10}]$
Medium high (MH)	$\bigcup\limits_{\lambda \in [0,\,1]} \lambda[\frac{\sqrt{24\lambda+25}}{10}, \frac{\sqrt{81-32\lambda}}{10}]$
High (H)	$\bigcup\limits_{\lambda \in [0,\,1]} \lambda[\frac{\sqrt{32\lambda+49}}{10}, \frac{\sqrt{100-19\lambda}}{10}]$
Very high (VH)	$\bigcup\limits_{\lambda \in [0,\,1]} \lambda[\frac{\sqrt{19\lambda+81}}{10}, 1]$

The power and details of the proposed approach for the goods return case study example are described as follows.

Step 1. A returner gives weights to O_1 and O_2, weights of C_{11}, C_{12} and C_{13} for O_1 and weights of C_{21}, C_{22} for O_2 respectively:

WO = {*Strongly not important, Strongly more important*}
WC$_1$ = {*Strongly not important, Strongly not important, Strongly more important*}
WC$_2$ = {*Strongly not important, Strongly more important*}

Step 2. The two objectives and their criteria are finalized:

WO = {*Strongly not important, Strongly more important*}
WC$_1$ = {*Strongly not important, Strongly not important, Strongly more important*}
WC$_2$ = {*Strongly more important, Strongly not important*}

Step 3. The returner provides relevant degrees of C_{ij} on A_k. ($k = 1, 2, 3$):

$AC_1^1 = \{AC_{11}^1, AC_{12}^1, AC_{13}^1\} = \{Medium\ high, Low, High\}$
$AC_2^1 = \{AC_{21}^1, AC_{22}^1\} = \{Low, High\}$
$AC_1^2 = \{AC_{11}^2, AC_{12}^2, AC_{13}^2\} = \{High, Low, Medium\ high\}$
$AC_2^2 = \{AC_{21}^2, AC_{22}^2\} = \{Low, High\}$
$AC_1^3 = \{AC_{11}^3, AC_{12}^3, AC_{13}^3\} = \{High, High, Medium\ high\}$
$AC_2^3 = \{AC_{21}^3, AC_{22}^3\} = \{Low, High\}$

Step 4. The weights proposed in Step 1 are normalized.

Because $\sum_{j=1}^{3} WC_{1j0}^R = 1.6$, $\sum_{j=1}^{2} WC_{2j0}^R = 1.3$, we got

$$WC_{11}^* = WC_{12}^* = \bigcup_{\lambda \in [0,\ 1]} \lambda \left[\frac{\sqrt{\lambda}}{16}, \frac{\sqrt{9-8\lambda}}{16} \right],$$

$$WC_{13}^* = \bigcup_{\lambda \in [0,\ 1]} \lambda \left[\frac{\sqrt{32\lambda + 49}}{16}, \frac{\sqrt{100 - 19\lambda}}{16} \right],$$

$$WC_{21}^* = \bigcup_{\lambda \in [0,\ 1]} \lambda \left[\frac{\sqrt{\lambda}}{13}, \frac{\sqrt{9-8\lambda}}{13} \right],$$

$$WC_{22}^* = \bigcup_{\lambda \in [0,\ 1]} \lambda \left[\frac{\sqrt{32\lambda + 4913}}{}, \frac{\sqrt{100 - 19\lambda}}{13} \right].$$

Step 5. Calculating the relevance degree OA_i^k of O_i on alternatives $\boldsymbol{A}_k, i = 1, 2$ and $k = 1, 2, 3$, we have

$$OA_1^1 = WC_1^* \times AC_1^1 = \sum_{j=1}^{3} WC_{1j}^* \times AC_{1j}^1$$

$$= \bigcup_{\lambda \in [0,\ 1]} \lambda \left[\frac{\sqrt{\lambda(24\lambda + 25)}}{160} + \frac{33\lambda + 49}{160}, \right.$$

$$\left. \frac{\sqrt{(9 - 8\lambda)(81 - 32\lambda)}}{160} + \frac{109 - 27\lambda}{160} \right]$$

$$OA_2^1 = WC_2^* \times AC_2^1 = \sum_{j=1}^{2} WC_{2j}^* \times AC_{2j}^1$$

$$= \bigcup_{\lambda \in [0,\ 1]} \lambda \left[\frac{2\sqrt{\lambda(32\lambda + 49)}}{130}, \frac{2\sqrt{(9 - 8\lambda)(100 - 19\lambda)}}{130} \right]$$

$$OA_1^2 = WC_1^* \times AC_1^2 = \sum_{j=1}^{3} WC_{1j}^* \times AC_{1j}^2$$

$$= \bigcup_{\lambda \in [0,\ 1]} \lambda \left[\frac{\sqrt{\lambda(32\lambda + 49)}}{160} + \frac{32\lambda + 49}{160} + \frac{\sqrt{(32\lambda + 49)(24\lambda + 25)}}{160}, \right.$$

$$\left. \frac{\sqrt{(9 - 8\lambda)(100 - 19\lambda)}}{160} + \frac{9 - 8\lambda}{160} + \frac{\sqrt{(100 - 19\lambda)(81 - 32\lambda)}}{160} \right]$$

$$OA_2^2 = WC_2^* \times AC_2^2 = \sum_{j=1}^{2} WC_{2j}^* \times AC_{2j}^1$$

$$= \bigcup_{\lambda \in [0,\ 1]} \lambda \left[\frac{2\sqrt{\lambda(32\lambda + 49)}}{130}, \frac{2\sqrt{(9 - 8\lambda)(100 - 19\lambda)}}{130} \right]$$

$$OA_1^3 = WC_1^* \times AC_1^3 = \sum_{j=1}^{3} WC_{1j}^* \times AC_{1j}^3$$

$$= \bigcup_{\lambda \in [0,\, 1]} \lambda \left[\frac{2\sqrt{\lambda\,(32\lambda + 49)}}{160} + \frac{\sqrt{(32\lambda + 49)\,(24\lambda + 25)}}{160}, \right.$$

$$\left. \frac{2\sqrt{(9 - 8\lambda)\,(100 - 19\lambda)}}{160} + \frac{\sqrt{(100 - 19\lambda)\,(81 - 32\lambda)}}{160} \right]$$

$$OA_2^3 = WC_2^* \times AC_2^3 = \sum_{j=1}^{2} WC_{2j}^* \times AC_{2j}^1$$

$$= \bigcup_{\lambda \in [0,\, 1]} \lambda \left[\frac{2\sqrt{\lambda\,(32\lambda + 49)}}{130}, \; \frac{2\sqrt{(9 - 8\lambda)\,(100 - 19\lambda)}}{130} \right]$$

Step 6. Normalizing the relevance degree OA_i^k of O_i on the alternatives A_k based on $OA^k = \{OA_1^k,\, OA_2^k, \ldots,\, OA_n^k\}, i = 1, 2$ and $k = 1, 2, 3$.

$$\overline{OA}_1^1 = \bigcup_{\lambda \in [0,\, 1]} \lambda \left[\frac{\sqrt{\lambda\,(24\lambda + 25)}}{160 \times 1.3115} + \frac{33\lambda + 49}{160 \times 1.3115}, \right.$$

$$\left. \frac{\sqrt{(9 - 8\lambda)\,(81 - 32\lambda)}}{160 \times 1.3115} + \frac{109 - 27\lambda}{160 \times 1.3115} \right]$$

$$\overline{OA}_2^1 = \bigcup_{\lambda \in [0,\, 1]} \lambda \left[\frac{2\sqrt{\lambda\,(32\lambda + 49)}}{130 \times 1.3115}, \; \frac{2\sqrt{(9 - 8\lambda)\,(100 - 19\lambda)}}{130 \times 1.3115} \right]$$

$$\overline{OA}_1^2 = \bigcup_{\lambda \in [0,\, 1]} \lambda \left[\frac{\sqrt{\lambda\,(32\lambda + 49)}}{160 \times 1.2678} + \frac{32\lambda + 49}{160 \times 1.2678} + \frac{\sqrt{(32\lambda + 49)\,(24\lambda + 25)}}{160 \times 1.2678}, \right.$$

$$\left. \frac{\sqrt{(9 - 8\lambda)\,(100 - 19\lambda)}}{160 \times 1.2678} + \frac{9 - 8\lambda}{160 \times 1.2678} + \frac{\sqrt{(100 - 19\lambda)\,(81 - 32\lambda)}}{160 \times 1.2678} \right]$$

$$\overline{OA}_2^2 = \bigcup_{\lambda \in [0,\, 1]} \lambda \left[\frac{2\sqrt{\lambda\,(32\lambda + 49)}}{130 \times 1.2678}, \; \frac{2\sqrt{(9 - 8\lambda)\,(100 - 19\lambda)}}{130 \times 1.2678} \right]$$

$$\overline{OA}_1^3 = \bigcup_{\lambda \in [0,\, 1]} \lambda \left[\frac{2\sqrt{\lambda\,(32\lambda + 49)}}{160 \times 1.3990} + \frac{\sqrt{(32\lambda + 49)\,(24\lambda + 25)}}{160 \times 1.3990}, \right.$$

$$\left. \frac{2\sqrt{(9 - 8\lambda)\,(100 - 19\lambda)}}{160 \times 1.3990} + \frac{\sqrt{(100 - 19\lambda)\,(81 - 32\lambda)}}{160 \times 1.3990} \right]$$

$$\overline{OA}_2^3 = \bigcup_{\lambda \in [0,\, 1]} \lambda \left[\frac{2\sqrt{\lambda\,(32\lambda + 49)}}{130 \times 1.3990}, \; \frac{2\sqrt{(9 - 8\lambda)\,(100 - 19\lambda)}}{130 \times 1.3990} \right].$$

Step 7. Calculating the relevance degree S_k of O on the alternatives A_k by using $S_k = \overline{OA}^k \times WO = \sum_{j=1}^{t_i} \overline{OA}_i^k \times WO_i, k = 1, 2, 3$.

$$S_1 = \overline{OA}^1 \times WO = \bigcup_{\lambda \in [0,\, 1]} \lambda \left[\frac{\sqrt{\lambda}}{10} \left(\frac{\sqrt{\lambda(24\lambda + 25)}}{160 \times 1.3115} + \frac{33\lambda + 49}{160 \times 1.3115} \right) \right.$$

$$+ \frac{\sqrt{32\lambda + 49}}{10} \times \frac{2\sqrt{\lambda(32\lambda + 49)}}{130 \times 1.3115}, \; \frac{\sqrt{9 - 8\lambda}}{10}$$

$$\times \left(\frac{\sqrt{(9 - 8\lambda)(81 - 32\lambda)}}{160 \times 1.3115} + \frac{109 - 27\lambda}{160 \times 1.3115} \right) + \frac{\sqrt{100 - 19\lambda}}{10}$$

$$\left. \times \frac{2\sqrt{(9 - 8\lambda)(100 - 19\lambda)}}{130 \times 1.3115} \right]$$

$$S_2 = \overline{OA}^2 \times WO = \bigcup_{\lambda \in [0,\, 1]} \lambda \left[\frac{\sqrt{\lambda}}{10} \left(\frac{\sqrt{\lambda(32\lambda + 49)}}{160 \times 1.2678} + \frac{32\lambda + 49}{160 \times 1.2678} \right. \right.$$

$$\left. + \frac{\sqrt{(32\lambda + 49)(24\lambda + 25)}}{160 \times 1.2678} \right) + \frac{\sqrt{32\lambda + 49}}{10} \times \frac{2\sqrt{\lambda(32\lambda + 49)}}{130 \times 1.2678},$$

$$\frac{\sqrt{100 - 19\lambda}}{10} \times \frac{2\sqrt{(9 - 8\lambda)(100 - 19\lambda)}}{130 \times 1.2678} + \frac{\sqrt{9 - 8\lambda}}{10}$$

$$\times \left(\frac{\sqrt{(9 - 8\lambda)(100 - 19\lambda)}}{160 \times 1.2678} + \frac{9 - 8\lambda}{160 \times 1.2678} \right.$$

$$\left. \left. + \frac{\sqrt{(100 - 19\lambda)(81 - 32\lambda)}}{160 \times 1.2678} \right) \right]$$

$$S_3 = \overline{OA}^3 \times WO = \bigcup_{\lambda \in [0,\, 1]} \lambda \left[\frac{\sqrt{\lambda}}{10} \left(\frac{2\sqrt{\lambda(32\lambda + 49)}}{160 \times 1.3990} \right. \right.$$

$$\left. + \frac{\sqrt{(32\lambda + 49)(24\lambda + 25)}}{160 \times 1.3990} \right) + \frac{\sqrt{32\lambda + 49}}{10} \times \frac{2\sqrt{\lambda(32\lambda + 49)}}{130 \times 1.3990},$$

$$\frac{\sqrt{100 - 19\lambda}}{10} \times \frac{2\sqrt{(9 - 8\lambda)(100 - 19\lambda)}}{130 \times 1.3990} + \frac{\sqrt{9 - 8\lambda}}{10}$$

$$\left. \times \left(\frac{2\sqrt{(9 - 8\lambda)(100 - 19\lambda)}}{160 \times 1.3990} + \frac{\sqrt{(100 - 19\lambda)(81 - 32\lambda)}}{160 \times 1.3990} \right) \right]$$

Step 8. The results S_k, $k = 1, 2, 3$ are normalized to be positive fuzzy numbers, and their ranges belong to closed interval $[0, 1]$. Positive distance and negative distance are then calculated respectively by

$$d_1^* = d(S_1,\ S^*) = \left(\int_0^1 \frac{1}{2} \left[\left(\frac{\sqrt{\lambda}}{10} \left(\frac{\sqrt{\lambda(24\lambda + 25)}}{160 \times 1.3115} + \frac{33\lambda + 49}{160 \times 1.3115} \right) \right. \right. \right.$$

$$\left. + \frac{\sqrt{32\lambda + 49}}{10} \times \frac{2\sqrt{\lambda(32\lambda + 49)}}{130 \times 1.3115} - 1 \right)^2$$

$$+ \left(\frac{\sqrt{9 - 8\lambda}}{10} \left(\frac{\sqrt{(9 - 8\lambda)(81 - 32\lambda)}}{160 \times 1.3115} + \frac{109 - 27\lambda}{160 \times 1.3115} \right) \right.$$

$$\left. \left. \left. + \frac{\sqrt{100 - 19\lambda}}{10} \times \frac{2\sqrt{(9 - 8\lambda)(100 - 19\lambda)}}{130 \times 1.3115} - 1 \right)^2 \right] d\lambda \right)^{\frac{1}{2}} = 0.80143$$

$$d_2^* = d(S_2,\ S^*) = 0.78983$$
$$d_3^* = d(S_3,\ S^*) = 0.81200$$

$$d_1^- = d(S_1, S-) = \left(\int_0^1 \frac{1}{2} \left[\left(\frac{\sqrt{\lambda}}{10} \left(\frac{\sqrt{\lambda(24\lambda + 25)}}{160 \times 1.3115} + \frac{33\lambda + 49}{160 \times 1.3115} \right) \right. \right. \right.$$

$$\left. + \frac{\sqrt{32\lambda + 49}}{10} \times \frac{2\sqrt{\lambda(32\lambda + 49)}}{130 \times 1.3115} - 0 \right)^2$$

$$+ \left(\frac{\sqrt{9 - 8\lambda}}{10} \left(\frac{\sqrt{(9 - 8\lambda)(81 - 32\lambda)}}{160 \times 1.3115} + \frac{109 - 27\lambda}{160 \times 1.3115} \right) \right.$$

$$\left. \left. \left. + \frac{\sqrt{100 - 19\lambda}}{10} \times \frac{2\sqrt{(9 - 8\lambda)(100 - 19\lambda)}}{130 \times 1.3115} - 0 \right)^2 \right] d\lambda \right)^{\frac{1}{2}}$$

$$= 0.26982$$
$$d_2^- = d(S_2,\ S-) = 0.27534$$
$$d_3^- = d(S_3,\ S-) = 0.25811$$

Step 9. After d_k^* and d_k^- of each alternative A_k ($k = 1,\ 2,\ 3$) are obtained, the closeness coefficient of each alternative is calculated as:

$$D_1 = \frac{1}{2} \left(d_1^* + (1 - d_1^-) \right) = \frac{1}{2} (0.80143 + (1 - 0.26982)) = 0.76581$$

$$D_2 = \frac{1}{2} \left(d_2^* + (1 - d_2^-) \right) = \frac{1}{2} (0.78983 + (1 - 0.27534)) = 0.75725$$

$$D_3 = \frac{1}{2} \left(d_3^* + (1 - d_3^-) \right) = \frac{1}{2} (0.81200 + (1 - 0.25811)) = 0.77695 .$$

As $D_3 = \max\{D_1, D_2, D_3\}$, the alternative A_3 is the best alternative for the returner, that is, the option maximally satisfies the business objectives for the particular goods return in the particular stage of reverse logistics chain.

5 Conclusions

There is a growing interest in exploiting reverse logistics models and developing decision support systems to enhance reverse logistics management [4]. Moreover, the interrelated relationship and dynamic feature in reverse logistics chain management require the capabilities of personalized multi-stage decision support. Uncertainty is inherent in the environment in which a reverse logistics chain propagates through a series of stages, and makes the chain management and control problems more complex. This study first analyses the characteristics of a reverse logistics chain and builds a set of corresponding relationships among goods returners, business objectives, alternatives and selection criteria. The paper then proposes a personalized multi-stage decision model and two sets of dynamic relationships among above decision compounds. Based on these results, a personalized fuzzy multi-criteria decision-making approach is developed. By using the approach, an alternative solution that meets maximally the business objectives under the preference of the logistics manager is selected to handle a goods return in reverse logistics.

The further study includes the development of a decision support system to implement the proposed approach. It will be expected to be applied in practice to enhance the efficiency and effectiveness of decision work for reverse logistics management problems. In order to validate the approach, a set of laboratory experiments will be further organized and more applications will be carried out. Also, the decision support system will be developed as online software and then embedded into e-logistics systems to support decision makers online choosing a suitable way to handle goods return problems in reverse logistics.

Acknowledgment

This research is partially supported by Australian Research Council (ARC) under discovery grants DP0557154 and DP0559213.

References

1. Ballou, R. (2004), Business Logistics/Supply Chain Management, Fifth edition, Pearson Prentice Hall.
2. de Brito, M.P. and Dekker, R. (2002), Reverse logistics – a framework, Econometric Institute Report 290, Erasmus University Rotterdam, Econometric Institute.
3. Blumberg, D. (1999), Strategic examination of reverse logistics and repair service requirements, needs, market size, and opportunities, Journal of Business Logistics, Vol. 20 No. 2, 141–159.
4. Dowlatshahi, S. (2000), Developing a theory of reverse logistics, Interfaces, Vol. 30, No. 3, 143–155.

5. Giuntini, R. and Andel, T. (1995), Advance with reverse logistics, Transportation and Distribution, Part 1, Vol. 36, No. 2, 93–98.

6. Giuntini, R. and Andel, T. (1995), Reverse logistics role models, Transportation and Distribution, Part 3, Vol. 36, No. 4, 97–98.

7. Gooley, T.B. (1998), Reverse Logistics: Five Steps to Success, *Logistics Management and Distribution Report*, Vol. 37 No. 6, 49–55.

8. Herrera, F. and Herrera-Viedma, E. (2000), Choice functions and mechanisms for linguistic preference relations, European Journal of Operational Research Vol. 120, 144–161.

9. Hutchinson, B. (2000), Return to sender, Logistics, May, 69–73.

10. Hu, T. L, Sheu, J.B., and Huang, K. H. (2002), A Reverse Logistics Cost Minimization Model for the Treatment of Hazardous Wastes, Transportation Research, Part E, 38E No. 6, 457–473.

11. Klausner, M. and Hendrickson, C. (2000), Reverse-logistics strategy for product take-back, Interfaces, Vol. 30, No. 3, 156–165.

12. Lambert, D.M. and Cooper, M.C. (2000), Issues in Supply Chain Management, Industrial marketing Management, Vol. 29, 65–83.

13. Pogorelec, J. (1998), Reverse logistics is doable, important, *Frontline Solutions*, Vol. 1 No. 10, 68–69.

14. Rice, D. (2002), Reverse Logistics Presents Opportunities, Challenges, Services News, Vol. 1, No. 2, 1–13.

15. Richardson, H.L. (2001), Logistics In Reverse, Industry Week 4/16/2001, Vol. 250, Issue 6, 37.

16. Rogers, D.S., Lambert, D.M., Croxton K.L. and Garcia-Dastugue, S. (2002), The Returns Management Process, The International Journal of Logistics Management, Vol. 13, No. 2, 1–18.

17. Rowley, J. (2000), The reverse supply-chain impact of current trends, Logistics and Transport Focus, Vol. 2 No. 6, 27–31.

18. Sakawa, M. (1993), Fuzzy sets and interactive multi-objective optimization, Plenum Press, New York.

19. Tecle, A. and Duckstein, L. (1992), A procedure for selecting MCDM techniques for forest resource management, Springer-Verlag, New York.

20. Zadeh, L.A. (1975), The concept of a linguistic variable and its application to approximate reasoning – part 1. Information Sciences Vol. 8, 199–249.

21. Zhang, G.Q., (1998), Fuzzy number-valued measure theory, Tsinghua University Press, Beijing.

22. Zhang G.Q. and Lu J. (2003), An integrated group decision-making method with fuzzy preference for alternatives and individual judgments for selection criteria, Group Decision and Negotiation, Vol. 12, No. 6, 501–515.

23. Zhang, G.Q. and. Lu, J (2004a), Using general fuzzy number to handle uncertainty and imprecision in group decision-making, in Ruan and Zeng (Ed.), Intelligent Sensory Evaluation: Methodologies and Applications, Springer, Northland.

24. Zhang, G.Q. and. Lu, J (2004b), A linguistic intelligent guide for method selection in multi-objective decision support systems, Accepted by *Information Science*

Industrial Engineering Applications

Fuzzy Process Control
with Intelligent Data Mining

Murat Gülbay and Cengiz Kahraman

Department of Industrial Engineering, Istanbul Technical University, Macka 34367, Istanbul, Turkey
gulbaym@itu.edu.tr, kahramanc@itu.edu.tr

Abstract. The quality-related characteristics cannot sometimes be represented in numerical form, such as characteristics for appearance, softness, color, etc. In this case fuzzy set theory can handle this problem. This chapter develops fuzzy control charts for linguistic data. Later, unnatural pattern analyses are made using the probability of a fuzzy event. Unnaturalness of the linguistic data is searched with an intelligent data mining procedure.

1 Introduction

The boundaries of classical sets are required to be drawn precisely and, therefore, set membership is determined with complete certainty. An individual is either definitely a member of the set or definitely not a member of it. This sharp distinction is also reflected in classical process control charts, where each process is treated as either "in control" or "out of control". However, most sets and propositions are not so neatly characterized. It is not surprising that uncertainty exists in the human world. To survive in our world, we are engaged in making decisions, managing and analyzing information, as well as predicting future events. All of these activities utilize information that is available and help us try to cope with information that is not. Lack of information, of course, produces uncertainty, which is the condition where the possibility of error exists. Research that attempts to model uncertainty into decision analysis is done basically through probability theory and/or fuzzy set theory. The former represents the stochastic nature of decision analysis while the latter captures the subjectivity of human behavior. When the data used to construct process control charts are incomplete, vague, or linguistic, classical process control charts fail to determine the nature of the process. Therefore, a fuzzy approach to process control charts are necessary to adopt.

Fuzzy sets were introduced in 1965 by Lotfi Zadeh with a view to reconcile mathematical modeling and human knowledge in the engineering sciences. Since then, a considerable body of literature has blossomed around the concept of fuzzy sets in an incredible wide range of areas, from mathematics and logics to traditional and advanced engineering methodologies. Applications are found in many contexts, from medicine to finance, from human factors to consumer products, from vehicle control to computational linguistics, and so on ... Fuzzy logic is now currently used in the industrial practice of advanced information technology.

Basically, when a point on the control chart is falling outside of the three-sigma control limits it shows an out of control situation. There are some interesting questions related to the computation of the probability that a chart will be out of control even all points on the chart are within three-sigma limits. Based on the expected percentages in each zone, sensitive run tests can be developed for analyzing the patterns of variation in the various zones on the control chart. The concept of dealing with probability of the data pattern on the control chart is known as "unnatural pattern analysis". Whenever a point is drawn on the control chart, the rules of accepting a pattern as unnatural should be examined. Analysis of unnatural patterns can be discovered through intelligent data mining.

Data mining (DM) is a non-trivial process of identifying valid, novel, potentially useful, and ultimately understandable patterns from data. The main data mining application areas are marketing, banking, retailing and sales, manufacturing and production, brokerage and securities trading, insurance, computer hardware and software, government and defense, airlines, health care, broadcasting, and law enforcement.

Intelligent data mining (IDM) is to use intelligent search to discover information within data warehouses that queries and reports cannot effectively reveal and to find patterns in the data and infer rules from them, and to use patterns and rules to guide decision-making and forecasting. Main tools used in intelligent data mining are case-based reasoning, neural computing, intelligent agents, and other tools (decision trees, rule induction, data visualization).

In recent years, the need to extract knowledge automatically from very large databases has grown. In response, the closely related fields of knowledge discovery in databases (KDD) and data mining have developed processes and algorithms that attempt to intelligently extract interesting and useful information from vast amounts of raw data. The term DM is frequently used to designate the process of extracting useful information from large databases. The term KDD is used to denote the process of extracting useful knowledge from large data sets. DM, by contrast, refers to one particular step in this process. Specifically, the data mining step applies so-called data mining techniques to extract patterns from the data. Additionally, it is preceded and followed by other KDD steps, which ensure that the extracted patterns actually correspond to useful knowledge. Indeed, without these additional KDD steps, there is a high risk of finding meaningless or uninteresting patterns [4, 8]. In other words, the KDD process uses data mining techniques along with any required pre- and post-processing to extract high-level knowledge from low-level data. In practice, the KDD process is interactive and iterative, involving numerous steps with many decisions being made by the user. DM techniques are essentially pattern discovery algorithms. Some techniques such as association rules are unique to data mining, but most are drawn from related fields such as databases, statistics, pattern recognition, machine learning, neurocomputing, and artificial intelligence.

The application of data mining to fuzzy process control has not yet been studied extensively. This chapter is organized as follows. In Sect. 2, the basics of process control charts are summarized. Next, fuzzy process control charts are explained in Sect. 3. Fuzzy Process Control with Intelligent Data Mining is developed in Sect. 4. Finally a numerical application in Sect. 5, and conclusions in Sect. 6 are given.

2 Process Control Charts

Based on the statistical variation of any process – control charts help focus on stability of a process. An essential element of producing a high quality product is insuring that the characteristics of that product remain constant over time. Statistical process control charts are widely used to determine whether a process is stable and to monitor that process over time. When the characteristics of interest can be measured (rather than simply observed), it is common to take periodic samples of measurements and then plot statistic such as the mean and range. It is used to determine how much variability in a process is due to random variation and how much is due to unique events/individual actions so that you know whether or not the process is in statistical control. A typical control chart plots the sample statistics together with upper and lower control limits shifted from center line (CL) as shown in Fig. 1.

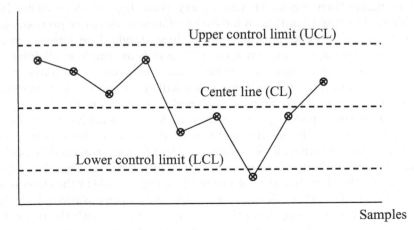

Fig. 1. A typical process control chart

Fluctuation of the points within the limits is due to variation built into the process such as design or preventative maintenance and can only be affected by changing that system. Fluctuation of the points outside of the limits comes from special causes such as people errors, unplanned outages, etc., which are not a part of the normal system or from an unlikely combination of steps. Special causes must be removed from the system to use the SPC effectively. Once this is done then the system can be described as "in control" and measurements can be taken at regular intervals to ensure that the process does not fundamentally change.

Based on the output of the process in consideration, SPC charts can be categorized into two groups. These are:

1. Variables control charts: SPC charts used to control characteristics of a product that can be measured on a continuous scale. An example of a variable would be the length or width of a product or part. Most commonly used variables control charts are X-Bar and R charts
2. Attributes control charts: SPC charts used to control which is an aspect or characteristic of a product that cannot be put on a linear scale. For example, a light bulb will either light or fail to light. "*Good/bad*" is an attribute, as is the number of defects. Examples of attributes control charts are p, np, c, and u charts.

If the process is stable, then the distribution of subgroup averages will be approximately normal. With this in mind, we can also analyze the *patterns* on the control charts to see if they might be attributed to a special cause of variation. To do this, we divide a normal distribution into zones, with each zone one standard deviation wide. Figure 2 shows the approximate percentage we expect to find in each zone from a stable process. Zone C is the area from the mean to the mean plus or minus one sigma, zone B is from plus or minus one to plus or minus two sigma, and zone A is from plus or minus two to plus or minus three sigma. Of course, any point beyond three sigma (i.e., outside of the control limit) is an indication of an out-of-control process. Since the control limits are at plus and minus three standard deviations, finding the one and two sigma lines on a control chart is as simple as dividing the distance between the grand average and either control limit into thirds, which can be done using a ruler. This divides each half of the control chart into three zones. The three zones are labeled A, B, and C as shown in Fig. 3.

Based on the expected percentages in each zone, sensitive run tests can be developed for analyzing the patterns of variation in the various zones. Reference [16] recommends four rules to identify patterns (and implicitly, points) in control charts as out-of-control. The first is the classical three-sigma rule; that is, the chart has at least one point falling outside of the three-sigma control limits. The other rules are: rule 2, two out of three consecutive points more than two sigma away from the centerline, zone A, (with the two points on the same side of the centerline); rule 3, four out of five consecutive points more than one sigma away from the centerline, zone B, (with all four on the

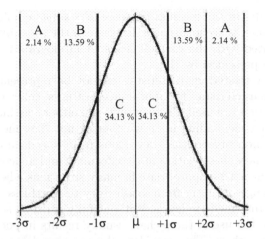

Fig. 2. Zones of normal distribution

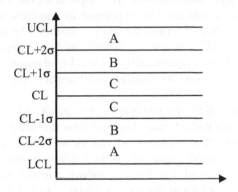

Fig. 3. Zones of a control chart

same side of the centerline); and rule 4, eight consecutive points on the same side of the centerline, zones $A + B + C$. One-sided probabilities of the rules 1, 2, 3, and 4 are calculated as 0.00135, 0.0015, 0.0027, and 0.0039, respectively.

Characteristics of interest for variables control charts can exactly be measured by the instruments and/or devices, but that for attributes cannot be measured directly and they consists of uncertainty. For this reason, fuzzy approaches to attributes control charts should be developed. In the next section, fuzzy process control charts are proposed for attributes control charts.

3 Fuzzy Process Control Charts

Even the first control chart was proposed during the 1920's by Shewhart, today they are still subject to new application areas that deserve further attention. The control charts introduced by [11] were designated to monitor processes for shifts in the mean or variance of a *single* quality characteristic. Further developments are focused on the usage of the probability and fuzzy set theories integrated with the control charts. A bibliography of control charts for attributes is presented by [17].

If the quality-related characteristics cannot be represented in numerical form, such as characteristics for appearance, softness, color, etc., then control charts for attributes are used. Product units are either classified as conforming or nonconforming, depending upon whether or not they meet specifications. The number of nonconformities (deviations from specifications) can also be counted. The binary classification into conforming and nonconforming used in the p-chart might not be appropriate in many situations where product quality does not change abruptly from satisfactory to worthless, and there might be a number of intermediate levels. Without fully utilizing such as intermediate information, the use of the p-chart usually results in poorer performance than that of the x-chart. This is evidenced by weaker detectability of process shifts and other abnormal conditions. To supplement the binary classification, several intermediate levels may be expressed in the form of linguistic terms. For example, the quality of a product can be classified by one of the following terms: *"perfect"*, *"good"*, *"medium"*, *"poor"*, or *"bad"* depending on its deviation from specifications appropriately selected continuous functions can then be used to describe the quality characteristic associated with each linguistic term.

In the literature, different procedures are proposed to monitor multinomial processes when products are classified into mutually exclusive linguistic categories. Reference [2] used fuzzy set theory as a basis for interpreting the representation of a graded degree of product conformance with quality standard. Reference [2] stressed that fuzzy economic control chart limits would be advantageous over traditional acceptance charts in that fuzzy economic control charts provide information on severity as well as the frequency of product nonconformance. References [10, 14] proposed an approach based on fuzzy set theory by assigning fuzzy sets to each linguistic term, and then combining for each sample using rules of fuzzy arithmetic and developed two approaches called *fuzzy probabilistic approach* and *membership approach*.

Apart from fuzzy probabilistic and fuzzy membership approach, [7] introduced modifications to the construction of control charts given by [13, 14]. Their study aimed at directly controlling the underlying probability distributions of the linguistic data, which were not considered by [10]. These procedures are reviewed by [18] and discussed by [9] and [1]. Reference [6] used triangular fuzzy numbers in the tests of control charts for unnatural patterns. Reference [3] proposed a neural fuzzy control chart for identifying process mean shifts. Reference [18] gave a review of statistical and fuzzy control charts based on categorical data. Reference [12] discussed different procedures of constructing control charts for linguistic data, based on fuzzy and probability theory. A comparison between fuzzy and probability approaches, based on the Average Run Length and samples under control, is conducted for real data.

Contrary to the conclusions of [10] the choice of degree of fuzziness affected the sensitivity of control charts.

Current fuzzy control charts are based on the fuzzy transformation from vague data to crisp data, and then, carried out as in the classical control charts. With the integration of the α-cut of fuzzy sets, [5] proposed α-cut fuzzy control charts.

3.1 Fuzzy p Control Charts

In classical p charts, products are distinctly classified as "*conformed*" or "*non-conformed*" when determining fraction rejected. In fuzzy p control charts, when categorizing products, several linguistic terms are used to denote the degree of being nonconformed product such as "*standard*", "*second choice*", "*third choice*", "*chipped*", and so on... A membership degree of being a non-conformed product is assigned to each linguistic term. Sample means for each sample group, M_j, are calculated as:

$$M_j = \frac{\sum_{i=1}^{t} k_{ij} r_i}{m_j} \tag{1}$$

where k_{ij} is the number of products categorized with the linguistic term i in the sample j, r_i is the membership degree of the linguistic term i, and m_j is the number of products in sample j. Center line, CL, is the average of the means of the n sample groups and can be determined by (2)

$$CL = \bar{M}j = \frac{\sum_{j=1}^{n} Mj}{n} \tag{2}$$

where n is the number of sample groups initially available. Since the CL is a fuzzy set, it can be represented by triangular fuzzy numbers (TFNs) whose fuzzy mode is CL, as shown in Fig. 4. Then, for each sample mean, $L_j(\alpha)$ and $R_j(\alpha)$ can be calculated using (3) and (4), respectively.

$$L_j(\alpha) = M_j \alpha \tag{3}$$
$$R_j(\alpha) = 1 - [(1 - M_j)\alpha] \tag{4}$$

Membership function of the $\bar{M}$, or CL, can be written as:

$$\mu_{M_j}(x) = \begin{cases} 0, & \text{if } x \leq 0 \\ \frac{x}{\bar{M}}, & \text{if } 0 \leq x \leq \bar{M} \\ \frac{1-x}{1-\bar{M}}, & \text{if } \bar{M} \leq x \leq 1 \\ 0, & \text{if } x \geq 1 \end{cases} \tag{5}$$

Control limits for α-cut is also a fuzzy set and can be represented by TFNs. Since the membership function of CL is divided into two components, then,

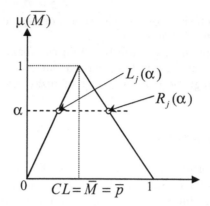

Fig. 4. TFN representation of $\bar{M}$ and M_j of the sample j

each component will have its own CL, LCL, and UCL. The membership function of the control limits depending upon the value of α is given below.

Control Limits(α)

$$
= \left\{ \begin{array}{l} \left\{ \begin{array}{l} CL^L = \bar{M}\alpha \\ LCL^L = \max\left\{ CL^L - 3\sqrt{\dfrac{(CL^L)(1-CL^L)}{\bar{n}}},\ 0 \right\} \\ UCL^L = \min\left\{ CL^L + 3\sqrt{\dfrac{(CL^L)(1-CL^L)}{\bar{n}}},\ 1 \right\} \end{array} \right\},\ \text{if } 0 \le M_j \le \bar{M} \\[3em] \left\{ \begin{array}{l} CL^R = 1 - \left[(1 - \bar{M}\alpha)\,\alpha \right] \\ LCL^R = \max\left\{ CL^R - 3\sqrt{\dfrac{(CL^R)(1-CL^R)}{\bar{n}}},\ 0 \right\} \\ UCL^R = \min\left\{ CL^R + 3\sqrt{\dfrac{(CL^R)(1-CL^R)}{\bar{n}}},\ 1 \right\} \end{array} \right\},\ \text{if } \bar{M} \le M_j \le 1 \end{array} \right.
\tag{6}
$$

where $\bar{n}$ is the average sample size (ASS). When the ASS is used, the control limits do not change with the sample size. Hence, the control limits for all samples are the same. A general illustration of these control limits is shown in Fig. 5.

For the variable sample size (VSS), $\bar{n}$ should be replaced by the size of the jth sample n_j. Hence, control limits change for each sample depending upon the size of the sample. Therefore, each sample has its own control limits. The decision that whether process is *in control* (1) or *out of control* (0) for both ASS and VSS is as follows:

α-level

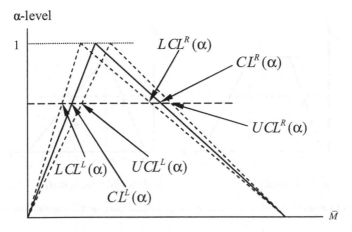

Fig. 5. Illustration of the α-cut control limits (ASS)

$$\text{Process Control} = \begin{cases} 1, & \text{if } LCL^L(\alpha) \le L_j(\alpha) \le UCL^L(\alpha) \wedge LCL^R(\alpha) \\ & \le R_j(\alpha) \le UCL^R(\alpha) \\ 0, & \text{otherwise} . \end{cases} \tag{7}$$

The value of α-cut is decided with respect to the tightness of inspection such that for a tight inspection, α values close to 1 may be used. As can be seen from Fig. 5, while α reduces to 0 (decreasing the tightness of inspection), the range where the process is *in control* (difference between UCL and LCL) increases.

3.2 Fuzzy c Control Charts

In the crisp case, control limits for number of nonconformities are calculated by the (8–10).

$$CL = \bar{c} \tag{8}$$

$$LCL = \bar{c} - 3\sqrt{\bar{c}} \tag{9}$$

$$UCL = \bar{c} + 3\sqrt{\bar{c}} \tag{10}$$

where $\bar{c}$ is the mean of the nonconformities. In the fuzzy case, each sample, or subgroup, is represented by a trapezoidal fuzzy number (a, b, c, d) or a triangular fuzzy number (a, b, d) as shown in Fig. 6. Note that a trapezoidal fuzzy number becomes triangular when $b = c$. For the ease of representation and calculation, a triangular fuzzy number is also represented as trapezoidal by (a, b, b, d) or (a, c, c, d). Center line, $\widetilde{CL}$, given in (8), is the mean of fuzzy samples, and it is shown as $(\bar{a}, \bar{b}, \bar{c}, \bar{d})$ where $\bar{a}, \bar{b}, \bar{c},$ and $\bar{d}$ are the arithmetic means of the $a, b, c,$ and d, respectively:

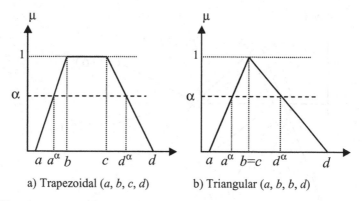

a) Trapezoidal (a, b, c, d) b) Triangular (a, b, b, d)

Fig. 6. Representation of a sample by trapezoidal and/or triangular fuzzy numbers

$$\widetilde{CL} = \left(\frac{\sum_{j=1}^{n} a_j}{n}, \frac{\sum_{j=1}^{n} b_j}{n}, \frac{\sum_{j=1}^{n} c_j}{n}, \frac{\sum_{j=1}^{n} d_j}{n} \right) = (\bar{a}, \bar{b}, \bar{c}, \bar{d}) \qquad (11)$$

where n is the number of fuzzy samples.

Since the $\widetilde{CL}$ is a fuzzy set, it can be represented by a fuzzy number whose fuzzy mode (multimodal) is the closed interval of $[\bar{b}, \bar{c}]$. $\widetilde{CL}$, $\widetilde{LCL}$, and $\widetilde{UCL}$ are calculated using (12–14).

$$\widetilde{CL} = (\bar{a}, \bar{b}, \bar{c}, \bar{d}) = (CL_1, CL_2, CL_3, CL_4) \qquad (12)$$

$$\widetilde{LCL}^\alpha = \widetilde{CL}^\alpha - 3\sqrt{\widetilde{CL}^\alpha} = (\bar{a}, \bar{b}, \bar{c}, \bar{d}) - 3\sqrt{(\bar{a}, \bar{b}, \bar{c}, \bar{d})}$$

$$= \left(\bar{a} - 3\sqrt{\bar{d}}, \bar{b} - 3\sqrt{\bar{c}}, \bar{c} - 3\sqrt{\bar{b}}, \bar{d} - 3\sqrt{\bar{a}} \right)$$

$$= (LCL_1, LCL_2, LCL_3, LCL_4) \qquad (13)$$

$$\widetilde{UCL} = \widetilde{CL} + 3\sqrt{\widetilde{CL}} = (\bar{a}, \bar{b}, \bar{c}, \bar{d}) + 3\sqrt{(\bar{a}, \bar{b}, \bar{c}, \bar{d})}$$

$$= \left(\bar{a} + 3\sqrt{\bar{a}}, \bar{b} + 3\sqrt{\bar{b}}, \bar{c} + 3\sqrt{\bar{c}}, \bar{d} + 3\sqrt{\bar{d}} \right)$$

$$= (UCL_1, UCL_2, UCL_3, UCL_4) \qquad (14)$$

An α-cut is a nonfuzzy set which comprises all elements whose membership is greater than or equal to α. Applying α-cuts of fuzzy sets (Fig. 4) values of a^α and d^α are determined by (15) and (16), respectively.

$$a^\alpha = a + \alpha(b - a) \qquad (15)$$

$$d^\alpha = d - \alpha(d - c) \qquad (16)$$

Using α-cut representations, fuzzy control limits can be rewritten as given in (17–19).

$$\widetilde{CL}^{\alpha} = (\bar{a}^{\alpha}, \bar{b}, \bar{c}, \bar{d}^{\alpha}) = (CL_1^{\alpha}, CL_2, CL_3, CL_4^{\alpha}) \tag{17}$$

$$\widetilde{LCL}^{\alpha} = \widetilde{CL}^{\alpha} - 3\sqrt{\widetilde{CL}^{\alpha}} = (\bar{a}^{\alpha}, \bar{b}, \bar{c}, \bar{d}^{\alpha}) - 3\sqrt{(\bar{a}^{\alpha}, \bar{b}, \bar{c}, \bar{d}^{\alpha})}$$

$$= (\bar{a}^{\alpha} - 3\sqrt{\bar{d}^{\alpha}}, \bar{b} - 3\sqrt{\bar{c}}, \bar{c} - 3\sqrt{\bar{b}}, \bar{d}^{\alpha} - 3\sqrt{\bar{a}^{\alpha}})$$

$$= (LCL_1^{\alpha}, LCL_2, LCL_3, LCL_4^{\alpha}) \tag{18}$$

$$\widetilde{UCL}^{\alpha} = \widetilde{CL}^{\alpha} + 3\sqrt{\widetilde{CL}^{\alpha}} = (\bar{a}^{\alpha}, \bar{b}, \bar{c}, \bar{d}^{\alpha}) + 3\sqrt{(\bar{a}^{\alpha}, \bar{b}, \bar{c}, \bar{d}^{\alpha})}$$

$$= \left(\bar{a}^{\alpha} + 3\sqrt{\bar{a}^{\alpha}}, \bar{b} + 3\sqrt{\bar{b}}, \bar{c} + 3\sqrt{\bar{c}}, \bar{d}^{\alpha} + 3\sqrt{\bar{d}^{\alpha}}\right)$$

$$= (UCL_1^{\alpha}, UCL_2, UCL_3, UCL_4^{\alpha}) \tag{19}$$

Results of these equations can be illustrated as in Fig. 7. To retain the standard format of control charts and to facilitate the plotting of observations on the chart, it is necessary to convert the fuzzy sets associated with linguistic

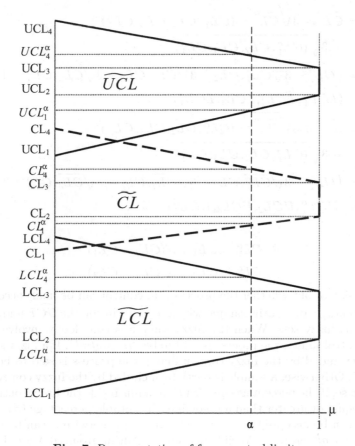

Fig. 7. Representation of fuzzy control limits

values into scalars referred to as *representative values*. This conversion may be performed in a number of ways as long as the result is intuitively representative of the range of the base variable included in the fuzzy set. Four ways, which are similar in principle to the measures of central tendency used in descriptive statistics, are *fuzzy mode*, *α-level fuzzy midrange*, *fuzzy median*, and *fuzzy average*. It should be pointed out that there is no theoretical basis supporting any one specifically and the selection between them should be mainly based on the ease of computation or preference of the user [14]. Conversion of fuzzy sets into crisp values results in loss of information in linguistic data. To retain information of linguistic data we prefer to keep fuzzy sets as themselves and to compare fuzzy samples with the fuzzy control limits. For this reason, a direct fuzzy approach (DFA) based on the area measurement is proposed for the fuzzy control charts. α-level fuzzy control limits, $\widetilde{UCL}^{\alpha}$, $\widetilde{CL}^{\alpha}$, and, $\widetilde{LCL}^{\alpha}$, can be determined by fuzzy arithmetic as shown in (20–22).

$$\widetilde{CL}^{\alpha} = (CL_1^{\alpha}, CL_2, CL_3, CL_4^{\alpha}) \tag{20}$$

$$\begin{aligned}
\widetilde{LCL}^{\alpha} &= \widetilde{CL}^{\alpha} - 3\sqrt{\widetilde{CL}^{\alpha}} = (CL_1^{\alpha}, CL_2, CL_3, CL_4^{\alpha}) \\
&\quad - 3\sqrt{(CL_1^{\alpha}, CL_2, CL_3, CL_4^{\alpha})} \\
&= \left(CL_1^{\alpha} - 3\sqrt{CL_4^{\alpha}}, CL_2 - 3\sqrt{CL_3}, CL_3 - 3\sqrt{CL_2}, CL_4^{\alpha} - 3\sqrt{CL_1^{\alpha}}\right) \\
&= (LCL_1^{\alpha}, LCL_2, LCL_3, LCL_4^{\alpha}) \tag{21}
\end{aligned}$$

$$\begin{aligned}
\widetilde{UCL}^{\alpha} &= \widetilde{CL}^{\alpha} + 3\sqrt{\widetilde{CL}^{\alpha}} = (CL_1^{\alpha}, CL_2, CL_3, CL_4^{\alpha}) \\
&\quad + 3\sqrt{(CL_1^{\alpha}, CL_2, CL_3, CL_4^{\alpha})} \\
&= \left(CL_1^{\alpha} + 3\sqrt{CL_1^{\alpha}}, CL_2 + 3\sqrt{CL_2}, CL_3 + 3\sqrt{CL_3}, CL_4^{\alpha} + 3\sqrt{CL_4^{\alpha}}\right) \\
&= (UCL_1^{\alpha}, UCL_2, UCL_3, UCL_4^{\alpha}) \tag{22}
\end{aligned}$$

where,

$$CL_1^{\alpha} = CL_1 + \alpha (CL_2 - CL_1) \tag{23}$$

$$CL_4^{\alpha} = CL_4 - \alpha (CL_4 - CL_3) \tag{24}$$

Decision about whether the process is in control can be made according to the percentage area of the sample which remains inside the $\widehat{UCL}$ and/or $\widehat{LCL}$ defined as fuzzy sets. When the fuzzy sample is completely involved by the fuzzy control limits, the process is said to be *"in-control"*. If a fuzzy sample is totally excluded by the fuzzy control limits, the process is said to be *"out of control"*. Otherwise, a sample is partially included by the fuzzy control limits. In this case, if the percentage area which remains inside the fuzzy control limits (β_j) is equal or greater than a predefined acceptable percentage (β), then the process can be accepted as *"rather in-control"*; otherwise it can be stated as *"rather out of control"*. Possible decisions resulting from DFA are illustrated in Fig. 8. Parameters for determination of the sample area outside the control

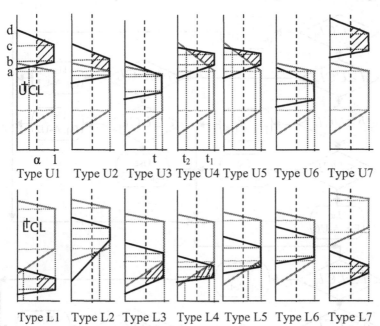

Fig. 8. Illustration of all possible sample areas outside the fuzzy control limits at α-level cut

limits for α-level fuzzy cut are LCL_1, $LCL_2, UCL_3, UCL_4, a, b, c, d$, and α. The shape of the control limits and fuzzy sample are formed by the lines of $\overline{LCL_1LCL_2}$, and $\overline{UCL_1UCL_2}, \overline{ab}, \overline{cd}$. A flowchart to calculate area of the fuzzy sample outside the control limits is given in Fig. 9. Sample area above the upper control limits, A_{out}^U, and sample area falling below the lower control limits, A_{out}^L, are calculated. Equations to compute A_{out}^U and A_{out}^L are given in Appendix A. Then, total sample area outside the fuzzy control limits, A_{out}, is the sum of the areas below fuzzy lower control limit and above fuzzy upper control limit. Percentage sample area within the control limits is calculated as given in (25).

$$\beta_j^\alpha = \frac{S_j^\alpha - A_{\text{out},j}^\alpha}{S_j^\alpha} \tag{25}$$

where S_j^α is the sample area at α-level cut.

DFA provides the possibility of obtaining linguistic decisions like "rather in control" or "rather out of control". Further intermediate levels of process control decisions are also possible by defining in stages. For instance, it may be defined as given below which is more distinguished.

$$\text{Process Control} = \begin{cases} \text{in control,} & 0.85 \le \beta_j \le 1 \\ \text{rather in control,} & 0.60 \le \beta_j < 0.85 \\ \text{rather out of control,} & 0.10 \le \beta_j < 0.60 \\ \text{out of control,} & 0 \le \beta_j < 0.10 \end{cases} \tag{26}$$

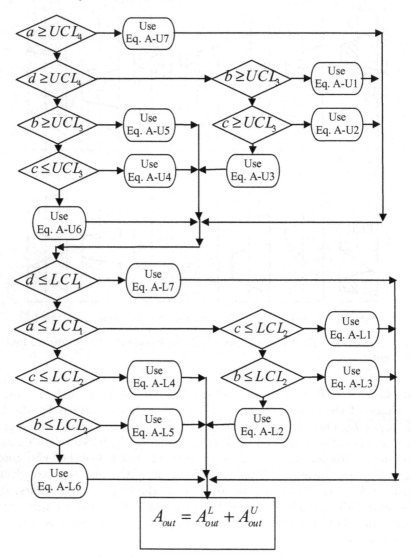

Fig. 9. Flowchart to cumpute the area outside the fuzzy control limits

4 Fuzzy Process Control with Intelligent Data Mining

Analysis of fuzzy unnatural patterns for fuzzy control charts is necessary to develop. Run rules are based on the premise that a specific run of data has a low probability of occurrence in a completely random stream of data. If a run occurs, then this must mean that something has changed in the process to produce a nonrandom or unnatural pattern. Based on the expected percentages in each zone, sensitive run tests can be developed for analyzing the patterns of variation in the various zones. The formula for calculating the probability of a fuzzy event A is a generalization of the probability theory:

$$P(A) = \begin{cases} \int \mu_A(x)P_x(x)dx, & \text{if } X \text{ is continuous} \\ \sum_i \int \mu_A(x_i)P_x(x_i), & \text{if } X \text{ is discrete} \end{cases} \tag{27}$$

where P_X denotes the probability distribution function of X. The membership degree of a sample to belong to a region is directly related to its percentage area falling in that region, and therefore, it is continuous. For example, a fuzzy sample may be in zone B with a membership degree of 0.4 and in zone C with a membership degree of 0.6. While counting points in zone B, that point is counted as 0.4.

Based on the Western Electric rules, the following fuzzy unnatural pattern rules can be defined.

Rule 1: Any fuzzy data falling outside the three-sigma control limits with a ratio (25) of more than predefined percentage (β) of sample area at desired α-level:

$$\mu_1 = \begin{cases} 0, & 0.85 \leq x \leq 1 \\ (x - 0.60)/0.25, & 0.60 \leq x \leq 0.85 \\ (x - 0.10)/0.50, & 0.10 \leq x \leq 0.60 \\ 1, & 0 \leq x \leq 0.10 \end{cases} \tag{28}$$

where x is the ratio of fuzzy data falling outside the three-sigma control limits.

Rule 2: A total membership degree *around* 2 from 3 consecutive points in zone A or beyond. Probability of a sample being in zone A (0.0214) or beyond (0.00135) is 0.02275. Let membership function for this rule be defined as follows:

$$\mu_2 = \begin{cases} 0, & 0 \leq x \leq 0.59 \\ (x - 0.59)/1.41, & 0.59 \leq x \leq 2 \\ 1, & 2 \leq x \leq 3 \end{cases} \tag{29}$$

Using the membership function above, fuzzy probability given in (27) can be rewritten as follows:

$$\int_0^3 \mu_2(x)P_2(x) = \int_0^{x_1} \mu_2(x)P_2(x) + \int_{x_1}^{x_2} \mu_2(x)P_2(x) + \int_{x_2}^3 \mu_2(x)P_2(x)$$

$$= \int_{x_1}^{x_2} \mu_2(x)P_2(x) + \int_{x_2}^3 \mu_2(x)P_2(x) \tag{30}$$

where,

$$P_x(x) = P_x\left(z \geq \frac{x - np}{\sqrt{npq}}\right) \tag{31}$$

To integrate the (30), membership function is divided into sections, each with a 0.05 width, and $\mu_2(x)P_x(x)$ values for each section are summed. For $x_1 = 0.59$ and $x_2 = 2$, the probability of the fuzzy event, rule 2, is determined as 0.0015, which corresponds to the crisp case of this rule. In the following rules, the membership functions are set in the same way.

Rule 3: A total membership degree *around* 4 from 5 consecutive points in zone C or beyond with the membership function (degree of unnaturalness) given below:

$$\mu_3 = \begin{cases} 0, & 0 \le x \le 2.42 \\ (x - 2.42)/1.58, & 2.42 \le x \le 4 \\ 1, & 4 \le x \le 5 \end{cases} \tag{32}$$

The fuzzy probability for this rule is calculated as 0.0027.

Rule 4: A total membership degree *around* 8 from 8 consecutive points on the same side of the centerline:

$$\mu_4 = \begin{cases} 0, & 0 \le x \le 2.54 \\ (x - 2.54)/5.46, & 2.54 \le x \le 8 \end{cases} \tag{33}$$

The fuzzy probability for the rule above is then determined as 0.0039. Probability of each fuzzy rule (event) above depends on the definition of the membership function which is subjectively defined with respect to the classical probabilities for unnatural patterns.

5 A Numerical Example for Fuzzy c Control Charts

Samples of 200 units are taken every 4 hours to control number of nonconformities. Data collected from 30 subgroups shown in Table 2 are linguistic such as *"approximately 30"* or *"between 25 and 30"*.

The linguistic expressions in Table 1 are represented by fuzzy numbers as shown in Table 3. These numbers are subjectively identified by the quality control expert who also sets $\alpha = 0.60$ and minimum acceptable ratio as $\beta = 0.70$. Quality control expert also set the acceptable membership degree of unnaturalness as 0.95, that is, when a sample refers to an unnatural sample with respect to any rule, it should refer a membership degree of unnaturalness more than 0.95 with respect to the membership function defined for that rule.

Using (5–7), $\widetilde{CL}$, $\widetilde{LCL}$, and $\widetilde{UCL}$ are determined as follows:

$$\widetilde{CL} = (18.13, 22.67, 26.93, 32.07)$$
$$\widetilde{LCL} = (1.15, 7.10, 12.65, 19.29)$$
$$\widetilde{UCL} = (30.91, 36.95, 42.50, 49.05)$$

Table 1. Number of nonconformities for 30 subgroups

No	Approximately	Between	No	Approximately	Between
1	30		16	40	
2		20–30	17		32–50
3		5–12	18	39	
4	6		19		15–21
5	38		20	28	
6		20–24	21		32–35
7		4–8	22		10–25
8		36–44	23	30	
9		11–15	24	25	
10		10–13	25		31–41
11	6		26		10–25
12	32		27		5–14
13	13		28		28–35
14		50–52	29		20–25
15		38–41	30	8	

Table 2. Fuzzy number (a, b, c, d) representation of 30 subgroups

No	a	b	c	d	No	a	b	c	d
1	25	30	30	35	16	33	40	40	44
2	15	20	30	35	17	28	32	50	60
3	4	5	12	15	18	33	39	39	43
4	3	6	6	8	19	12	15	21	38
5	32	38	38	45	20	23	28	28	36
6	16	20	24	28	21	28	32	35	42
7	3	4	8	12	22	14	18	28	33
8	27	36	44	50	23	24	30	30	34
9	9	11	15	20	24	20	25	25	31
10	7	10	13	15	25	25	31	41	46
11	3	6	6	10	26	7	10	25	28
12	27	32	32	37	27	3	5	14	20
13	11	13	13	15	28	23	28	35	38
14	39	50	52	55	29	17	20	25	29
15	28	38	41	45	30	5	8	8	15
	Average					18.13	22.67	26.93	32.07

Table 3. Fuzzy zones calculated for the example problem

Zone	a	b	c	d
UCL^{α}	34.53	36.95	42.50	45.12
$+2\,\sigma$	29.97	32.19	37.31	39.74
$+1\,\sigma$	25.41	27.43	32.12	34.37
CL^{α}	20.85	22.67	26.93	28.99
$-1\,\sigma$	15.47	17.48	22.17	24.43
$-2\,\sigma$	10.10	12.29	17.41	19.87
LCL^{α}	4.72	7.10	12.65	15.31

Applying α-cut of 0.60, values of $\widetilde{CL}^{\alpha=0.60}$, $\widetilde{LCL}^{\alpha=0.60}$, and $\widetilde{UCL}^{\alpha=0.60}$ are calculated. (10–12)

$$\widetilde{CL}^{\alpha=0.60} = (20.85, 22.67, 26.93, 28.99)$$

$$\widetilde{LCL}^{\alpha=0.60} = (4.72, 7.10, 12.65, 15.31)$$

$$\widetilde{UCL}^{\alpha=0.60} = (34.53, 36.95, 42.50, 45.12)$$

Membership functions in (28, 29, 32), and (33) are used for the rules 1–4 as μ_1, μ_2, μ_3, and μ_4, respectively. These membership functions set the degree of unnaturalness for each rule. As an example, when a total membership degree of 1.90 is calculated for the rule 2, its degree of unnaturalness is determined from μ_2 as 0.9291. Probabilities of these fuzzy events are calculated using normal approach to binomial distribution.

In order to make calculations easy and mine our sample database for unnaturalness a computer program is coded using Fortran 90 programming language. Table 4 gives total membership degrees of the fuzzy samples and those of unnaturalness in zones.

As can be seen from Table 4, sample 14 is beyond the $\pm 3\sigma$ limits and shows an out of control situation. Considering samples 14, 15, and 16 for the rule 2, total membership degree is calculated as 2, which refers an unnatural pattern with unnaturalness degree of 1. Then, only the last sample, sample 16, is marked as unnatural pattern and necessary actions should be taken at this stage. The mining is the restarted with sample 17. With the sample 18, total membership degree for rule 2 reaches to 1.61 and degree of unnaturalness determined from μ_2 is 0.72. Since minimum acceptable membership degree of unnaturalness for this problem is set to 0.95, it is not treated as an unnatural pattern. Sample 10 denotes unnaturalness with respect to the rule 2 applied to the lower side of the control chart. There is no sample indicating unnaturalness with a degree more than 0.95 according to the rules 3 and 4.

Table 4. Total membership degrees of the fuzzy samples and degree of unnaturalness in zones

Sample No	Beyond ±3σ	In or Above Fuzzy CL			In or Below Fuzzy CL		
		Rule 2	Rule 3	Rule 4	Rule 2	Rule 3	Rule 4
1	0.00	0.24	1	1	0	0	0.06
2	0.00	0.04	0.38	0.77	0.03	0.36	0.75
3	0.14	0	0	0	0.86	0.86	0.86
4	0.32	0	0	0	0.68	0.68	0.68
5	0.00	1	1	1	0	0	0
6	0.00	0	0	0.54	0.05	0.73	1
7	0.42	0	0	0	0.58	0.58	0.58
8	0.13	0.87	0.87	0.87	0	0	0
9	0.00	0	0	0	1	1	1
10	0.00	0	0	0	1 ($\mu^* = 1$)	1	1
11	0.26	0	0	0	0.74	0.74	0.74
12	0.00	0.96	1	1	0	0	0
13	0.00	0	0	0	1	1	1
14	**1.00**	0	0	0	0	0	0
15	0.00	1	1	1	0	0	0
16	0.00	**1 ($\mu^*=1$)**	1	1	0	0	0
17	0.39	0.61	0.61	0.61	0	0	0
18	0.00	1 ($\mu^* = 0.72$)	1	1	0	0	0
19	0.00	0	0.03	0.28	0.42	0.87	1
20	0.00	0.05	1	1	0	0	0.42
21	0.00	0.99	1	1	0	0	0
22	0.00	0	0.22	0.61	0.13	0.52	0.91
23	0.00	0.17	1	1	0	0	0.11
24	0.00	0	0.2	1	0	0.11	1
25	0.00	0.9	1	1	0	0	0
26	0.00	0	0.01	0.24	0.57	0.86	1
27	0.12	0	0	0	0.88	0.88	0.88
28	0.00	0.53	1	1	0	0	0.13
29	0.00	0	0.03	0.63	0.02	0.61	1
30	0.00	0	0	0	1	1	1

* unnatural sample with the corresponding degree of unnaturalness defined by the membersip functions for each rule.

6 Conclusions

In this chapter, fuzzy process control charts for attributes have been developed. Well-known Western Electric rules for examining unnaturalness are fuzzified using probability of fuzzy events and searched with data mining. A linguistic data of 30 samples have been used for illustration purposes. For larger data sets, unnaturalness can be mined using the same procedure. There are other unnatural pattern rules defined in the literature. These rules can also be examined under fuzziness. Some new rules can be added to the existing rules. When rules for unnaturalness are defined for longer runs, the usage of an intelligent data mining procedure is inevitable.

Appendix

Equations to compute sample area outside the control the limits.

$$A_{\text{out}}^U = \frac{1}{2} \left[(d^\alpha - UCL_4^\alpha) + (d^t - UCL_4^t) \right] (\max(t - \alpha, 0))$$
$$+ \frac{1}{2} \left[(d^z - a^z) + (c - b) \right] (\min(1 - t, 1 - \alpha)) \tag{A-U1}$$

where,

$$t = \frac{UCL_4 - a}{(b - a) + (c - b)} \text{ and } z = \max(t, \alpha)$$

$$A_{\text{out}}^U = \frac{1}{2} \left[(d^\alpha - UCL_4^\alpha) + (c - UCL_3) \right] (1 - \alpha) \tag{A-U2}$$

$$A_{\text{out}}^U = \frac{1}{2} (d^\alpha - UCL_4^\alpha)(\max(t - \alpha, 0)) \tag{A-U3}$$

$$A_{\text{out}}^U = \frac{1}{2} \left[(c - UCL_3) + (d^z - UCL_4^z) \right] (\max(1 - t, 1 - \alpha)) \tag{A-U4}$$

where

$$t = \frac{UCL_4 - d}{(UCL_4 - UCL_3) - (d - c)} \text{ and } z = \max(t, \alpha)$$

$$A_{\text{out}}^U = \frac{1}{2} \left[(d^{z_2} - UCL_4^{z_2}) + (d^{t_1} - UCL_4^{t_1}) \right]$$
$$\times (\min(\max(t_1 - \alpha, 0), t_1 - t_2))$$
$$+ \frac{1}{2} \left[(d^{z_1} - a^{z_1}) + (c - b) \right] (\min(1 - t_1, 1 - \alpha)) \tag{A-U5}$$

where

$$t_1 = \frac{UCL_4 - a}{(b - a) + (UCL_4 - UCL_3)},$$

$$t_2 = \frac{UCL_4 - d}{(UCL_4 - UCL_3) - (d - c)}$$

$$z_1 = \max(\alpha, t_1), \text{ and } z_2 = \max(\alpha, t_2)$$

$$A_{\text{out}}^U = 0 \tag{A-U6}$$

$$A_{\text{out}}^U = \frac{1}{2} \left[(d^\alpha - a^\alpha) + (c - b) \right] (1 - \alpha) \tag{A-U7}$$

$$A_{out}^L = \frac{1}{2}[(LCL_1^\alpha - a^\alpha) + (LCL_1^t - a^t)](\max(t - \alpha, 0))$$
$$+ \frac{1}{2}[(d^z - a^z) + (c - b)](\min(1 - t, 1 - \alpha)) \qquad \text{(A-L1)}$$

where

$$t = \frac{d - LCL_1}{(LCL_2 - LCL_1) + (d - c)} \text{ and } z = \max(\alpha, t)$$

$$A_{out}^L = \frac{1}{2}[(d^\alpha - a^\alpha) + (c - b)](1 - \alpha) \qquad \text{(A-L2)}$$

$$A_{out}^L = \frac{1}{2}(LCL_1^\alpha - a^\alpha) + (LCL_2 - b)](1 - \alpha) \qquad \text{(A-L3)}$$

$$A_{out}^L = \frac{1}{2}[(LCL_1^{z_2} - a^{z_2}) + (LCL_1^{t_1} - a^{t_1})]$$
$$\times (\min(\max(t_1 - \alpha, 0), t_1 - t_2))$$
$$+ \frac{1}{2}[(d^{z_1} - a^{z_1}) + (c - b)](\min(1 - t, 1 - \alpha)) \qquad \text{(A-L4)}$$

where

$$t_1 = \frac{d - LCL_1}{(LCL_2 - LCL_1) + (d - c)}, t_2 = \frac{a - LCL_1}{(LCL_2 - LCL_1) + (b - a)}$$
$$z_1 = \max(\alpha, t_1), \text{and } z_2 = \max(\alpha, t_2)$$

$$A_{out}^L = \frac{1}{2}[(LCL_4^z - a^z) + (LCL_2 - b)](\min(1 - t, 1 - \alpha)) \qquad \text{(A-L5)}$$

where

$$t = \frac{a - LCL_1}{(LCL_2 - LCL_1) - (b - a)}, \text{and } z = \max(\alpha, t)$$

$$A_{out}^L = 0 \qquad \text{(A-L6)}$$

$$A_{out}^L = \frac{1}{2}[(d^\alpha - a^\alpha) + (c - b)](1 - \alpha) \qquad \text{(A-L7)}$$

References

1. Asai, K. (1995), Fuzzy Systems for Management, IOS Press.
2. Bradshaw, C.W. (1983), A fuzzy set theoretic interpretation of economic control limits, *European Journal of Operational Research*, 13(4), 403–408.
3. Chang S.I., Aw, C.A. (1996), A neural fuzzy control chart for detecting and classifying process mean shifts, *International Journal of Production Research*, 34 (8), 2265–2278.
4. Fayyad, U. (1998), Mining Databases: Towards Algorithms for Knowledge Discovery, *IEEE Computer Society*, Special Issue on Mining Databases, Vol. 21, No. 1, pp. 39–48.

5. Gülbay, M., Kahraman, C., Ruan, D. (2004), α-cut fuzzy control charts for linguistic data, *International Journal of Intelligent Systems*, 19, 12, pp. 1173–1195.

6. Kahraman, C., Tolga, E., Ulukan, Z. (1995), Using triangular fuzzy numbers in the tests of control charts for unnatural patterns, in Proceedings of *INRIA / IEEE Conference on Emerging Technologies and Factory Automation*, October, 10-13, Paris-France, Vol. 3, pp. 291–298.

7. Kanagawa, A., Tamaki, F., Ohta, H. (1993), Control charts for process average and variability based on linguistic data, *International Journal of Production Research*, 2, 913–922.

8. Klemettinen, M., Mannila, H., Toivonen, H. (1997), A datamining methodology and its application to semi-automatic knowledge acquisition, *In DEXA'97 Workshop*, September 1–5, 1997. IEEE Computer Society Press.

9. Laviolette, M., Seaman, J.W., Barrett, J.D., Woodall, W.H. (1995), A probabilistic and statistical view of fuzzy methods, with discussion., *Technometrics*, 37, pp. 249–292.

10. Raz, T., Wang, J.-H. (1990), Probabilistic and membership approaches in the construction of control charts for linguistic data, *Production Planning and Control*, 1, 147–157.

11. Shewhart, W.A. (1931), Economic control of quality of manufactured product, *D. Van Nostrand,In.c*, Princeton, N.J.

12. Taleb, H., Limam, M. (2002), On fuzzy and probabilistic control charts, *International Journal of Production Research*, 40 (12), 2849–2863.

13. Wang, J.-H., Raz, T. (1988), Applying fuzzy set theory in the development of quality control charts, *International Industrial Engineering Conference proceedings*, Orlando, FL, 30–35.

14. Wang, J.-H., Raz, T. (1990), On the construction of control charts using linguistic variables, *International Journal of Production Research*, 28, 477–487.

15. Wang, J.-H., Chen, C.-H. (1995), Economic statistical *np*-control chart designs based on fuzzy optimization, *International Journal of Quality and Reliability Management*, 12(1), 88–92.

16. Western Electric Company, Bonnie B. Small, Chairman of the Writing Committee (1956), *StatisticalQuality Control Handbook*, Indianapolis, IN: AT&T Technologies (Select Code 700–444, P.O. Box 19901, Indianapolis 46219).

17. Woodall, W. (1997), Control charts based on attribute data: bibliography and review., *Journal of Quality Technology.*, 29, pp. 172–183.

18. Woodall, W., Tsui, K.-L., Tucker, G.L. (1997), A review of statistical and fuzzy control charts based on categorical data, *Frontiers in Statistical Quality Control 5 (Heidelberg, Germany: Physica-Verlag)*, pp. 83–89.

Accelerating the New Product Introduction with Intelligent Data Mining

Gülçin Büyüközkan and Orhan Feyzioğlu

Department of Industrial Engineering, Galatasaray University Ortaköy 34357
İstanbul – Turkey
{gbuyukozkan,ofeyzioglu}@gsu.edu.tr

Abstract. New product development (NPD) is a vital activity for companies. It is also a very risky process since every development stage involves a high degree of uncertainty and the success of each step depends on how previous steps are successfully accomplished. Hence, there is a clear need to evaluate new product initiatives systematically and make accurate decisions under uncertainty. Another actual concern for the companies is the time pressure to launch a significant number of new products due to the high competition in the market. In this chapter, we first present the available assessment models and techniques that help the evaluators to reduce their risks under uncertainty in NPD process. Then, we propose an integrated framework which is based on fuzzy logic, neural networks and multi criteria decision making and which enables us to make appropriate decisions while accelerating the decision process. We are especially interested in two first stages of new product decision-making: the choice of a new product idea ("go"/"no go" decision) and the choice of the right implementation order of the selected product ideas. We show that this two-staged intelligent approach allows practitioners to roughly and quickly separate good and bad product ideas by making use of previous experiences, and then to analyze in details a more shortened list rigorously.

1 Introduction

New product development (NPD) is the process by which an organization uses its resources and capabilities to create a new product or improve an existing one. Today, product development is seen as "among the essential processes for success, survival, and renewal of organizations, particularly for firms in either fast-paced or competitive markets" ([3] p. 344). Markets are generally perceived to be demanding higher quality and higher performing products, in shorter and more predictable development cycle-times and at lower cost [35]. In order to obtain best performance from NPD, the efficient and effective management of the product development process is vital. However, new product failure rates are substantial and the cost of failure is large, and this

Gülçin Büyüközkan and Orhan Feyzioğlu: *Accelerating the New Product Introduction with Intelligent Data Mining*, Studies in Computational Intelligence (SCI) **5**, 337–336 (2005)
www.springerlink.com   © Springer-Verlag Berlin Heidelberg 2005

makes successful NPD rather a complicating task to be exercised with caution [19].

The NPD process is structured around well-defined phases; each phase encloses many decision points, where management decides about the future of the project. The decision maker must take into account the customers' needs, the company's strategies as well as technological opportunities and the company's resources, and deduce the goals based on these factors for a successful NPD. With the support of a successful management system, an enterprise must be able to determine right products or features to be developed, the right time to develop and launch, the right amount of development investments, its effective implementation, etc. As it can be easily understood, no NPD operation can be accomplished without effective and timely decision-making.

An important cornerstone of the new product management is the idea selection and new product project launch decision. Several researchers have suggested that it is difficult for managers to end NPD projects once they are begun [7, 10]. Historical cases suggest that firms can make two types of erroneous decisions when evaluating their new product ideas. First, they might decide to pursue a potentially unsuccessful new product idea. Second, they might decide not to develop a potentially successful new product. In either case, firms accrue big losses, while the former leads to investment loses the latter leads to missed investment opportunities [39]. For this reason, here we focus especially on increasing the accuracy of the necessary decisions before a new product project is launched.

Similar to all decision problems, NPD decisions contain considerable amount of uncertainty causing elements, which confuse the decision-maker to reach the targeted performance. Uncertainty arises from multiple sources including technical, management and commercial issues, both internal and external to the project. It is also widely recognized and accepted that successful management of uncertainty is intimately associated with project success, as the proactive project manager constantly seeks to steer the project towards achievement of the desired objectives [22]. Then, it is critical to use a structured approach that can minimize the risks caused by the uncertainty for NPD projects. In this work, we propose an integrated approach based on fuzzy logic, neural networks and multi criteria decision making (MCDM) to make more rational selection decisions.

The rest of the chapter is organized as follows. In the next section, we briefly expose the uncertainty factors affecting the NPD process. In the third section, we present different decision-making techniques yet available to reduce the risks under uncertainty in NPD. The next two sections give the details on how to incorporate intelligent techniques in NPD and the proposed approach based on intelligent data mining. Finally, the last section contains some concluding remarks and perspectives.

2 Risks under Uncertainty in NPD Process

New product development is a very risky and uncertain process. Risk is defined as the exposure to loss/gain, or the probability of occurrence of loss/gain multiplied by its respective magnitude. Events are said to be certain if the probability of their occurrence is 100% or totally uncertain if the probability of occurrence is 0%. In between these extremes the uncertainty varies quite widely. On projects it is necessary to define one or a number of objective functions to represent the project under consideration and then measure the likelihood of achieving certain target values for them. Examples of such functions include capital expenditure, completion time and so on. Risk management involves modeling the project's objective functions against project variables, which include such variables as cost and quantities of input resources, external factors, etc. Since the project variables are often stochastic in nature and dynamic (i.e., exhibiting varying degrees of uncertainty over time) it is quite natural that the objective functions will also exhibit uncertainty. Project uncertainty is the probability that the objective function will not reach its planned target value [24].

It can be observed that different approaches exist in the literature to define and analyze the uncertainty in NPD projects. [17] combined three dimensions of uncertainty as technical, market and process. They rated and categorized uncertainty along each dimension as being either low or high. For technical uncertainty, when uncertainty is low, the technologies used in the development of the project are well known to the organization and relatively stable. When technical uncertainty is high, technologies used in the development of the project are neither existent nor proven at the start of the project, and/or are rapidly changing over time. For market uncertainty, when uncertainty is low the organization has good market data on both customers and competitors, and product is being sold through familiar channels of distribution. When market uncertainty is high, the organization has little information regarding who the customer is, how the market is segmented and what are the needed channels of distribution. For process uncertainty, when uncertainty is low the engineering, marketing, and communications (both internal and external) processes used in this project are well tested, stable, and embedded in the organization. When process uncertainty is high, a significant portion of any or all of the engineering, marketing, and communications processes are relatively new, unstable, or evolving.

Similarly, [38] identified three levels of uncertainty that confront companies operating in rapidly changing markets. First, potential customers cannot easily articulate needs that a new technology may fulfill. Consequently, NPD managers are uncertain about the market opportunities that a new technology offers. Second, NPD managers are uncertain about how to turn the new technologies into new products that meet customer needs. This uncertainty arises, not only from customers' inability to articulate their needs, but also from managers' difficulties in translating technological advancements into product features and benefits. Finally, senior management faces uncertainty about how much capital to invest in pursuit of rapidly changing markets as well as when to invest.

Reference [36] identified three main risk categories for engineering projects: "completion risks" group formed by technical, construction and operational risks, "market related risks" group formed by demand, financial and supply risks and finally, "institutional risks" group formed by social acceptability and sovereign risks.

We refer also to the recent work of [42] where NPD risks from uncertainty are organized into three general categories such as technical risks, commercial risks and NPD personnel. If we analyze NPD from different perspectives, we can precise risk structure in a more detailed manner. As an example, we can allocate product positioning, pricing and customer uncertainties to marketing; organizational alignment and team characteristics uncertainties to organizations; concept, configuration and performance uncertainties to engineering design; supplier, material, design of production sequence and project management uncertainties to operations management.

Efficient and effective NPD requires the appropriate management of all these uncertainty sources. While considering the decision points in whole NPD process, we expect to minimize the side effects of uncertainties described previously and to increase the effectiveness of the decisions. Numerous decision tools and techniques have been developed to assist managers in making better screening decisions in an uncertain environment. Some of them are summarized in the next section.

3 Risk Analysis Tools and Techniques in NPD

The balance between opportunities and risks has got to be carefully maintained for the performance of the NPD project. There are several, or many, tools and techniques, which are applicable to risk analysis in NPD projects [8, 9, 11, 13, 14, 18, 23, 27, 31, 47, 49]. The summary of them is given as follows.

Probabilistic Models: These include Monte Carlo Simulation and decision trees [48]. Monte Carlo analysis uses the process of simulation to achieve a range of solutions to a problem. Decision tree is a diagram that provides a structured approach to decision making that incorporates uncertainty of outcome and expected revenues.

Options Pricing Theory: It is being proposed as a mean of understanding what level of research investment is justified for a particular project. It treats each stage of the new product project much like purchasing an option on a future investment [16].

Scoring Models and Checklists: Here, projects are rated and scored on a variety of qualitative questions (in some cases, the project score becomes the criterion for project prioritization) [20]. The questions or items often capture proven drivers of new product success such as product advantage, market attractiveness, and synergy with the base business (leverages core competencies), familiarity, etc. [37].

Behavioral Approaches: These are tools designed to bring managers to a consensus in terms of which projects to undertake, and include methods such as the Delphi method that is a qualitative forecasting method which uses a panel of experts [48]. They are particularly useful for the early stages, where only qualitative information is available.

Analytical Hierarchy Process (AHP): These are decision tools based on paired comparisons of both projects and criteria [45]. Software tools such as Expert Choice® enable a team of managers to arrive at the preferred set of projects in a portfolio [54], with relative ease.

Sensitivity Analysis: It examines how the optimal solution and the optimal objective value are affected from the changes of the uncertainty parameters (values and probabilities) that are considered to be important [40].

Scenario Analysis: This technique has been widely preferred and used by many decision makers. Here, a combination of possible values of the uncertainty parameters are assumed regarding to different point of views (e.g., pessimistic, neutral and optimistic), and the resulting scenario is solved. By solving the problem repeatedly for different scenarios and studying the solutions obtained, the decision maker observes sensitivities and heuristically decides on an appropriate solution.

Fuzzy Logic: It deals with problems in which a source of vagueness is involved [53]. In general, the probability concept is related to the frequency of occurrence of events, captured by repeated experiments whose outcomes are recorded, while the fuzzy sets provide the appropriate framework to evaluate the possibility of events rather than their probability [18].

Artificial Intelligence: It is a discipline that is concerned with the study and creation of computer systems that exhibit some form of intelligence. Intelligence is a system that can learn new concepts and tasks; reason and draw useful conclusions about the world around us; understand a natural language; and perceive and comprehend a visual scene [41]. Typical research areas of artificial intelligence include problem solving and planning, expert systems, natural language processing, robotics, computer vision, neural networks, genetic algorithms and machine learning [29]. Case-based reasoning, rough set theory and intelligent agent are the recent emerging areas [42].

These techniques can be used exclusively or in a hybrid way. We must note that there is no best technique. Each of them has some advantages and also disadvantages. For example, the decision tree method is easy to understand

where the risk is interpreted as probability and not as a discount rate. In the same time, the risk estimates easily biased and difficult to estimate accurately. The method lacks flexibility since decision points occur continuously and not always at discrete junctions. If too many possibilities are considered, then "tree" becomes a "bush." As another example, Monte Carlo simulation has the advantage to analyze a greater number of scenarios and to estimate the probabilities of these scenarios. But it has also some drawbacks: probability distributions for individual variables and variable correlations may be difficult to calculate. To reflect reality, more variables have to be added which makes the model more complicated and difficult to understand. Moreover, the project value due to the managerial flexibility is not calculated. For these reasons we think that the extent to which different techniques for the NPD idea evaluation can be used will depend upon the nature of the project, the information availability, the company's culture and several other factors. This is clear from the variety of techniques, which are theoretically available, and the extent to which they have been used in practice. In any case, no matter which technique is selected by a company, it should be implemented, and probably adapted, according to the particular needs of that company.

In this study, where we analyze the new product idea evaluation, we propose an intelligent decision-making procedure based on neural networks, fuzzy logic and MCDM that will be described in details in the next section.

4 Use of Intelligent Techniques for New Product Idea Selection

As stated before, being able to consistently and rationally evaluate and justify go/no-go decision-making for each NPD project becomes extremely desirable from both top management as well as project manager's point of view. When there are numerous ideas generating sources, it is almost impossible to rate all new product ideas in a very detailed way and in a reasonable amount of time. In this study, we propose to use a two-stage intelligent decision making approach to accelerate the NPD process and to improve the efficiency of the decisions in an environment of uncertainty. The research in the intersection area of artificial intelligence and NPD is comparatively new. For a comprehensive overview of the application of the related techniques in NPD, we refer the interested readers to [42, 56]. We note that, [56] identified neural networks and genetic search as the predominant techniques for the initial phases of NPD process.

The proposed two-staged new product idea selection approach allows practitioners to roughly and quickly separate good and bad product ideas by making use of previous experiences, and then to analyze in details a more shortened list. The first stage consists of a technique that merges neural networks and fuzzy logic. Artificial Neural Networks (ANN) [21, 34]

- make use of the way that the human brain learns and functions,
- possess the ability to learn from examples,
- have the ability to manage systems from their observed behavior rather than from a theoretical understanding,
- have the capacity to treat large amount of data and capturing complex interactions among the input variables, and thus reducing the development time by learning underlying relationships.
  Meanwhile fuzzy logic [28, 53, 57]
- is used to deal with imprecise linguistic concepts or fuzzy terms,
- allows us to make rational decisions in an environment of uncertainty, fuzziness and imprecision without loosing the richness of verbal judgment,
- is highly suitable for approximate reasoning by incorporating fuzzy rules.
  So it is likely that substantial improvements on NPD idea selection decisions can be made by merging the ANN and fuzzy set theory. The characteristics of such hybrid architecture can be described as follows:
- It realizes an *automatic procedure* for obtaining in the same time both the consequents and antecedents of a set of fuzzy rules starting from a system's set of input (previous new product idea evaluations) and output data (ideas' grades). Moreover, they allow us to appropriately modify the shape of the membership functions.
- It requires a *small number of parameters* with respect to the number of connections in a Multilayer Perceptron (MLP). Besides, the number of neurons in such architecture is wholly determined by that of membership functions chosen for each new product evaluation input variables.
- It allows us *to incorporate the knowledge* of an expert regarding the choice of new product idea input-output topology.
- It leads us to determine a system model, which is *easily comprehensible*, unlike the model obtained with an MLP. In fact, neural networks reach their own limits precisely because the knowledge acquired by a neural network consists in a set of interconnection weights, which is not simply interpretable. Instead, a fuzzy rules system is always *transparent* in the sense that a practitioner can easily read the knowledge base of the fuzzy system and interpret its behavior when faced by a given new product.

The second stage of the proposed approach is based on MCDM, particularly fuzzy AHP method, which allows a more accurate description of the evaluation and decision making process. Among the different MCDM methods, AHP is the most widely used and easily understandable one [45, 54]. Other researchers also have noted the usefulness of AHP for new product screening [7, 32, 33, 51, 52]. The methodology allows decision makers to model a complex problem like a new product idea selection in a structure showing relationships of the goal, objectives and alternatives. The goal of selecting the best new product idea is defined as a statement of the overall objectives. Therefore, the definition of the goal is that it is the idea that best meets the objectives. With AHP, it is also possible to have a set of ideas that would become the "best choice." AHP allows for decision makers to pull information together for one idea, assess pros and cons for that idea, weight that idea against others using a variety of measurement techniques and finally communicate the decision through synthesis of the new product ideas in relation to the goal. Fuzzy AHP is a natural extension of the traditional AHP where decision makers do not require to express their assessments through crisp values but rather they use fuzzy numbers which is more suitable when uncertainty is high. This is especially true for a decision process like new product idea selection where there are also many qualitative attributes to rate subjectively. Recently, [6] suggested an integrated decision making approach for NPD under uncertainty and they used the fuzzy AHP method to select new product development strategies, which minimize project uncertainties. Fig. 1 illustrates the simplistic view of our proposed two-stage approach

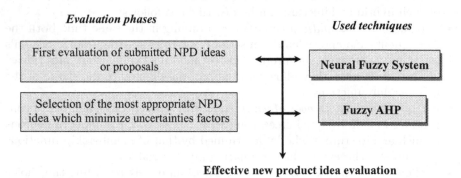

Fig. 1. Proposed intelligent decision-making approach

5 An Intelligent Data Mining Approach for New Product Idea Selection

5.1 Neural Fuzzy System

In this study, new product ideas generated individually or by groups of individuals have been collected by a formal system. The preprocessing of ideas is left to an intelligent neuro-fuzzy inference system, which is trained with precedent decisions. This type of system has clearly an unbiased nature and posses an internal mechanism that can learn the viewpoint of the company management towards products ideas by making use of the extracted rules. This will also reduce the needed effort to make decisions when the number of applications is large.

For our fuzzy inference system (FIS), the input space for the mapping is the information provided by past ideas evaluations and the output space is the status of the idea (i.e., "good" or "bad"). Regarding to NPD, most of the time evaluations are based on a scoring system with determined evaluation criteria. Therefore, translating if necessary these new products' criteria scores to eligibility percentages, one can build the input database for FIS. The mapping then provides a basis from which decisions can be made, or patterns discerned. The details of the FIS are given in [1]. Neural network techniques aid the fuzzy modeling procedure to learn information about a data set, and compute the membership function parameters that best allow the associated FIS to track the given input/output data. ANFIS (adaptive network-based fuzzy inference system) is a class of adaptive networks that are functionally equivalent to FIS [25]. Using a given input/output data set, ANFIS constructs a FIS whose membership function parameters are adjusted using either a back propagation algorithm or a hybrid-learning algorithm. Therefore, using ANFIS, fuzzy systems can learn from the modeling data.

The architecture of ANFIS is a feed-forward network that consists of five layers [25]. Figure 2 shows the equivalent ANFIS architecture for a two-input Sugeno-type fuzzy inference system.

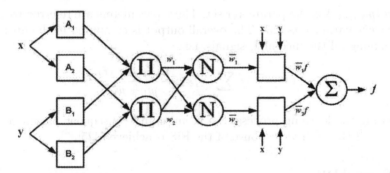

Fig. 2. ANFIS architecture for a two inputs, two rules Sugeno FIS

A rule in the first order Sugeno FIS has the form:

$$\text{If } x \text{ is } A_i \text{ and } y \text{ is } B_i \text{ then } f_i = p_i x + q_i y + r_i$$

The output of a node in the first layer specifies to which degree a given input, x, satisfies a quantifier, A, i.e., the function of the node i in this layer is a membership function for the quantifier, A_i, of the form:

$$O_i^1 = \mu_{A_i}(x) . \tag{1}$$

Each membership function has a set of parameters that can be used to control that membership function. For example, a Gaussian membership function that has the form

$$\mu_{A_i}(x) = \exp\left[-\left(\frac{x - c_i}{\sigma_i}\right)^2\right] \tag{2}$$

and has two parameters, c_i and σ_i. Tuning the values of these parameters will vary the membership function, which means a change in the behavior of the FIS. Parameters in this layer are referred to as premise parameters [25].

In the second layer, the output of a node represents a firing strength of a rule. The node generates the output (firing strength) by multiplying the signals that come on its input,

$$w_i = \mu_{A_i}(x) \times \mu_{B_i}(y) . \tag{3}$$

The function of a node in the third layer is to compute the ratio between the ith rule's firing strength to the sum of all rules' firing strengths:

$$\bar{w}_i = \frac{w_i}{w_1 + w_2} \tag{4}$$

where $\bar{w}_i$ is referred to as the normalized firing strength [25]. In the fourth layer, each node has a function of the form:

$$O_i^4 = \bar{w}_i f_i = \bar{w}_i \left(p_i x + q_i y + r_i \right) \tag{5}$$

where $\{p_i, q_i, r_i\}$ is the parameter set. These parameters are referred to as the consequent parameters [25]. The overall output is computed in the fifth layer by summing all the incoming signals, i.e.,

$$O^5 = f = \sum_i \bar{w}_i f_i = \frac{w_1 f_1 + w_2 f_2}{w_1 + w_2} \tag{6}$$

During the learning process, the premise and consequent parameters are tuned until the desired response of the FIS is achieved [25].

5.2 Fuzzy AHP

We propose to use the fuzzy AHP method in the second stage of our study. Traditional AHP [44, 45] uses the principle of comparative judgments to construct pair wise comparisons of the relative importance of elements at some given level of a criteria hierarchy with respect to shared criteria or property at the level above, giving rise to a weight matrix. Priorities are synthesized from the second level down from multiplying local priorities by the priority of their corresponding criterion in the level above and adding for each element in a level according to the criterion it affects. The construction of comparison matrices requires the relative importance among attributes and options being expressed as precise numbers on a standard scale (usually from 1 to 9) where the degree of the preference is proportional with the magnitude of the chosen number. However, precise numbers fail to contain the subjectivity and vagueness in such decision-making. Comparisons between alternatives always contain ambiguity and multiplicity of meaning. Moreover, human assessment on qualitative attributes is always subjective and thus imprecise. We overcome this difficulty of modeling the uncertainty of human assessment by using the fuzzy AHP methodology. The fuzzy AHP approach allows a more accurate description of the decision making process.

The earliest work in fuzzy AHP appeared in [50], which compared fuzzy ratios described by triangular membership functions. Logarithmic least square was used to derive the local fuzzy priorities. Later, using geometric mean, [4] determined fuzzy priorities of comparison, whose membership functions were trapezoidal. By modifying the Van Laarhoven and Pedrycz method, [2] presented a more robust approach to the normalization of the local priorities. In a recent study, [15] used a fuzzy extension of the AHP method in the project selection.

A fuzzy number is a special fuzzy set $F = \{(x, \mu_F(x)), x \in R\}$, where x takes its value on the real line, $R : -\infty < x < +\infty$ and $\mu_F(x)$ is a continuous mapping from R to the closed interval $[0,1]$ which represents the membership degree of x. A triangular fuzzy number denoted as $\tilde{a} = (l, m, u)$, where $l \leq m \leq u$, has the following triangular type membership function:

$$\mu_{\tilde{a}}(x) = \begin{cases} 0, & \text{if } x \leq l \text{ or } x > u \\ \dfrac{x - l}{m - l}, & \text{if } l < x \leq m \\ \dfrac{u - x}{u - m}, & \text{if } m < x \leq u \end{cases}$$

It can also be characterized alternatively as

$$\tilde{a}^{\alpha} = [l^{\alpha}, u^{\alpha}] = [(m - l)\alpha + l, u - (u - m)\alpha]$$

for all $\forall \alpha \in [0,1]$. Then, main operations like addition or multiplication can be accomplished by usual interval arithmetic. Here, we use triangular fuzzy numbers $\tilde{1}$ to $\tilde{9}$ as a superior means of representing pair wise comparisons in the AHP judgment matrix and improve the conventional nine-point scaling scheme. These numbers together with their corresponding membership functions are defined in Fig. 3.

A comparison matrix $\tilde{R}$ is constructed for the n-new product idea selection problem, in which pair wise comparisons are assumed to be triangular fuzzy numbers $\tilde{a}_{ij}$ for all $i, j \in \{1, 2, \ldots, n\}$, such that

$$\tilde{R} = \begin{bmatrix} (1,1,1) & \tilde{a}_{12} & \tilde{a}_{13} & \cdots & \tilde{a}_{1(n-1)} & \tilde{a}_{1n} \\ 1/\tilde{a}_{12} & (1,1,1) & \tilde{a}_{23} & \cdots & \tilde{a}_{2(n-1)} & \tilde{a}_{2n} \\ \vdots & \vdots & \vdots & \vdots & \vdots & \vdots \\ 1/\tilde{a}_{1(n-1)} & 1/\tilde{a}_{2(n-1)} & 1/\tilde{a}_{3(n-1)} & \cdots & (1,1,1) & \tilde{a}_{(n-1)n} \\ 1/\tilde{a}_{1n} & 1/\tilde{a}_{2n} & 1/\tilde{a}_{3n} & \cdots & 1/\tilde{a}_{(n-1)n} & (1,1,1) \end{bmatrix}$$

The triangular fuzzy number $\tilde{a}_{ij} = (l_{ij}, m_{ij}, u_{ij})$ is obtained for each lowest level decision criterion and alternative idea simply by weighted average of different evaluators

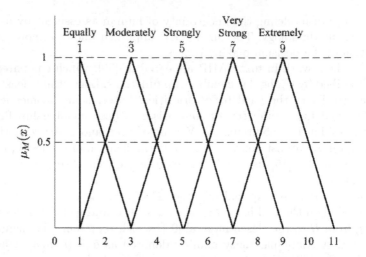

Fig. 3. Membership functions of triangular fuzzy numbers

$$\tilde{a}_{ij} = \tilde{w}_1 \otimes \tilde{a}_{ij1} \oplus \tilde{w}_2 \otimes \tilde{a}_{ij2} \oplus \ldots \oplus \tilde{w}_K \otimes \tilde{a}_{ijK} \text{ for all } i,j \in \{1,2,\ldots,n\} \quad (7)$$

where $\tilde{w}_k$ is the weight of the evaluator k and there are K evaluators. $\tilde{w}_k$ can be a fuzzy number or just the crisp number $(1/K)$ if one would prefer to treat decision makers equally. Note that, each $\tilde{a}_{ijk} = (l_{ijk}, m_{ijk}, u_{ijk})$ is selected among the triangular fuzzy numbers given in Fig. 3. Equation (8) can also be used to aggregate assessments.

$$\tilde{a}_{ij} = \left(\min_k (l_{ijk}), \left(\prod_{k=1}^{K} m_{ijk} \right)^{1/K}, \max_k (u_{ijk}) \right) \quad (8)$$

Next, the fuzzy eigenvector of the matrix $\tilde{R}$ is estimated. According to [45], the right principal eigenvector of the matrix expresses the importance of the alternatives. The fuzzy eigenvalue, $\tilde{\lambda}$, is a fuzzy number solution to $\tilde{R}\tilde{x} = \tilde{\lambda}\tilde{x}$ where $\tilde{x}$ is a non-zeros $n \times 1$ fuzzy vector. Using interval arithmetic, this is equivalent to

$$[a_{i1l}^{\alpha} x_{1l}^{\alpha}, a_{i1u}^{\alpha} x_{1u}^{\alpha}] \oplus \ldots \oplus [a_{inl}^{\alpha} x_{nl}^{\alpha}, a_{inu}^{\alpha} x_{nu}^{\alpha}] = [\lambda_l^{\alpha} x_{il}^{\alpha}, \lambda_u^{\alpha} x_{iu}^{\alpha}]$$

where $\tilde{a}_{ij}^{\alpha} = [a_{ijl}^{\alpha}, a_{iju}^{\alpha}]$, $\tilde{x}_i^{\alpha} = [x_{il}^{\alpha}, x_{iu}^{\alpha}]$ and $\tilde{\lambda}^{\alpha} = [\lambda_l^{\alpha}, \lambda_u^{\alpha}]$ for $0 ¡ \alpha \leq 1$ and all $i,j = 1, 2, \ldots, n$.

The degree of satisfaction for the matrix $\tilde{R}$ is estimated by the index of optimism μ. Larger value of μ indicates higher degree of optimism. Optimism index is the convex combination defined as [30]

$$\hat{a}_{ij}^{\alpha} = \mu a_{iju}^{\alpha} + (1-\mu) a_{ijl}^{\alpha}, \; \forall \mu \in [0,1]. \quad (9)$$

When α and μ is fixed, the following matrix can be obtained.

$$\tilde{R} = \begin{bmatrix} 1 & \hat{a}_{12}^{\alpha} & \cdots & \hat{a}_{1(n-1)}^{\alpha} & \hat{a}_{1n}^{\alpha} \\ \hat{a}_{21}^{\alpha} & 1 & \cdots & \hat{a}_{2(n-1)}^{\alpha} & \hat{a}_{2n}^{\alpha} \\ \vdots & \vdots & \vdots & \vdots & \vdots \\ \hat{a}_{(n-1)1}^{\alpha} & \hat{a}_{(n-1)2}^{\alpha} & \cdots & 1 & \hat{a}_{(n-1)n}^{\alpha} \\ \hat{a}_{n1}^{\alpha} & \hat{a}_{n2}^{\alpha} & \cdots & \hat{a}_{n(n-1)}^{\alpha} & 1 \end{bmatrix}$$

Then, the eigenvector of $\tilde{R}$ corresponding to its maximal eigenvalue can be computed. After a normalization, the importance of alternatives for the given criterion is obtained. The procedure explained so far is repeated for all the lowest level criteria and also others in between criteria. In other words, all alternatives have to be compared one with another for each lowest level criterion so as to find their ratings and meanwhile the importances of criteria have to be determined again by pair wise comparison for each hierarchical level.

Finally, if we denote $x' = \{x_1', x_2', \ldots, x_n'\}$ as the adjusted performance measures of new product ideas, then we can construct the following fuzzy decision matrix

$$\tilde{Z} = \begin{array}{c} \\ A_1 \\ A_2 \\ \vdots \\ A_n \end{array} \begin{array}{c} C_1 \ C_2 \ C_3 \ C_4 \\ \begin{bmatrix} x_{11}' & x_{12}' & \cdots & x_{1m}' \\ x_{21}' & x_{22}' & \cdots & x_{2m}' \\ \vdots & \vdots & \vdots & \vdots \\ x_{n1}' & x_{n2}' & \cdots & x_{nm}' \end{bmatrix} \end{array}$$

where A and C stand for alternative and criterion, respectively. Note that x_{ij}' is the performance measure of the alternative i for the criteria j. Then, the importance of each alternative is obtained by multiplying each criterion weight v_j (calculated as a crisp number) by the related alternative's performance. In other words, we require

$$p_i = v_1 \times x_{i1}' + v_2 \times x_{i2}' + \ldots + v_m \times x_{im}' \text{ for all } i \in \{1, 2, \ldots, n\}$$

This calculation process continues level by level in the hierarchical structure until the finite performance of alternatives can be obtained.

5.3 Algorithmic Form of Proposed Approach

To summarize our approach, the necessary steps are given in an algorithmic form as follows.

Step 1. Accumulation of the new product ideas through selected collecting techniques (i.e., forms, contest, web, etc.).

Step 2. Rating of individual ideas in percentage for all evaluation criteria by the marketing team.

Step 3. Determination of the input membership functions and related parameters by exercising neural networks techniques on rating data.

Step 4. Building the fuzzy inference system with adjusted membership functions of the previous step.

Step 5. Using the inference system as needed to accept/reject ideas.

Step 6. Given uncertainty factors, each individual member of the expert team is required to evaluate pre-selected ideas by using the linguistic terms.

Step 7. Aggregation of expert results to figure out the right implementation order.

We apply Steps 3–4 if necessary after the idea pool update. The application of this proposed methodology to a specific toy manufacturing firm has been recently reported by [5].

6 Final Remarks and Perspectives

In this study, we aim to improve the quality of decision-making in NPD under uncertainty and to higher the level of success of associated activities by introducing a new iterative methodology. First we describe uncertainty factors affecting the NPD process and cite the essential methods for the decision maker to reduce these factors. Then, we emphasize the motivation behind our approach, which incorporates fuzzy logic, neural networks and MCDM for the new product idea selection.

We believe that the application of our method will be a good practice in terms of the aggregation and purification of the subjective judgments and to clarify the big picture, which is covered by risks and uncertainties. Moreover, it is generic in a sense that although in different sectors, companies exercising similar vast new product idea selection process and having a scoring system can adopt it quite easily. However, we have to also underline two limitations of this study:

- The methodology is proposed to the companies that had already a successful scoring system and want to computerize and speed up the selection process. Without a reliable historical database, the neural network cannot be trained and the FIS can only be equipped with theoretical understanding. This can lead to inconsistent results.
- We underline that the approach is not applicable in all cases. In other words, the method is structured especially for companies/sectors where many new product ideas are stimulated and there is a need for a more efficient evaluation procedure for the initial selection. There is a need for intellectual capital evaluation for high innovative and creative, very few new product developing or highly R&D oriented companies.

No matter which evaluation technique is used, a long period of time is always necessary to observe the results of such a strategic level decision. Additionally, a product success is not only depending on catching the best idea but also on how to manage subsequent development and launch processes. We keep trying to understand the sources of conflict and possible improvements on the approach.

A practical reality is that environmental factors and customer's tendencies towards new products change over time and previously selected genuine ideas cannot be adequate for the actual period. It is then advised to practitioners to update the database in a way that old ideas are discarded (e.g., dating five or more years old) and new ones are added. It is clear that the update frequency highly depends on the targeted market segment.

Based on this work, our future extension is to investigate other decision phases in NPD and to provide similar approaches to enrich the available literature. We will evaluate in a more detailed form, the influence of other methods on the final quality and accuracy of decisions. We would also try to enhance our decision support system with new techniques to enable managers comparing different solutions and making more rigorous decisions.

References

1. Bezdek, J.C., Dubois, D., and H. Prade, (edit.), 1999, Fuzzy sets approximate reasoning and information systems. Kluwer Academic Publishers Group, Netherlands.
2. Boender, C.G.E., de Grann, J.G., and F.A. Lootsma, 1989, Multi criteria decision analysis with fuzzy pair wise comparison, Fuzzy Sets and Systems, 29, 133–143.
3. Brown, S.L., and K.M. Eisenhardt, 1995, Product development: past research, present findings and future directions, Academy of Management Review, 20, 343–378.
4. Buckley, J.J., 1985, Fuzzy Hierarchical Analysis, Fuzzy Sets and Systems, 17, 233 247.
5. Büyüközkan, G., and O. Feyzioğlu, 2004a, A new approach based on soft computing to accelerate new product introduction, Computers in Industry, 54 (2), 151–167.
6. Büyüközkan, G., and O. Feyzioğlu, 2004b, A fuzzy-logic-based decision-making approach for new product development, International Journal of Production Economics, 90 (1), 27–45.
7. Calantone, R.J., Di Benedetto, C.A. and J.B. Schmidt, 1999, Using the analytic hierarchy process in new product screening, Journal of Product Innovation Management, 16 (1), 65–76.
8. Cohen, M.A., Eliashberg, J., and T.-H. Ho, 1997, New Product Development: The Performance and Time-To-Market Tradeoff, Journal of Product Innovation Management, 14 (1), 65-66.
9. Coldrick, S., Longhurst, P., Ivey, P. and J. Hannis, 2004, An R&D options selection model for investment decisions, Technovation, Article In Press, Available through www.sciencedirect.com
10. Cooper, R.G., 1994, Perspective: third generation new product processes, Journal of Product Innovation Management, 11, 3–14.

11. Cooper, R.G., 2001, Winning at New Products: accelerating the process from idea to launch, Pereus Publishing, New York, Third Edition.
12. Carbonell-Foulquié, P., Munuera-Alemán, J.L. and Rodríguez-Escudero, A.I., 2004, Criteria employed for go/no-go decisions when developing successful highly innovative products, Industrial Marketing Management, 33 (4), 307–316.
13. Davila, T., 2000, An empirical study on the drivers of management control systems' design in new product development, Accounting, Organizations and Society, 25, 383–409.
14. Doctor, R.N., Newton, D.P., and A. Pearson, 2001, Managing uncertainty in research and development, Technovation, 21 (2), 79–90.
15. Enea, M., and T. Piezza, 2004, Project Selection by Constrained Fuzzy AHP, Fuzzy Optimization and Decision Making, 3, 39–62.
16. Faulkner, T., 1996, Applying options thinking to R&D valuation, Research and Technology Management, 39, 50–57.
17. Fox, J., Gann, R., Shur, A., Glahn, L., and B. Zaas, 1998, Process uncertainty: a new dimension for new product development, Engineering Management Journal, 10, 19–27.
18. Garavelli, A.C., Gorgoglione, M., and B. Scozzi, 1999, Fuzzy logic to improve the robustness of decision support systems under uncertainty, Computers and Industrial Engineering, 37, 477–480.
19. Goldenberg, J., Lehmann, D.R., and D. Mazursky, 2001, The idea itself and the circumstances of its emergence as predictors of new product success, Management Science, 47 (1), 69–84.
20. Hall, D.L., and A. Naudia, 1990, An interactive approach for selecting R&D project, IEEE Transactions on Engineering Management, 37, 126–133.
21. Hammerstrom, D., 1993, Neural networks at work, IEEE Spectrum Computer Applications, 30 (6), 26–32.
22. Hillson, D., 2002, Extending the risk process to manage opportunities, International Journal of Project Management, 20 (3), 235–240.
23. Infanger, G., 1994, Planning Under Uncertainty. International Thomson Publishing, MA.
24. Jaafari, A., 2001, Management of risks, uncertainties and opportunities on projects: time for a fundamental shift, International Journal of Project Management 19, 89–101.
25. Jang, J.S.R., 1993, ANFIS: Adaptive-Network-Based Fuzzy Inference System, IEEE Transaction on Systems, Man and Cybernetics, 23 (3), 665–685.
26. Jovanović, P., 1999, Application of sensitivity analysis in investment project evaluation under uncertainty and risk, International Journal of Project Management, 17 (4), 217–222.
27. Klir, G.J., and B. Yuan, 1995, Fuzzy Sets and Fuzzy Logic: Theory and Applications. Prentice-Hall, Englewood Cliffs, NJ.
28. Krishnamoorthy, C.S. and S., Rajeev, 1996, Artificial Intelligence and Expert Systems for Engineers, CRC Press, Inc., Doca Raton.
29. Lee, A.R., 1999, Application of modified fuzzy AHP method to analyze bolting sequence of structural joints. UMI Dissertation Services, A Bell & Howell Company.
30. Li, S., 2000, The development of a hybrid intelligent system for developing marketing strategy, Decision Support Systems, 27, 395–409.
31. Lilien, G.L., and A. Rangaswamy, 1997a, Marketing Engineering: Computer-Assisted Marketing Analysis and Planning, Addison-Wesley, MA.

32. Lilien, G.L., and A. Rangaswamy, 1997b, Software tools for new product development, Journal of Marketing Research, 34, 177–184.
33. Lin, C., and C. Lee, 1996, Neural Fuzzy Systems. Prentice Hall, Englewood Cliffs, NJ.
34. Maffin, D., and P. Braiden, 2001, Manufacturing and supplier roles in product development, International Journal of Production Economics, 69, 205–213.
35. Miller, R., Lessard, D., 2001. Understanding and managing risks in large engineering projects. International Journal of Project Management 19, 437–443.
36. Montoya-Weiss, M.M., and R.J., Calantone, 1994, Determinants of new product performance: A review and meta analysis, Journal of Product Innovation Management, 11, 397–417.
37. Mullins, J.W., and D.J. Sutherland, 1998, New product development in rapidly changing markets: an exploratory study, Journal of Product Innovation Management, 15, 224–236.
38. Özer, M., 2005, Factors which influence decision making in new product evaluation, European Journal of Operational Research, 163 (3), 784–801.
39. Parnell, G.S., Jackson, J.A., Burk, R.C., Lehmkuhl, L.J., Engelbrecht, J.A., 1999, R&D concept decision analysis:Using alternate futures for sensitivity analysis, Journal of Multi-Criteria Decision Analysis, 8, 119–127.
40. Patterson, D.W., 1990. Introduction to Artificial Intelligence and Expert Systems. Prentice Hall, Englewood Cliffs.
41. Rao, S. S., Nahm, A., Shi, Z., Deng, X., and A., Syamil, 1999. Artificial intelligence and expert systems applications in new product development – a survey, Journal of Intelligent Manufacturing, 10 (3), 231–244.
42. Riek, R.F., 2001, From experience: Capturing hard-won NPD lessons in checklists, Journal of Product Innovation Management, 18, 301–313.
43. Saaty, T.L., 1977, A scaling method for priorities in hierarchical structures, Journal of Mathematical Psychology, 15, 234–281.
44. Saaty, T.L., 1980, The Analytic Hierarchic Process. McGraw-Hill International, New York.
45. Schmidt, J.B., and R.J. Calantone, 1998, Are really new product development projects harder to shut down? Journal of Product Innovation Management, 15 (2), 111–123.
46. Shipley, M.F., de Korvin, A., and O. Khursheed, 2001, A fuzzy logic-based decision model to satisfy goals for successful product/service introduction, European Journal of Operational Research, 135 (1), 209–219.
47. Souder, W.E., and T. Mandakovic, 1986, R&D project selection models, Research Management, 29, 36–42.
48. Trittle, G.L., Scriven, E.F.V., and A.R. Fusfeld, 2000, Resolving uncertainty in R&D portfolios, Research and Technology Management, March-April, 47–55.
49. Van Laarhoven, P.J.M., and W. Pedrycz, 1983, A fuzzy extension of Saaty's priority theory, Fuzzy Sets and Systems, 11, 229–241.
50. Wind, J., and V. Mahajan, 1981, Designing product and business portfolios, Harvard Business Review, 59, 155–165.
51. Wind, Y., and T.L. Saaty, 1980, Marketing applications of the analytic hierarchy process, Management Science, 26, 641–658.
52. Zadeh, L.A., 1965, Fuzzy sets, Information and control, 8, 338–353.
53. Zahedi, F., 1996, The analytic hierarchy process – A survey of the method and its applications, Interfaces, 16, 108.

54. Zaremba, M.B. and G., Morel, 2003, Integration and control of intelligence in distributed manufacturing, Journal of Intelligent Manufacturing, 14 (1), 25–42.
55. Zimmermann, H.J., 1999, Practical Applications of Fuzzy Technologies. Kluwer Academic Publishers, Massachusetts.

Integrated Clustering Modeling with Backpropagation Neural Network for Efficient Customer Relationship Management

Tijen Ertay[1]* and Bora Çekyay[2]

[1] Istanbul Technical University Faculty of Management, Management Engineering Department, Macka, 34367, Istanbul, Turkey
[2] Istanbul Technical University Faculty of Management, Industrial Engineering Department, Macka, 34367, Istanbul, Turkey

1 Introduction

The rapid progress in digital data acquisition and storage technology has lead to the fast growing tremendous and amount of data stored in databases, data warehouses, or other kinds of data repositories such as the World Wide Web [40].

The advent of the network world induced by the rapid development of the Internet and the accompanying adoption of the Web has promoted the changes to create greater business opportunities and to reach customers more easily. This situation is required the capability to both generate and collect data and this capability has been expanded enormously and provides us with huge amounts of data. Although valuable information may be hiding behind the huge amounts of data, it makes difficult for human to extract them without powerful tools. *Data Mining (DM)* emerged to extract knowledge from huge volumes of data with the help of computing device during the late 1980s. *DM* has become a research area with increasing importance with the amount of data greatly increasing [5, 7, 9, 27].

Due to its interdisciplinary nature, data mining has received contributions from many disciplines such as databases, machine learning, statistics, information retrieval, data visualization, and parallel and distributed computing etc. The field of *DM* is a new discipline in engineering and computer science to address these new opportunities and challenges. Industrial Engineering (IE), with the diverse areas it encompasses, presents unique opportunities for the

[1]*Corresponding Author: ertay@itu.edu.tr

Tijen Ertay and Bora Çekyay: *Integrated Clustering Modeling with Backpropagation Neural Network for Efficient Customer Relationship Management*, Studies in Computational Intelligence (SCI) **5**, 355–354 (2005)
www.springerlink.com

application of *DM* and for the development of new concepts and techniques in this field. Specific techniques used in *DM* applications include market basket analysis, memory based reasoning, cluster detection, link analysis, decision trees, neural networks and genetic algorithms.

Reference [3] defines data mining as the exploration and analysis, by automatic or semiautomatic means of large quantities of data in order to discover meaningful patterns and rules. Reference [12] defines data mining as the process of discovering interesting knowledge from large amounts of data stored either in data bases, data warehouses, or other information repositories. Reference [13] considers data mining as the analysis of observational data set to find unsuspected relationships and to summarize the data in novel ways that are both understandable and useful to the data owner. DM business applications can be found in a diverse group of business including for example, banks [32], healthcare [20], insurance [15].

Taxonomy of DM tasks can be broadly divided into five categories: dependency analysis, class identification, concept description, and data visualization. The taxonomy reflects the emerging role of data visualization as a separate DM task, even as it is used to support other data mining tasks. Different DM tasks are grouped into categories depending on the type of knowledge extracted by the tasks. The identification of patterns in large data set is the first step to gaining useful marketing insights and making critical marketing decisions. The DM tasks generate an assortment of customer and market knowledge, which form the core of knowledge management process. Knowledge management is prerequisite for e-business and its increasing customer centric focus. DM is usually used to answer two main types of application questions: to generate predictions on basis of available data; to describe behavior captured in the data [8].

To answer the first type of question, classification can be used as one of the most popular approaches. Classification and clustering are prime targets of most empirical research of the real world and of DM. These aim to group entities or objects into classes so that there is maximum intra-class proximity between members, and maximum inter-class proximity between members and maximum inter-class distinction among groups. Commonly, a clustering model provides a representation scheme in the form of a data structure, an index to calculate similarity and a grouping technique. There are many different methods, which may be used data predict the appropriate class for situations. Among the most popular one are logistic regression, case-base reasoning, neural networks, decision trees, and rule induction [17].

Especially, neural networks have been applied to almost all types of data mining applications including forecasting [29], credit approval problems [19], target marketing [25], cost comparison of assembly systems [30, 31], development of maintenance policies [2].

In general, development of good neural network applications can be very time consuming and requires the building, training and testing of many different network structures to arrive at an efficient model. Neural networks are characterized by learning capability, the ability to improve performance over time. A closely related feature is that of generalization, relating to the recognition of new objects, which are similar but not identical to previous ones. A typical neural network consists of a number of simple processing elements called neurons. Each neuron is connected to other neurons by means of directed information being used by the net to solve a problem. Neural networks are usually modeled into one input layer, one or several hidden layers and one output layer [34].

The most popular neural network method for practical applications is bacpropagation algorithm. In spite of its simple structure, the presence of one or more hidden units, together with a non-linear activation function, gives it the ability to solve many complex problems. A backpropagation neural network (BP) is the most common multi-layer network estimated to be used in percent 80 of all applications.

The research presented in this study emanates from the apparent use of neural network modeling for customer segmentation and the apparent primary advantages of self-organizing mapping (SOM) neural networks over cluster analysis. SOM network has been considered as one of the most popular unsupervised competitive neural network learning models, for clustering and visualization in a number of real world problems [18].

The primary objectives of the research are to consider the use of SOM NN for segmenting customers and to analyze the predictive ability of BP NN for classifying customers from follow-up surveys by using the output provided by SOM NN. Data collected by the *Group Lens Research Project* at the University of Minnesota (http://www.cs.umn.edu/research/GroupLeans/data) are used in this research. The rest of this study is organized as follows. Section 2 provides an overview on customer relationship management. Section 3 includes the backgrounds and the methodology to be used in this study. Section 4 provides the results of modeling process based on the used methodology. Section 5 is related to the conclusions of this study.

2 Customer Relationship Management: An Overview

Among business practitioners and marketing scientists today, there is an increasing interest in customer relationship management (CRM). The focus of CRM is to forge closer and deeper relationships with customers, "being willing and able to change your behavior toward an individual customer based on what the customer tells you and what else you know about the customer" [28].

In this manner, CRM is an enterprise wide business strategy designed to optimize profitability, revenue and customer satisfaction by organizing the enterprise around customer segments, fostering customer-satisfying behaviors and linking processes from customers through suppliers [24]. It requires a clear focus on the service attributes that represent value to the customer and create loyalty [14].

CRM deals with the mechanics of building relationships, including data capture and analysis and business process simplification. Proponents of CRM argue that may benefits can be derived from implementing CRM; including sales functionality by developing customer profiles and history; customer service support through warranty management, tracking and problem resolution; cross selling and upselling higher – margin products or services to targeted customer segments; and attracting additional customers by offering personalized service such as direct mail outs. Another key benefit claimed by the CRM industry is that customers are segmented and communication programs are developed to retain the most profitable customers. With CRM, emphasis is placed on selling more products and services through data mining to determine the types of customers that would be most likely to buy a particular product. This is achieved by developing sophisticated predictive models that assess a segment's propensity to purchase product based on the purchasing behavior of individuals with similar demographic other profiles. Segmentation is product focused rather than customer focused and does not consider any element of the emotional connection with the customer. The relationship is driven entirely by historical behaviors. CRM does have the capability to allow companies to better understand customer purchasing behavior, or at least that portion that is captured in the system, and to determine the type of communications that should be undertaken with the customer [1].

CRM is essentially a two-stage concept. The task of the first stage is to master the basics of building customer focus. This means moving from a product orientation to a customer orientation and defining market strategy from outside-in and not from inside-out. The focus should be on customer needs rather than product features. The second stage includes company's development of customer orientation by integrating CRM across the entire customer experience chain, by leveraging technology to achieve real-time customer management, and by constantly innovating their value proposition to customer. The goal of CRM is to identify a customer, understand and predict the customer-buying pattern, identify an appropriate offer, and deliver it in a personalized format directly to the customer. CRM means that companies manage relationships with individual customers with the aid of customer databases and interactive and mass customization technologies. The adoption of CRM has been enhanced by developments in information and communication technology (e.g., Database Technology, E-commerce and the Internet) [28, 35].

The first step in CRM is based on information processing about company's customers through internal customer data or the purchased data from outside sources. An enterprise data warehouse is a critical component of a successful CRM. By using an efficient enterprise data warehouses, companies can invest in the customers that are potentially valuable for the company, but also minimize their investments in non-valuable customers. Hence, customers are categorized by the enterprise in terms of their different values and served with different relationship strengthening practices. In other words, on the turnover of each customer or customer profitability can be used as segmentation variables to distinguish between valuable and nonvaluable customers. Besides, considering customers' demographic information, it makes possible the segmentation according to the customer values. This indicates what direction customers' preferences to move. In this situation, it can be said that the *customer value* (CV) provides a good framework for applying data mining to CRM. On the "input" side of data mining, the customer value tells what is likely to be interesting. In general, marketers say there are three ways to increase a customer value – increase their use (or purchases) of products they already have – sell them more or higher margin products – keep the customers for a longer period of time. However, to obtain information on the potential value of a customer, analysts need on the customer's purchasing behavior at their own company, as well as at other companies in the market. Usually, companies only have data on customers' purchasing behavior at their own company in their customer information file. Hence, models are needed to predict the potential value of a customer based on the purchasing behavior and any available socio-demographic data. Behavioral clustering and segmentation help drive strategic marketing initiatives, while sub-segments based on demographic lifestyle or value characteristics could also be determined and used for tactical marketing.

The process of establishing a traditional static CRM begins with the identification of customers of a retailer and the construction of a customer purchase data mart from customer history of purchases within a certain time period. Once the customer purchase data mart has been built on the enterprise intranet, the summarized data such as the time of purchase, quantity, frequency, and product's rating are extracted from the customer information of the data mart for mining customer purchase behavior patterns. According to the calculated CV, with the aid of the SOM, the customers can be divided into a set of segments of similar customers with similar CV. The last component of a CRM is campaign execution and tracking. Implementation of decisions made as a result of data mining is based on campaign execution and tracking. Campaign execution is based on recommendation about company's products for the define segments. Tracking is also based on getting users' opinions about the promotions made for the company's products. These are the processes and systems that allow the user to develop and deliver targeted messages in a test-and-learn environment.

3 Backgrounds and Methodology

The first part of this section will be related to a discussion on customer value and a segmentation method for CRM that uses customer value. Next, it will be considered the methodology that combines SOM being one of the most popular unsupervised competitive neural network learning models for clustering and backpropagation neural networks (BP) for predictive classifying the customers from follow-up surveys by using the output provided by a SOM.

3.1 Customer Value Concept

Customer Value (CV) has turned out to be a very important concept in marketing strategy and research in spite of the fact that the growing body of knowledge about the construct is fragmented. Costumer value measurement is a strategic marketing tool to clarify a company's proposition to its customers, thus creating a differential superior offering compared with the competition. The tool assesses a company's performance in comparison with its main competitors as perceived by former, present, and potential customers. Reference [39] has proposed that customer value was a customer's perceived preference for and evaluation of those product attributes, attribute performances, and consequences arising from use that facilities achieving the customer's goals and purposes in use situations. There are some areas of consensus about customer value. First, customer value is inherent in or linked to the use of a product. Second, customer value is something perceived by customers rather than determined by sellers. Third, these perceptions include a trade-off between what the customer receives, and what he/she gives up in order to acquire and use a product. Despite the increasing attention being focused on customer value, extant definitions of the construct are somewhat ambiguous, because they typically rely on other terms such as utility, worth, benefits and quality which are too often not well defined [26]. In general, definitions take a rather narrow perspective, because value frequently is measured as attribute-based desires or preferences, which are influencing purchase. Besides, researchers have assessed extensively the construct of value in marketing. Reference [10] proposed that value can be approached in three different concepts: values, desired values and value judgments. Value is defined as centrally held enduring core beliefs, desired end states, or higher order goals of the individual customer or customer organization that guide behavior. Desired customer value, to the contrary, are the customers' perceptions of what they want to have happen in a specific kind of use situation, with the help of a product or service offering, to accomplish a desired purpose or goal [38]. Value judgment is "the customer's assessment of the value that has been created for them by a supplier given the trade-offs between all relevant benefits and sacrifices in a specific-use situation". For this situation, value judgment is the most important for a definition of customer-perceived value.

The most important success factor for a film is the ability to deliver better customer value than the competition. Good customer value can be achieved only when product quality, service quality, and value-based prices are in harmony and exceed customer expectations. Maximizing customer value must flow from a firm's culture, beliefs, values, management style, reward systems, and structure. As a concept, customer value is fairly simple. But since it is

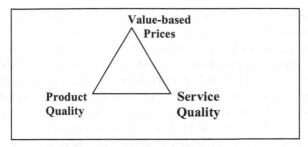

Fig. 1. Customer Value Triad

defined by the customer, it is inherently ambiguous. The customer value triad as seen in Fig. 1 consists of only three things: product quality, service quality and value based prices. For this reason, customer value is created when customer expectations in any one of the three areas, the firms have not delivered good customer value. Only when all three are in harmony will customer value be maximized. Therefore a firm must implement an organized system to capture customer expectations of product quality, service quality and reasonable price. The first subject is that customer expectations must be formed relative to competitor alternatives.

A firm can steadily improve product and service quality, and reduce price and still convey less value than competitors. The competitors may simply be improving faster. The second subject is that customer expectations are dynamic. Therefore, determining customer expectations of product quality, service quality and price relative to the competition should be part of a continuous, ongoing program. The third subject is that product and service quality must extend throughout the channel. Upstream, suppliers must adhere to higher and higher quality standards. Downstream, all of the channel intermediaries must be coordinated to deliver maximum value to the end user. Maximizing customer value would achieved by achieving superior product and service quality at a price competitive with the competition ([23] pp. 15–24). Besides, the expected customer value is ultimately determined by the comparison of expected benefits to expected sacrifice. Decreasing the sacrifice while holding benefits constant also enhances customer value. Increasing the benefits faster than the sacrifice enhances customer value too ([23] p. 117).

3.2 Self-Organizing Neural Network for Customer Segmentation

In CRM, managers develop specific strategies for different segments of their customer base. Segmentation proposes that groups of customers with similar needs and purchasing behaviors are likely to demonstrate a more homogenous response to marketing programs that target specific consumer groups. A critical issue to successful market segmentation is the selection of segmentation (clusters) variables. Segmentation variables can be expanded to general variables and product specific variables [36]. General variables have been used in many researches because these ones are intuitive and easy to operate [11]. Besides, [22] proposed artificial neural network clustering method incorporating both clusters and segment discriminate analysis to estimate the relationship between customer demographics. Clustering is prime target of most empirical research of the real world and of data mining. That aims to group entities or objects into classes so that there is maximum intra-class proximity between members and maximum inter-class distinction among groups. Customer segmentation based on general variables is more intuitive and easier to conduct than product specific variables. However, the assumption that customers with similar demographics will exhibit similar purchasing behavior is doubtful. Each customer pursues personalized products and services even within groups with similar demographics in order to present uniqueness and identity. This makes customer's purchase patterns difficult to determine using only general variables. Self-Organizing Map (SOM) can be used for market segmentation. The versatile properties of SOM make it a valuable tool in data mining. SOM is designed to do unsupervised clustering of data; i.e. given training patterns that contain inputs only, the SOM assigns output units that represent clusters to inputs. The SOM not only assigns cluster centers to units, but also it tends to group similar centers on units close together, thus giving on impression of the relationship of the clusters and the topological behavior of the network. Clustering customers via a SOM is divided into two phases. On phase is to train a SOM using the customer value patterns. The other is to map input customer value patterns to output customer segments. When an input pattern is imposed on the network, an output node is selected from among all the output nodes as having the smallest Euclidean distance between the presented input pattern vector and its weight vector. This output unit is declared the winner in the competition among all the neurons in the output layer. Only the winning neuron generates an output signal from the output layer. All the other neurons have a zero output signal. As learning involves adjustment of weight vectors, only the neurons within the neighborhood of the winning neuron are allowed to learn with this particular input pattern. The neighborhood means the physical proximity of the neurons to the winning neuron. Learning for the nodes within the neighborhood is carried out by adjusting their weights closer to the input vector. The size of the neighborhood is initially chosen to be large enough to include all units in the output layer. As learning proceeds, the size of the neighborhood is progressively reduced to a pre-defined limit. Thus during these stages, fewer neurons have their weights adjusted closer to the input vector. Data variables representing continuous data (e.g. rating of costumer) are scaled to assume a value between zero and one, while dichotomy variables (e.g. male/female) are recoded to assume values of zero or one. In this manner each data is considered by the network as binary-valued input. In order to create the training and test sets, the data set is randomly, nearly sub-divided so that percent 70 of the data is allocated to the training set, percent 20 to the test set, percent 10 to the validation set. This portion is based on the principle that the size of the validation set must procure a balance between obtaining a sufficient sample size to evaluate both the training and test tests. Reference [16] In this study, The MATLAP 6.0 software package will be used for training the selected SOM neural network model.

3.3 Backpropagation Neural Network for Predictive Clustering

Backpropagation NNs have been extremely popular for their unique learning capability [37] and have been shown to perform well in different applications in our previous research such as medical application [33] and game playing [6] A typical backpropagation neural network consists of a three layer structure: input layer nodes, output layer nodes, hidden layer nodes. In this second stage of methodology, in order to specify an output variable, which is required for supervised learning, each respondent belonging to a specific segment is mapped back to the original data set and represented the dependent variable for the training of the BP neural network. Backpropagation networks are fully connected, layered, feed-forward models. Activations flow from the input layer through the hidden layer, then to the output layer. A backpropagation network typically starts out with a random set of weights. The network adjusts its weights each time it sees an input-output pair. Each pair is processed at two stages, a forward pass and backward pass. The forward pass involves presenting a sample input to the network and letting activitations flow until they reach the output layer. During the backward pass, the network's actual output is compared with the target output and error estimates are computed for the output units. The weights connected to the output units are adjusted to reduce the errors (a gradient descent method). The error estimates of the output units are then used to derive error estimates for the units in the hidden layer. Finally, errors are propagated back to the connections stemming from the input units. The backpropagation network updates its weights incrementally until the network stabilizes. The algorithm details can be shown in [4] and [37]. In this study, we followed the standard neural network architecture because it provides comparable results to the optimal architecture and works well as a benchmark for comparison. In this study, a BP NN model is trained using the demographic variables as inputs and twelve variables each representing a segment as outputs. The primary aim of this stage is to develop a trained model, which learns the pattern in the data and has the ability to generalize the pattern for predictive purposes. The accuracy of the trained model is validated by the validation patterns and tested on a portion of the data set to determine its ability to generalize. The size of the training, validation and test sets used for the SOM neural network modeling is also applied to the BP NN model. A validation set is used for visual comparative purposes during training. The accuracy of the model is verified using a test set of data not included for the purposes of training or validation. The selected sigmoid transfer function requires that each data point be scaled between 0 and 1. For this reason, each of the input variables is scaled to primarily enable the network to learn the relevant patterns. The sigmoid function scales the data by channel to match the range of the first hidden transfer function and the desired data to match the range of the output transfer function. A transfer function is used to prevent outputs from reaching very large values, which can "impair" neural networks. The BP NN training is monitored by using two stopping rules. The second rule is related to the acceptable error rate as measures of overall accuracy and applied whenever the output error term indicated signs of over-training.

4 Finding of the Modeling Procedure

4.1 Data Preparation

In this study, the data was collected through the Movie Lens web site (movie-lens.umn.edu) during the seven month period from September 19th, 1997 through April 22nd, 1998. The historical data consists of 100000 ratings from 943 users (customer) on 1682 movies with every user having at least 20 ratings and simple demographic information for the users such as age, gender, occupation, and zip code is included. The purpose of data preparation is to integrate, select and transform the data from one or more data bases into the data required for the proposed methodology. For this reason, it will be sampled a reference set that has enough rating information to discover similar user patterns. First, the data from web environment have been transferred to ACCESS program. Rating values that the user has given to the films, the demographic information related to the user, the contents of the films have been transferred to each separate file. Later, how many the users to go to the determined film kinds, the rating values given to these films and how many the films to be gone are extracted by using the interrogation commend. In order to accelerate the interrogation, it has been prepared as table sheet the information to be used for interrogation. A data preparation example is shown in Fig. 2. This information will be used for calculation of the customer value in the following section. The above extracted information has been combined with the extracted demographic information for data mining. Sexual ones from demographic information are coded as "1" for male and "0" for female. Occupation information can be also coded as follows.

$$X_i^j = \begin{cases} 1 & \text{if person } i \text{ has occupation } j, \\ 0 & \text{otherwise} \end{cases} \quad i = 1, 2, \ldots, n \; j \in Q$$

n : The number of selected person
Q : The set of all occupations

Nine genre types and 308 people gone and voted to these film types are selected. The selected people are considered each occupation, age, and sexual groups.

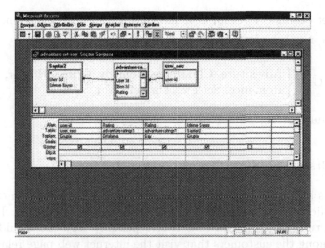

Fig. 2. A Data Preparation Example

4.2 Evaluation Metric: Calculating Customer Value

Value is the customer's overall assessment of the utility of a product based on perceptions of what is received and what is given. In other words, value in business markets is the perceived worth in monetary units of the set of economic, technical, service and social benefits received by customer firm in exchange for the price paid for a product, taking into consideration the available suppliers' offerings and prices. Buyers' perceptions of value represent a trade off between the quality and benefits they perceive in the product relative to the sacrifice they perceive by paying the price [39].

Most customer value importance scaling techniques used in practice today, such as rating scales, ranking scales and derived importance through regression and conjoint analysis can handle only a relatively few value dimensions at one time. In this study, we evaluate customer value with two viewpoints. Equation (1) evaluates the customer value having been developed for this study.

$$
\begin{aligned}
CV_i^j &= \text{Preference rate of } j. \text{ person for } i. \text{ film type} * \text{rate of the} \\
&\quad \text{being met expectation} \\
CV_i^j &= \frac{\text{The number of films of } i. \text{ type being seen by } j. \text{ person}}{\text{The number of films seen by } j. \text{ person}} \\
&\quad \cdot \frac{\text{Expectation being met}}{\text{Total expectation}}
\end{aligned}
$$

$$
CV_i^j = \frac{\left\| \alpha_i^j \right\|}{\sum\limits_{k \in G} \left\| \alpha_i^k \right\|} \cdot \frac{\sum\limits_{k \in \alpha_i^j} r_i^k}{\mu \cdot \left\| \alpha_i^j \right\|}, \quad i = 1, 2, \ldots, 308, \quad j \in G \tag{1}
$$

r_i^j: rating value that i. person gives j. film

α_i^j: the set of films of j. film type having been seen by i. person.

$\left\| \alpha_i^j \right\|$: the number of members in α_i^j

$$G = \left\{ \begin{array}{l} \text{Adventure, Children's, Comedy, Crime, Horror,} \\ \text{Romance, Sci-Fic, Thriller, War} \end{array} \right\}$$

μ : Maximum rating value (5 for this study).

4.3 Results of the SOM Neural Network Model

First, it has been clustered the customers according to the customer value, the demographic variables, the occupations, the sexual information. Preliminary the SOM studies, it appears that twelve possibly segments could be distinguished among the customers that visit the internet web page related to data collected by the Group Lens Research Project at the University of Minnesota. It is considered the ones to be in the different states from the point of view of the customer value of the segments in the determination of the cluster numbers. The following parameters and values are used for training the selected SOM neural network model. Ordering phase learning rate 0.9; ordering phase steps 1000; tuning phase learning rate 0.02; topology function adopts *"hextop"* and distance function uses *"mandist."* Topology and distance functions are selected to the best being seen distinction for segments. Table 1 indicates the centroid values of the segments according to the customer values for genre. Also Table 2 indicates the centroid values of the segments according to the demographic variables.

Table 1. The centroid values to the customer values of the segments

	Genre (Customer Value)								
Cluster No	Adventure (I)	Children's (II)	Comedy (III)	Crime (IV)	Horror (V)	Romance (VI)	Sci-Fi (VII)	Thriller (VIII)	War (IX)
1	0.25	0.18	0.37	0.21	0.07	0.39	0.19	0.24	0.20
2	0.24	0.13	0.25	0.23	0.05	0.28	0.17	0.28	0.30
3	0.27	0.14	0.31	0.20	0.09	0.32	0.22	0.22	0.33
4	0.30	0.14	0.29	0.27	0.13	0.26	0.26	0.28	0.31
5	0.26	0.15	0.35	0.19	0.05	0.46	0.18	0.23	0.32
6	0.29	0.10	0.30	0.19	0.10	0.30	0.25	0.23	0.34
7	0.26	0.08	0.29	0.23	0.06	0.33	0.25	0.28	0.29
8	0.21	0.12	0.32	0.22	0.10	0.23	0.23	0.24	0.19
9	0.22	0.17	0.30	0.25	0.12	0.30	0.22	0.31	0.20
10	0.25	0.18	0.29	0.22	0.08	0.33	0.25	0.24	0.26
11	0.34	0.13	0.27	0.24	0.08	0.25	0.28	0.25	0.32
12	0.14	0.09	0.27	0.27	0.11	0.19	0.29	0.30	0.20

Table 2. The centroid values to the demographic values of the segments

Cluster No	Age	Sex													Demographic Centroid Values								
			1	2	3	4	5	6	7	8	9	10	11	12	13	14	15	16	17	18	19	20	21
1	30.14	0.48	0.00	0.00	0.00	0.01	0.00	0.00	0.00	0.01	0.00	0.00	0.00	0.00	0.00	0.00	0.00	0.00	0.00	0.00	0.49	0.00	0.00
2	43.00	0.27	0.03	0.00	0.00	0.02	0.00	0.00	0.00	0.81	0.00	0.04	0.01	0.00	0.00	0.00	0.00	0.05	0.01	0.01	0.03	0.00	0.00
3	46.31	1.00	0.00	0.00	0.00	0.00	0.00	0.00	0.00	0.02	0.00	0.27	0.33	0.00	0.00	0.00	0.00	0.34	0.04	0.00	0.00	0.00	0.00
4	35.29	1.00	0.33	0.00	0.00	0.00	0.00	0.33	0.33	0.00	0.00	0.00	0.00	0.00	0.00	0.00	0.00	0.00	0.00	0.00	0.00	0.00	0.00
5	36.57	0.05	0.03	0.01	0.00	0.82	0.03	0.01	0.01	0.01	0.00	0.00	0.00	0.02	0.02	0.02	0.01	0.00	0.00	0.01	0.02	0.00	0.01
6	46.71	0.93	0.01	0.01	0.01	0.80	0.01	0.00	0.00	0.02	0.00	0.01	0.01	0.01	0.00	0.02	0.00	0.00	0.01	0.05	0.00	0.04	0.00
7	35.71	1.00	0.00	0.00	0.25	0.00	0.00	0.00	0.00	0.00	0.00	0.00	0.00	0.00	0.00	0.00	0.01	0.00	0.00	0.36	0.00	0.36	0.00
8	26.69	1.00	0.00	0.00	0.00	0.00	0.00	0.00	0.01	0.00	0.00	0.00	0.00	0.00	0.00	0.00	0.48	0.00	0.00	0.01	0.47	0.01	0.00
9	34.71	0.00	0.00	0.00	0.00	0.00	0.00	0.04	0.06	0.00	0.11	0.04	0.19	0.00	0.08	0.00	0.11	0.02	0.06	0.06	0.00	0.02	0.19
10	31.81	0.00	0.00	0.33	0.00	0.00	0.00	0.00	0.00	0.00	0.00	0.00	0.00	0.33	0.00	0.33	0.00	0.00	0.00	0.00	0.00	0.00	0.00
11	34.23	0.99	0.00	0.33	0.00	0.00	0.00	0.00	0.00	0.00	0.00	0.00	0.00	0.33	0.00	0.33	0.00	0.00	0.00	0.00	0.00	0.00	0.00
12	30.32	1.00	0.00	0.00	0.00	0.00	0.29	0.00	0.00	0.00	0.03	0.00	0.00	0.00	0.15	0.00	0.00	0.00	0.23	0.00	0.00	0.00	0.29

1: administrator, 2: artist, 3: doctor, 4: educator, 5.engineer, 6: entertainment, 7: executive, 8: healthcare, 9: homemaker, 10: lawyer, 11: librarian, 12: marketing, 13: none, 14: other, 15: programmer, 16: retired, 17: salesman, 18: scientist, 19: student, 20: technician, 21: writer

Segment 1: In general, this cluster includes the women, which are of the occupations of the students and administrator and the average 30 years old. The order of the film types (genre) from point of view of the customer value has been determined as VI-III-I-VIII.

Segment 2: In general, this cluster includes the women, which are of the occupation of the healthcare and the average 43 years old. The order of the film types (genre) from the point of view of the customer value has been determined as IX-VI-VIII-III.

Segment 3: In general, this cluster includes the men, which are of the occupations of librarian, lawyer, and retired and the average 46 years old. The order of the film types (genre) from the point of view of the customer value has been determined as IX-VI-III-I.

Segment 4: In general, this cluster includes the men, which are of the occupations of administrator, entertainment and executive and the average 35 years old. The order of the film types (genre) from the point of view of the customer value has been determined as IX-I-III-VIII.

Segment 5: In general, this cluster includes the women, which are of the occupation of educator and the average 36 years old. The order of the film types (genre) from the point of the customer value has been determined as VI-III-IX-I.

Segment 6: In general, this cluster includes the men, which are of the occupation of educator and the average 36 years old. The order the film types (genre) from the point of view of the customer value has been determined as VI-III-IX-I.

Segment 7: In general, this cluster includes the men, which are of the occupation of doctor, scientist, technician, and the average 35 years old. The order of the film types (genre) from the point of view of the customer value has been determined as VI-IX-III-VIII.

Segment 8: In general, this cluster includes the men, which are of the occupation of programmer, student, and the average 26 years old. The order of the film types (genre) from the point of view of the customer value has been determined as III-VIII-VI-VII.

Segment 9: In general, this cluster includes the women, which are of the occupation of artist, marketing and other and the average 32 years old. The order of the film types (genre) from the point of view of the customer value has been determined as VIII-VI-III-IV.

Segment 10: In general, this cluster includes the women, which are of the occupation of artist, marketing and other and the average 32 years old. The order of the film types (genre) from the point of view of the customer value has been determined as VI-III-IX-I.

Segment 11: In general, this cluster includes the men, which are of the occupation of artist, marketing and other and the average 34 years old. The order of the film types (genre) from the point of view of the customer value has been determined as I-IX-VII-III.

Segment 12: In general, this cluster includes the men, which are of the occupation of engineer, salesman and writer and the average 30 years old. The order of the film types (genre) from the point of view of the customer value has been determined as VIII-VII-III-IV.

4.4 Results of Backpropagation Neural Network Model

It has been designed a BP neural network model to predict the segments determined in SOM model according to the demographic values of the new customers. This model required for the development of BP neural network considers the variables that are also used for the SOM neural network modeling process. In order to determine output variables, which is required for supervised learning, each respondent belonging to a specific segment is mapped back to the original data set and represented the output variable for the training of the BP neural network. A BP neural network model is trained using the 23 variables as inputs and 12 variables each representing a segment as outputs. This model is shown in Fig. 3.

The aim of this stage is to develop a trained model, which learns the pattern in the data and gets the ability to generalize the pattern for predictive purposes. The accuracy of the trained model is validated by unseen data and tested on a portion of the data set to determine its ability to generalize. A validation set is used for comparative purposes during training. The accuracy of the model is verified using the test data not used for training or validation. In this study, the data set belonging to 260 people was used for training.

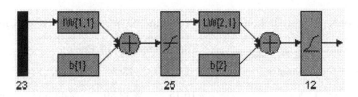

Fig. 3. BP Neural Network Model

Also the data set related to 48 people was used for testing. The one intermediate (hidden) layer with 25 neurons used for the model specification is a function of the non-linearity of the data. The tansig function has been used as the activity one for the hidden layer because of indicating the better generalization for training. The logsig function has been also used as the activity one for the output layer because of being 0–1 intervals values. The neuron number in the hidden layer has been expanded one by one and continued as to the acceptable error level. The accuracy of BP neural network model is determined by how well the training set can learn the pattern in data when compared to the test data set, which comprises unseen data. Figure 4 indicates the error graph of the training and validation sets for BP NN.

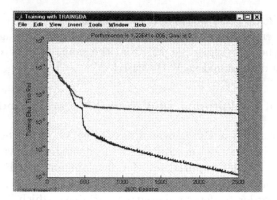

Fig. 4. The error graph of BP-NN model

In Fig. 4, the training set and validations set converge as the error between the trained data and the validation data starts to decrease. The BP NN training is monitored by using stopping rule if the total number of iterations reached 1000. The overall accuracy for BP NN model could also be considered using the unseen 161 people's data. BP NN has been simulated by these people's demographic information and the segments in which the people are located have been surmised. According to these predictions, the film's types having the highest customer value have been recommended to the person and the accuracy of these recommendations related to the first three preferences has been determined as shown in Table 3.

Table 3. The prediction accuracy of segments according to BP-NN model

Film Preference of Cluster	Film Preference of Person		
	1	1&2	1&2&3
1	%16.15	%32.30	%43.50
1&2	%39.75	%58.40	%67.70
1&2&3	%75.77	%85.70	%89.44
1&2&3&4	%89.44	%97.52	%99.38

When the first four film's types with the highest customer value are recommended to the related person for each segment, at least one of the first three film's type on the person's preferences could be predicted accurately. For this reason, it can be said that the more accurate segmentation requires more detailed data and more definite recommendations are based on more accurate predictions.

5 Conclusion

The requirement for the detailed knowledge on film industry and the preference of the customer and the need to cope with the limitations in analyzing non-linear relationship are the aim of this research. Knowledge discovery is a complicated process of extracting useful information from raw data. It includes many steps such as data warehousing; target data section, data cleaning, preprocessing, and transformation. In a majority of the real cases, the knowledge discovery process is iterative and interactive in nature. At the core of the knowledge discovery process, there are the data mining methods for extracting patterns from data. These methods can have different goals and may be applied successively to achieve the desired result in the knowledge discovery process. For example, SOM neural network models based on artificial intelligence technology can be developed to create clusters based on combinations of natural characteristics within a set of customer data (e.g. information in this study). In addition, BP-NN models could be also used to predict the segmentation of new customers as part of an already existing segment. This study demonstrates the usefulness of Artificial Neural Networks as means of grouping the respondents reached according to the survey in web page and predicting the segmentation of new respondents. Profiling the clusters and tracking the dynamic behavioral changes of segments can be considered in the future research. Besides, additional inputs required for the more segmentation and calibration of BP-NN model should be obtained through a process of data enrichment. The knowledge generated by the predictive ability of the BP-NN model could induce greater effectiveness in film industry and financial efficiency. The application of the BP-NN could generate further knowledge, which may be used to order the composition of the film recommendation with the changing behavior of customers. Using the BP-NN model to generate knowledge related to high customer value segments could increase revenue for the local economy and assist with the development of film industry's management strategies. A competitive advantage could be gained by researching the differences in behavior of customer segments that go and do not go to a film. The matching of cluster profiles of high value customers that go to a film to similar profiles of customers that do not go to the film by using the BP-NN model could procure a target market opportunity.

References

1. Barnes, J.G., 2000. Secrets of Customer Relationship Management: It's All About How You Make Them Fell, Mc Graw-Hill.
2. Bellandi, G., Dulmin, R., Mininno, V., 1998. Failure rate neural analysis in the transport sector. International Journal of Operations & Production Management, 18(8), 778–793.
3. Berry, M.J.A., Linoff, G., 1997. Data mining techniques for marketing, sales and customer support, New York, Wiley.

4. Bishop, C.M., 1995. Neural Networks for Pattern Recognition, Oxford University Press, New York.
5. Changchien, S.W., Lu, T., 2001. Mining association rules procedure to support on-line recommendation by customer and products fragmentation. Expert Systems with Applications, 20, 325–335.
6. Chen, H., Buntin, P., She L., Sutjahjo, S., Sommer, C., Neely, D., 1994. Expert prediction, symbolic. IEEE Expert, 9(6), 21–27.
7. Chiang, D., Chow, L.R., Wang, Y., 2000. Mining time series data by a fuzzy linguistic summary system. Fuzzy Sets and Systems, 112, 419–432.
8. Edelstein, H., 1997. Mining for gold. Information Week, 4/21/1997.
9. Fayyads, U.M., Ratetsky-Shapiro, G., Smyth, P., 1996. The KDD processes for extracting useful knowledge from volumes of data. Communications on the ACM, 39(11), 27–34.
10. Flint, D.J., Woodruff, R.B., Gardial, S.F., 1997. Customer Value Change in Industrial Marketing Relationships-A Call for New Strategies and Research. Industrial Marketing Management, 26(2), 163–175.
11. Hammond, K., Ehrenberg, A.S.C., Goodhandt, G.J., 1996. Market segmentation for competitive brands. European Journal of Marketing, 30(12), 39–49.
12. Han, J., Kamber, M., 2001. Data Mining: Concepts and Techniques. Academic Press, New York.
13. Hand, D., Mannila, H., Smyth, P., 2001. Principles of Data Mining. London, England: MIT Press.
14. Handen, L., 2000. Putting CRM to work: The rise of the relationship. In S.A. Brown (Ed.), Customer relationship management: A Strategic imperative in the world of e-Business, 7–18, Toronto, Wiley.
15. Iannotte, D., 2000. Up-front analysis helps control losses. National Underwriter, 104(32), 23.
16. Kaastra, I., Body, M., 1996. Designing a neural network for forecasting financial and economic time series. Neurocomputing, 10, 215–236.
17. Kennedy, R.L., Lee, Y., Van Roy, B., Reed C.D., Lippman, R.P., 1997. Solving data mining problems through pattern recognition, New York, Wiley.
18. Kohonen, T., Hynninen, J., Kangas, J., Laaksonen, J., 1996. SOM-PAK: The Self-organizing Map Program Package. Technical Report A31, Helsinki University of Technology, Laboratory of Computer and Information Science, FIN-02150.
19. Lee, T.H., Jung, S., 1999. Forecasting credit worthiness: Logistic vs. artificial neural net. The Journal of Business Forecasting, 18(4), 28–30.
20. Milley, A., 2000. Healthcare and data mining. Health Management Technology, 21(8), 44–45.
21. Moody, J., Utans, J., 1995. Architecture selection strategies for neural networks application to corporate bound rating, in: A. Refenes (Ed.), Neural Networks in the Capital Markets, Wiley, Chichester, 277–300.
22. Natter, M., 1999. Conditional market segmentation by neural networks: a Monte Carlo Study. Journal of Retailing and Consumer Services, 6(4), 237-2-48.
23. Naumann, E., 1994. Creating Customer Value: The Path to sustainable competitive advantage. Thomson Executive Press, Cincinnati, Ohio, International Thomson Publishing.
24. Nicolett, M., Andren, E., Gilbert, M., 2000. Challenges of aggregating and managing catalog content. Gartner Group report, April 12; (Online) available: http://gartner4.gertnerweb.com:80/gg/purchase/0/00/877/63/doc/00087763/.

25. O'Brien, T.V., 1994. Neural nets for direct marketers. Marketing Research, 6(1), 47–49.
26. Parasuraman, A., 1997. Reflections on Gaining Competitive Advantage Through Customer Value. Academy of Marketing Science Journal, 25(2), 154–161.
27. Park, S.C., Piramuthu, S., Shaw, M.J., 2001. Dynamic rule refinement in knowledge-based data mining systems. Decision Support Systems, 31, 205–222.
28. Peppers, D., Rogers, M., 1999. The One to One Manager: Real-World Lessons in Customer Relationship Management. Doubleday, New York.
29. Poli, I., Jones, R.D., 1994. A Neural netmodel for prediction. Journal of the American Statistical Association, 89(425) 117–121.
30. Schalkoff, R.J., 1997. Artificial Neural Networks. McGraw-Hill, New York.
31. Shtup, A., Versano, R., 1999. Estimating the cost of steel pipe bending, a comparison between neural networks and regression analysis. International Journal of Production Economics, 62(3), 201–207.
32. Tillett, L.S., 2000. Banks mine customer data. Internetweek 830, 45–48.
33. Tolle, K.M., Chen, H., Chow, H., 2000. Estimating drug/plasma concentration levels by applying neural networks to pharmacokinetic data sets, Decision Support Systems, Special Issue on Decision Support for Health Care in a New Information Age, 30(2), 139–152.
34. Turban, E., Trippi, R.R., 1996. Neural Network Fundamentals for financial analysts In R.R. Trippi & E. Turban (Eds.), Neural Networks in Finance and Investment, 3–24, New York, McGraw-Hill.
35. Verhoef, P.C., Donkers, B., 2001. Predicting customer potential value an application in the insurance industry. Decision Support Systems, 32, 189–199.
36. Wedel, S., Kamura, W., 1997. Market segmentation: Conceptual and Methodological foundations. Boston, Kluwer.
37. Widrow, B., Rumelhart, D.E., Lehr, M.A., 1994. Neural Networks: Applications in industry, business and science, communications of the ACM, 37, 93–105.
38. Woodruff, R.B., Gardial, S.F., 1996. Know Your Customer: New Approaches to Customer Value and Satisfaction. Blackwell Publishers, Cambridge, MA.
39. Woodruff, R.B., 1997. Customer Value: The Next Source for Competitive Advantage. Journal of the Academy of Marketing Science, 25(2), 139–153.
40. Zhou, Q., 2003. Adaptive Knowledge Discovery Techniques for Data Mining, Ph.D. Thesis, University of Otago, New Zealand.

Sensory Quality Management and Assessment: from Manufacturers to Consumers

Ludovic Koehl*, Xianyi Zeng*, Bin Zhou*,#, and Yongsheng Ding#

*GEMTEX Lab., ENSAIT 9, rue de l'Ermitage, 59070 Roubaix
ludovic.koehl@ensait.fr
#College of Information Sciences and Technology Donghua University,
1882 Yan-An West Road, Shanghai 200051, P.R. China

Abstract. This paper presents an intelligent technique based method for analyzing and interpreting sensory data provided by multiple panels for the evaluation of industrial products. In order to process the uncertainty existing in these sensory data, we first transform all sensory data into fuzzy sets on a unified scale using the 2-tuple fuzzy linguistic model. Based on these normalized data sets, we compute the dissimilarities or distances between different panels and between different evaluation terms used by them, defined according to the degree of consistency of data variation. The obtained distances, expressed with crisp numbers, are turned into fuzzy numbers for a better physical interpretation. Thus, these fuzzy distances permit to characterize in an easier way the evaluation behaviour of each panel and the quality of the evaluation terms used. Also, based on soft computing techniques and the dissimilarity between terms, we develop procedures for interpreting terms of one panel using those of another panel and a model for setting the relationships between the physical product features and the evaluation terms. Then, we introduce a new method to forecast the consumer preference from the sensory evaluation provided by an expert panel. This general approach has been applied to two kinds of industrial products concerning both cosmetic and textile industries.

Key words: Sensory Evaluation, quality, assessment, fuzzy linguistic model, dissimilarity, distance, fuzzy distance, interpretation

1 Introduction

In many industrial sectors such as food, cosmetic, medical, chemical, and textile, sensory evaluation is widely used for determining the quality of end products, solving conflicts between customers and manufacturers, developing new products, and exploiting new markets adapted to the consumer's preference [1, 2, 3]. In the [2], sensory evaluation is defined as a scientific discipline used to evoke, measure, analyze, and interpret reactions to the characteristics of products as they are perceived by the senses of sight, smell, taste, touch,

Ludovic Koehl et al.: *Sensory Quality Management and Assessment: from Manufacturers to Consumers*, Studies in Computational Intelligence (SCI) **5**, 375–373 (2005)
www.springerlink.com

and hearing. In general, sensory evaluation can be described as "under pre-defined conditions, a group of organized individuals evaluate some products with respect to certain given target". Consequently, there are four basic factors in sensory evaluation: evaluation product, evaluation panel, evaluation target and evaluation environment. According to the difference cases of these factors, we can divide sensory evaluation into two levels [4]: (1) design-oriented sensory evaluation; and (2) market-oriented sensory evaluation. Design-Oriented Sensory Evaluation (DOSE) is done by a trained panel composed of experienced experts or consultants inside the enterprise for judging industrial products on a number of analytical and non-hedonic linguistic descriptors in a controlled evaluation environment, such as an evaluation laboratory. The evaluation target of design-oriented sensory evaluation is to obtain the basic sensory attributes of products to improve the quality of product design and development. Market-Oriented Sensory Evaluation (MOSE) is given by untrained consumer panels using analytical and hedonic descriptors according to their preference on the products to be evaluated in an uncontrolled evaluation environment, such as supermarkets. The evaluation target of market-oriented sensory evaluation is to obtain the preference degree of consumers in order to forecast the market reaction to the evaluated product. Afterwards, the key issue is to compute data provided by a DOSE in order to forecast the consumers' preference (B2C: Business to Consumer) assuming the data are obtained for a precise context and/or end-use for the industrial product, since sensory evaluation is context dependant.

Sensory evaluation of industrial products leads to a set of linguistic terms, named subjective evaluation, strongly related to consumer's preference but difficult to be normalized due to their uncertainty and imprecision (see Fig. 1). As such, this evaluation restricts the scientific understanding of product characteristics for those who wish to design high quality product by engineering means. Hence, a great number of researchers tried to develop objective evaluation systems by physical measurements in order to replace sensory evaluation, e.g. [5, 6]. In practice, these objective evaluation systems are often expensive and lead to precise numerical data describing indirectly products but their

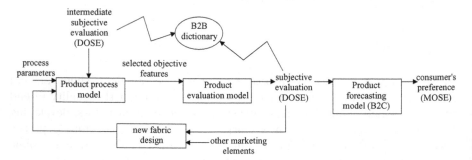

Fig. 1. Different steps for the industrial product sensory evaluation

interpretation for the product quality related to consumer's preference has to be exploited. In this chapter (Sect. 4.4), we present a fuzzy based method for modeling the relationships between the sensory evaluation provided by panels and the physical features measured on appropriate apparatus, which is called objective evaluation. The published model is named 'Product evaluation model'. Compared with physical measures, sensory evaluation as a result of a human measurement is more efficient for quality determination and it can not be, for a long term, completely replaced by objective evaluation. This method can be applied to set up the product process model (see Fig. 1).

In sensory evaluation, the main difficulties can be summarized as follows:

1. For an individual, the evaluation of a sample (in numerical score or linguistic expression) gives a relative result depending on the comparison with the other samples. This score is significant only for one specific collection of products and for one particular individual. It is not normalized in a general background.
2. The terms used by different individuals in an evaluation are not normalized. Even if they use a common term, its significance is not necessarily the same for them.
3. The scales and the upper and lower bounds used by different individuals are often different, which should be unified to the same scale so that the aggregation of all sensory data can be done.

In our previous work [7], we propose an approach based on a linguistic 2-tuple model [8] for the formalization and the analysis of sensory data. This approach permits to initially normalize and aggregate sensory data inside each panel, i.e. a group of evaluators and compute the dissimilarities or distances between different panels and between different terms used according to the degree of consistency of relative data variation. The quality of each panel and evaluation terms can be estimated from these dissimilarity criteria. However, the physical meaning of the corresponding results is not easy to be interpreted and distances for different panels are not easy to be compared because the definition of the distance is related to a specific evaluation space, whose number of evaluation terms is not the same with respect to another evaluation space.

We propose here two procedures in order to provide a systematic interpretation to sensory data obtained by different panels and to forecast the consumer preference using a training set of products with known DOSE. The first procedure permits to interpret the values of the dissimilarities between panels and between evaluation terms. In this procedure, each evaluation score is generated by a random variable distributed between its lower and upper bounds uniformly and then the scores of each panel on all terms for all products to be evaluated constitute a random evaluation matrix. The statistical distribution of the dissimilarity between two panels or between two terms can be obtained from the corresponding random evaluation matrices using the equations given in Sect. 4.1. For specific panels or terms, the fuzzy values or linguistic values of distances between them can be calculated according to these statistical distributions. This interpretation will be very important in the understanding of the behavior of panels and terms used in the evaluation.

The second procedure permits to interpret the relationship between terms used by different panels. This relationship is recurrently determined using a genetic algorithm with penalty strategy. It can be considered as a dictionary for the understanding between different panels. Using this dictionary, an evaluation term used by one panel can be transformed into one or several terms used by another panel. It will be very helpful for solving commercial conflicts between producers and consumers at the level of understanding of evaluation terms (B2B: Business to Business). The general proposed approach tries to allow manufacturers to reduce cost and time for designing new products and thus to become more reactive and competitive for the market demands and requirements. Using suitable tools, the manufacturers will be able to predict the ability for a product to become a success on a specified market. The final challenge is to be able to tune the process parameter for producing the appropriate product which will fit the market demand. Additional components, such as marketing elements (price, retailer location, and so on, ...), have also to be taken into account.

In order to illustrate the effectiveness of our proposed approach, we apply it to sensory data provided by two sets of industrial products. The first set corresponds to fabrics designed for apparel: T-shirts. The hand evaluation for those 43 knitted cotton samples are obtained from 4 sensory panels in France and China. Each panel uses its own terms and evaluation bounds different from the others. Based on the proposed method, we compare the behaviors on fabric hand evaluation between textile professionals and students and between French and Chinese consumers in order to make the adaptive design of textile products to consumer's preference. The second set of data is related to the cosmetic industry. It includes 8 lotions with varying performance according to their interaction with the skin or the human feeling they express at the first contact.

2 Description and Formalization of Sensory Data

The concepts of sensory evaluation used in this chapter can be formalized as follows.

$P = \{P_1, P_2, \ldots, P_r\}$: the set of r panels, each panel, $P_i = \{I_{i1}, I_{i2}, \ldots, I_{i,h(i)}\}$, being composed of $h(i)$ individuals evaluating the hand feeling of fabric samples.

$A = \{a_{ik} | i = 1, 2, \ldots, r; \; k = 1, 2, \ldots, m(i)\}$: the set of linguistic terms used for the fabric hand evaluation. For the panel P_i, it uses $m(i)$ terms, i.e. $A_i = \{a_{i1}, a_{i2}, \ldots, a_{i,m(i)}\}$. For one or several panels, different linguistic terms can be correlated between them, but in general they can never be replaced one by another. It is also possible for two terms used by different panels to be identical, but they are not necessarily equivalent at semantic level.

$T = \{t_1, t_2, \ldots, t_w, \ldots, t_n\}$: the set of n industrial products to be evaluated. The relationships among different terms of the evaluation, the behaviors of different panel members can be studied from these samples.

$E_i = \{A_i; E_{i1}, E_{i2}, \ldots, E_{i,h(i)}\}$: the evaluation space for the panel $P_i/(i \in \{1, 2, \ldots, r\})$.

E_{ij}: the evaluation matrix $n \times m(i)$ of the individual I_{ij}. Each element of this matrix, $e_{ij}(k, 1)$ represents a relative numerical score or a granular linguistic expression given by I_{ij} for evaluating the sample t_k on the term $a_{il}/ (k \in \{1, 2, \ldots, n\}, l \in \{1, 2, \ldots, m(i)\})$. It is obtained by a classification procedure for the whole samples of T.

The sensory data, provided by the consumers, are obtained through a survey and the consumers' panel can be considered as a free profiling panel. The consumers fill a questionnaire form. For each question, they select the most appropriate grade (intensity) according to their feeling. For example, if the question about the product concerns its 'softness', the consumer's answer can be chosen from the following grades of softness: {not Soft, slightly Soft, Soft, very Soft, extremely soft}. Since the survey is repeated for many people, the consumer evaluation matrix is computed so that each element denotes the population percentage who thinks the product satisfies the considered grade compared to the term and for all the products of T. Thus, a consumer panel (MOSE) can be regarded as a regular panel of P, but with only one member $(h(i) = 1)$, with a number of terms that corresponds to the number of terms used in the survey and with an evaluation matrix (scores) which corresponds to the population percentage. We also define the vector $X = (x_1, \ldots, x_w, \ldots, x_n)^T$ which concerns the consumers' preference about all the products which belong to T. Through a survey, they express their preference for one product over the all set. Thus, the component x_w of X corresponds to the consumers' rate who appreciate the w-th product over the $(n-1)$-th other products. As a mother of fact, the more consumers are enrolled in the survey, the more precise is the market preference evaluation. The sum for all the $x'_w s$, as the expression of the consumers rate who rank the w-th product at the top level, is then equal to 1: $\sum_{w=1}^{n} x_w = 1$

3 Linguistic 2-Tuple for Finding the Optimal Unified Scale

In sensory evaluation, results given by different individuals on different attributes or terms may have different scales. This is because the sensitivity of each individual to the samples to be evaluated, strongly related to his personal experience and the corresponding experimental conditions, is often different from others. Moreover, these sensory data may be in a numerical form or a granular linguistic form. So it is necessary to develop a suitable unified scale in order to normalize and aggregate these data.

The 2-tuple fuzzy linguistic model [8] can be used for unifying multi-granular linguistic information without loss of information. Using this model, sensory data provided by different individuals on different attributes can be normalized on the common optimal scale. However, in [8], the transformation is carried out between levels of a linguistic hierarchy only. Here, we generalize the 2-tuple model to be used in transforming among arbitrary scales and several quantitative criteria permitting to select the most suitable scale for all individuals and all attributes or terms.

3.1 2-Tuple Fuzzy Linguistic Model

In the panel P_i, for each individual I_{ij} and each term a_{il}, the corresponding sensory data varying between 0 and $g(i,j,l) = \max\{e_{ij}(k,l)|k = 1,2,\ldots,n\}$ can be transformed into a fuzzy set of $g(i,j,l) + 1$ modalities denoted by $U_{ijl} = \{u_1, u_2, \ldots, u_{g(i,j,l)}\}$ as shown in Fig. 2.

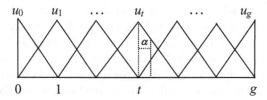

Fig. 2. Fuzzy set U_{ijl} composed of $g + 1$ modalities

For simplicity, $g(i,j,l)$ is denoted by g for unambiguous cases. We consider that any evaluation score of I_{ij} for the term a_{il} is included between 0 and g and it can be represented by a 2-tuple (u_t^g, α^g) with $\alpha^g \in [-0.5, 0.5]$.

For any $\beta \in [0, g]$, the 2-tuple that expresses the equivalent information is obtained using the following function:

$$\Delta(\beta) = \begin{cases} u_t & t = \text{round}(\beta) . \\ \alpha = \beta - t\alpha \in [-0.5, 0.5) \end{cases} \text{ and } \Delta^{-1}(u_t, \alpha) = t + \alpha = \beta .$$

For the panel P_i, the evaluation results of the individuals I_{ij}'s can be aggregated by transforming all the corresponding fuzzy sets to be on a unified scale.

Let $ug(i,l)$ be the value of the unified scale for all the individuals of P_i on the term a_{il}. For each individual I_{ij}, any evaluation score (u_t^g, α^g) on this term can be transformed into a new 2-tuples:

$$(u_s^{ug}, \alpha^{ug}) = \Delta\left(\frac{\Delta^{-1}(u_t^g, \alpha^g) \cdot ug}{g}\right) \tag{1}$$

This transformation can be denoted by the function $s = Tr(t, g, ug)$.

3.2 Obtaining the Optimal Common Scale

In order to best aggregate the sensory evaluation of multiple individuals in the same panel, we have to find an optimal value of the unified scale ug for all the individuals of P_i.

For the panel P_i, its optimal unified scale can be calculated according to the two following principles:

1. The sensory data given by the individuals I_{ij} should cover all the modalities of the unified scale, i.e., any $u_s^{ug}(s \in \{0, 1, 2, \ldots, ug\})$ should correspond to at least one data.
2. The variation or the trend of the sensory data should not change very much with the transformation of the scale.

The sensory data of I_{ij} for evaluating n samples on the term a_{il} before the transformation are $\{e_{ij}(1,l), e_{ij}(2,l), \ldots, e_{ij}(n,l)\}$. After the transformation, these data become $\{s_{ij}(1,l), s_{ij}(2,l), \ldots, s_{ij}(n,l)\}$, where

$$s_{ij}(k,l) = Tr(e_{ij}(k,l), g, ug) \quad \text{for } k = 1, 2, \ldots, n.$$

According to the first principle, we first calculate the number of data for each modality q of the unified scale ug, i.e.,

$$N\_\text{mod}_i(l,q) = \sum_{j=1}^{h(i)} \sum_{k=1}^{n} equ(s_{ij}(k,l), q),$$

with

$$equ(p,q) = \begin{cases} 1 & \text{if } p = q \\ 0 & \text{otherwise} \end{cases}$$

The criterion of coverage of the unified scale is then defined by

$$\text{Cover}_i(l) = \min \{N\_\text{mod}_i(l,q) \,|\, q = 0, 1, 2, \ldots, ug\}.$$

According to this criterion, ug should be selected so that $\text{Cover}_i(l)$ is as big as possible. If this value is 0, it means that there exists at least one non-significant modality u_s^{ug} on the unified scale ug.

According to the second principle, the difference of the trend between two data sets $\{e_{ij}(1,l), e_{ij}(2,l), \ldots, e_{ij}(n,l)\}$ and $\{s_{ij}(1,l), s_{ij}(2,l), \ldots, s_{ij}(n,l)\}$ should be as small as possible. So, the corresponding criterion is defined by

$$\text{Trend}_i(l) = \min \{\text{trend}_{ij}(l) \,|\, j = 1, \ldots, h(i)\},$$

with

$$\text{trend}_{ij}(l) = \frac{2}{n(n-1)} \sum_{k_1 < k_2} c_{ij}(k_1, k_2, l),$$

and

$$c_{ij}(k_1, k_2, l) = \begin{cases} 1 & \text{if } (e_{ij}(k_1, l) - e_{ij}(k_2, l))(s_{ij}(k_1, l) - s_{ij}(k_2, l)) > 0 \\ 0 & \text{otherwise} \end{cases}.$$

According to this criterion, ug should be selected so that the value of $\text{Trend}_i(l)$ is as big as possible. If two data sets $\{e_{ij}(1, l), e_{ij}(2, l), \ldots, e_{ij}(n, l)\}$ and $\{s_{ij}(1, l), s_{ij}(2, l), \ldots, s_{ij}(n, l)\}$ for any individual I_{ij} vary completely with the same trend, the value of $\text{Trend}_i(l)$ will be maximized and there is no information lost in the data set $\{s_{ij}(k, l)\}$ on the unified scale ug.

The optimal value of ug can be obtained by maximizing the linear combination of these two criteria as follows:

$$\max \{\text{Cover}_i(l) + \rho \cdot \text{Trend}_i(l)\} \tag{2}$$

where ρ is a positive constant adjusting the ratio of these two criteria.

As the optimal value of ug is obtained, for each individual I_{ij} and each term a_{il}, the optimal unified evaluation score (u_s^{og}, α^{og}) can be obtained by transforming sensory data into a new 2-tuples by using (1). The sensory data for all individuals of this panel can be aggregated on this optimal scale.

3.3 Aggregating Sensory Data on the Desired Domain

On the common optimal scale, the transformed evaluation scores on term a_{il} for all individuals can be aggregated using an averaging operator. The aggregated evaluation result of P_i for one sample on the term a_{il} can be calculated by

$$(\bar{u}_s^{ug}, \bar{\alpha}^{ug}) = \Delta \left(\frac{\sum_{j=1}^{h(i)} \Delta^{-1}(u_{s_j}^{ug}, \alpha_{s_j}^{ug})}{h(i)} \right)$$

where $(u_{s_j}^{ug}, \alpha_{s_j}^{ug})$ is the transformed 2-tuples of the evaluation scores of the individual I_{ij} for the same sample on the term a_{il}.

In the same way, all the aggregated evaluation scores of P_i for different terms of A_i are transformed to be on the unique unified scale and can be expressed by a matrix of $n \times m(i)$ 2-tuples, denoted by S_i (evaluation space for P_i). Each element of S_i is a 2-tuples denoted by $(u_{s_i(k,l)}, \alpha_{s_i(k,l)})$ for $k = 1, 2, \ldots, n$ and $l = 1, 2, \ldots, m(i)$. For simplicity, this 2-tuples is replaced by $s_i(k, l)$ if the weak influence of $\alpha_{s_i(k,l)}$ can be neglected. The matrix $S_i = (S_{i1} \ S_{i2} \cdots S_{i,n})^T$ includes n vectors, each of them representing the evaluation results for one sample. These vectors will be used in the next section for the analysis and interpretation of panels and term.

4 Analysis and Interpretation of Sensory Data

In this section, we first define a number of criteria, according to the normalized matrices $S_i = (s_i(k, l)), i \in \{1, \ldots, r\}$ obtained from Sect. 3, in order to analyze the performance of the evaluators and the used description terms.

4.1 Dissimilarity between Individuals and between Terms

The sensory data of two panels P_a and P_b constitute two evaluation spaces S_a and S_b. The dissimilarity between P_a and P_b cannot be defined using classical methods, which compute distances between the vectors S_{ak} and S_{bk} ($k \in \{1, \ldots, n\}$) because these two vectors are not in the same space. So a new dissimilarity criterion between two individuals P_a and P_b has been defined in [7].

In this definition, the dissimilarity criterion takes into account the degree of consistency of relative variations of two different sensory data sets. If the internal relative variations of these two data sets are close each other, and then the dissimilarity between the corresponding panels is small. Otherwise, this dissimilarity is great. Formally, this dissimilarity is defined by

$$D_{ab} = \frac{2}{n(n-1)} \sum_{i<j} d_{ab}(i,j) \tag{3}$$

It depends on the following elements:

1. The dissimilarity between P_a and P_b related to the relative variation between fabric samples t_i and t_j: $d_{ab}(i,j) = |vr_a(i,j) - vr_b(i,j)|$.
2. The relative variations between t_i and t_j for P_a and P_b:

$$vr_a(i,j) = \frac{1}{\sqrt{m(a)}} \|S_{ai} - S_{aj}\|$$

$$vr_b(i,j) = \frac{1}{\sqrt{m(b)}} \|S_{bi} - S_{bj}\|.$$

The definition of D_{ab} permits to compare, between these two panels, the relative variations on the samples of T. The dissimilarity between two panels reaches its minimum only when the internal variations of their sensory data are identical.

The dissimilarity D_{ab} can be considered as a distance between P_a and P_b because it satisfies the following three conditions:

(i) $D_{aa} = 0$
(ii) $D_{ab} = D_{ba}$
(iii) $D_{ab} + D_{bc} \geq D_{ac}$

The two first conditions (i)&(ii) can be easily proved from the definition of the dissimilarity. The proof of the third condition (iii) is given as follows.

If $vr_a(i,j) \geq vr_b(i,j) \geq vr_c(i,j)$, then $d_{ab}(i,j) + d_{bc}(i,j) = vr_a(i,j) - vr_c(i,j) = d_{ac}(i,j)$.

If $vr_a(i,j) \geq vr_c(i,j) \geq vr_b(i,j)$, then $d_{ab}(i,j) + d_{bc}(i,j) = vr_a(i,j) - vr_b(i,j) + vr_c(i,j) - vr_b(i,j) \geq vr_a(i,j) - vr_c(i,j) = d_{ac}(i,j)$.

For any other conditions, the inequality $d_{ab}(i,j) + d_{bc}(i,j) \geq d_{ac}(i,j)$ also holds and then we have $D_{ab} + D_{bc} \geq D_{ac}$. So, the three conditions of distance are completely satisfied by the dissimilarity defined previously.

Another criterion is developed in order to compare two panels according to the sensitivity of data for the evaluation of the samples of T [7]. The sensitivity of P_a is defined by

$$SSB_a = \frac{2}{n(n-1)} \sum_{i<j} vr_a(i,j) \tag{4}$$

where, $vr_a(i,j)$ characterizes the relative variation of the sensory data given by P_a from the sample t_i to t_j. If the value of SSB_a is bigger than that of SSB_b, then we consider that P_a is more sensitive to the samples of T than P_b. However, this does not mean that P_a is more efficient than P_b.

In the same way, we also define the dissimilarity or distance between terms used by the same panel and by different panels [7]. This criterion permits to study the redundancy of the terms used by each panel. In general, the bigger the dissimilarity between any two terms used by a panel is, the more efficient the evaluation results are.

4.2 Fuzzy Dissimilarity or Fuzzy Distance

The criteria of dissimilarity and sensitivity for panels and terms defined in Sect. 4.1 are significant only for comparison because we do not know how to physically interpret the absolute values of these criteria. We do not know if a slight variation of such a criterion is physically important or not. In order to give a physical interpretation to the results calculated from dissimilarity criteria and sensitivity criteria, we transform these numerical values into fuzzy numbers, whose membership functions are generated according to the probability density distributions of the corresponding random matrices. The detailed procedure is given as follows and, according to the above section, P_a and P_b denote two panels a and b:

Step 1: For fixed values n, $m(a)$ and $m(b)$, generating two random matrices S_a (dimension: $n \times m(a)$) and S_b (dimension: $n \times m(b)$), whose elements obey the uniform distribution between lower and upper bounds of normalized evaluation scores, i.e. 0 and ug.

Step 2: Computing the values of dissimilarity and sensitivity D_{ab}, SSB_a and SSB_b according to the equations in Sect. 4.1.

Step 3: Repeat Step 1 and Step 2 several times in order to obtain the probability density distributions for D_{ab}, SSB_a and SSB_b (see Fig. 3).

Step 4: We then divide equally the area of each distribution into 5 parts. According to these divided areas, we generate 5 fuzzy sub-sets for each of D_{ab}, SSB_a and SSB_b: {*very small, small, medium, large, very large*}. The corresponding membership functions can be determined from these 5 fuzzy numbers.

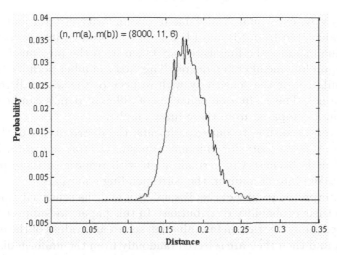

Fig. 3. Distance Distribution Function (D_{ab}) with uniformly selected random evaluation scores

Figure 3 gives the probability distribution for the dissimilarity D_{ab} between two panels P_a and P_b using 11 and 6 terms respectively as Fig. 4 shows how we turn a crisp number of dissimilarity (or sensitivity) into a fuzzy one, which makes easier the understanding. The membership functions corresponding to the five fuzzy values equally dividing the area of this distribution are given in Fig. 4. From these membership functions, we can see that the dissimilarity D_{ab} is sensitive only in the interval of [0.11, 0.30], in which three fuzzy values small (S), medium (M) and large (L) are asymmetrically distributed. A value of D_{ab} smaller than 0.161 is considered as very small (VS) and a value of D_{ab} larger than 0.207 as very large (VL).

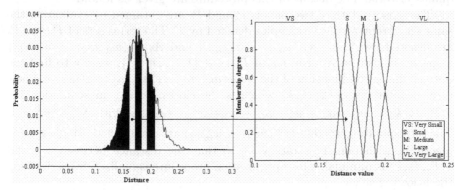

Fig. 4. From a distance crisp number to a fuzzy distance number

In this way, each numerical value of dissimilarity criteria and sensitivity criteria, calculated from (3) and (4) can be transformed into a fuzzy number whose value includes the linguistic part taken from the previous 5 terms and the corresponding membership degree. This fuzzy number permits to interpret the dissimilarity or the sensitivity with respect to the whole distribution of random values. The evaluation behaviors of different panels can be effectively analyzed and compared from these fuzzy numbers.

Moreover, according to our experiments, the distributions of the dissimilarity and the sensitivity for different values of $n, m(a)$ and $m(b)$ are rather similar. This is because the normalization with respect to these parameters has been taken into account in the corresponding equations.

The interpreted results of the dissimilarity and the sensitivity are strongly related to their probability distributions. In this paper, we suppose that there does not exist any restriction in evaluation scores and values of the elements of S_a and S_b and then they are selected randomly from the uniform distribution. If some restriction exists in evaluation scores, the probability distributions of the dissimilarity and the sensitivity will change accordingly and new membership functions of the corresponding fuzzy values should be generated in order to guarantee the correctness of the interpreted results.

This principle of interpretation using fuzzy distances can also be applied to the analysis of terms used by the same panel and different panels.

4.3 Relationships Settings between Linguistic Terms

In industrial applications, there exists a strong need for interpreting evaluation terms of one panel using those of another panel. In this paper, we propose a genetic algorithm based procedure to do so. This procedure can be considered as a dictionary of terms for different panels and it is helpful for solving commercial conflicts between sensory panels related to the understanding of quality criteria. The details of this procedure are given as follows.

The sensory data of two panels P_a and P_b are obtained by evaluating the same set of representative samples denoted by T. The terms sets of P_a and P_b are denoted by $A_a = \{a_{a1}, a_{a2}, \ldots, a_{a,m(a)}\}$ and $A_b = \{a_{b1}, a_{b2}, \ldots, a_{b,m(b)}\}$ respectively. For each term a_{ak} of P_a ($k \in \{1, \ldots, m(a)\}$), we try to find the optimal linear combination of the terms $a_{b1}, a_{b2}, \ldots, a_{b,m(b)}$ to generate a new term denoted by $a(P_a, P_b, k)$ which is the closest to a_{ak} in semantics, i.e.
$$a(P_a, P_b, k) = w_1^k \cdot a_{b1} + w_2^k \cdot a_{b2} + \cdots + w_{m(b)}^k \cdot a_{b,m(b)} \text{ with } \sum_{i=1}^{m(b)} w_i^k = 1.$$
The corresponding weights $\{w_1^k, w_2^k, \ldots, w_{m(b)}^k\}$ are determined using a genetic algorithm with penalty strategy [9] so that the distance between a_{ak} and $a(P_a, P_b, k)$ is minimal. This optimization procedure is realized by performing the following steps:

Step 1: Finding the term of P_b the closest to a_{ak}

Computing the distance between a_{ak} and each term of P_b: $a_{b1}, a_{b2}, \ldots, a_{b,m(b)}$ using the method presented in Sect. 4.1. The corresponding values are denoted by $\{D_1, D_2, \ldots, D_{m(b)}\}$. Selecting the term a_{bx} so that D_x (the distance between a_{bx} and a_{ak}) is the smallest of $\{D_1, D_2, \ldots, D_{m(b)}\}$.

Step 2: Building the support set of terms of P_b related to a_{ak}

Building a set of new terms of P_b: $\{a_{b1}', \ldots, a_{b,x-1}', a_{b,x+1}', \ldots, a_{b,m(b)}'\}$ by adding the normalized evaluation scores of a_{bx} for all samples of T to those of each term a_{bi} ($i \in \{1, \ldots, b(m)\}$ and $i \neq x$). Computing the distance between a_{ak} and each of these new terms and denoting the corresponding results as $\{D_1', \ldots, D_{x-1}', D_{x+1}' \cdots D_{m(b)}'\}$. If $D_j' < D_x$ ($j \in \{1, \ldots, m(b)\}$ and $j \neq x$), we consider that a_{bj} has a contribution to the decrease of the distance between a_{bx} and a_{ak} and then the corresponding weight $w_j^k > 0$. If $D_i' > D_x$, we consider that a_{bj} has no contribution to the decrease of the distance between a_{bx} and a_{ak} and then the corresponding weight $w_j^k = 0$. Therefore, we define the support set of terms of P_b related to a_{ak} by $A_b^k = \{a_{bh_1}, a_{bh_2}, \ldots, a_{bh_q}\}$ with $a_{bh_1} = a_{bx}$ and $a_{bh_j} \in A_b$ and $w_{h_j}^k > 0$ for $j \in \{2, \ldots, q\}$.

In this case, all terms of P_b satisfying the condition $w_j^k = 0$ are deleted from A_b and only the relevant terms having contributions to the construction of the new term $a(P_a, P_b, k)$ are preserved in the support set A_b^k. This step can largely reduce the computing complexity of Step 3.

Step 3: Building the term $a(P_a, P_b, k)$ that is the closest to a_{ak}

By applying a genetic algorithm, we compute the optimal weights of the relevant terms of A_b^k in order to construct the term $a(P_a, P_b, k)$ that is the closest to a_{ak} in semantics. It is an optimization problem with constraints because the sum of the weights should be equal to 1 and each weight should be no smaller than 0. In this case, we use the penalty strategy [9] in the genetic algorithm. The detail for this algorithm is given as follows.

Procedure for computing the weights of the relevant terms using a Genetic Algorithm:

Begin

Coding and initializing the population of weights $W_b^k(t) = (w_{b h_1}^k(t) w_{h_2}^k(t) \cdots w_{h_q}^k(t))$ ($t \leftarrow 0$)

IF $W_b^k(t)$ satisfies the constraints ($\sum_{j=1}^q w_{h_j}^k = 1$ and $w_{h_j}^k \geq 0$ for $j \in \{1, \ldots, q\}$)

    THEN Evaluate the fitness by fitness function A
    ELSE Evaluate the fitness by fitness function B
End_IF

While Not satisfying stop conditions

Do
  Random Selection Operation
  Crossover Operation

Mutation Operation

Updating $W_b^k(t)$ $(t \leftarrow t+1)$ for generating the next population of weights

IF $W_b^k(t)$ satisfies the constraints

THEN Evaluate the fitness by fitness function A

ELSE Evaluate the fitness by fitness function B

End_IF

End_While

End

The fitness function A is defined by $D_{ab}(a(P_a, P_b, k), a_{ak}) = \frac{2}{n(n-1)} \sum_{i<j} d_{ab}(a(P_a, P_b, k), a_{ak}, i, j)$ with $a(P_a, P_b, k) = (1 - \sum_{j=2}^{q} w_j^k) \cdot a_{bh_1} + \sum_{j=2}^{q}(w_j^k \cdot a_{bh_j})$ $\left(a_{bh_j} \in A_b^k\right)$.

under the constraints $\begin{cases} 1 - \sum_{j=2}^{q} w_j^k > 0 \\ w_j^k \geq 0 \quad (2 \leq j \leq q) \end{cases}$.

The concepts D_{ab} and d_{ab} are computed according to the definitions in Sect. 4.1.

We also define the fitness function B with penalty factor ρ as follows:

$$D_{ab}(a(P_a, P_b, k), a_{ak}) = \frac{2}{n(n-1)} \sum_{i<j} d_{ab}(a(P_a, P_b, k), a_{ak}, i, j) + \rho$$

where $\rho = \gamma \sum_{j=2}^{q} w_j^k$. γ is the parameter of penalty.

Figure 5 gives a practical example which recurrently computes the weights of the relevant terms of P_b related to a_{ak} using the genetic algorithm running for 100 generations. The evolution of the best value and the averaged value of the fitness function shows that the algorithm converges to its optimum after 20 populations. The best linear combination of the terms of P_b related to the term a_{ak} of P_a is then obtained.

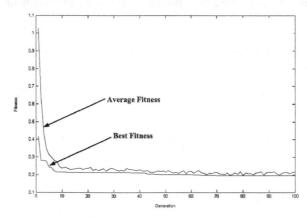

Fig. 5. Evolution of the fitness function for computing the weights (one example)

4.4 Forecasting the Consumers' Preference

Usually, the number of descriptors used by experts is greater than those of consumers. This lies in the fact the consumers' knowledge is more basic. But even if their experience on the product quality evaluation is poorer, their feeling can be included in the results of experts or trained panels. Thus, it means that relationships can be found between linguistic terms used by experts or trained panels and terms used by the consumers. Since the consumer evaluation matrix is computed so that each element denotes the population percentage, we need to adapt the relative variation between fabric samples t_i and t_j to compute the distances between DOSE and MOSE panels. The relative variation $vr_{\mathrm{MOSE}}(i, j)$ is now estimated by considering the averaged scores, computed with the grades values for a specific descriptor (see Table 1). For each consumer's term and each grade, we assign an absolute score according to the number of grades and the range $[0, 1]$, since the sensory data are all normalized between 0 and 1. The average is then obtained by multiplying the population rate and the absolute score. For the considered term, we obtain a score which tends toward 1, if the major part of the consumers agrees with the descriptor. Then, using the procedure described in Sect. 4.3, we compute the relationships between the consumers' sensory evaluation and the terms used by DOSE panels. In the same way, we reproduce the same technique for extracting the relationships between the consumers' preference (vector X) and the DOSE-linguistic descriptors. These last relationships lead to a vector of weights, which characterizes the optimal linear combination of the $m(i)$ DOSE-terms for explaining the consumers' preference X: $(w_1^k, w_2^k, \ldots, w_{m(i)}^k)^T$ where k equals 1, because the consumers' preference is there considered as only one descriptor and no more as a set of linguistic terms. Our target is to estimate what could be the consumers' preference for an additional $(n + 1)$-th product when its sensory evaluation is performed by the DOSE. Assuming the x_i's vary in the range $[0, 1]$, to find the preference x_{n+1}, we construct an array of all the preference vectors with an additional component which lies in $[0, 1]$ and represents the possible preference. Then, we compute all the distances between the preference vector with additional components and the optimal linear combination of DOSE-terms for the new product. The predicted preference corresponds to the minimum of those distances, because it is the nearest distance between the optimal linear combination which models the preference and the forecasted preference. The Fig. 6 shows the curve obtained with varying consumers' preference. In this example, the nearest distance between the DOSE and the preference is about 13, which means that 13% of the consumers are expected to appreciate the product.

5 Product Evaluation and Process Models

In the product evaluation model (see Fig. 1), we have numerical input variables (selected physical features). Those input parameters, measured with

Table 1. Example of consumer sensory evaluation for a set of samples T

		Not 'soft'	Very Little 'soft'	Slightly 'soft'	'soft'	Very 'soft'	Extremely 'soft'
	Absolute scores	0	0.2	0.4	0.6	0.8	1
Product #1	Population rate (%)	9	16	18	27	27	3
	Score ([0, 1])	0.09	0.16	0.18	0.27	0.27	0.03
average = 0.09x0 + 0.16x0.2 + 0.18x0.4 + 0.27x0.6 + 0.27x 0.8 + 0.03x1 = **0.512**							
...	...			...			
Product #n	Population rate (%)	1	7	5	45	32	10
	Score ([0, 1])	0.01	0.07	0.05	0.45	0.32	0.10
average = 0.01x0 + 0.07x0.2 + 0.05x0.4 + 0.45x0.6 + 0.32x 0.8 + 0.10x1 = **0.66**							

appropriate devices, are chosen for their influence on the product quality. The Abe's method is then applied for extracting fuzzy rules directly from these numerical data [10] and to build the model between the physical measurements on the products and the sensory evaluation. This method permits to obtain a good compromise between precision, robustness and interpretability.

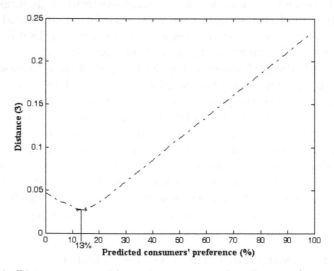

Fig. 6. Distance curve with varying consumers' preferences (percentage)

The Abe's method is briefly described below. At first, the universe of the output is divided into a number of intervals. By putting the input data into different classes according to the output intervals, we define two kinds of regions in the input space: activation hyperboxes and inhibition hyperboxes. For a given class of input data, an activation hyperbox contains all data belonging to this class and an inhibition hyperbox inhibits the existence of data for this class. Inhibition hyperboxes can be located by finding overlaps between neighboring activation hyperboxes. In these located inhibition hyperboxes can be defined new activation and inhibition hyperboxes for the next level. This procedure is repeated until overlaps are solved (see Fig. 7).

In this procedure, the fuzzy rules are defined by activation and inhibition hyperboxes (see Fig. 7). We select a suitable Gaussian function as membership function and calculate the output value using Sugeno's defuzzification method.

By comparison with other methods of fuzzy rules extraction which assume the space of input variables is partitioned into a number of fixed regions, this procedure generates more accurate fuzzy partition and fuzzy rules.

In practice, the fuzzy rules extracted using Abe's method are less efficient when the number of input variables is too great with respect to the quantity of available data. It is the case in many industrial applications. For solving this problem, we use Principle Component Analysis (PCA) [11] to reduce the number of input variables before starting the procedure of fuzzy rules extraction. By using this technique, the lower dimensional input space used in the Abe's method is obtained from the projection of the original high dimensional space. Its principle is given below.

PCA performs a linear transformation of an input variable vector for representing all original data in a lower-dimensional space with minimal information lost.

The principle of the product evaluation model for one descriptor is shown in Fig. 7. A very slight model internal parameters adjustment is required for any other descriptor and the same procedure can be repeated in order to build the product process model described in Fig. 1.

In the example described below (Sect. 6), experts obtain 7 levels for "Soft" and for the whole fabric samples. Their evaluation scores are taken as output data of the model. The data measured on 11 selected physical features are taken as input data after the projection of the original space into the two-dimensional subspace using PCA. Fuzzy rules are then directly extracted from these input-output learning data (see Fig. 8).

6 Application

In order to highlight and to illustrate the effectiveness of the above approach, we apply it to sensory data on fabric hand evaluation provided by 2 sensory panels in France, including a fashion design expert (FE) and a group of trained

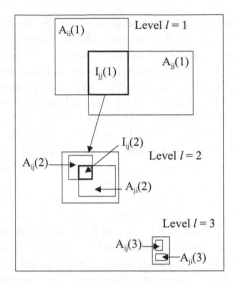

Fig. 7. Fuzzy rules extraction by generating Activation and Inhibition hyperboxes in the input space (2-D)

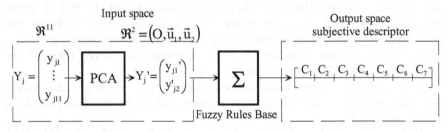

Fig. 8. Product evaluation model for one descriptor with 11 product features

students (FTS) and 2 sensory panels in China, including a group of textile experts (CE) and a group of trained students (CTS). The fabric set is composed of 43 knitted cotton samples produced using 3 different spinning processes. These samples can be then classified into 3 categories: Carded, Combed and Open-End, corresponding to different touch feelings. For FE, 11 evaluation terms have been used and they include "soft", "smooth", "tight", "slippery", "floppy", "compact", "hollow", "pleasant", "fresh", "dense", "flexible". For FTS, they use 4 terms, including "smooth", "slippery", "soft", "tight". For CE and CTS, the terms used are not normalized and they vary with evaluator. In general, each CE uses 6 or 7 terms such as "soft, "slippery", "flexible", "texture", "elasticity", "thickness" and each CTS uses 3 or 4 terms such as "soft", "slippery", "flexible".

6.1 Analysis at the Level of Evaluators

Applying the methods presented in Sects. 4.1 and 4.2 to analyze and interpret the evaluation scores given by these 4 evaluation panels, we obtain the corresponding crisp and fuzzy values of dissimilarity or distance and sensitivity in Table 1.

From Table 2, we can notice that the averaged distances between French and Chinese experts and between professional experts and students are very small. It means that the general evaluation on fabric hand is related to neither the cultural background nor the professional background. Moreover, since the experts' sensitivities are medium or large and the students' sensitivities are very small or small, it means that experts are more sensitive in the evaluation of products.

6.2 Analysis in the Level of Terms

Using the methods in Sects. 4.1 and 4.2, we also calculate the crisp and fuzzy distances between evaluation terms used by each panel. The averaged results are given in Table 3. From Table 3, we can notice that the crisp values of distances between different terms are very large or large for French experts (FE) and very small or small or large for the other panels. This means that French experts define more suitable terms for describing fabric hand and understand better their meaning in fabric hand evaluation. The overlap between two different terms is not important.

The results of crisp and fuzzy distances between different panels on the term "soft" are given in Table 4 and Table 5 respectively.

Table 2. Averaged distances and sensitivities for different panels

			FE	FTS	CE		CTS		
	Crisp#		0	0.1563	0.1284		0.1717		FE
	Fuzzy set		VS	VS	VS		VS		
	Crisp#		0.1563	0	0.1456		0.1692		FTS
Distances	Fuzzy set		VS	VS	VS		VS		
	Crisp#		0.1284	0.1456	0		0.1622		CE
	Fuzzy set		VS	VS	VS		VS		
	Crisp#		0.1717	0.1692	0.1622		0		CTS
	Fuzzy set		VS	VS	VS		VS		
	Crisp#			0.4219	0.3635		0.4187		0.3977
Sensitivities	Fuzzy set	S	0;2833	VS	S	0.5459	VS	0.4900
		M	0;7167		M	0.4531	S	0.5100

Table 4 and Table 5 show that the distances between different panels on the term "soft" are sometimes rather important (M, L, VL) although most of the fuzzy values of these distances are very small. The same phenomenon

Table 3. The averaged crisp and fuzzy distances between terms

PROCESS		FE	CE		FTS	CTS	
Carded	Crisp#	0.2937	0.2546		0.1844	0.289	
	Fuzzy Set	VL	VS		VS	L	0.792
						VL	0.208
Combed	Crisp#	0.2911	0.291		0.1585	0.2869	
	Fuzzy Set	VL	L	0.167	VS	L	0.967
			VL	0.833		VL	0.033
Open-End	Crisp#	0.3708	0.3779		0.2774	0.3289	
	Fuzzy Set	VL	VL		VS	S	0.144
						M	0.856

can be observed for the other terms such as "slippery" and "smooth". This means that one evaluation term is often semantically interpreted in different ways by different panels. This remark has been validated by some industrial companies. In these companies, there exist conflicts between suppliers and consumers on quality criteria expressed in linguistic terms. A dictionary is then needed for the understanding of evaluation terms between different professional populations.

The performance of each evaluator or panel can be characterized by the following criteria: 1) dissimilarity between terms he/she uses; 2) total number of terms used; 3) sensitivity to the products be evaluated and 4) stability or capacity of reproduction of evaluation scores. According to the previous analysis on the sensory evaluation data provided by our panels, we can see that the experts specialized in textile technology, especially the French expert (FE) are more efficient than the other panels in the evaluation of fabric hand. A sensory panel can also be trained according to these 4 criteria.

Table 4. The values of crisp distances between different evaluators on the common term "soft"

Process	Carded			Combed			Open-End		
Evaluators	CE	FTS	CTS	CE	FTS	CTS	CE	FTS	CTS
FE	0.1343	0.2123	0.1743	0.2262	0.2311	0.2661	0.102	0.21	0.1803
CE		0.2061	0.201		0.1325	0.1643		0.1803	0.1973
FTS			0.2189			0.1574			0.2874

6.3 Interpretation of the Relationship between Terms used by Different Panels

We use the method in Sect. 4.3 to interpret each term used by the panel of French trained students (FTS), the panel of Chinese experts (CE) and the

Table 5. The values of fuzzy distances between different evaluators on the common term "soft"

Evaluators	Carded				Combed				Open-End				
	CE	FTS		CTS	CE		FTS		CTS	CE	FTS		CTS	
FE	VS	VS 0.813 / S 0.187		VS	S 0.417 / M 0.583		S 0.417 / M 0.583		VL	VS	VS 0.031 / S 0.969		VS 0.969 / S 0.031	
CE		VS 0.813 / S 0.187		VS			VS		VS		VS 0.969 / S 0.031		VS 0.344 / S 0.656	
FTS				VS 0.125 / S 0.875					VS				L 0.1 / VL 0.9	

panel of Chinese trained students (CTS) by those of the French Expert (FE). For simplicity, we only discuss the case of $P_a = $ FTS and $P_b = $ FE in this section. The 11 terms used by FE corresponds to $A_b = \{a_{b1}, a_{b2}, \ldots, a_{b,11}\}$ and the 4 terms of FTS to $A_a = \{a_{a1}, a_{a2}, a_{a3}, a_{a4}\}$. After applying the genetic algorithm with penalty strategy presented previously, we obtain the optimal linear combination of the terms of FE related to each term of FTS. The corresponding weights w_j^k's of these linear combinations are shown in Table 6.

Table 6. The weights of optimal linear combinations of terms of FE related to those of FTS

Terms Of FTS	Terms of FE										
	Soft	Smooth	Tight	Slippery	Floppy	Compact	Hollow	Pleasant	Fresh	Dense	Flexible
Smooth	0.6	0	0	0	0.1	0	0.3	0	0	0	0
Slippery	1	0	0	0	0	0	0	0	0	0	0
Soft	0.3	0	0	0	0.3	0	0.4	0	0	0	0
Tight	0	0.1	0	0	0	0	0.1	0	0	0.4	0.4

Table 6 permits to interpret the relationship between terms used by FTS and FE. Under this relationship, the terms used by FTS can be approximately expressed by linear combinations of the terms of FE. For example, the term *"soft"* used by FTS can be approximately expressed as a linear combination of three terms of FE: *"soft", "floppy" and "hollow"*, i.e.

$$soft\_FTS \approx 0.33 \cdot soft\_FE + 0.32 \cdot floppy\_FE + 0.35 \cdot hollow\_FE$$

Table 7 gives the distances between the terms of FTS and their corresponding optimal linear combinations of terms of FE, obtained using the procedure

Table 7. Evolution of distances between terms of FTS and FE

Term of FTS	Distance between the term of FTS and the closest term of FE			Distance between the term of FTS and the optimal linear combination of terms of FE	
	Crisp#	*Fuzzy set*		*Crisp#*	*Fuzzy set*
Smooth	0.211	VS	0.431	0.156	VS
		S	0.569		
Slippery	0.161	VS		0.161	VS
Soft	0.212	VS	0.367	0.131	VS
		S	0.638		
Tight	0.237	M	0.501	0.165	VS
		L	0.499		

in Sect. 4.3. These results show that the obtained optimal linear combinations are very close to the terms of FTS related to the original terms of FE. Each term of FTS corresponds to several terms of FE.

For the term "slippery" of FTS, there is only one term in the support set of terms of FE. So, the procedure described in Sect. 4.3 can not decrease the closest distance between terms of FTS and FE. For the other three terms, the procedure implemented in Sect. 4.3 can effectively decrease the closest distance between terms of FTS and FE. From the optimal linear combinations, we obtain new terms of FE much closer to those of FTS than the original terms of FE.

6.4 Interpretation of the Relationship between Terms used by Different Panels

In the modeling of fabric evaluation, we first use Principle Component Analysis (PCA) to obtain the reduced data and then extract fuzzy rules using the Abe's method. The corresponding results are given as follows.

The two components obtained from PCA are $\lambda_1 = 4.43$ and $\lambda_2 = 1.16$ with 46.9% and 16.7% of explanation respectively. For testing the effectiveness of the model, we remove at each time one sample from the learning base and we apply the Abe's algorithm with the remaining data for extracting the fuzzy rules base. By taking the removed sample as testing data, we compare the output estimated from the model and the real output of the removed sample. The corresponding results are shown in Tables 8 & 9.

From Table 8, we can see that the difference between the real output and the output estimated by the fabric evaluation model is rather small. For the descriptor "Soft", the maximum evaluation error doesn't exceed 2 marks over 7. For example, "0" means that the model gives the same score as the experts and "1" means that there exists only one level of difference between the model's evaluation and the expert's evaluation. The model also gives good results for the other samples and the other descriptors. In average,

Table 8. Fabric evaluation model results for the descriptor "soft"

Removed Sample	# Rules	# Level(s)	Real Output	Estimated Output	Error
4_1	8	2	7	5.83	1
10_1	8	2	7	4.95	2
14_1	8	2	6	3.50	2
14_2	7	1	4	4.50	1
16_1	8	2	6	4.25	2
16_2	8	2	4	3.74	0
22_1	8	2	6	3.50	2
22_2	8	2	4	4.02	0
24_1	7	2	5	3.27	2
24_2	8	2	4	3.51	0
26_1	8	2	3	3.44	0
26_2	8	2	3	3.71	1
28_1	8	2	3	3.85	1
28_2	8	2	2	2.89	1
31_1	8	2	2	2.49	0
31_2	8	2	2	2.71	1
34_1	8	2	1	1.65	1
34_2	8	2	1	1.20	0
				average	**0.94**

by considering 9 linguistic terms all together for describing the touch handle of the whole samples set, the evaluation error is about 1 point (1.11 − Table 9). This represents the fabric evaluation model accuracy.

6.5 An Example of Consumers' Preference Forecasting

To illustrate the consumers' preference prediction, we collected one set of sensory data related to the cosmetic industry. It includes 8 lotions with varying performance according to their interaction with the skin or the human feeling they express at the first contact. 19 terms are required by the experts for describing the lotions quality. The consumers expressed their feeling with two linguistic terms: the softness and the touch feeling. They ranked also the lotions according to their preference. The running of the procedure described above in Sect. 4.3 leads to the weights set of Fig. 9. Then, applying the method of Sect. 4.4., we are able to predict the consumers' preference (see Table 10). The forecasting works quite well even if sometimes the error raises 15% over 100%. This drawback is due to the size of the training data set base which contains only 8 samples. For testing the methods, we used 7 samples and tried to predict the preference of the 8-th.

Table 9. Results of fabric evaluation model for 9 linguistic terms used by a French Experts panel

	Tight	Shiftless	Smooth	Compact	Weak	Pleasant	Fresh	Heavy	Pliant
Removed Sample				Class Error					
4_1	1	1	2	1	4	2	2	0	1
10_1	1	1	1	1	1	0	2	1	2
14_1	1	1	1	1	0	1	1	0	1
14_2	1	1	1	1	4	1	1	1	2
16_1	1	1	1	1	0	2	1	0	2
16_2	2	2	1	3	0	3	1	1	2
22_1	1	0	1	1	0	1	0	0	1
22_2	1	1	1	1	3	2	1	0	2
24_1	0	3	1	0	0	2	1	0	1
24_2	2	1	1	3	0	1	1	0	2
26_1	1	1	1	0	2	0	1	2	0
26_2	1	2	1	0	1	0	1	0	3
28_1	0	2	1	0	0	0	1	1	0
28_2	2	1	1	0	1	0	1	0	3
31_1	2	3	1	1	0	0	1	2	1
31_2	1	1	0	2	3	0	1	2	3
34_1	1	1	1	2	0	0	1	2	1
34_2	2	1	1	2	1	0	1	1	3
Total average:				errors (average):					
1.11	**1.17**	**1.33**	**1.00**	**1.11**	**1.11**	**0.83**	**1.06**	**0.72**	**1.66**

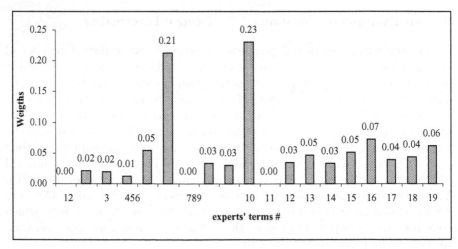

Fig. 9. Weights values for explaining the relationships between consumers' preference about the softness and experts' terms

Table 10. Difference between forecasted and real consumers' preference for 8 lotions

Expressed (%)	Forecasted (%)	Error (%)
14	17	3
9	10	1
16	7	9
16	19	3
18	7	11
4	20	16
10	9	1
13	18	5

7 Conclusions

This paper presents a general method for analyzing and interpreting sensory data given by different panels. The 2-tuple linguistic model is used for normalizing and aggregating sensory data of different individuals inside each panel on an optimal unified scale. The dissimilarity criteria and the sensitivity criteria are transformed into fuzzy numbers in order to obtain a suitable physical interpretation, leading to a better understanding of the quality of panels and evaluation terms. Also, we propose a procedure permitting to interpret terms of one panel using the linear combination of terms of another panel. The optimal weights of this linear combination are obtained using a genetic algorithm with penalty strategy. This procedure is particularly significant for solving commercial conflicts related to the understanding of product quality criteria expressed in linguistic terms. The proposed method has been successfully applied to the analysis and the interpretation of the sensory data on fabric hand evaluation provided by four panels.

This paper deals mainly with the introduction of intelligent methods for both formalizing sensory data, which are expressed by human being, and modeling the relationships between these sensory data and objective measures operated on the fabrics. It gives promising results for assessing the sensory quality of industrial products from manufacturers to consumers. Another contribution in this paper is that we used PCA to project original higher dimensional data into a lower dimensional subspace before starting the procedure of fuzzy rules extraction. In general, a fuzzy model is efficient only when the number of input variables is small enough with respect to the number of learning data. We have to reduce the number of input variables if we can not measure more numerical data.

References

1. H. Stone, J.L. Sidel, Sensory Evaluation Practice, Academic Press Inc., 1993.
2. G.B. Dijksterhuis, Multivariate Data Analysis in Sensory and Consumer Science, Food & Nutrition Press Inc., Trumbull, Connecticut, USA, 1997.
3. J.R. Piggot, E.A. Hunter, Evaluation of assessor performance in sensory analysis, Italian Journal of Food Science, vol. 11, no. 4, pp. 59–64, 1999.
4. Bin Zhou, Xianyi Zeng, Ludovic Koehl, Yongsheng Ding, A 2-level Model for Description of Fabric Hand Sensory Evaluation, Int. Conf. World Textile Conference – 4th AUTEX Conference, Roubaix, France, June 22–24, 2004.
5. Kawabata S. and Niwa M., Objective measurement of fabric hand, Modern Textile Characterization Methods (Eds. M. Raheel and M. Dekker), 1996, 329–354.
6. Hu J., Vhen W., and Newton A., A psychophysical model for objective fabric hand evaluation: An application of Steven's law, *J. the Textile Institute*, 1993, 84(3): 354–363.
7. X. Zeng, Y. Ding and L. Koehl, A 2-tuple fuzzy linguistic model for sensory fabric hand evaluation, in Intelligent sensory evaluation, D. Ruan and X. Zeng, Eds. Berlin: Springer, 2004, pp. 217–234.
8. F. Herrera, L. Martizez, A model based on linguistic 2-tuples for dealing with multigranular hierarchical linguistic contexts in multi-expert decision-making, IEEE Trans. Systems, Man, and Cybernetics – Part B: Cybernetics, vol. 31, no. 2, pp. 227–234, 2001.
9. F. Back, Evolutionary Algorithms in Theory and Practice, Oxford University Press, New York, 1996.
10. S. Abe and M. Lan (1995). Fuzzy rules extraction directly from numerical data for function approximation, IEEE Trans. SMC 25(1), 1995, pp. 119–129.
11. K. Fukunaga (1990), Introduction to Statistical Pattern Recognition, 2nd ed. San Diego, CA: Academic

Simulated Annealing Approach for the Multi-objective Facility Layout Problem

Umut R. Tuzkaya[1], Tijen Ertay[2]*, and Da Ruan[3]

[1] Yildiz Technical University, Mechanical Faculty, Department of Industrial Engineering, Besiktas, 34349, İstanbul, Turkey
[2] Istanbul Technical University, Faculty of Management, Management Engineering Department, Macka, 34367, Istanbul, Turkey
[3] Belgian Nuclear Research Centre (SCK•CEN), Boeretang 200, B-2400 Mol, Belgium

1 Introduction

In general, facility layout problems are occurred if there are changes in requirements of space, people and equipments. When the changes of requirements are shown frequently, the problem goes toward a dynamic structured problem. Designing the facility layout that can respond to requirements related to changes in economic situation of the scope, variability of demand in terms of volume or variety, and changes in production technologies is a comprehensive and complicated study.

To assure this dynamic facility structure over the last decades, computational tools have been integrated to the manufacturing systems as originated computer integrated manufacturing (CIM). The CIM provides computational assistance, control, and high degree of integration in the automation at all levels of the manufacturing industry [1]. The included technology in CIM intensifies the use of distributed computer networks, data processing techniques, and artificial intelligence. Thanks to the above-mentioned technology, data gathering becomes more efficient. Obtaining the valid and required data in voluminous databases to use in a convenient model for solving a facility layout problem is as important as constituting the right model.

Because the data available in a large volume in this system make their evaluation with the traditional methods difficult, a new generation of techniques and tools are required to assist humans in intelligently analyzing voluminous data for pieces of useful knowledge. The field of Knowledge Discovery in Databases (KDD) and Data Mining (DM) has recently emerged as a new discipline

---

* Corresponding ertay@itu.edu.tr

Umut R. Tuzkaya et al.: *Simulated Annealing Approach for the Multi-objective Facility Layout Problem*, Studies in Computational Intelligence (SCI) **5**, 401–400 (2005)
www.springerlink.com © Springer-Verlag Berlin Heidelberg 2005

in engineering and has integrated Database Management System (DBMS) and artificial intelligence technologies to assist humans in analyzing large quantities of data [14]. *Data Mining (DM)* emerged to extract knowledge from huge volumes of data with the help of computing device during the late 1980s and has become a research area with increasing importance with the amount of data greatly increasing [3, 4, 7, 23].

Also, the valid data that will be used in the layout problems may be discovered in huge databases belonging to Enterprise Resources Planning (ERP) system that include the data related to production quantities, production mix, product sequences, bill of materials and schedules. By using data mining tools, the valid data can be obtained to use in some related model.

Even though this information is used for the layout model, finding the most efficient layout solution is not possible generally. The reason is that the layout problems are Multi-objective Combinatorial Optimization (MOCO) problems. A solution technique for this kind of problems is Multi-objective Simulated Annealing (MOSA), of which one is discussed in this study. The following study includes a background research related to data mining, facility layout problems which are considered as multi-objective problems and simulated annealing algorithms as solution technique for multi-objective layout problems. In Sect. 3, the constituted framework starts with preparing the useful data by data mining tools for the MOSA algorithm. After running the algorithm, the efficient layout solutions set is obtained for evaluating and decision-making. In Sect. 4, the framework is applied for a case study and the results are evaluated in terms of feasibility.

2 Background

2.1 Data Mining

In today's world, tremendous growth in capability of both generating and collecting data has been seen. Developing in data storage technologies has allowed transforming this large amount of stored data. Although extensive and voluminous data is handled, it is difficult to find the hiding useful data without powerful tools. In order to relieve such a data rich but information poor plight, a new discipline named data mining (DM) emerged, which devotes itself to extracting knowledge from huge volumes of data, with the help of the ubiquitous modern computing device [35].

There are different definitions of DM in the literature. Reference [2] defined DM as the exploration and analysis, by automatic and semiautomatic means, of large quantities of data in order to discover meaningful patterns and rules. Reference [9] defined DM: a core step that is included by entire process of knowledge discovery in databases (KDD). Also KDD can be defined as an exploration and interpretation of the patterns to be able to determine the useful, novel and required knowledge in data. According to [8], DM is an application, under human control, of low-level induction algorithms that are used to extract patterns from data in specific categories. While DM is surely part of technology in the broad sense, we use it here as an approach to knowledge generation. DM models range from simple, parametric equations derived from linear techniques to complex, non-linear models derived from nonlinear techniques. Non-linear DM methods create complex algorithmic models that do not lend themselves to simple mathematical representation. In general, non-linear techniques can find subtle patterns in the data that cannot be deciphered with linear techniques; they do so at the expense of model interpretability. Specific techniques used in DM applications include market basket analysis, memory based reasoning, cluster detection, link analysis, decision trees and rule induction, neural networks, and genetic algorithms, etc. These techniques are related to classification, clustering, and graphical models. The primary goals of these techniques used for knowledge discovery are prediction and description. Prediction involves using variables or fields in the database to predict unknown or future values of other variables or attributes. For example, some of layout problem characteristics for this study, such as the size of departments, the adjacency scores among the departments, production quantities for products, etc., can predict of huge databases belonging to ERP. Also description focuses on finding human-interpretable patterns describing the data, such as finding patterns for "a good layout." For a good description, in this study, a simulated annealing approach accepted as one of non-linear methods like genetic algorithms and based on the nearest neighbor relation has been used.

2.2 Facility Layout Problems

As a well-known fact, a facility layout problem is one of the most frequently confronted field where much research effort is spent and numerous approaches and methodologies are developed for different layout models. When some assumptions are considered as facilities need equal areas instead of the real shapes and sizes, the facility layout problem is generally formulated as a quadratic assignment problem (QAP) [15] of allocating facilities to discrete locations on a grid with the objective of minimizing a given cost function. In addition to QAP, the facility layout problem has been modeled as [16]:

- quadratic set covering problem,
- linear integer programming problem,
- mixed integer programming problem,
- graph theoretic problem.

Many published results obtained with these exact (optimal) models can be seen in literature. Some optimal algorithms like branch and bound and cutting plane algorithms are used to solve these exact models. However, because optimal methods are limited in the number of departments, the substantial researchers have been developing sub-optimal algorithms that can solve much larger problems. These are called as heuristic algorithms that can be used for construction or improvement of a layout design and solving large sized problems in a reasonable time. Due to the combinatorial nature of the facility layout problem, heuristic approaches will be convenient. The mostly known meta-heuristics that can be applied to various models are simulated annealing, tabu search and genetic algorithms. The recent studies about facility layout problem, which includes this solution approach, can be found in [1] and [18].

Also, facility layout problems are considered as multi-objective problems and their solutions require consideration of conflicting objectives. Usually, a facility design must provide minimum handling cost, penalty values and maximum flexibility, and quality etc. at the same time. Solving a combinatorial optimization problem with a single objective has difficulties because of its NP-hard complexity, so using a heuristic method for a multi-objective optimization will be appropriate to obtain an efficient solution set. The multi-objective simulated annealing method [34], which approximates to the set of efficient solutions of a multi-objective combinatorial optimization (MOCO) problem, can be considered for multi-objective facility layout problems.

2.3 Multi-objective Problems and Metaheuristic Procedures

Many real-world applications naturally call for Multi-objective Combinatorial Optimization (MOCO), because discrete variables and conflicting objectives are often inherent [34]. Although many multi-objective optimization procedures are developed, they have a number of shortcomings. Examples of these solving techniques are surrogate worth trade-off method [25, 26], goal programming and attainment methods [6], min-max approach [20, 21] and fuzzy aggregation method [24]. But later, multi-objective metaheuristics have become very popular and alternative for the above mentioned techniques. The increasing acceptance of these algorithms is due to their ability to [28]: (1) find multiple solutions in a single run; (2) work without derivatives; (3) converge speedily to Pareto-optimal solutions with a high degree of accuracy; and (4) handle both continuous function and combinatorial optimization problems with ease.

Mostly, genetic algorithms and simulated annealing techniques are used to solve the multi-objective problems. Reference [13] suggested a simple multi-objective evolutionary algorithm. Reference [11] proposed a method that employs a novel genetic local search for MOCO. Reference [17] mentioned convergence of evolutionary algorithms and loosing Pareto optimal solutions. Later, [36] suggested several methods to overcome these problems.

Also, simulated annealing which is known as an easy and good solution providing method has been adapted for multi-objective problems by [27] and [31]. Later, [32] generated the MOSA algorithm completely and applied it to a MOCO problem. Also the MOSA algorithm is used for a bi-criteria problem by [30]. Some of these and other improved versions will be discussed in Sect. 2.4.2.

2.4 Multi Objective Simulated Annealing Approach

Simulated Annealing Algorithm: an Overview

The name of the algorithm comes from the analogy to the behavior of physical systems by melting a substance and lowering its temperature slowly until it reaches the stable state (physical annealing). This physical annealing was first simulated by [19]. Thirty years later, [12] established an analogy between optimizing an objective function of a combinatorial optimization problem and annealing of a solid. In the same paper, the simulated annealing algorithm was first purposed as a generally applicable stochastic technique that applied to solve large combinatorial optimization problems.

The algorithm starts with an initial solution, which can be a random solution or the current state of a system. Then, the system is perturbed at random to a new solution in the neighborhood of the original one. If there is a reduction in an objective function for a minimization process, the current solution is replaced by the generated one; else the replacement by a generated neighbor is accepted with a certain probability of;

$$p(\Delta S) = e^{\frac{-\Delta S}{k_b T}} \tag{1}$$

where ΔS is the change in the objective function, T is a control parameter corresponding to the temperature in the analogy and k_b is Boltzmann's constant. The simulated annealing (SA) algorithm attempts to avoid entrapment in a local optimum by accepting small increases in the objective function value. This is realized by controlling the probability of acceptance p (ΔS) through the temperatures. A pseudo-code of the simulated annealing procedure is given below [22].

> **Input:** A problem instance
> **Output:** A (sub-optimal) solution

1. Generate an initial solution at random and initialize the temperature T.
2. **While** $(T > 0)$ **do**
   (a) **While** (thermal equilibrium not reached) **do**
       (i) Generate a neighbor state at random and evaluate the change in energy level ΔS.
       (ii) If $\Delta S < 0$ update current state with new state.
       (iii) If $\Delta S >= 0$ update current state with new state
          with probability $e^{\frac{-\Delta S}{k_b T}}$ where $\kappa_\beta$gs a constant.
   (b) Decrease temperature T according to annealing schedule.
3. Output the solution having the lowest energy.

   Due to SA is an easy and versatile technique, it is widely used for both function and combinatorial optimization problem. Also several implementations of the SA algorithm have been purposed for the facility layout problems which in generally have multiple objectives.

Simulated Annealing Procedures for Multi Objective Problems

As mentioned above, various simulated annealing algorithms based on multi-objective optimization algorithms have been generated by many researchers. For a good overview of this subject, we refer to the study by [28].

– The method of Suppapitnarm and Parks (SMOSA)

Reference [29] used the concept of archiving the Pareto-optimal solutions coupled with return to base strategy. The probability scalarizing approach of which the weight vector is not used in acceptance criteria is used. Scalarizing approaches will be explained in Sect. 3.2.

In this approach, a constrained problem is converted to an unconstrained one with the help of a penalty function. A penalty value is defined to take the constrained violation into account. This value is added or subtracted depending on the problem, which is a minimization or maximization, respectively.

– The method of Ulungu and Teghem (UMOSA)

The UMOSA algorithm of [32, 33] is an interactive version of MOSA and considers three different cases when a move from a present position to a new position is realized. These cases can occur as (1) the move is an improving one; (2) causes improvements in certain objectives and deterioration in other objectives; and (3) a deteriorating move in all the objectives. The UMOSA algorithm takes into consideration all these issues by using a criterion scalarizing approach that will be discussed in Sect. 3.2. This approach aims to project a multi-dimensional objective space into a mono dimensional space.

– The Pareto simulated annealing (PSA)

Reference [5] proposed PSA by combining the single objective simulated annealing and genetic algorithms to provide efficient solutions. The outcome of the multi-objective PSA is the set of solutions which are neither better nor worse solutions. Each of them is called as non-dominated Pareto optimal solutions with respect to the multi-objective. In PSA, objective weights are used in the multi-objective rules for an accepting probability in order to assure dispersion of generating solutions over the whole set of efficient solutions. Increasing or decreasing of the probability of improving values for a particular objective can be controlled by changing the objective weights.

– Multi-objective Simulated Annealing Using Constraints Violation in Acceptance Criterion (WMOSA)

Many simulated annealing based multi-objective optimization algorithms do not handle constraints without using a separate technique. The WMOSA algorithm attempts to handle the constraints in the main algorithm by using a weight vector in the acceptance criterion.

A detailed summary and the algorithms of all these approaches for the multi-objective simulated annealing can be found in [28].

3 An Applied Framework
for Multi-objective Layout Problems

3.1 Data Collection and Preparation

Facility layout design and evaluation is often a challenging and time consuming procedure due to its inherent multiple objective nature and its data
collection process. An effective facility layout design procedure necessitates
the consideration of qualitative data, e.g., penalty values that depend on the
relationship between departments and a flexibility rate related to volume or
variety, as well as quantitative data such as handling cost and material handling vehicle utilization etc.

The compulsory data in designing a facility layout is the frequency of
the trips between facilities (departments, cells or machines etc.) for handling
the materials, work-in-process (WIP), finished goods etc. This is a tangible
data that can be obtained by observations and motion time studies for a time
period. Instead of this kind of hard solution, the required data for the proposed
methodology can be prepared by integrating, selecting, and transforming the
data from the database or databases which belong to Enterprise Resources
Planning (ERP) system.

In production planning module of an ERP system, Master Production
Scheduling (MPS) is prepared according to the demand forecasts and orders.
Considering the MPS and integrating it with the Bill of Materials (BOM)
and inventory status, Material Requirement Planning (MPR) is run to get
the order receipts. The data related to order receipts are reached in database
that includes transactional data. Also BOM shows the required quantities of
materials and subassembly works for each product and operation plans (work
routes) show the manufacturing process (sequence) including the machines on
which the products will be processed.

By selecting these data for a time period e.g. month, week, or day and
integrating them with queries, the data of total trips between machines or
work centers can be obtained.

3.2 Using Simulated Annealing Algorithm
for Multi-objective Combinatorial Optimization

In the multi-objective problems, it is difficult to determine if the realized move
(new solution) is an improving or deteriorating one when all the objectives are
considered. In this study, the Multi-Objective Simulated Annealing (MOSA)
method [34] is used to overcome this difficulty. Before giving the details of
the method, critical situations and strategies related with this difficulty are
discussed. A list of the related notations is given below as the symbols that
are used in the MOSA algorithm.

Nomenclature

K number of the criteria

$z_k(X)$ kth objection function value for the solution X.

λ_k weight of the kth objective

p accepting probability

ΔS change in the objective function values

T annealing temperature (controlling parameter)

T_i initial temperature

T_f final temperature

α cooling factor

$rd(\ )$ random number

X_0 initial solution

X_f final solution

X_n current solution in the nth iteration of the outer loop

X_m current best solution

As mentioned in Sect. 2.4.2, in the UMOSA method, when a neighboring solution Y is compared with the current solution X_n in iteration n, there are three possible situations. The situations related to moving from X_n to Y are (1) improvement in all objectives, (2) improvement in some objectives and deterioration in remains and (3) deterioration in all objectives. Also three strategies can be considered for the three situations as summarized in Fig. 1. Strategy 1 is the accepting solution Y with the probability $p = 1$ when the situation 1 is occurred ($\forall k\ \Delta z_k \leq 0$ in a minimization problem). Strategy 3 is the accepting solution Y with a calculated p probability and is used for the situation 3. It is possible to calculate p value with "probability scalarizing approach."

In the probability scalarizing approach, two solutions X_n and Y are compared in terms of the each objective function separately. Then the obtained probabilities for each individual objective are aggregated. The decision rule can be shown in the following relation:

$$\pi_k = \begin{cases} \exp\left(-\dfrac{\Delta z_k}{T_n}\right) & \text{if } \Delta z_k > 0 \\ 1 & \text{if } \Delta z_k \leq 0 \end{cases} \tag{2}$$

In the aggregating phase, many scalarizing function can be considered. The most widely used ones are the product and the min:

$$p = \prod_{k=1}^{K} (\pi_k)^{\lambda_k} \quad \text{or} \quad p = \min_{k=1,\dots,K} (\pi_k)^{\lambda_k} \tag{3}$$

When the situation 2 is considered, it is possible to apply all the three strategies. However, applying the strategy 1 only assures the "diversity" because all the solutions are accepted as improving ones even only one of the objectives is improving. In the same manner, applying the strategy 3 assures only "intensification" because all the solutions are accepted deteriorating ones even only one of the objectives is deteriorating.

In meta-heuristics, it is necessary to assure both diversity and intensification. Therefore the strategy 2 assumes that the move is sometimes an improving one and sometimes a deteriorating one depending on the value of Δz_k. Thanks to "the criterion scalarizing approach" in the strategy 2, both diversity and intensification will be hold. This approach aims to reduce multi-criteria to a single criterion. After the single-criterion is obtained, the classical rules can be applied easily. The scalarizing function can be stated as $s(.) = s(Z(.), \lambda)$ which performs the single criterion project. The essence of this approach is to get weighted objectives by using the λ weight vector and obtain a single objective value. In this way, the change in the single objective function value between the neighboring solution Y and the current solution X_n (ΔS) equals to $s(Z(Y), \lambda) - s(Z(X_n), \lambda)$. Now the decision rule can be shown as follows:

$$p = \begin{cases} 1 & \text{if } \Delta s \leq 0 \\ \exp\left(-\frac{\Delta s}{T_n}\right) & \text{if } \Delta s > 0 \end{cases} \qquad (4)$$

A number of scalarizing functions may be considered for the different situations. However the mostly used and well known ones are "the weighted sum" and "the weighted Chebyshev norm." Here only "the weighted sum" will be considered. For the others, we refer to [27]. The scalarizing function can be expressed as follows:

$$s(Z, \lambda) = \sum_{k=1}^{K} \lambda_k z_k$$

$$\sum_{k=1}^{K} \lambda_k = 1,$$

$$\lambda_k > 0, \quad \forall k \qquad (5)$$

In the MOSA algorithm all the iterations are conducted as the same as the classical SA algorithm by using T, T_i, T_f, p and α. But, the differences are occurred because of the multiple objectives. To deal with this problem, a criterion scalarizing approach is embedded to the algorithm. Also, instead of obtaining an optimal solution as in a single objective SA, MOSA generates a potentially efficient solution set, which is not dominated by another one. To assure this, a wide diversified set of weights is considered. Different weight vectors $\lambda^{(l)}, l \in L$ are generated where $\lambda^{(l)} = (\lambda_k^{(l)}, k = 1, \ldots, K)$ with $\lambda_k^{(l)} \geq 0 \; \forall k$ and $\sum_{k=1}^{K} \lambda_k^{(l)} = 1$, $\forall l \in L$. This set of weights is uniformly generated. A pseudo code for the MOSA algorithm is given below:

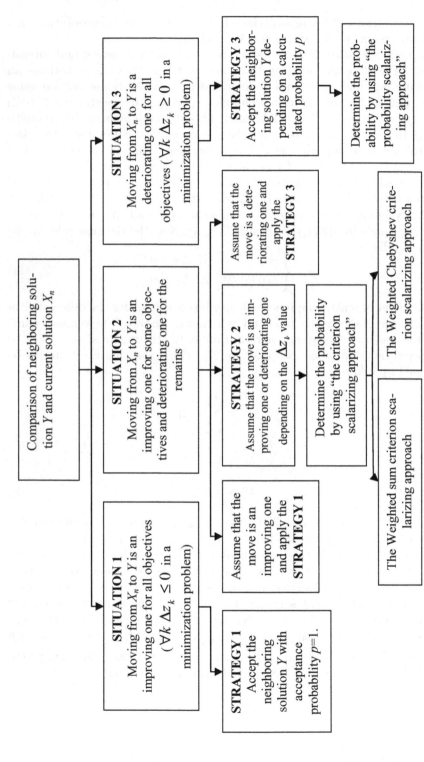

Fig. 1. Summary of probable situations and strategies for comparison of neighboring solutions

Initialization
n=1;
Draw at random an initial solution X_0;
$X_f = X_0$;
$X_n = X_0$;
$X_m = X_0$;
$T = T_i$;
Evaluate $z_k (X_0)$ $\forall k$;
While $(T >= T_f)$ do;
    While (all generated neighboring solutions have not been analyzed) do;
    $Y = $ a neighboring solution of X_n;
    calculate $s(z(Y), \lambda)$;
    if $(s(z(Y), \lambda) < s(z(X_m), \lambda))$ then;
      $X_m = Y$;
    end if;
    calculate $\Delta S = s(z(Y), \lambda) - s(z(X_n), \lambda)$;
    $prob = \min(1, \exp(-\Delta S/T)$;
    if $(rd\ () <= prob)$ then;
      $X_f = Y$;
  end if;
  end while;
  $n = n + 1$;
  $X_n = X_f$;
  $T = T.\ \alpha$;
  end while;
  if $(s\ (z\ (X_m), \lambda)\ ¡\ s\ (z(X_f), \lambda))$ then;
    $X_f = X_m$;
end if;
write the solution X_f;
end.

This algorithm can be considered in the proposed framework together with the data queried by SQL from ERP databases. At the end of the framework, the generated solution set is proposed to the decision maker for choosing an appropriate one to her/him.

4 An Application of the Framework for a Case Study

The proposed framework is applied to a pipe clam, anchors and hanging & fixing systems manufacturing facility, which has the layout shown in Fig. 2. Layout problem, the required data preparation, using the MOSA algorithm and evaluation of the results are detailed below.

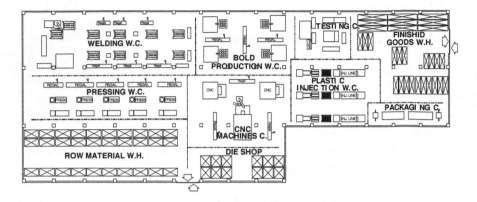

Fig. 2. The current layout of the facility

4.1 Problem Definition and Data Preparation

The aim of the application is to redesign the facility layout to remove a new place. Ten work centers are tried to assign to convenient sites. The problem seems as a Quadratic assignment problem (QAP) that will be solved with MOSA heuristic. The used data in problem is related to numbers of trips between work centers and the distances of the sites that facilities (work centers) will be assigned to them. Also qualitative data that are determined based on relationship of work centers are penalty values. This kind of data are derived from distance data: if the specific two work centers which have to be close to each other depending on their relationship importance are not close, then the distance between them is multiplied with a constant penalty value.

The number of trips between facilities is determined by the following way that is mentioned in Sect. 3.1. The data obtained from the database of the ERP system are BOM and operation plans, which can be seen in Fig. 3. The circles represent the sub-assembly products and raw materials and the squares represent the machines in which the product will be processed.

The Fig. 4 shows the query that is realized to obtain the trips in *Query Analyzer Module of SQL Server 2000*. The material's name, the operation levels in BOM, work center numbers, unit handling quantities and other data related to processing times can be seen in the query.

By using these source data, the required data about trips are gathered for a mount period. Table 1 shows names of work centers, distances between them under the diagonal and trips between them above the diagonal.

4.2 Multi-objective Simulated Annealing

The gartered data is used to get a robust layout from the potentially efficient layout alternatives set by using the MOSA algorithm. The pseudo code given by the MOSA algorithm in Sect. 3.2 is coded in *Microsoft Visual C++* ®6.0.

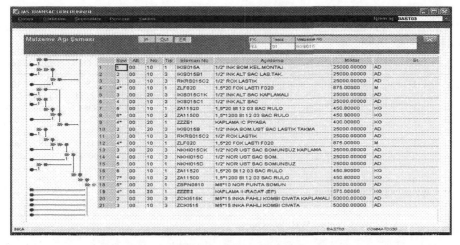

Fig. 3. A screen related to BOM and Operations plans

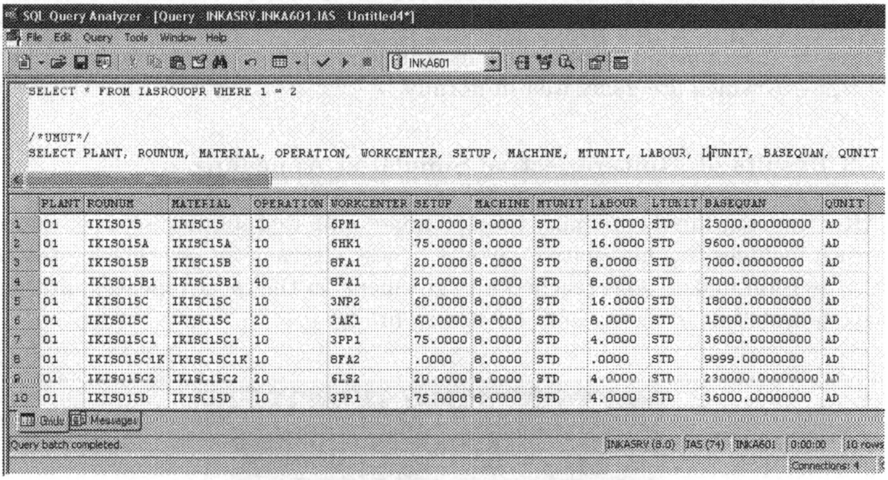

Fig. 4. A query screen example for the required data

There are two objectives that are tried to minimize. The first one is related to material handling costs that is tried to satisfy by putting closer the work centers which have intensive trips between them. The second one is minimizing the penalty values, which are occurred when the two particular departments are not close to each other but have to be. This closeness necessity can be related to some quality problems. In this study, the plastic injection work center and finished goods warehouse must be close to the packaging center. To assure this, constant penalty values are multiplied by the distances between these departments.

Table 1. Number of trips and distances between work centers

	Welding WC	Bolt Production WC	Testing Center	Finished Goods WH	Pressing WC	CNC Machines C	Plastic Injection WC	Packaging C	WC Raw	Die Shop
	1	2	3	4	5	6	7	8	9	10
1		13725	2450	4573	17330	2342	2985	12784	15368	325
2	1		1120	3256	12563	1756	2230	9284	18450	1176
3	2	1		9045	2987	236	720	3749	2341	0
4	3	2	1		6879	3215	133	36551	0	0
5	1	2	3	4		5448	3970	14189	23785	1905
6	2	1	2	3	1		4550	6036	5198	390
7	3	2	1	2	2	1		4520	7543	546
8	4	3	2	1	3	2	1		1276	0
9	2	3	4	5	1	2	3	4		0
10	3	2	3	4	2	1	2	3	1	

C: Center, WC: Work center, WH: Warehouse

The convenient parameter values are determined according to Heragu (1997) as $T_i = 900.00$, $T_f = 1.00$ and $\alpha = 0.90$. The program is run many times to get non-dominated efficient solutions with different weight vectors $\lambda^{(l)}$, $l \in L$ which are generated uniformly.

4.3 Results of Multi-objective Simulated Annealing

After the program is run many times, a non-dominated solution set is obtained which includes five alternative layouts. These alternatives are represented to the decision makers and reached an agreement on the most repeated alternative layout that is shown in $C++$ screen in Fig. 5.

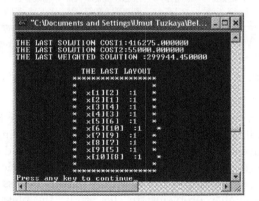

Fig. 5. $C++$ screen that shows the layout result

In an initial solution all the facilities were assigned to the same numbered sites, but Fig. 5 shows the changes as the facility 1 is assigned to the site 2, the

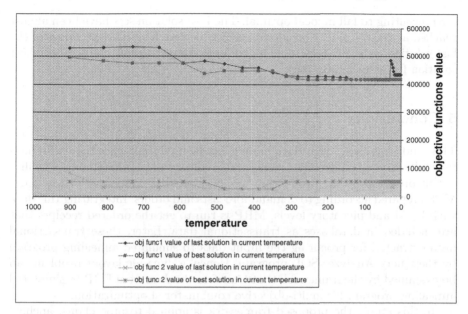

Fig. 6. Changes of objective functions separately while temperature decreases

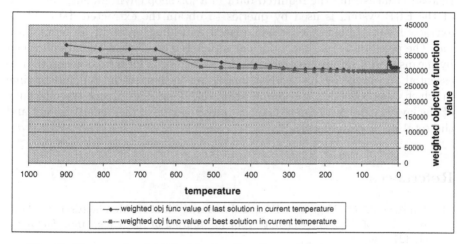

Fig. 7. Change of weighted objective functions while temperature decreases

facility 2 is assigned to the site 1 etc. At the end of the iterations, the last and best layout has the values of 416275, 55000, and 299944.45 for the objective function 1, objective function 2 and weighted objective function, respectively.

Figure 6 and Fig. 7 show the changes of values related to each objective function and their weighted objective function while temperature T is decreasing. It can be seen that some worse solutions are accepted by the algorithm for preventing to fall in local optimal. The best solution sets have been almost reached at 200 T-value. Considering each objective function, the values of the last solution generally take some higher ones comparing with those of the best solution for assuring diversity.

5 Conclusions

This study has shown that data mining can be used to prepare the useful data for multi-objective layout problems. The required data in designing a facility layout procured by the production model of an ERP system, which includes MPS prepared according to demand forecasts and orders. Integrating this data with BOM and inventory levels, MRP is run to get the ordered receipts that are included in databases as transactional data. Later, these transactional data extracted for procuring the input data to simulated annealing program by the Query Analyzer SQL program. Hence, a dynamic layout problem can be examined by the framework that transfers the data from ERP to simulated annealing program for multi-objective combinatorial optimization.

In this study, the proposed framework is applied to pipe clams, anchors and hanging & fixing systems manufacturing facility having 10 work centers. Instead of measuring the required data in a job shop environment, databases of the ERP system is used by queries to obtain the extracted data. These extracted data transformed to simulated annealing program as input data file. The program is run in a sufficient number. The efficient layouts obtained by means of run set are presented to decision makers for a final decision. The essential advantage of this study is providing to shorten the concluding time in which data collection and preparation take long time. Hence the efficiency of the layout can be considered continually in terms of the penalty values and the total distances objects.

References

1. Alvarenga, A.G., Negreiros, F.J., and Mestria, M. 2000. Metaheuristic methods for a class of the facility layout problem. *Journal of Intelligent Manufacturing.* 11, 421–430.
2. Berry, M.J.A., and Linoff, G. 1997. Data Mining Techniques for Marketing, Sales and Customer Support. New York: Wiley.
3. Changchien, S.W. and Lu, T. 2001. Mining association rules procedures to support on-line recommendation by customer and products fragmentation. *Expert Systems with Applications.* 20, 325–335.
4. Chiang, D., Chow, L.R. and Wang, Y. 2000. Mining time series data by a fuzzy linguistic summary system. *Fuzzy Sets and Systems.* 112, 419–432.
5. Czyzak, P. and Jaszkiewicz, A. 1998. Pareto simulated annealing – A metaheuristic technique for multiple objective combinatorial optimization. *Journal of Multi-Criteria Decision Analysis.* 7, 34–47.

6. Dhingra, A. K. 1987. Optimal apportionment of reliability and redundancy in series systems under multiple objectives. *IEEE Trans. Reliability.* 41, 621–623.
7. Fayyads, U.M., Piatetsky-Shapiro, G. and Smyth, P. 1996a. The KDD processes for extracting useful knowledge from volumes of data. *Communications on the ACM.* 39 (11), 27–34.
8. Fayyads, U.M., Djorgovski, S.G., and Weir, N. 1996b. Automating the analysis and cataloging of sky surveys, Advances in Knowledge Discovery and Data Mining. Cambridge MA:MIT Press, 471–493.
9. Frawley, W.J., Piatesky,_Shapiro, G., and Matheus, C.J. 1991. Knowledge discovery in databases: an overview. In Fayyad, G. Piatesky-Shapiro and P. Smyth. Knowledge Discovery in Databases. 1–27. Cambridge MA:AAAI/MIT Press.
10. Heragu, S. 1997. Facility Design. PWS Publishing Company. Boston.
11. Jaszkiewicz, A. 2002. Genetic local search for multiple objective combinatorial optimization. *European Journal of Operations Research.* 1(137), 50–71.
12. Kirkpatrick, S., Gelatt, C.D., Jr., and Vecchi, M.P. 1983. Optimization by simulated annealing. *Science.* 220 (4598): 671–679.
13. Knowles, J.D. and Corne, D.W. 2000. Approximating the non-dominated front using the Pareto archived evolutionary strategy. *Evolutionary Computation.* 8(2), 142–172.
14. Koonce, D.A. and Tsai, S.C. 2000. Using data mining to find patterns in genetic algorithm solutions to a job shop scheduling. *Computers & Industrial Engineering.* 38, 361–374.
15. Koopmans, T.C. and Beckmann, M. 1957. Assignment problems and the location of economic activities. *Econometrica* 25, 1, 53–76.
16. Kusiak, A. and Heragu S. 1987. The facility layout problem. *European Journal of Operational Research.* 29, 229–251.
17. Mahfoud, S.W. 1995. Niching methods for genetic algorithms. Ph.D. dissertation, University of Illinois Urbana-Champaign.
18. Mavridou, T.D. and Pardalos, P. M. 1997. Simulated annealing and genetic algorithms for the facility layout problem: A survey. *Computational Optimization and Applications.* 7, 111–126.
19. Metropolis, N., Rosenbluth, M. Rosenbluth, and Teller, A. 1953. Equation of state calculations by fast computing machines. *Journal of Chemical Physic,* 21, 1953, 1087–1092.
20. Mishra K.B. and Sharma, U. 1991a. An efficient approach for multiple criteria redundancy optimization problems. *Microelectronics and Reliability.* 31(2/3), 303–321.
21. Mishra, K.B. and Sharma, U. 1991b. Multicriteria optimization for combined reliability and redundancy allocation in systems employing mixed redundancies. *Microelectronics and Reliability.* 31(2/3), 323–335.
22. Pardalos, P.M., Pitsoulis, L.S., Mavridou, T.D. and Resende, M.G.C. 1995. Parallel search for combinatorial optimization: genetic algorithms, simulated annealing, Tabu search and GRASP. IRREGULAR 1995: 317–331.
23. Park, S.C., Piramuthu, S., Shaw M.J. 2001. Dynamic rule refinement in knowledge-based data mining systems. *Decision Support Systems.* 31, 205–222.
24. Ravi, V., Reddy, P.J. and Zimmermann, H.J. 2000. Fuzzy global optimization of complex system reliability. *IEEE Trans. Fuzzy Systems.* 8(3), 241–248.
25. Sakawa, M. 1981. Optimal reliability design of a series-parallel system by a large-scale multiobjective optimization method. *IEEE Trans Reliability.* R-30, 173–174.

26. Sakawa, M. 1982. Interactive multiobjective optimization by sequential proxy multiobjective technique (SPOT). *IEEE Trans Reliability.* R-31, 461–464.

27. Serafini P. 1985. Mathematics of Multiobjective Optimization. *CISM courses and lecturers* (Vol. 289) Berlin: Springer Verlag.

28. Suman B. 2004. Study of simulated annealing based algorithms for multi-objective optimization of a constrained problem. *Computers Chemical Engineering.* Article in Press.

29. Suppapitnarm, A., Seffen, K.A., Parks, G.T. and Clarkson, P.J. 2000. Simulated annealing: an alternative approach to true multiobjective optimization. *Engineering Optimization,* 33(1), 59–85.

30. Tuyttens, D., Teghem, J., Fortemps. P.H., and Nieuwenhuyze, K.V. 2000. Performance of the MOSA method for the bi-criteria assignment problem. *J. Heuristics.* 6, 295–310.

31. Ulungu, L.E. and Teghem J. 1994. Multiobjective combinatorial optimization problems: A survey. *Journal of Multicriteria Decision Analysis.* 3, 83.

32. Ulungu, L.E., Teghem J., and Fortemps, P. 1995. Heuristics for multiobjective combinatorial optimization problems by simulated annealing. In J. Gu, G. Chen, Q. Wei, & S. Wang (Eds.), *MCDM: Theory and applications* (p. 269). Windsor: Sci-Tech.

33. Ulungu, L.E., Teghem, J. and Ost, C. 1998. Interactive simulated annealing in a multiobjective framework: application to an industrial problem. *J. Oper. Res. Soc.* 49, 1044–1050.

34. Ulungu, E.L., Teghem, J., Fortemps, P.H. and Tuyttens, D. 1999. MOSA method: a tool for solving multiobjective combinatorial optimization problems. *Journal of Multi-criteria Decision Analysis.* 8: 221–236.

35. Zhou, Z.H. 2003. Three perspective of data mining. *Artificial Intelligence.* 143, 139–146.

36. Zitzler, E. Laumanns, M., Thiele, L. 2002. SPEA2: Improving the strength Pareto evolutionary algorithm for multiobjective optimization. EUROGEN 2001, *Evolutionary Methods for Design, Optimization and Control with Applications to Industrial Problems.* Athens, Greece, September 12–21.

Self-Tuning Fuzzy Rule Bases
with Belief Structure

Jun Liu[1], Da Ruan[2], Jian-Bo Yang[3], and Luis Martinez Lopez[4]

[1] School of Computing and Mathematics, University of Ulster at Jordanstown,
   Newtownabbey BT37 0QB, Northern Ireland, UK
   j.liu@ulster.ac.uk
[2] Belgian Nuclear Research Centre (SCK•CEN), Boeretang 200, 2400 Mol,
   Belgium
[3] Manchester Business School (East), P.O. Box 88, The University of Manchester,
   Manchester M60 1QD, UK
[4] Dept. of Computer Sciences, University of Jaén, E-23071, Jaén, Spain

Abstract. A fuzzy rule-based evidential reasoning (FURBER) approach has been
proposed recently, where a fuzzy rule-base designed on the basis of a belief structure
(called *a belief rule base*) forms a basis in the inference mechanism of *FURBER*.
This kind of rule-base with both subjective and analytical elements may be difficult
to build in particular as the system increases in complexity. In this paper, a learning
method for optimally training the elements of the belief rule base and other knowledge representation parameters in *FURBER* is proposed. This process is formulated
as a nonlinear multi-objective function to minimize the differences between the output of a belief rule base and given data. The optimization problem is solved using
the optimization tool provided in MATLAB. A numerical example is provided to
demonstrate how the method can be implemented.

Key words: uncertainty, fuzzy logic, belief rule-base, evidential reasoning, optimization, MATLAB, safety estimate

1 Introduction

In real world applications, intrinsically vague information may coexist with
conditions of "lack of specificity" originating from evidence not strong enough
to completely support a hypothesis but only with degrees of belief or *credibility*
[8]. One realistic way to deal with imprecision is to use linguistic assessments
instead of numerical values. Fuzzy logic approaches [24] and [25] employing
fuzzy IF-THEN rules can model the qualitative aspects of human knowledge
and reasoning process without employing precise quantitative analysis. This
actually provides a tool for working directly with the linguistic information,
which are commonly used in engineering system analysis (e.g., representing

Jun Liu et al.: *Self-Tuning Fuzzy Rule Bases with Belief Structure*, Studies in Computational
Intelligence (SCI) **5**, 419–418 (2005)
www.springerlink.com

risk factors and carrying out safety assessments) [1, 3, 7, 16]. Dempster-Shafer (D-S) theory of evidence [6, 13] based on the concept of *belief function* is well suited to modeling subjective credibility induced by partial evidence [15]. It also provides appropriate methods for computing belief functions for combination of evidence. Besides, the D-S theory also shows great potentials in multiple attribute decision analysis (MADA) under uncertainty, where an evidential reasoning (ER) approach for MADA under uncertainty has been developed, on the basis of a distributed assessment framework and the evidence combination rule of the D-S theory [18, 19, 20, 21, 22].

Accordingly, it seems reasonable to extend the fuzzy logic framework to cover credibility uncertainty as well. To combine fuzzy logic and D-S models to deal with fuzziness and incompleteness in safety analysis, a framework for modelling the safety of an engineering system using a FUzzy Rule-Based Evidential Reasoning (FURBER) approach has been recently proposed [9], which is based on a generic Rule-base Inference Methodology using the Evidential Reasoning approach (RIMER) proposed in [23]. Within this framework, a fuzzy rule-base designed on the basis of a belief structure, called *a belief rule base*, is used to capture uncertainty and non-linear relationships between the parameters, and the inference of the rule-based system is implemented using the evidential reasoning algorithm [21].

A belief rule base forms a basis in the inference mechanism of *FURBER*, which is a framework for representing expert knowledge but it is difficult to determine its elements entirely subjectively, in particular for a large scale rule base with hundreds of rules. Also, a change in a rule weight or an attribute weight may lead to significant changes in the performance of a belief rule base. Moreover, the form of fuzzy membership function in the antecedent of the rule still remains an important factor for the system performance.

As such, there is a need to develop a method that can generate an optimal belief rule base using expert judgments as well as statistical data. In this paper, a learning method for optimally training the elements of the belief rule base and other knowledge representation parameters in *FURBER* is proposed. This process is formulated as a *nonlinear objective function* to minimize the differences between the output of a belief rule base and given data and is solved using the optimization tool provided in MATLAB.

In addition, for some linguistic terms (such as *good, fair, intelligent, smart*, or *beautiful*, etc.), which do not have clearly defined bases, it is too subjective to clearly define them. If they appear in the consequent part of a rule, it is better to draw a conclusion which has the same linguistic values as the one in the consequent but with different degree of confidence. We do not try to change the linguistic values used in the consequent of the rule, but a degree of confidence of this conclusion is added. The optimization approach proposed in the present paper provides a more flexibility for optimally training, i.e., it is not required that the training sample should be numerical data pairs, which can be subjective judgment using linguistic values with belief. Consequently, the reasoning process can be approximately imitated by the optimized *FURBER* (e.g., assigning weights to antecedent attributes and/or by adjusting belief degrees in the consequents of rules in a systematic manner). This is one of the prominent features of the belief rule-base.

The rest of this paper is organized as follows. *FURBER* is briefly reviewed in Sect. 2. The optimization method for constructing belief rule-base in *FURBER* is proposed in Sect. 3. A numerical example is illustrated in Sect. 4. Conclusions are drawn in Sect. 5.

2 FURBER

This section reviews the *FURBER* framework [9].

2.1 Fuzzy Rule-Base with the Belief Structure

Fuzzy logic systems are knowledge-based or rule-based ones constructed from human knowledge in the form of fuzzy *IF-THEN* rules. For example, the following is a fuzzy *IF-THEN* rule for safety analysis [9]:

IF Failure Rate of a hazard is *frequent AND* Consequent Severity is *catastrophic AND* Failure Consequent Probability is *likely THEN safety estimate* is *Poor*.

To take into account the belief degrees of a rule, attribute weights and rule weights, fuzzy rules can be extended in the following way. In general, assume that the T antecedent parameters, $U_1, \ldots, U_T$ can be described by J_i linguistic terms $\{A_{ij}; j = 1, \ldots, J_i\}, i = 1, \ldots, T$, respectively. One consequent variable can be described by N linguistic terms, i.e., $D_1, D_2, \ldots, D_N$. Suppose that the rule-base is given by $R = \{R_1, R_2, \ldots, R_L\}$, the kth rule can be represented as follows:

$$R_k: \ IF U \text{ is } A^k \ THEN D \text{ with belief degree } \beta^k, \text{ with a rule weight } \theta_k \text{ and}$$
$$\text{attribute weights } \delta_1, \ldots, \delta_T \tag{1}$$

where U represents the antecedent attribute vector $(U_1, \ldots, U_T)$, A^k the packet antecedents $\{A_1^k, \ldots, A_T^k\}$, where $A_i^k (\in \{A_{ij}; j = 1, \ldots, J_i\})$ is a linguistic term corresponding to the ith attribute in the kth rule with $i = 1, \ldots, T$. D is the consequent vector $(D_1, \ldots, D_N)$, and β^k the vector of the belief degrees $(\beta_{1k}, \ldots, \beta_{Nk})$ for $k \in \{1, \ldots, L\}$ with $\sum_{i=1}^{N} \beta_{ik} \leq 1$. This is the vector form of a *belief rule*, β_{ik} measures the degree to which D_i is the consequent if the input activates the antecedent A^k in the kth rule for $i = 1, \ldots, N; k = 1, \ldots, L$. L is the number of rules in the rule-base. If $\sum_{i=1}^{N} \beta_{ik} = 1$, the output assessment or the kth rule is said to be complete; otherwise, it is incomplete. The rule base in (1) is referred to as a *belief rule base*.

A belief rule base given in (1) represents functional mappings between antecedents and consequents with uncertainty. It provides a more informative

Table 1. A belief rule expression matrix

Belief Output	Input					
	A^1 (w_1)	A^2 (w_2)	$\ldots$	A^k (w_k)	$\ldots$	A^L (w_L)
D_1	β_{11}	β_{12}	$\ldots$	β_{1k}	$\ldots$	β_{1L}
$\vdots$	$\vdots$	$\vdots$	$\ldots$	$\vdots$	$\ldots$	$\vdots$
D_i	β_{i1}	β_{i2}	$\ldots$	β_{ik}	$\ldots$	β_{iL}
$\vdots$	$\vdots$	$\vdots$	$\ldots$	$\vdots$	$\ldots$	$\vdots$
D_N	β_{N1}	β_{N2}	$\ldots$	β_{Nk}	$\ldots$	β_{NL}

and realistic scheme for uncertain knowledge representations. Note that the degrees of belief $\beta_{ik}(i = 1, \ldots, N; k = 1, \ldots, L)$, and the weights could be assigned directly by experts or more generally they may be trained and updated using dedicated learning algorithms if a priori or *up-to-date* information regarding the input and output of a rule-based system is available. Once such a belief rule-base is established, the knowledge contained in the belief rule base can be used to perform inference for given input. The rule base can be summarized using a belief rule expression matrix shown in Table 1.

2.2 Fuzzy Rule-Base Inference Mechanism Based on the Evidential Reasoning Approach

In the matrix, w_k is the activation weight of A^k, which measures the degree to which the kth rule is weighted and activated. The degree of activation of the kth rule w_k is calculated as:

$$\omega_k = \left(\theta_k * \prod_{i=1}^{T_k} (\alpha_i^k)^{\bar{\delta}_i} \right) \Bigg/ \left(\sum_{i=1}^{L} \left[\theta_i * \prod_{i=1}^{T_k} (\alpha_l^i)^{\bar{\delta}_l} \right] \right) \tag{2}$$

where $\bar{\delta}_i = \delta_i/(\max_{i=1,2,3} \{\delta_i\})$. α_i^k $(i = 1, \ldots, T_k)$ is the individual matching degree to which the input for U_i belongs to A_i^k of the ith individual antecedent in the kth rule, T_k is the number of antecedents involved in the kth rule.

For a given real input vector $a = (a_1, \ldots, a_{T_k}) = \mu_{A_i^k}(a_i)$, here is the fuzzy membership function of the linguistic term A_i^k. Fuzzy membership functions can be applied in different forms depending on the system. The straight-line membership functions can be used due to its advantage of simplicity, such as the triangular membership function and trapezoidal membership function. Continuous and differentiable Gaussian function is used in this paper, i.e.,

$$\mu_{A_i^k}(a_i) = \exp\left(-\frac{1}{2} \left(\frac{a_i - c_i^k}{\sigma_i^k} \right)^2 \right), \tag{3}$$

where c_i^k is the central value of fuzzy membership function and σ_i^k is the variance at the central value.

Having represented each rule as (1), the ER approach can be directly applied to combine rules and generate final conclusions as follows. First, transform the degrees of belief β_{jk} for all $j = 1, \ldots, N, k = 1, \ldots, L$ into basic probability masses and then aggregate all the packet antecedents of the L rules to generate the combined degree of belief in each possible consequent D_j in D using the evidential reasoning (ER) algorithm [20, 21]. The final conclusion generated by aggregating the L rules, which are activated by the actual input vector $\boldsymbol{a} = \{a_t; t = 1, \ldots, T\}$ can be represented as follows

$$\{(D_j, \beta_j), j = 1, \ldots, N\} \tag{4}$$

The ER Recursive Algorithm used in [21, 22] has been equivalently transformed into the ER overall analytical algorithm [17]. Using this overall algorithm, the overall combined degree of belief β_j in D_j is generated as follows:

$$\beta_j = \frac{\mu * \left[\prod_{k=1}^{L}\left(\omega_k\beta_{j,k} + 1 - \omega_k \sum_{j=1}^{N}\beta_{j,k}\right) - \prod_{k=1}^{L}\left(1 - \omega_k\sum_{j=1}^{N}\beta_{j,k}\right)\right]}{1 - \mu * \left[\prod_{k=1}^{L}(1 - \omega_k)\right]} \tag{5}$$

where $j = 1, \ldots, N$, and

$$\mu = \left[\sum_{j=1}^{N}\prod_{k=1}^{L}\left(\omega_k\beta_{j,k} + 1 - \omega_k\sum_{j=1}^{N}\beta_{j,k}\right) - (N-1)\prod_{k=1}^{L}\left(1 - \omega_k\sum_{j=1}^{N}\beta_{j,k}\right)\right]^{-1}$$

Notice that it is the beliefs used in the belief structure and the activation weights that determine the actual performance of inference. The degree to which the final output can be affected is determined by the magnitude of the activation weight and the belief degrees in each rule. Therefore, the performance of inference can be improved if the following parameters in (5) are adjusted by autonomous learning.

(1) Rule weight $\theta_k (k = 1, \ldots, L)$ and attribute weights $\delta_1, \delta_2, \delta_3$;
(2) The degrees of belief $\beta_{ik}(i = 1, \ldots, N; k = 1, \ldots, L)$;
(3) The central value of fuzzy membership function c_i^k and the variance σ_i^k at the central value.

Notice that there are some constraint conditions on each parameter in the above formulation, which are described in the optimization formulation in the following section.

3 Optimal Learning Method for Belief Rule Bases in FURBER

3.1 Multiple Variable Constrained Nonlinear Optimization Problem

In this section, the learning algorithm is to be incorporated in the context *FURBER* model whose function is to search for optimal belief rule matrix and other knowledge representation parameters simultaneously.

Based on the formulation of the system output, the learning method includes a *constrained nonlinear optimization* problem. This objective can be formulated as the minimization of a nonlinear programming problem expressed as follows:

$$\min . f(\boldsymbol{P})$$
$$\text{s.t. } A(\boldsymbol{P}) = 0$$
$$B(\boldsymbol{P}) \geq 0 \tag{6}$$

where $f(\boldsymbol{P})$ is the objective function, $\boldsymbol{P}$ is the parameter vector or matrix of the system, $A(\boldsymbol{P})$ are the equality functions and $B(\boldsymbol{P})$ are the inequality functions respectively.

The optimization starts with the pre-selected initial values of parameters. Then the nonlinear algorithm is used to iteratively adjust the parameters, until the objective function (6) is minimized. In the proposed method, a *nonlinear objective function* is formulated to minimize the differences between the output of a belief rule base and given data. Parameter specific limits and partial expert judgments are formulated as constraints. These so determined parameters are the optimal settings of the *FURBER*.

In the learning process we must use observations on the input and output to determine the parameters. In the following, we shall assume that we have available a collection of observation pairs $(\boldsymbol{x}, y)$, where $\boldsymbol{x}$ is an input vector, and y is the corresponding output. Besides the ability to handle various types of input information, as one may see from the output of the *FURBER*, it also provides the flexibility of the output status, i.e., the qualitative output forms (subjective judgment).

Notice that the output of *FURBER* is actually a distribution assessment instead of a single numerical score, which provides a panoramic view about the output status, from which one can see the variation between the original output and the revised output on each linguistic term. A distribution is easier to understand and flexible to represent output information than a single average value. Especially it is very useful in the case that the outputs are too subjective to quantify them. In fact, this subjective judgment with belief is quite popular in such area as diagnosis, classification, prediction etc. For instance, a physician wishes to predict the likelihood of a new patient's falling ill given his symptoms by analyzing a database of previous patient symptoms and outcomes. Teachers wish to determine which students need extra help given their aptitude test score by analyzing a database of past student test scores and performance. Manufactures wish to determine which parts will fail under stress by analyzing a database of manufacturing parameters and previous part failures and so on.

Hence, for example, for some linguistic terms (such as *good, fair, intelligent, smart,* or *beautiful,* etc.), which do not have clearly defined bases, it is too subjective to clearly define them. If they appear in the consequent part of a rule, it is better to draw a conclusion which has the same linguistic values as the one in the consequent but with a different degree of confidence, i.e., we do not try to change the linguistic values used in the consequent of the rule, but a degree of confidence of this conclusion is added. The optimization approach proposed in the present paper provides more flexibility for optimal training, i.e., it is not required that the training sample should be numerical data pairs, which can be subjective judgment using linguistic values with belief.

3.2 The Optimization Algorithm Based on the Output in the Form of the Subjective Judgment

In this case, a training set composed of M input-output pairs $(x_m, y_m)(m = 1, \ldots, M)$, where y_m can be a subjective judgment, i.e., a distributed assessment on the linguistic value with belief. Notice that the single judgment as one linguistic value can be regarded as a special case of the distribution assessment.

In this paper the Matlab Optimization Toolbox is applied [5]. Since the function to be minimized and the constraints are both continuous, for this case based on the output in the form of the subjective judgment is to solving problems with multi-objective functions using Fminimax in Matlab [5] while the output is given as the expert judgment using a belief distribution. Based on (4), i.e., the final conclusion generated by aggregating the L rules, which are activated by the actual input can be represented as follows

$$\{(D_j, \beta_j), j = 1, \ldots, N\} \tag{7}$$

So each $(D_j, \beta_j)(j \in \{1, \ldots, N\})$ can be regarded as one component of the multi-objective function vector. The function to be minimized is continuous. This multi-objective function is solved using Fminimax function in MATLAB referred to as the *minimax problem* defined as follows:

$$\min_{Q} \max_{\{\xi_j\}} \{\xi_j(Q); j = 1, \ldots, N\} \tag{8}$$

where

$$\xi_j(Q) = \frac{1}{M} \sum_{m=1}^{M} (\beta_j(m) - \hat{\beta}_j(m))^2, j = 1, \ldots, N \tag{9}$$

Here Q is the tuning parameter vector. $\beta_j(m)$ is given by (5) for the mth input in training set. M is the number of points in the training set, $\hat{\beta}_j(m)$ is the expected belief corresponding to the individual consequent $D_j.(\beta_j(m) - \hat{\beta}_j(m))$ is the residual at the mth point. The tuning parameters are beliefs, weights and parameters of fuzzy membership function, without utilities. Equation (8) is an N-objective and multi-variable nonlinear optimization problem. The constraint conditions are given as follows:

$$0 \leq \beta_{jk} \leq 1, j = 1, \ldots, N; k = 1, \ldots, L \qquad (10a)$$

$$\sum_{j=1}^{N} \beta_{jk} = 1 \qquad (10b)$$

$$0 \leq \delta_i \leq 1, i = 1, \ldots, T \qquad (10c)$$

$$0 \leq \theta_k \leq 1, k = 1, \ldots, L \qquad (10d)$$

$$lb_{MF} \leq c_{ij} \leq ub_{MF}, i = 1, \ldots, 3; j = 1, \ldots, J_i \qquad (10e)$$

$$c_{ij} \leq c_{ik} \text{ if } j \leq k, j, k = 1, \ldots, J_i \qquad (10f)$$

$$0 \leq \sigma_{ij}, i = 1, \ldots, 3; j = 1, \ldots, J_i \qquad (10g)$$

lb_{MF} and ub_{MF} are the bounds of the universal courses of fuzzy membership function. Here, $\sum_{j=1}^{N} \beta_{jk} = 1$, i.e., the optimized rule-base should be complete. Therefore, each generation of the optimization algorithm is used to get the minimal mean square error. There is no nonlinear constraint here, all the constraints are linear. Arrange the problem into the standard form of MATLAB, and use the FMINCON function to solve the problem.

Minimax method [5, 11, 12] is also called *ideal point method* that is to minimize a worst case objective function. In other words, the purpose of minimax formulation strategy is to minimize the maximum relative deviation of the objective function from its minimum objective function value. All the objective functions are evaluated qualitatively according to their functional importance and are assigned weights, $\boldsymbol{\omega} = (\omega_{1,\ldots}, \omega_N)$ indicating the designer's subjective preference. Usually $0 \leq \omega_j \leq 1$, $\sum_{j=1}^{N} \omega_j = 1$. The group of multiple objective function is separated and the optimum is $\boldsymbol{P}^*$ in feasible solution space corresponding to each objective function vector $\xi^*(\boldsymbol{Q}) = \{\xi_j^*(\boldsymbol{Q});\ j = 1, \ldots, N\}$. Specify $\xi_j^*(\boldsymbol{Q}) = \xi_j^*$. So the objective function can be formulated as follows:

$$\min_{\boldsymbol{Q}} \max_{j=1,\ldots,N} \{\varphi_j(\boldsymbol{Q})\} \qquad (11)$$

with

$$\varphi_j(\boldsymbol{Q}) = \frac{\xi_j(\boldsymbol{Q}) - \xi_j^*}{\xi_j^*}, \xi_j^* > 0, j = 1, \ldots, N \qquad (12)$$

The computational steps of the minimax method of a multi-objective optimization problem in (11) are summarized below:

Step 1. Based on the multi-objective function, solve the following single-objective optimization problem with the conventional methods individually (e.g., FMINCON).

$$\min_{\boldsymbol{Q}} \xi_j(\boldsymbol{Q})$$

$$\text{s.t. } 10(a) \sim 10(g) \qquad (13)$$

where $j = 1, \ldots, N$, $\xi_j(\boldsymbol{Q})$ is given by (9), and there are N single-objective optimization problem to solve. Suppose that the optimum is $\boldsymbol{Q}^*$ in feasible solution space and its corresponding function value $\xi^* = \{\xi_j^*; j = 1, \ldots, N\}$.

Step 2. To gain the designer's preference of the objectives, a relative weight vector $\omega = (\omega_1,...,\omega_N)$ is given. Here each component (D_j, β_j) of the multi-objective function is regarded as having the same weight, i.e., equally important.

Step 3. The multi-objective function can be formulated as follows:

$$\min_{Q} \max_{j=1,...,N} \left\{ \omega_j \cdot \frac{\xi_j(Q) - \xi_j^*}{\xi_j^*} \right\}$$

$$\text{s.t.,} 10(a) \sim 10(g) \tag{14}$$

Step 4. Arrange the problem into the standard form of MATLAB, and use the MATLAB function, FMINIMAX, to calculate the result.

This optimization process proposed is a kind of iterative process, the iterative process continues until the mean square error becomes smaller than a specified tolerance which can be determined by the expert within the relevant application context.

4 A Numerical Example

The *FURBER* framework has been applied in [9] modeling system safety of an offshore and marine engineering system: floating production storage offloading (FPSO) system [4, 10], specially focus on collision risk between FPSO and a shuttle tanker due to technical failure during a tandem offloading operation.

4.1 Problem Description and the Optimization Algorithms

In this section, an example within the same application framework in [9] is used to demonstrate how the optimization method can be implemented in safety analysis. As an illustration, we only consider the safety assessment related to controllable pitch propeller (CPP) failure to demonstrate the procedure involved in the optimization of belief rule base and other knowledge representation parameters in *FURBER*.

The three fundamental parameters used to assess the safety level of an engineering system on a subjective basis are the *failure rate* (**FR**), *consequence severity* (**CS**) and *failure consequence probability* (**FCP**). Subjective assessments (using linguistic variables instead of ultimate numbers in probabilistic terms) are more appropriate for safety analysis as they are always associated with great uncertainty, especially for a novel system with high level of innovation. These linguistic assessments can become the criteria for measuring safety levels. The typical linguistic variables used to describe **FR, CS, FCP** of a particular element may be described as follows [8, 14]:

FR describes failure frequencies in a certain period, which directly represents the number of failures anticipated during the design life span of a particular system or an item. To estimate the **FR**, one may choose to use such linguistic terms as *"very low," "low," "reasonably low," "average," "reasonably frequent," "frequent,"* and *"highly frequent."*

CS describes the magnitude of possible consequences, which is ranked according to the severity of failure effects. One may choose to use such linguistic terms as *"negligible," "marginal," "moderate," "critical,"* and *"catastrophic."*

FCP defines the probability that consequences happened gives the occurrence of the event. One may choose to use such linguistic terms as *"highly unlikely," "unlikely," "reasonably unlikely," "likely," "reasonably likely,"* and *"Definite."*

For the detailed definitions of these parameters we refer to [8, 14].

Safety estimate is the only output fuzzy variable used in this study to produce safety evaluation for a particular cause to technical failure. This variable is also described linguistically, which is described and determined by the above parameters. In safety assessment, it is common to express a safety level by degrees to which it belongs to such linguistic variables as *"poor," "fair," "average,"* and *"good"* that are referred to as safety expressions. These linguistic terms do not have clearly defined bases and it is difficult to clearly quantitatively define them. They appear in the consequent part of a rule, it is better to draw a conclusion which has the same linguistic values as the one in the consequent but with different degree of confidence, i.e., a degree of confidence of this conclusion is added. This subjective judgment with belief is popular and useful in such area as safety/risk classification.

Twenty-seven fuzzy rules from a total of 245 rules [9, 14] are extracted and used in our example, which are described in Appendix. They are used as the initial belief rule base in the learning process. Here the linguistic terms for describing **FR** are supposed to be (*average, frequent, highly frequent*), for **CS** (*moderate, critical, catastrophic*), and for **FCP** (*likely, highly likely, definite*) respectively. The definitions of their linguistic terms refer to [9, 14], and the corresponding initial membership function are given by Gaussian function as shown in Fig. 1 in Sect. 4.2.

Here it is assumed that each input parameter may be fed to the proposed safety model in a single deterministic value although there are other input forms possible to address the inherent uncertainty associated with the data as discussed in [9, 14].

Notice that here the training data are given in subjective output forms. So the corresponding optimization formulation is given in (8) for subjective judgment form, where $L = 27, T = 3, N = 4$. The computation steps are given in Sect. 3. The FMINIMAX function is used to solve the problem.

4.2 Experiment Results Base on Distributed Assessment with Belief

For the learning purpose, a set of 14 data is used. What we would like to retrieve from the learning are the tuned and refined beliefs, weights, and parameters of fuzzy membership functions. Here we partition the available data into a training set and a test set commonly tests generalization. The training set is used for parameter training purposes; and, once a goal performance measure value is achieved, the corresponding approximation error on the test data is measured. In the example, the output is given by the subjective judgment, i.e., a distribution assessment with belief.

We use this example as an illustration, i.e., 7 training data for parameter estimation (Table 2). 7 remaining data for test purpose (Table 4). The initial belief matrix and the rule weights are supposed to be given by Expert as shown in [9]. The initial attribute weights are all equal to 1.

Table 3 shows the comparison between the expected distributed assessment and the learning results based on distributed assessment with belief from expert. In this example, the error tolerance for each sub-objective minimization is set to 0.00001 and the maximum iteration is set to 100, and the error tolerance for multi-objective minimization is set to 0.0001 and the maximum iteration is set to 100 to avoid an endless loop in the learning process.

Table 2. 7 training input vaues of three safety related antecedent attributes

Number	Antecedents		
	FR	CS	FCP
1	7.75	8.25	7.6
2	7	8	7.25
3	8	8.5	7
4	7	7	5.5
5	6.5	8	7.5
6	7.15	7.95	7.25
7	7	8.5	7

In Table 3 (in Table 5 as well), G represents "Good;" A represents "Average;" F represents "Fair;" P represents "Poor," respectively.

The initial belief structure of the rule base is given by experts shown in Appendix. The initial fuzzy membership function and the optimized fuzzy membership function are shown in Fig. 1 and Fig. 2 respectively.

After the tuning, a testing data consisting of 7 records (Table 4) is used. The expected and actual values of the output are listed in Table 5.

As one may see from Table 3, there is a big difference between the initial and expected outputs. The big difference is due to the fact that the initial outputs are obtained when the belief rule matrix and the weights have not been tuned. Although the difference is not great, that may be due to the small size of the rule base, however, it would be difficult to obtain such relatively accurate belief rule matrix while the system increases in complexity.

Table 3. Training results comparison based on distributed assessment with belief

	Safety Estimate (Distribution Assessment with Belief)			
	Expected Values of Output			
Number	G	A	F	P
1	0	0.0123	0.3641	0.6236
2	0	0.0033	0.3090	0.6876
3	0	0.0057	0.3735	0.6208
4	0	0.0373	0.7802	0.1825
5	0	0.0640	0.4165	0.5195
6	0	0.0013	0.4179	0.5808
7	0	0.0047	0.6151	0.3802
	Trained Output Values			
Number	G	A	F	P
1	0.0098	0.3462	0.6342	0.0098
2	0.0419	0.4510	0.4652	0.0419
3	0.0060	0.3374	0.6506	0.0060
4	0.0016	0.7677	0.2292	0.0016
5	0.0502	0.4689	0.4307	0.0502
6	0.0359	0.4364	0.4918	0.0359
7	0.0395	0.3503	0.5707	0.0395
	Initial System Output Values Before Training			
Number	G	A	F	P
1	0	0	0.8056	0.1944
2	0	0	0.8310	0.1690
3	0	0	0.7982	0.2018
4	0	0	0.9768	0.0232
5	0	0	0.7808	0.2192
6	0	0	0.8473	0.1527
7	0	0	0.7218	0.2782

Especially, as shown in Table 3 and Table 5, after the parameter optimization, the performance of the system is perfectly achieved. One may notice that the testing results in the above Table indicate that 90% of them are already correct within the specified tolerance 0.01 (except test 6). The error including test 6 would be decreasing while the number of training data is increasing.

As a result, a belief-rule-base matrix, and other knowledge representation parameters can be built from partial knowledge about the output and can be then refined by learning, or it can begin with an empty rule base (randomly generated) that is filled by creating rule-bases from the training data. The optimized belief rule base is given in Table A1 of Appendix.

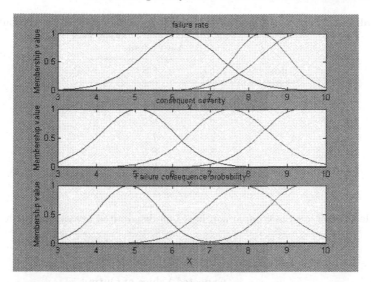

Fig. 1. The initial fuzzy membership function

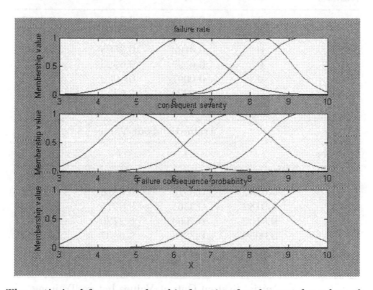

Fig. 2. The optimized fuzzy membership function for the case based on the subjective output

Table 4. 7 test input vaues of three safety related antecedent attributes

		Antecedents	
Number	FR	CS	FCP
1	7.5	8	6
2	7.5	7.2	7.1
3	7.5	8.5	7
4	7	7	6
5	8	8.5	7.5
6	7.25	6.75	7
7	7.95	8.25	7.9

Table 5. Test results comparison based on distributed assessment with belief

	Safety Estimate (Distribution Assessment with Belief)			
	Expected Values of Output			
Number	G	A	F	P
1	0	0.0041	0.6142	0.3817
2	0	0.0080	0.3694	0.6226
3	0	0.0102	0.3595	0.6303
4	0	0.0097	0.6926	0.2977
5	0	0.0097	0.3930	0.5973
6	0	0.0200	0.5733	0.4067
7	0	0.0256	0.2688	0.7056
	Trained Output Values			
Number	G	A	F	P
1	0.0062	0.6956	0.2920	0.0062
2	0.0182	0.3818	0.5818	0.0182
3	0.0182	0.3561	0.6075	0.0182
4	0.0102	0.7069	0.2726	0.0102
5	0.0059	0.3386	0.6497	0.0059
6	0.0278	0.4116	0.5328	0.0278
7	0.0060	0.3498	0.6383	0.0060
	Initial System Output Values before Training			
Number	G	A	F	P
1	0	0	0.9602	0.0398
2	0	0	0.8684	0.1316
3	0	0	0.8084	0.1916
4	0	0	0.9632	0.0368
5	0	0	0.7756	0.2244
6	0	0	0.8050	0.1950
7	0	0	0.7799	0.2201

5 Conclusion

A learning method for optimally training a fuzzy rule base with the belief structure in FURBER for engineering system safety analysis is proposed. This learning method provides practical support to construct flexible and reliable belief rule bases, which can optimally imitate complex reasoning processes and represent nonlinear or nonsmooth relationships using both human knowledge and numerical data. The unique feature of the new method is that the output data can be judgmental, which makes the learning process flexible and practical in decision making.

Acknowledgements

This work forms part of the projects supported by the UK Engineering and Physical Sciences Research Council (EPSRC) under Grant No: GR/S85498/01 and Grant No: GR/S85504/01.

Appendix

Table A1. The optimized belief rule base

Rule Number	Trained Belief Structure of the Rule			
	Good	Average	Fair	Poor
1	0.0020	0.0001	0.9434	0.0545
2	0.0155	0.0250	0.5058	0.4537
3	0.0013	0.0003	0.8364	0.1620
4	0.0108	0.0038	0.9852	0.0001
5	0.2144	0.1123	0.0230	0.6503
6	0.0004	0.0000	0.0000	0.9996
7	0.0604	0.0073	0.9323	0.0000
8	0.2699	0.1107	0.0256	0.5939
9	0.0005	0.0000	0.0000	0.9995
10	0.0000	0.0000	0.9996	0.0004
11	0.0002	0.0002	0.9820	0.0177
12	0.0002	0.0015	0.0002	0.9981
13	0.0231	0.0010	0.9717	0.0042
14	0.0323	0.0135	0.0043	0.9499
15	0.0635	0.0215	0.0163	0.8986
16	0.0537	0.0186	0.0002	0.9274
17	0.2584	0.1029	0.0013	0.6374
18	0.0916	0.0391	0.0033	0.8660
19	0.0000	0.0000	1.0000	0.0000
20	0.0000	0.0001	0.9977	0.0022
21	0.0000	0.0000	0.0000	1.0000
22	0.0030	0.0001	0.9959	0.0010
23	0.0450	0.0159	0.0010	0.9381
24	0.0796	0.0312	0.0011	0.8882
25	0.0066	0.0025	0.0000	0.9908
26	0.0780	0.0271	0.0009	0.8941
27	0.0865	0.0287	0.0003	0.8845

For example, the first rule (the third row) in this table represents the following rule:

Rule # 137: *IF the* **failure rate** *is average AND the* **consequence severity** *is catastrophic AND the* **failure consequence probability** *is likely THEN the* **safety estimate** *is* {(*good*, 0.0020), (*average*, 0.001), (*fair*, 0.9434), (*poor*, 0.0545)}

RULE BASE using in the case study which is extracted from the rule-base established in [14]

1. Rule # 137: *IF the* **failure rate** *is average AND the* **consequence severity** *is catastrophic AND the* **failure consequence probability** *is likely THEN the* **safety estimate** *is fair*
2. Rule # 139: *IF the* **failure rate** *is average AND the* **consequence severity** *is catastrophic AND the* **failure consequence probability** *is highly likely THEN the* **safety estimate** *is poor*
3. Rule # 140: *IF the* **failure rate** *is average AND the* **consequence severity** *is catastrophic AND the* **failure consequence probability** *is definite THEN the* **safety estimate** *is poor*
4. Rule # 193: *IF the* **failure rate** *is frequent AND the* **consequence severity** *is moderate AND the* **failure consequence probability** *is likely THEN the* **safety estimate** *is fair*
5. Rule # 195: *IF the* **failure rate** *is frequent AND the* **consequence severity** *is moderate AND the* **failure consequence probability** *is highly likely THEN the* **safety estimate** *is fair*
6. Rule # 196: *IF the* **failure rate** *is frequent AND the* **consequence severity** *is moderate AND the* **failure consequence probability** *is definite THEN the* **safety estimate** *is poor*
7. Rule # 200: *IF the* **failure rate** *is frequent AND the* **consequence severity** *is critical AND the* **failure consequence probability** *is likely THEN the* **safety estimate** *is fair*
8. Rule # 202: *IF the* **failure rate** *is frequent AND the* **consequence severity** *is critical AND the* **failure consequence probability** *is highly likely THEN the* **safety estimate** *is poor*
9. Rule # 203: *IF the* **failure rate** *is frequent AND the* **consequence severity** *is critical AND the* **failure consequence probability** *is definite THEN the* **safety estimate** *is poor*
10. Rule # 207: *IF the* **failure rate** *is frequent AND the* **consequence severity** *is catastrophic AND the* **failure consequence probability** *is likely THEN the* **safety estimate** *is poor*
11. Rule # 209: *IF the* **failure rate** *is frequent AND the* **consequence severity** *is catastrophic AND the* **failure consequence probability** *is highly likely THEN the* **safety estimate** *is poor*

12. Rule # 210: *IF the **failure rate** is frequent AND the **consequence severity** is catastrophic AND the **failure consequence probability** is definite THEN the **safety estimate** is poor*

13. Rule # 214: *IF the **failure rate** is highly frequent AND the **consequence severity** is negligible AND the **failure consequence probability** is likely THEN the **safety estimate** is fair*

14. Rule # 216: *IF the **failure rate** is highly frequent AND the **consequence severity** is negligible AND the **failure consequence probability** is highly likely THEN the **safety estimate** is fair*

15. Rule # 217: *IF the **failure rate** is highly frequent AND the **consequence severity** is negligible AND the **failure consequence probability** is definite THEN the **safety estimate** is fair*

16. Rule # 221: *IF the **failure rate** is highly frequent AND the **consequence severity** is marginal AND the **failure consequence probability** is likely THEN the **safety estimate** is fair*

17. Rule # 223: *IF the **failure rate** is highly frequent AND the **consequence severity** is marginal AND the **failure consequence probability** is highly likely THEN the **safety estimate** is fair*

18. Rule # 224: *IF the **failure rate** is highly frequent AND the **consequence severity** is marginal AND the **failure consequence probability** is definite THEN the **safety estimate** is fair*

19. Rule # 228: *IF the **failure rate** is highly frequent AND the **consequence severity** is moderate AND the **failure consequence probability** is likely THEN the **safety estimate** is fair*

20. Rule # 230: *IF the **failure rate** is highly frequent AND the **consequence severity** is moderate AND the **failure consequence probability** is highly likely THEN the **safety estimate** is fair*

21. Rule # 231: *IF the **failure rate** is highly frequent AND the **consequence severity** is moderate AND the **failure consequence probability** is definite THEN the **safety estimate** is poor*

22. Rule # 235: *IF the **failure rate** is highly frequent AND the **consequence severity** is critical AND the **failure consequence probability** is likely THEN the **safety estimate** is fair*

23. Rule # 237: *IF the **failure rate** is highly frequent AND the **consequence severity** is critical AND the **failure consequence probability** is highly likely THEN the **safety estimate** is poor*

24. Rule # 238: *IF the **failure rate** is highly frequent AND the **consequence severity** is critical AND the **failure consequence probability** is definite THEN the **safety estimate** is poor*

25. Rule # 242: *IF the **failure rate** is highly frequent AND the **consequence severity** is catastrophic AND the **failure consequence probability** is likely THEN the **safety estimate** is poor*

26. Rule # 244: IF the **failure rate** is *highly frequent* AND the **consequence severity** is *catastrophic* AND the **failure consequence probability** is *highly likely* THEN the **safety estimate** is *poor*

27. Rule # 245: IF the **failure rate** is *highly frequent* AND the **consequence severity** is *catastrophic* AND the **failure consequence probability** is *definite* THEN the **safety estimate** is poor

References

1. Bell P.M. and Badiru A.B. (1996), Fuzzy modelling and analytic hierarchy processing to quantify risk levels associated with occupational injuries – Part I: The development of fuzzy-linguistic risk levels, *IEEE Transactions on Fuzzy Systems*, Vol. 4, No. 2, pp. 124–131.
2. Binaghi E. and Madella P. (1999), Fuzzy Dempster–Shafer reasoning for rule-based classifiers, *International Journal of Intelligent Systems*, Vol. 14, pp. 559–583.
3. Bowles J.B. and Pelaez C.E. (1995), Fuzzy logic prioritisation of failures in a system failure mode, effects and criticality analysis, *Reliability Engineering and System Safety*, Vol. 50, pp. 203–213.
4. Chen H. and Moan T. (2002), Collision risk analysis of FPSO-tanker offloading operation, *21st International Conference on Offshore Mechanics & Arctic Engineering*, 23–28 June 2002, Oslo, Norway.
5. Coleman T., Branch M.A, and Grace A. (1999), *Optimization Toolbox – for Use with Matlab*, The Mathworks Inc.
6. Dempster A.P. (1968), A generalization of Bayesian inference, *Journal of the Royal Statistical Society, Series B*, Vol. 30, pp. 205–247.
7. Duckstein L. (1994), Elements of fuzzy set analysis and fuzzy risk, In *Decision Support Systems in Water Resources Management*, H.P. Nachtnebel (Ed.), pp. 410–430. Paris: UNESCO Press.
8. Liu J., Yang J.B., Wang J. and Sii H.S. (2003), Review of uncertainty reasoning approaches as guidance for maritime and offshore safety-based assessment, *J. of UK Safety and Reliability Society*, Vol. 23, No. 1, pp. 63–80.
9. Liu J., Yang J.B., Wang J., Sii H.S., Wang Y.M. (2004), Fuzzy rule-based evidential reasoning approach for safety analysis, *International Journal of General Systems*, **33** (2–3), 183–204.
10. McCaul J.R. (2001), Special Report – Floating production systems, *Oil & Gas Journal*, June, 11.
11. Marler R.T. and Arora J.S. (2004), Survey of multi-objective optimization methods for engineering, *Struct Multidisc Optim.*, Vol. 26, pp. 369–395.
12. Miettinen K. (1999), *Nonlinear Multiobjective Optimization*. Boston: Kluwer Academic Publishers
13. Shafer G. (1976), *A Mathematical Theory of Evidence*. Princeton, N.J.: Princeton University Press.
14. Sii H.S. and Wang J. (2002), *Safety assessment of FPSOs – The process of modelling system safety and case studies*, Report of the project – *"The Application of Approximate Reasoning Methodologies to Offshore Engineering Design"* (EPSRC GR/R30624 and GR/R32413), Liverpool John Moores University, UK.
15. Smets P. (1988), Belief function, in *Non-Standard Logics for Automated Reasoning* (Smet P., Mamdani E.H., Dubois D., and Prade H. eds.), London, Academic Press, pp. 253–277.

16. Wang J., Yang J.B. and Sen P. (1995), Safety analysis and synthesis using fuzzy set modelling and evidential reasoning, *Reliability Engineering and System Safety*, Vol. 47, No. 3, pp. 103–118.

17. Wang Y.M., Yang J.B. and Xu D.L. (2003), Environmental impact assessment using the evidential reasoning approach, accepted by *European Journal of Operational Research.*

18. Yang J.B. and Singh M.G. (1994), An evidential reasoning approach for multiple attribute decision making with uncertainty, *IEEE Transactions on Systems, Man, and Cybernetics*, Vol. 24, No. 1, pp. 1–18.

19. Yang J.B. and Sen P. (1994), A general multi-level evaluation process for hybrid MADM with uncertainty, *IEEE Transactions on Systems, Man, and Cybernetics*, Vol. 24, No. 10, pp. 1458–1473.

20. Yang J.B. (2001), Rule and utility based evidential reasoning approach for multi-attribute decision analysis under uncertainties. *European Journal of Operational Research,* Vol. 131, pp. 31–61.

21. Yang J.B. and Xu D.L. (2002a), On the evidential reasoning algorithm for multiple attribute decision analysis under uncertainty. *IEEE Transactions on Systems, Man, and Cybernetics-Part A*, Vol. 32, No. 3, 289–304.

22. Yang J.B. and Xu D.L. (2002b), Nonlinear information aggregation via evidential reasoning in multi-attribute decision analysis under uncertainty. *IEEE Transactions on Systems, Man, and Cybernetics Part A: Systems and Humans*, Vol. 32, No. 3, pp. 376–393.

23. Yang J.B., Liu J., Wang J., and Sii H. S. (2003), A generic rule-base inference methodology using the evidential reasoning approach – RIMER, *IEEE Transactions on Systems, Man, and Cybernetics*, accepted and in press.

24. Zadeh, L. A. (1965), Fuzzy sets, *Information and Control*, Vol. 8, pp. 338–353.

25. Zadeh L.A. (1975), The concept of a linguistic variable and its applications to approximate reasoning. Part I, *Information Sciences*, Vol. 8, pp. 199–249; Part II, *Information Sciences*, Vol. 8, pp. 301–357; Part III, *Information Sciences*, Vol. 9, pp. 43–80.

A User Centred Approach to Management Decision Making

L.P. Maguire, T.A. McCloskey, P.K. Humphreys, and R. McIvor

School of Computing and Intelligent Systems, Faculty of Engineering, University of Ulster, Magee Campus, Northern Ireland, UK
lp.maguire@ulster.ac.uk

Abstract. The management decision making process is becoming increasingly complicated as more detailed and extensive data is available in this information age. The ability of human decisions makers to consistently analyse huge volumes of data and to do so in a repeatedly identical manner is questionable. This uncertainty about consistency and the cost in time and money creates the need for an artificial intelligent system. The system must be capable of processing large quantities of data and imitate the human decision making process but in a more consistent and cost effective manner. The authors employ a user centred fuzzy system, which is based on a hierarchical system, employing scalable fuzzy membership functions. The hierarchical structure of the system is self adjusting to facilitate the particular business problem and user's decision making process. The fuzzy membership functions are scaled to reflect the human precedence given to particular data in the decision making process. The proposed system supports decision making in any data intense management decision making processes. Two case studies are presented, "The supplier selection process" and "The corporate acquisition process". The development of this system is intended to illustrate that a fuzzy system can aid management in the most complicated management decision making processes.

Key words: User-centred decision making; fuzzy reasoning; acquisitions analysis; supplier selection

1 Introduction

A methodology for management decision making based on fuzzy logic is presented. The system employs scalable fuzzy membership functions that implements human priorities in the management decision making process. Fuzzy logic provides a method by which human reasoning can be emulated and decisions can be made with vague and imprecise information. The manipulation of the magnitude of the fuzzy membership functions will enable the authors to employ human priorities on the system to varying degrees and at varying stages of the decision making process. A hierarchical fuzzy system is presented

L.P. Maguire et al.: *A User Centred Approach to Management Decision Making*, Studies in Computational Intelligence (SCI) **5**, 439–437 (2005)
www.springerlink.com   © Springer-Verlag Berlin Heidelberg 2005

that is adjustable to suit all management decision making factors and their degree of importance to the management decision making process. The hierarchical fuzzy system presented in this paper enables the user to implement preference and priorities at varying levels on the system. This facilities the creation of a suitable system for the user, that reflects their requirements in the management decision making process and allows the user to assign preference and varying priority levels. The user will be prompted to identify the particular decision making process undertaken and the relevant factors and sub-factors associated with this process. The factors and sub-factors will be assigned relevant linguistic priorities by the user. This enables a robust system to be created that replicates the general priorities that are considered within the human decision making process of the business.

The first management decision making process undertaken is "The supplier selection process" with particular focus on the inclusion of environmental performance in the supplier selection process. Pressure from governments, institutions and consumers [20] has forced many companies to improve their environmental performance [1, 3]. Environmentally conscious firms, mainly large companies, are developing environmental programmes aimed at organising their supply chains [7, 13]. It is now widely acknowledged that environmental issues must be considered as strategic in a growing number of industries because of market pressures and the threat of environmental regulations [22, 36]. The inclusion of environmental performance in the supplier selection process introduces additional data into an already extensive process. This is viewed as ideal opportunity for the introduction of a computational system that is a user centred approach to management decision making.

The second management decision making process undertaken is "The corporate acquisition system". Changes in the global environment of many industries have impacted upon how individual firms in those sectors must compete to survive, [14]. A strategy that companies have adopted to meet the international challenge has been international expansion via acquisitions [23]. Identifying suitable companies in the vastly diverse and escalating markets requires considerable research effort from the company's financial experts, whose time represents a considerable investment [28]. This increased cost and workload also provides the opportunity for a computational system to replace or assist in the acquisition process. This chapter illustrates the potential benefits of the management decision making system by providing results obtained for both the supplier selection and corporate acquisition decision making processes.

2 The Development of a Fuzzy Based System

2.1 Fuzzy Inference Method

The fuzzy inference method used in this system is the Takagi-Sugeno-Kang (TSK) which was introduced in 1985 [30, 31, 32]. The TSK method was selected rather than the Mamdani's fuzzy inference method [19], as it is more computationally efficient and it works well with optimisation and adaptive techniques [8].

2.2 Input Membership Function

The membership functions for the inputs to each fuzzy system are determined by three major factors:

- The total range of all the membership functions, the universe of discourse. The system determines how each input in the database performs in relation to the strongest and weakest inputs.
- The second factor is how the data is dispersed between the strongest and weakest inputs.
- The last is the priority level given to the input in the system; this will determine the maximum degree of membership possible for each membership function.

2.3 Membership Range

The range of the membership functions in any input is determined by the strongest and weakest value retrieved in the input data. All the input data is normalised with the strongest input value set as 1 (x-axis) and the weakest value set as 0 (x-axis). The authors have selected five membership functions across each universe of discourse although increasing the number of membership functions may improve the model accuracy but it will increase computational demands. The five membership functions have been termed "Very Poor", "Poor", "Average", "Good" and "Very Good". Assuming that the input is equally dispersed, the membership functions were evenly divided across the range. Using these membership functions, each supplier in the sector under analysis is assigned a membership function based upon its position in the range. The degree of membership of each function would relate to the shape of the membership function used, in this case a triangular shaped membership function. The Triangular curve is a function of a vector x, and depends on three scalar parameters a, b, and c as given by:

$$f(x; a, b, c) = \begin{cases} 0, & x \le a \\ \frac{x-a}{b-a}, & a \le x \le b \\ \frac{c-x}{c-b}, & b \le x \le c \\ 0, & c \le x \end{cases} \tag{1}$$

2.4 Width of Individual Membership Functions to Cover Data Dispersion

If data is dispersed evenly across the membership range then the fuzzy membership functions are divided evenly over the range with partition of unity.

However from analysis of the data it was apparent that a small number of suppliers were present at the extremes of the membership range and that the data was not evenly dispersed across the membership range. In these circumstances the fuzzy membership functions are altered. In the range were data is concentrated the width of the fuzzy membership functions is narrowed and in the areas of sparse data the width of the membership function is widened. This widening and narrowing of the membership functions attempts to create an even distribution of companies in each membership function. In order to calculate how the membership functions are narrowed or widened for each membership functions three points are found in the range. The three points correspond to the "b" parameter or the peaks of $mf2, mf3$ and $mf4$. The "b" parameter of $mf1$ and $mf5$ are set to 0 and 1 respectively. The other three "b" parameters or peaks are calculated using the following formulas:

$$mf2(b) = \frac{\sum_{i=1}^{n} x_i}{2(n)} \tag{2}$$

$$mf3(b) = \frac{\sum_{i=1}^{n} x_i}{n} \tag{3}$$

$$mf4(b) = \left(\frac{1 - \frac{\sum_{i=1}^{n} x_i}{n}}{2} \right) + \frac{\sum_{i=1}^{n} x_i}{n} \tag{4}$$

Where n = number of inputs for x.

2.5 Priority Levels and Scaling

The system hierarchy is presented to the user with all inputs identified to each fuzzy system. The user is then prompted to assign a priority level for each input in relation to the other inputs of that fuzzy system. These priorities are set once on the system and do not require amendment unless the user's business or perspective changes. The authors in have selected five levels of priority within the system:

- Very High Priority
- High Priority
- Medium Priority
- Low Priority
- Very Low Priority

The standard membership function allows a degree of membership from 0 to 1. The proposed scaling of the membership functions replaces this membership function for each input with a scaled membership function. This scaling changes the membership functions in accordance to the priority level given to the input. The calculated scaling values of this system are as follows:

Very High Priority	1.0
High Priority	0.8
Medium Priority	0.6
Low Priority	0.4
Very Low Priority	0.2

The triangular function defined in equation (1) is altered to enable the degree of membership of a function to be changed. The triangular curve is still a function of the vector "x", but now depends on four scalar parameters a, b, c and d. The "d" parameter determines the maximum degree of membership for the membership function. The triangular function is given by:

$$f(x; a, b, c) = \begin{cases} 0, & x \leq a \\ \frac{(x-a)d}{b-a}, & a \leq x \leq b \\ \frac{(c-x)d}{c-b}, & b \leq x \leq c \\ 0, & c \leq x \end{cases} \tag{5}$$

This scaling determines how influential a particular input can be; as the degree of membership for the input is limited this limiting factor determines how influential the input is on the output of its fuzzy system. This scaling determines the influential levels of inputs to outputs throughout the whole system.

2.6 Rules for the Fuzzy Systems

Each fuzzy system produces an output from their respective inputs. This output is determined by the rules employed by the fuzzy system. The combination of rules that are fired and the firing strength of the rule determine the output from the fuzzy system. The number of rules defined in this system is a dependant on the number of membership functions (p) in each input (total of n inputs).

$$\text{The number of rules} = p^n \tag{6}$$

A fuzzy hierarchical system proposed, enables the management team to dictate the overall hierarchy of the system dictating hierarchical levels and groupings. This enables the system to reflect the associative groupings and priorities present within the user's decision making process. A level can contain one or more factors but if more than one factor exists then their relationship must be defined with a series of rules. A level with one or more factors combines their outputs so a single output is present for each level of the hierarchy. A benchmark value dictates what percentage of the output from one level will proceed to the next level. Outputs which can be considered as highly unlikely solutions can be discarded at each level. The overall structure of the system is illustrated in Fig. 1 for the supplier selection process. This illustrates how only suppliers that meet a defined benchmark will proceed to the next level, this

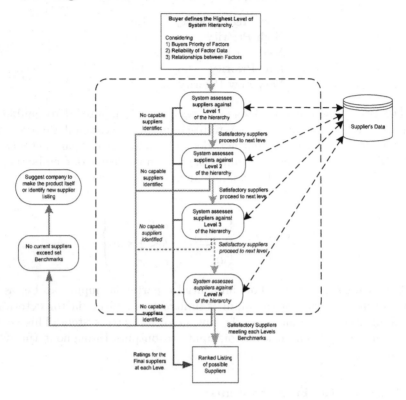

Fig. 1. System Overview

reduces the need to process data for suppliers that are obviously unsuitable and would save time investigating supplier data.

A more detailed look at the fuzzy system developed for the environmental factor and how the ranking system is implemented for this will establish the basic building block of the system. The environmental factor is a self contained fuzzy system which contains other sub-fuzzy systems that represent the other levels of inputs present in the supplier selection process. As shown in Fig. 2 the system uses the sub-factors as inputs to the "Factor fuzzy system", the criteria as inputs to the "Sub-Factor fuzzy system" and the sub-criteria as inputs to the "Criteria fuzzy System". The output of the lower fuzzy systems becomes an input to the higher fuzzy system at all levels as can be seen in Fig. 2. The system uses dynamic scaling of the fuzzy membership functions to prioritise the inputs to each fuzzy system and to enable the degree of influence held by each input to be altered. Each input within the "Environmental factor" contributes to a specified degree to the overall output of the "Environmental factor". The degree of influence for each input is set within each fuzzy system at each level and once set does not require adjustment unless buyer's position

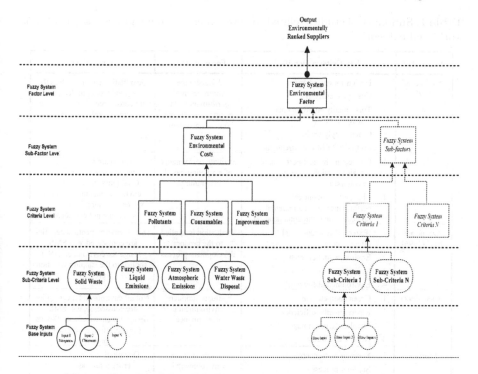

Fig. 2. Supplier Selection Overview Fuzzy Hierarchical Systems

and priorities change. If the buyer does not wish to set any priorities then the system can be set to equal priorities at all levels.

3 Case Study 1: The Supplier Selection System

Historically, several methodologies have been developed for evaluating, selecting and monitoring potential suppliers [9, 35] that take into account factors dealing with, for example, quality, logistics and cost. However, none of these methodologies have considered the importance of environmental factors, such as, life cycle analysis or design for environment in the decision-making process. Further evidence collected by [5] of US and European firms, supports this view, indicating that few companies included environmental attributes in evaluating suppliers. Due to the increasing realisation of the importance of integrating environmental factors into assessing supplier's performance, a number of researchers have begun to identify some possible environmental indicators and criteria [4, 17, 21, 34]. Early case studies indicate that companies have applied different approaches to deal with environmental issues and initially there was no generic model for incorporating environmental criteria into the supplier

Table 1. Summary of studies related to developing environmental assessment frameworks and categories

Researcher	Key Criteria identified	Focus of study	Limitations
Sarkis et al (1996)	1. Design for environment 2. Life cycle analysis 3. Total quality environmental management 4. Green supply chain 5. ISO 14000 EMS requirements	Evaluate the environmental performance of a company's existing operation system	Quantitative factors such as the emission level of pollutants are not considered. Not applied to the supplier selection process.
Noci (1995)	1. Change in physical performance, e.g. air emissions, energy consumption 2. Change in economical performance, e.g. incremental revenues, environmental taxation	Evaluate performance of recycling-based programmes	Criteria are not applied to supplier selection process. Qualitative criteria such as environmental management system and supplier's 'green' image are not considered.
Azzone and Noci (1996)	1. 'External' environmental effectiveness 2. Environmental efficiency 3. 'green' image 4. Environmental flexibility	Evaluation is applied to the product development process	Not all environmental categories are considered, e.g. EMS, design for environment. Not applied to the supplier selection process.
Noci (1997)	1. Green competencies 2. Environmental efficiency 3. Supplier 'green' image 4. Net life cycle cost	Evaluate suppliers' environmental performance	Not all environmental categories are considered, e.g. EMS, design for environment. Details of the selection process are not provided.
Enarrson (1998)	1. Supplier as company 2. Supplier process 3. Product 4. Transportation	Evaluate suppliers' environmental performance	Quantitative environmental criteria such as energy consumption, waste emission levels are not considered. Procedures for selecting suppliers are not provided.

selection process. The key work by researchers on developing environmental frameworks and their limitations are summarised in Table 1.

Within the literature a more systematic approach has been developed by the identification of several environmental categories and criteria. The system is created in a generic form as the supplier selection process is often a very personal process. The system is a combination of a number of self contained fuzzy systems with each system receiving a number of fuzzy or numerical inputs and providing a defuzzified output. This output can then be used to rank the supplier or as input to a further fuzzy system. Each factor under consideration in the supplier selection process requires a fuzzy system; the overall system presented in this paper considers seven supplier selection factors as illustrated in Fig. 3.

The detail involved in the entire system is too vast for this paper and so only the environmental factor has been illustrated. A similar process exists for each of the other six factors identified in Fig. 3 as part of the supplier selection process. The "Environmental Issues" factor has many sub-factors,

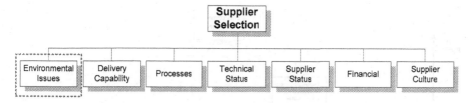

Fig. 3. Supplier Selection Factors

which have been established through consolidating and classifying the factors and sub-factors from the as reviewed earlier. This gives rise to the proposed environmental factors and sub-factors shown in Fig. 4.

Each factor and sub-factor is a self contained fuzzy system and therefore is interchangeable within the overall hierarchy of the system, enabling the user to establish a system that best reflects their decision making process. The importance the buyer places on a particular factor or sub-factor will determine its priority setting, this can be considered as applying weights to each factor

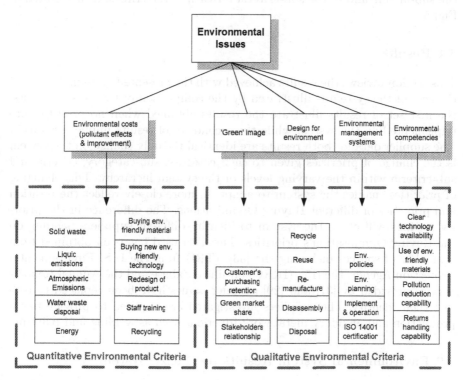

Fig. 4. Environmental framework for incorporating environmental factor into the supplier selection process

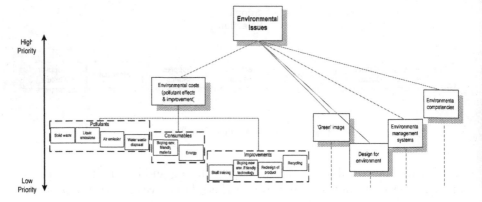

Fig. 5. Environmental Factor, Sub-factors, Criteria and Sub-Criteria

or sub-factor. The reliability of data [6, 11, 12, 18] can also be considered as a weighting element and can be combined with level of importance to form a weighting for each factor. This process is extended to criteria for each of the sub-factors and to the sub-criteria for each of the criteria as illustrated in Fig. 5.

3.1 Results

This section reviews the results achieved with the presented system. The effectiveness of the system is illustrated by the comparison of two sets of results, the first set of results illustrate the results obtained when no priorities are implemented on the system while the second set of results implies priorities. The supplier data for both results are identical the only change on the system is the change of priorities given to each category, sub-category, criteria and sub-criteria within the varying levels of the system hierarchy. This changing of priorities enables the system to adapt, to more closely reflect the position and priorities of different Buying Organisations. The difference in the priority settings will cause the system to identify different suppliers pending on the Buying Organisation's priorities. The data used has been obtained from a number of sources including the [33], (TRI) from the U.S. Environmental Protection Agency, the [15], (IRRC) on-line database and the Annual Reports Service provided by [37] and [29] supplier names have been replaced to protect their identity. The suppliers investigated are required to provide metal production and auxiliary metal work functions such as welding and painting.

3.2 Environmental System Results per Fuzzy System

The results presented are for Buying Organisation 1 which has set priorities and for Buying Organisation 2 which has no set priorities. The results show

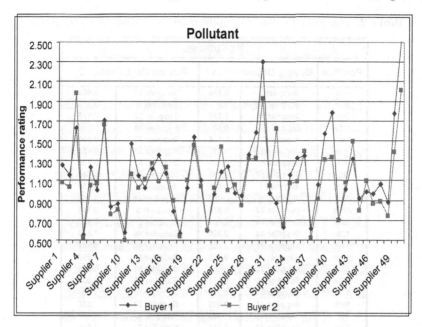

Fig. 6. Pollutant Output for Buying Organisation 1 & 2

the output from the four fuzzy systems that account for the four sub-criteria Solid Waste, Liquid Emissions, Air Emissions and Water Waste from the criteria Pollutants. The Pollutant fuzzy system results are shown in Fig. 6 and Table 2, other results are not shown due to limitation of space.

Table 2 identifies the suitability of suppliers based on the "Pollutant" factor for both Buying Organisation 1 and 2. The most suitable supplier for Buying Organisation 1 is Supplier 4 while Supplier 10 has been identified as the most suitable supplier for Buying Organisation 2. Supplier 10 while being first for Buying Organisation 2 only appears in third position for Buying Organisation 1 appearing behind Supplier 18. This can be identified with the fact that Supplier 10's strong attributes, sub-factors "Liquid Emissions" and "Water Waste" have been decreased to Very Low Priority and Low Priority for Buying Organisation 1. While in sub-factor "Air Emissions" the priority is increased to Very High Priority and Supplier 10 is beaten both by Supplier 4 and Supplier 18. Similar movements can be seen throughout the two listings and are accounted for by the changing in priority levels. Doing well in a very high priority will positively influence a supplier's case for selection but will not be the only determining factor. This system achieves a balance that enables an input to be more influential but without complete control. This enables a simulated human reasoning were one aspect may influence a decision more but not to the extent that it overrides the influence of other aspects.

Table 2. Top Fifty Percent of Suppliers in the Pollutant Fuzzy System

Pollutants				
Position	Buying Organisation 1		Buying Organisation 2	
1st	Supplier 4	0.556	Supplier 10	0.504
2nd	Supplier 18	0.567	Supplier 4	0.520
3rd	Supplier 10	0.575	Supplier 37	0.525
4th	Supplier 22	0.600	Supplier 18	0.540
5th	Supplier 37	0.618	Supplier 22	0.600
6th	Supplier 33	0.632	Supplier 33	0.659
7th	Supplier 41	0.705	Supplier 41	0.700
8th	Supplier 17	0.790	Supplier 48	0.746
9th	Supplier 8	0.835	Supplier 8	0.758
10th	Supplier 9	0.867	Supplier 44	0.794
11th	Supplier 32	0.876	Supplier 9	0.806
12th	Supplier 48	0.883	Supplier 27	0.853
13th	Supplier 44	0.920	Supplier 46	0.866
14th	Supplier 27	0.952	Supplier 47	0.887
15th	Supplier 46	0.965	Supplier 17	0.897
16th	Supplier 23	0.968	Supplier 38	0.910
17th	Supplier 26	0.974	Supplier 25	1.001
18th	Supplier 31	0.977	Supplier 12	1.029
19th	Supplier 45	0.989	Supplier 23	1.030
20th	Supplier 6	1.003	Supplier 2	1.037
21st	Supplier 42	1.014	Supplier 21	1.040
22nd	Supplier 19	1.025	Supplier 31	1.046
23rd	Supplier 13	1.027	Supplier 5	1.048
24th	Supplier 38	1.055	Supplier 26	1.060
25th	Supplier 47	1.068	Supplier 6	1.069

The three outputs from the Pollutant Fuzzy system, Consumables Fuzzy System and Improvements Fuzzy Systems become inputs to the Environmental Cost Fuzzy System this system gives a crisp rating for the sub-factor Environmental cost that is used as an input for fuzzy system on the next level of the hierarchy.

3.3 Environmental System Output Fuzzy System

The final stage of the fuzzy hierarchy for the Environmental Factor is the Output Fuzzy System which summates the outputs from all the sub-factors. The inputs to the Output Fuzzy System are also scaled membership function, the scaling depending on the priority level given to each individual sub-factor. The results obtained from the Output Fuzzy System provide a rating for each supplier which indicated how suitable it would be for a particular Buying Organisation. For Buying Organisation 1 and 2 the suppliers identified for selection are presented in Fig. 7 and Table 3.

The company with the lowest output value is considered to be the most suitable supplier for selection for the particular Buying Organisation. As

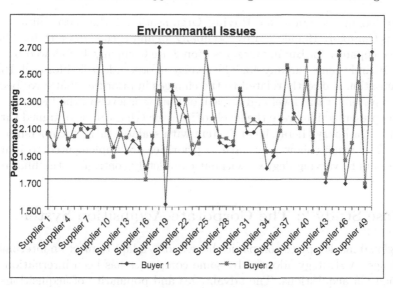

Fig. 7. Comparing Supplier Output Ratings for Buying Organisation 1 and 2

Table 3. Top Fifty Percent of Suppliers in Environmental Issues

Environmental Issues				
Position	Buying Organisation 1		Buying Organisation 2	
1st	Company 19	1.520	Company 49	1.666
2nd	Company 49	1.646	Company 16	1.695
3rd	Company 46	1.670	Company 43	1.738
4th	Company 43	1.678	Company 19	1.781
5th	Company 16	1.779	Company 46	1.837
6th	Company 34	1.782	Company 11	1.860
7th	Company 35	1.871	Company 35	1.901
8th	Company 23	1.888	Company 41	1.902
9th	Company 13	1.892	Company 44	1.904
10th	Company 44	1.917	Company 34	1.904
11th	Company 15	1.930	Company 2	1.949
12th	Company 11	1.933	Company 23	1.951
13th	Company 28	1.939	Company 24	1.957
14th	Company 2	1.941	Company 47	1.965
15th	Company 4	1.947	Company 29	1.974
16th	Company 29	1.947	Company 4	1.992
17th	Company 17	1.958	Company 28	1.996
18th	Company 47	1.964	Company 15	1.999
19th	Company 27	1.967	Company 13	2.000
20th	Company 14	1.984	Company 7	2.007
21st	Company 41	2.000	Company 27	2.007
22nd	Company 24	2.008	Company 5	2.010
23rd	Company 31	2.041	Company 17	2.017
24th	Company 32	2.042	Company 12	2.019
25th	Company 1	2.045	Company 1	2.028

illustrated in the graph for Buying Organisation 1 the most suitable supplier is Company 19 followed by Company 49 and then Company 46. The top three companies for Buying Organisation 2 are Company 49 followed by Company 16 and then Company 43. The change in order identified for each Buying Organisation is directly related to the changes in priorities that have been set on the system at this level and at each of the lower levels of the hierarchy. The results presented illustrate the levels of influence that can be obtained through the use of a Fuzzy Hierarchical System with scalable fuzzy membership functions. The results show how natural priorities are implemented to influence the results to varying degrees without completely controlling the final result.

4 Case Study 2: The Corporate Acquisition System

In a global market place companies have had to adapt to meet an international challenge. A strategy adopted by some companies has been international expansion via acquisitions. The advantages and popularity of acquisitions as a policy for corporate development are well established [16, 26]. The need for expert knowledge to determine an appropriate company to acquire has been complicated by the sheer size of the global market place. The costs associated with this in relation to time and personnel have created the need for a computerised expert system to be developed. The corporate acquisition system endeavours to show how the proposed fuzzy based system can assist in the identification of a company for acquisition.

The system obtains a precise set of aims and objectives from the "Acquiring Organisation" and the motivation they have in considering the acquisition of another company. Then utilising this information the system creates a hierarchical structure on which the system is based. Within this hierarchical system, financial categories and ratios are identified that are deemed relevant to the particular "Acquiring Organisation" and their aims and objectives. In addition a quality input may be added by the "Acquiring Organisation" that is not reflected in financial statistics such as the quality of a product, quality of management, quality of in-house machinery or even the perceived quality of a brand by the public.

An overview of the corporate acquisition system is shown in Fig. 8. The model represents the basic hierarchical nature of the decision making process for corporate acquisition. The focal point for the corporate acquisition system is "Stage 3" in the acquisition process.

The number of financial categories and ratios required in the system and how they are placed within the hierarchy determines the structure of the system. This in conjunction with the priorities given to each financial category and ratio determine the output of the system. For the purposes of this case study four categories have been identified: Profitability, Liquidity, Efficiency, and Financial Strength and Gearing. Within each category financial ratios are selected that are considered to offer the best information about that particular

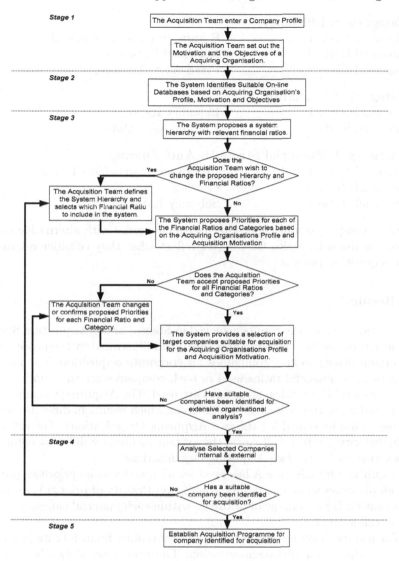

Fig. 8. Overview of Acquisition System

category. The financial categories and ratios selected for this demonstration are:

- **Category 1: Profitability**
  Financial Ratio 1:   Gross Profit Percentage
  Financial Ratio 2:   Net Profit Percentage
  Financial Ratio 3:   Profit per Employee

- **Category 2: Efficiency**
  Financial Ratio 1:   Return on Capital Employed
  Financial Ratio 2:   Net Asset Turnover
  Financial Ratio 3:   Fixed Asset Turnover

- **Category 3: Liquidity**
  Financial Ratio 1:   Liquidity Ratio
  Financial Ratio 2:   Current Asset Ratio

- **Category 4: Financial Strength And Gearing**
  Financial Ratio 1:   Return on Shareholders Funds
  Financial Ratio 2:   Gearing
  Financial Ratio 3:   Solvency Ratio

Different companies may chose alternative categories with alternative combinations of financial ratios that better reflect what they consider relevant in their acquisition process.

4.1 Results

This section reviews the results achieved in Stage 3 of the Acquisition System. The merits of the Acquisition system are again illustrated by the processing of two organisations seeking companies for corporate acquisition. The financial categories and financial ratios used in both companies are the same, and the same, two level hierarchical structures are used. The Acquiring Organisation's profile and aims and objectives are different which results in different priority settings being identified for the two Acquiring Organisations. The difference in the priority settings will cause the system to identify different companies for acquisition for the two Acquiring Organisations.

Acquiring Organisation A has been set with no financial priorities (i.e. setting all priorities to an equal Very High Level Priority of priority). Acquiring Organisation B has a range of priorities within each financial category and for the overall system.

The results show the output from the Profitability financial category which has been chosen for this demonstration. The results are shown in graphical form so comparison of results can be easily achieved. Due to the nature of the membership functions the important fact is not in what position a company achieves although this is a good indicator; it is the magnitude of difference that is achieved in comparison to the other companies and the position in relation to the average of all the companies. The average is used to determine the position of all the membership functions therefore a companies position in relation to the average ratio will determine which membership functions it will fall under. The results from the Profitability Fuzzy System are analysed. The output results from the Profitability Fuzzy System are shown in Fig. 9 and Table 4.

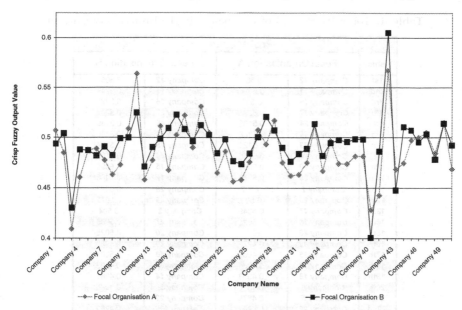

Fig. 9. Profitability Output for Acquiring Organisation A and B

Table 4 identifies the order of the companies from the Profitability Fuzzy System for both Acquiring Organisation A and B. As can be seen in the table the general suitability of companies for the two Acquiring Companies is different. This is due to the changed priorities within the system.

The strongest prospect for acquisition for both Acquiring Organisations is Company 42, but it is a stronger prospect for Acquiring Organisation B. This is due to the fact that Net Profit is given a Very High Priority in Acquiring Organisation B and Company 42 is the best company in this financial ratio by a strong margin. Company 11 which is returned in second position for both Acquiring Organisations is in 39th position in the Net Profit ratio but still holds second position when Net Profit is given a Very High Priority. This is due to the fact that although Net Profit is Very High Priority it is not the only determiner and because Company 11 has done well in the other financial ratios, 2nd in Gross Profit ratio and 1st in the Profit per Employee ratio it hold its second position. Company 16 which obtained third position for Acquiring Organisation B, is in 14th position for Acquiring Organisation A. It achieves this rise in position due to the fact that the Net Profit ratio for Company 16 is in 5th position and has a value of 0.6734 well above the average of 0.493. Company 16 does not do as well in the other two categories but as they are of a lower priority they have less influence. The fact that Company 16 has done better than other companies that are above it in the Net Profit ratio such as Company 44, 27 and 15 is how it secures third position for Acquiring Organisation B. This illustrated how doing well in a high priority financial

Table 4. Top Fifty Percent of Companies in the Profitability Category

Profitability				
Position	Focal Organisation A		Focal Organisation B	
1st	Company 42	0.567	Company 42	0.6051
2nd	Company 11	0.5636	Company 11	0.5249
3rd	Company 19	0.531	Company 16	0.5227
4th	Company 17	0.5221	Company 27	0.5207
5th	Company 28	0.5169	Company 49	0.5139
6th	Company 49	0.5148	Company 33	0.5136
7th	Company 33	0.5121	Company 19	0.5124
8th	Company 14	0.5116	Company 44	0.5106
9th	Company 10	0.5089	Company 15	0.5095
10th	Company 26	0.5075	Company 17	0.5085
11th	Company 1	0.507	Company 28	0.5073
12th	Company 20	0.505	Company 45	0.5072
13th	Company 47	0.5049	Company 2	0.504
14th	Company 16	0.5025	Company 47	0.5035
15th	Company 46	0.5	Company 20	0.5025
16th	Company 45	0.4972	Company 26	0.5005
17th	Company 35	0.4967	Company 10	0.5001
18th	Company 27	0.4934	Company 9	0.4992
19th	Company 18	0.4903	Company 14	0.4988
20th	Company 6	0.4889	Company 38	0.4985
21st	Company 5	0.4879	Company 22	0.4984
22nd	Company 22	0.4863	Company 39	0.4982
23rd	Company 2	0.4849	Company 36	0.4976
24th	Company 48	0.4843	Company 18	0.4965
25th	Company 38	0.4814	Company 37	0.4958

ratio will heavily influence a company's case for acquisition but will not be the only determining factor. As in the case with Company 11 the combination of doing well in a number of the lower priority financial ratios also heavily influenced the company's case for acquisition. This system achieves a balance that enables a financial ratio to be more influential but without complete control.

4.2 Acquisition System Output Fuzzy System

The final stage of the fuzzy hierarchy is the Output Fuzzy System which summates the outputs from all the Financial Category Fuzzy Systems. The inputs to the Output Fuzzy System are also scaled membership function, the scaling depending on the priority level given to each individual Financial Category in the context of the entire system. The results obtained from the Output Fuzzy System provide a rating for each company which indicated how suitable it would be for corporate acquisition for a particular Acquiring Organisation. The system or user can then select a number of the top companies identified for further analysis.

For Acquiring Organisations A and B the companies identified for corporate acquisition are presented in Fig. 10 and Table 5. The company with

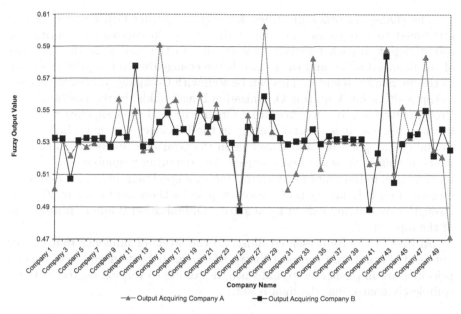

Fig. 10. Company Output Ratings for Acquiring Company A and B

the highest output value is considered to be the most suitable company for acquisition for the particular Acquiring Organisation. As illustrated in the graph for Acquiring Organisation A the most suitable company for acquisition is Company 27 followed by Company 42 and then Company 14. The top three companies for Acquiring Organisation B are Company 42 followed by Company 14 and then Company 10.

Table 5. Top Ten Companies System Output

	System Output			
Position	Focal Organisation A		Focal Organisation B	
1st	Company 27	0.544	Company 42	0.5425
2nd	Company 42	0.5394	Company 14	0.5401
3rd	Company 14	0.5389	Company 10	0.5378
4th	Company 11	0.5374	Company 28	0.5377
5th	Company 28	0.5374	Company 17	0.5376
6th	Company 19	0.5365	Company 35	0.5375
7th	Company 44	0.5365	Company 18	0.5372
8th	Company 17	0.5363	Company 37	0.5372
9th	Company 47	0.5362	Company 33	0.5371
10th	Company 46	0.5361	Company 32	0.537

The change in order identified for each Acquiring Organisation can be attributed to two issues; the first is the change in companies identified as the strongest in each financial category due to the change in the priorities of the financial ratios and the second is the change due to the priorities given to the financial categories. This can be seen with Company 42 which has risen to top position for Acquiring Organisation B as it holds the top position in the Profitability Category for Acquiring Organisation B by a significant margin but also because it is around average in the Efficiency Category, high in the Liquidity Category and above average in the Gearing Category. Company 27 which is identified as the best company for Acquiring Organisation A has its real strength in Liquidity in which it is top by a good margin but for Acquiring Organisation B Liquidity is of a very low priority therefore Company 27's real strength is not of any benefit to Acquiring Organisation B and so it drops out of the top ten.

Again the results presented illustrate the levels of influence that can be obtained through the use of a Fuzzy Hierarchical System and how natural priorities are implemented to influence the results to varying degrees without completely controlling the final result.

5 Conclusions

Companies are investing a considerable amount in both financial and employee resources to varying business problems, which are becoming ever more complicated by the detail and mass of data available. This chapter has looked at two of these problems and identified a system that assists in the evaluation of suppliers in the supplier selection process and the identification of companies in the corporate acquisition process. A user centred approach has been achieved that adequately reflects the position and priorities of any company selecting a supplier or acquiring a company. The major benefit of this system is that in a computational inexpensive manner the proposed system is capable of implementing a range of user priorities that influence to varying degrees the system output. This hierarchical fuzzy system with scalable fuzzy membership function can as seen by the results, successfully impart user priorities onto the system. Enabling the system to be gently or strongly influence by specific data in the selection process. This provides a method of applying the prioritised influences involved in the human decision making process. The system attempts to emulate the priorities adhered to by a companies own experts, but on a larger scale and in a more timely and cost effective manner. The results presented in this chapter illustrate the varying degrees of influence that have been exerted on the system in different business problems and how the system has successfully emulated the selection process in both cases. The results demonstrate an accurate reflection of the results that are obtained from a manual analysis of the same data by a human expert. From the results obtained, it can be concluded that the approach is promising, for implementing management decision making processes.

References

1. Azzone, G., & Bertele, U. (1994). Exploiting green strategies for competitive advantage. Long Range Planning, 27 (6), 69–81.
2. Azzone, G., & Noci, G. (1996). Measuring the environmental performance of new products: an integrated approach. International Journal Prod, 3 (11), 3055–3078.
3. Azzone, G., Bianchi, R., Mauri, R., & Noci, G. (1997). Defining, operating environmental strategies: programmes and plans within Italian industries. Environmental Management and Health, 8 (1), 4–19.
4. B & Q. (1993). How green is my hammer. B & Q's Environmental Review. B & Q, Eastleigh, UK.
5. Birou, L., & Fawcett, S. (1994). Supplier involvement in integrated product development. International Journal of Physical Distribution and Logistics Management, 24 (5), 4–14.
6. Bowen, F., Cousins, P.D., Lamming, R.C., Faruk, A. Horses for Courses: Explaining the gap between the theory and practice of green supply'. International Journal for Green Management. Vol 35, Autumn, (2002), pp. 41–60
7. Carter, J., & Narasimhan, R. (1996). A Comparison of North America and European future purchasing trends. International Journal of Purchasing and Materials Management, Spring, 12–22.
8. Cherkassky V., Fuzzy Inference Systems: A Critical Review, Computational Intelligence: Soft Computing and Fuzzy-Neuro Integration with Applications, Kayak O. et al (Eds.), Springer Verlag, Germany, pp. 177–197, (1998).
9. Ellram, L. (1987). The supplier selection decision in strategic partnership. Journal of Purchasing and Materials Management, 26 (3), 8–14.
10. Enarsson, L. (1998). Evaluation of suppliers: how to consider the environment. International Journal of Physical Distribution and Logistics Management, 28 (1), 5–17.
11. Faruk, A., Lamming, R., Cousins, P. & Bowen, F Developing an Assessment for Environmental Supply Chain Analysis. Business Strategy and the Environment. (2002)
12. Faruk, A.C., Lamming, R.C., Cousins, P.D. and Bowen, F.E 'Streamlined integrated life Cycle Assessment: A Tool for Supply Chain Managers in Pursuit of Environmental Soundness' Journal of Industrial Ecology. Vol. 5 no. 2. (2001)
13. Gupta, M. (1995). Environmental Management and its impact on the operations function. International Journal of Operations and Production Management, 15 (8), 34–51.
14. Harrison, B., Lean and Mean: The Changing Landscape of Corporate Power in the Age of Flexibility, Basic Books, New York, (1994).
15. Investor Responsibility Research Centre, (2004), Database, https://oa.irrc.com/
16. Johnson G. and Scholes K., Exploring Corporate Strategy, Financial Times: Prentice Hall, Sixth Edition, (2002).
17. Lamming, R., & Hampson, J. (1996). The Environment as a supply chain management issue. Journal of Management, 7 (Special Issue), 45–62.

18. Lamming, R.C. and Cousins, P.D. "Developing and Evaluating Supply Relationships: Putting Theory into Practice" "DILF Orientering" (Dansk Indkoebs og Logistik Forum – Danish Purchasing and Logistics Forum), October, (2002), pp. 18–23
19. Mamdani E. H. and S. Assilian, An Experiment in Linguistic Synthesis with a Fuzzy Logic Controller, International Journal of Man-Machine Studies, Vol. 7, No. 1, (1975), pp. 1–13.
20. McAleer, E., McIvor, R., Humphreys, P. and McCurry, L. (2000). What multinationals corporations with manufacturing plants in Northern Ireland and the Republic of Ireland are demanding from their suppliers. Journal of Small Business and Enterprise Development, 7 (4), 363–373.
21. McIntyre K., Smith H., Henham A., & Pretlove J., Environmental performance indicators for integrated supply chain: the case of Xerox Ltd, Supply Chain Management, 3(3), 1998, pp. 149–156.
22. Murphy, J. and A. Gouldson, 2000. Environmental policy and industrial innovation: integrating environment and economy through ecological modernisation. Geoforum 31 (1): pp. 33–44.
23. Newton, J.K. Acquisitions : A Directional Policy Matrix Approach, Long Range Planning, Vol. 14, No. 6, (1981), pp. 51–57.
24. Noci, G. (1995). Supporting decision making on recycling based investment. Business Strategy and the Environment, 4 (2), 62–72.
25. Noci, G. (1997). Designing green vendor rating systems for the assessment of a supplier's environmental performance. European Journal of Purchasing and Supply Management, 2, 103–114.
26. Rappaport A., Strategic Analysis for More Profitable Acquisitions, Harvard Business Review, Vol. 57, No. 4, pp. 111–122 (July-August, 1979).
27. Sarkis, J., Nehaman, G., & Priest, J. (1996). A Systemic evaluations model for environmentally friendly conscious business practices and strategy. Proceeding of the 1996 IEEE International Symposium on Electronics and the Environment (pp. 281–286).
28. Schoenberg R. and Reeves R., What Determines Activity within an Industry, European Management Journal, Vol. 17, No. 1, (1999), pp. 93–98.
29. StockHouse.com Annual Reports Service, (2004), http://www.stockhouse.com.au/wilinks_directory/index.asp
30. Sugeno M, "Industrial applications of Fuzzy Control", Elsevier Science Pub. Co., (1985).
31. Sugeno M. and Kang G.T, Structure identification of fuzzy model, Fuzzy Sets and Systems, Vol. 28, (1988) pp. 15–33.
32. Sugeno M. and Yasukawa T., A fuzzy-logic-based approach to qualitative modelling, IEEE Trans. Fuzzy Systems, vol. 1, (1993), pp. 7–31.
33. Toxic release inventory, (2004), The U.S. Environmental Protection Agency, http://www.epa.gov/triexplorer/
34. Wathey, D., O'Reilly, M., 2000. ISO 14031: a practical guide to developing environment performance indicators for your business. The Stationary Office, London.
35. Weber, C., Currenet, J., & Benton, W. (1991). Vendor selection criteria and methods. European Journal of Operational Research, 50 (1), 2–18.

36. Welford, R., & Gouldson, A. (1993). Environmental Management and Business Strategy. London: Pitman.
37. WILink Annual Reports Service, (2004), http://ft.ar.wilink.com/(ujse2345-hjmrdoz10g0iegff)/index.aspx

Techniques to Improve Multi-Agent Systems for Searching and Mining the Web

E. Herrera-Viedma[1], C. Porcel[1], F. Herrera[1], and L. Martínez[2] and A.G. Lopez-Herrera[1]

[1] Dept. of Computer Science and Artificial Intelligence, University of Granada, 18071 – Granada, Spain
viedma@decsai.ugr.es,cporcel@invest.ugr.es,herrera@decsai.ugr.es, agabriel@ugr.es
[2] Dept. of Computer Science, University of Jaén, 23071 - Jaén, Spain
martin@ujaen.es

Abstract. Nowadays, an abundant amount of information is created and delivered over electronic media. The information gathering in the Internet is a complex activity and Internet users need tools to assist them to find the information required. Web multi-agent systems assist the users by gathering from the Internet the information that best satisfies their specific needs.

In this paper, we analyze some techniques that applied together could provide major advances in the design of these Web multi-agent systems in order to improve their performance: i) information filtering tools and ii) fuzzy linguistic modelling. Then, we present a model of a fuzzy linguistic multi-agent system for searching and mining the Web that is designed using some filtering tools and a particular kind of fuzzy linguistic modelling, called multi-granular fuzzy linguistic modelling, which is useful when we have different label sets to assess the information.

Key words: Web, fuzzy linguistic modelling, information filtering, information gathering, intelligent agents

1 Introduction

The exponential increase of Web sites and documents is contributing to that Internet users not being able to find the information they seek in a simple and timely manner. Users are in need of tools to help them cope with the large amount of information available on the Web [22, 23]. Therefore, techniques for searching and mining the Web are becoming increasingly vital.

A multi-agent system is one in which a number of agents cooperate and interact with each other in a distributed environment. On the Web the activity of a multi-agent system consists of assisting Internet users in information gathering processes by means of distributed intelligent agents in order to find

E. Herrera-Viedma et al.: *Techniques to Improve Multi-Agent Systems for Searching and Mining the Web*, Studies in Computational Intelligence (SCI) **5**, 463–461 (2005)
www.springerlink.com

the fittest information to their information needs. In a typical multi-agent system, the agents work together to achieve a global objective based on distributed data and control. Multi-agent systems have been widely used in Web applications [4, 24, 25].

In this paper we study two techniques that applied together can contribute to achieve major advances in the design of Web multi-agent systems in order to improve their performance:

- *Information Filtering Tools:* A promising direction to improve the information access on the Web concerns the way in which it is possible to filter the great amount of information available across the Web. Information filtering is a name used to describe a variety of processes involving the delivery of information to people who need it. Operating in textual domains, *filtering systems* or *recommender systems* evaluate and filter the great amount of information available on the Web to assist people in their search processes [28].

- *Fuzzy Linguistic Modelling:* The great variety of representations and evaluations of the information existing in Internet is the main obstacle to the communication among the agents and between agents and user from what is very important the design of appropriate communication protocol. The problem becomes more noticeable when users take part in the process. This reveals the need for more flexibility in the communication among agents and between agents and users. To solve this problem we propose the use of *fuzzy linguistic modelling* [13, 14, 29] to represent and handle flexible information by means of linguistic labels.

Firstly, we revise the main aspects and models of information filtering tools, as for example, the content-based filtering tools and the collaborative filtering tools. Then, we revise different approaches of fuzzy linguistic modelling to represent the information in the information gathering process of a Web multi-agent system, as for example, the ordinal fuzzy linguistic modelling [9, 13], the 2-tuple fuzzy linguistic modelling [14, 16], the multi-granular fuzzy linguistic modelling [12, 15] and the unbalanced fuzzy linguistic modelling [10, 11]. And finally, we present a model of fuzzy linguistic multi-agent systems which is designed using both information filtering tools and a multi-granular fuzzy linguistic modelling.

The paper is structured as follows. Section 2 revises the information filtering techniques. Section 3 analyzes different approaches of fuzzy linguistic modelling. Section 4 presents the new model of fuzzy linguistic multi-agent system for gathering information on the Web. Finally, some concluding remarks are pointed out.

2 Information Filtering Tools

Information gathering on the Internet is a complex activity. Finding the appropriate information, required for the users, on the World Wide Web is not a simple task. This problem is more acute with the ever increasing use of the Internet. For example, users who subscribe to internet lists waste a great deal of time reading, viewing or deleting irrelevant e-mail messages. To improve the information access on the Web the users need tools to filter the great amount of information available across the Web. *Information Filtering* (IF) is a name used to describe a variety of processes involving the delivery of information to people who need it. It is a research area that offer tools for discriminating between relevant and irrelevant information by providing personalized assistance for continuous retrieval of information.

IF systems are characterized by [8]:

- applicable for unstructured or semi-structured data (e.g. web documents, e-mail messages),
- based on user profiles,
- handle large amounts of data,
- deal primarily with textual data and
- their objective is to remove irrelevant data from incoming streams of data items.

We can find some of the above features in Information Retrieval (IR) systems, but IF differs from traditional IR in that the users have long information needs that are described by means of user profiles, rather than ad-hoc needs that are expressed as queries posed to some IR system. Traditionally IR develops storage, indexing and retrieval technology for textual documents. A user describes his information need in the form of a query to the IR system and the system attempts to find items that match the query within a document store. The information need is usually very dynamic and temporary, i.e., a user issues a query describing an immediate need. Furthermore, information retrieval systems tend to maintain a relatively static store of information. Unlike IR systems, IF systems generally operate on continuous information streams, and always maintain a profile of the user interests needs throughout many uses of the system. As a result, IF systems tend to filter information based on more long-term interests.

Traditionally, these IF systems or recommender systems have fallen into two main categories [26]. *Content-based filtering systems* filter and recommend the information by matching user query terms with the index terms used in the representation of documents, ignoring data from other users. These recommender systems tend to fail when little is known about user information needs, e.g. when the query language is poor. *Collaborative filtering systems* use explicit or implicit preferences from many users to filter and recommend documents to a given user, ignoring the representation of documents. These recommender systems tend to fail when little is known about a user, or when he/she has uncommon interests [26]. In these kind of systems, the users' information preferences can be used to define user profiles that are applied as filters to streams of documents; the recommendations to a user are based on another user's recommendations with similar profiles. Many researchers think that the construction of accurate profiles is a key task and the system's success will depend to a large extent on the ability of the learned profiles to represent the user's preferences [27]. Several researchers are exploring hybrid content-based and collaborative recommender systems to smooth out the disadvantages of each one of them [1, 2, 7, 26].

2.1 Approaches in the Design of IF Systems

In this section we present two major approaches followed in the design and implementation of IF systems, that is, the *statistical approach* and *the knowledge based approach* [8].

2.1.1. Statistical Approach

This kind of IF systems represent the user profiles as a weighted vector of index terms. To filter the information, the system implements a statistical algorithm that computes the similarity of a vector of terms that represents the data item being filtered to a user profile. The most common algorithm used is the Correlation or the Cosine measure between the user profile and the document vector.

The filtering activity is followed by a relevance feedback phase. Relevance feedback is a cyclic process whereby the user feeds back into the system decisions on the relevance of retrieved documents and the system then uses these evaluations to automatically update the user profile.

2.1.2 Knowledge Based Approach

IF systems that follow the knowledge based approach utilize Artificial Intelligence techniques, such as production rules, neural networks and evolutionary genetic algorithms, to represent user profiles and to implement the filtering and the learning (feedback) phases.

- **Rule based IF systems:** These IF systems use rules to represent user profiles, where each rule can represent a user information need or pattern of information filtering. For example, in e-mail messages, rules can be defined and applied to fields that appear in the message header (e.g. subject or sender). The rules may contain instructions on how to handle a message, depending on the values of these fields. The rules allow us either to filter out the data item or to treat it as relevant. For example, if the sender of an e-mail does not appear in a certain predefined list, the message gets a low relevance rank.
- **Neural network based IF systems:** A neural network is an interconnected assembly of simple processing elements, units or nodes, whose functionality is roughly based on the animal neuron. The processing ability of the network is stored in inter-unit connection weights, obtained by

a process of adaptation to, or learning from, a set of training patterns. The weights are supposed to adapt when the net is shown examples from training sets. Neural networks can also be applied in IF systems, where a user profile is representing a user's concept with unseen associations, that adapts from training.

- **Evolutionary genetic algorithms based IF systems:** Evolutionary genetic based techniques borrow their model from the Darwinian concept of the natural process of survival. Nature selects the most fit individuals to survive, and genetic patterns are passed by the individuals down through generations. The changes take place by recombining the genetic codes of pairs of individuals. These features allow us to apply an evolutionary and genetic approach in IF systems. The analogy in information filtering makes use of the vector space model to represent documents. In this model, a gene would be represented as a term, an individual as a document in the vector space, and the community as a profile. An appropriate objective function is introduced as the survival process, to decide whether to update the profile.

2.2 On the Acquisition of User Data

Another topic that we must have in mind when we design a IF system is the method to gather user information. In order to discriminate between relevant and irrelevant information for a user, we must have some information about this user, i.e. we must know the user preferences. Information about user preferences can be obtained in two different ways [8], *implicit* and *explicit* *mode*, although these ways not be mutually exclusive.

The implicit approach is implemented by inference from some kind of observation. The observation is applied to user behavior or to detecting a user's environment (such as bookmarks or visited URL). The user preferences are updated by detecting changes while observing the user.

The other approach, the *explicit* approach, interacts with the users by acquiring feedback on information that is filtered, that is, the user expresses some specifications of what they desire. This approach is the most common.

3 Fuzzy Linguistic Modelling

There are situations in which the information cannot be assessed precisely in a quantitative form but may be in a qualitative one. For example, when attempting to qualify phenomena related to human perception, we are often led to use words in natural language instead of numerical values, e.g. when evaluating the *comfort* or *design* of a car, terms like *good*, *medium* or *bad* can be used. In other cases, precise quantitative information cannot be stated because either it is unavailable or the cost for its computation is too high and an "approximate value" can be applicable, eg. when evaluating the speed of a car, linguistic terms like *fast, very fast* or *slow* can be used instead of numeric values.

The use of Fuzzy Sets Theory has given very good results for modelling qualitative information [29]. *Fuzzy linguistic modelling* is a tool based on the concept of *linguistic variable* [29] to deal with qualitative assessments. It has proven to be useful in many problems, e.g., in decision making [13], quality evaluation [21], models of information retrieval [17, 18], etc.

In this section, we revise four different approaches of fuzzy linguistic modelling which can provide a different support to represent the linguistic information managed in the communication processes developed by the multi-agent systems:

1. *Ordinal fuzzy linguistic modelling* [13, 9], which is defined to eliminate the excessive complexity of the traditional fuzzy linguistic modelling [29].
2. *2-tuple fuzzy linguistic modelling* [14, 16], which is is defined to improve the performance of the ordinal fuzzy linguistic approach.
3. *Multi-granular fuzzy linguistic modelling* [12, 15], which is defined to deal with situations in which the linguistic information is assessed on different label sets.
4. *Unbalanced fuzzy linguistic modelling* [10, 11], which is defined to deal with situations in which the linguistic information is assessed on an unbalanced label set, that is, a non-symmetrical and non-uniform label set.

3.1 The Ordinal Fuzzy Linguistic Modelling

Ordinal fuzzy linguistic modelling [9, 13] is a very useful kind of fuzzy linguistic approach proposed as an alternative tool to the traditional fuzzy linguistic modelling [29] which simplifies the computing with words process as well as linguistic aspects of problems. It is defined by considering a finite and totally ordered label set $S = \{s_i\}, i \in \{0, \ldots, g\}$ in the usual sense, i.e., $s_i \geq s_j$ if $i \geq j$, and with odd cardinality (7 or 9 labels). The mid term represents an assessment of "approximately 0.5", and the rest of the terms being placed symmetrically around it. The semantics of the label set is established from the ordered structure of the label set by considering that each label for the pair (s_i, s_{g-i}) is equally informative. For example, we can use the following set of seven labels to represent the linguistic information:

$$S = \{s_0 = N, \ s_1 = VL, \ S_2 = L, \ s_3 = M, \ s_4 = H, \ s_5 = VH, \ s_6 = P\} .$$

Additionally, a fuzzy number defined in the $[0, 1]$ interval can be associated with each linguistic term. A way to characterize a fuzzy number is to use a representation based on parameters of its membership function. The linguistic assessments given by the users are just approximate ones, some authors consider that linear trapezoidal membership functions are good enough to capture the vagueness of such linguistic assessments. The parametric representation is achieved by the 4-tuple (a, b, c, d), where b and d indicate the interval in which

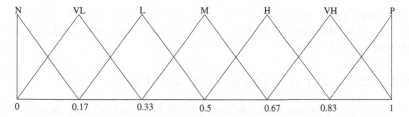

Fig. 1. A set of seven linguistic terms with its semantics

the membership value is 1, with a and c indicating the left and right limits of the definition domain of the trapezoidal membership function. A particular case of this type of representation are the linguistic assessments whose membership functions are triangular, i.e., $b = d$, then we represent this type of membership functions by a 3-tuple (a, b, c). An example may be the following set of seven terms (Fig. 1):

$$s_0 = Null(N) = (0, 0, .17) \qquad s_1 = VeryLow(VL) = (0, .17, .33)$$
$$s_2 = Low(L) = (.17, .33, .5) \qquad s_3 = Medium(M) = (.33, .5, .67)$$
$$s_4 = High(H) = (.5, .67, .83) \qquad s_5 = VeryHigh(VH) = (.67, .83, 1)$$
$$s_6 = Perfect(P) = (.83, 1, 1).$$

In any linguistic modelling we need management operators of linguistic information. An advantage of ordinal fuzzy linguistic modelling is the simplicity and quickness of its computational model. It is based on symbolic computation [9, 13] and acts by direct computation on labels by taking into account the order of such linguistic assessments in the ordered structure of labels. Usually, the ordinal fuzzy linguistic model for computing with words is defined by establishing i) a negation operator, ii) comparison operators based on the ordered structure of linguistic terms, and iii) adequate aggregation operators of ordinal fuzzy linguistic information. In most ordinal fuzzy linguistic approaches the negation operator is defined from the semantics associated to the linguistic terms as

$$NEG(s_i) = s_j \mid j = g - i \,;$$

and there are defined two comparison operators of linguistic terms:

1. *Maximization operator:* $MAX(s_i, s_j) = s_i$ if $s_i \geq s_j$; and
2. *Minimization operator,* $MIN(s_i, s_j) = s_i$ if $s_i \leq s_j$.

Using these operators it is possible to define automatic and symbolic aggregation operators of linguistic information, as for example the LOWA operator [13] and the LWA operator [9].

We must point out that in [5] we apply satisfactorily the ordinal fuzzy linguistic modelling to model the communication processes in the design of a Web multi-agent system.

3.2 The 2-Tuple Fuzzy Linguistic Modelling

The *2-tuple fuzzy linguistic modelling* [14, 16] is a kind of fuzzy linguistic modelling that mainly allows to reduce the loss of information typical of the ordinal fuzzy linguistic modelling. Its main advantage is that the linguistic computational model based on linguistic 2-tuples can carry out processes of computing with words easier and without loss of information. To define it we have to establish the 2-tuple representation model and the 2-tuple computational model to represent and aggregate the linguistic information, respectively.

Let $S = \{s_0, \ldots, s_g\}$ be a linguistic term set with odd cardinality ($g + 1$ is the cardinality of S), where the mid term represents an assessment of approximately 0.5 and with the rest of the terms being placed symmetrically around it. We assume that the semantics of labels is given by means of triangular membership functions represented by a 3-tuple (a, b, c) and consider all terms distributed on a scale on which a total order is defined $s_i \leq s_j \Longleftrightarrow i \leq j$. In this fuzzy linguistic context, if a symbolic method [9, 13] aggregating linguistic information obtains a value $\beta \in [0, g]$, and $\beta \notin \{0, \ldots, g\}$, then an approximation function is used to express the result in S.

Definition 1. *[14] Let β be the result of an aggregation of the indexes of a set of labels assessed in a linguistic term set S, i.e., the result of a symbolic aggregation operation, $\beta \in [0, g]$. Let $i = round(\beta)$ and $\alpha = \beta - i$ be two values, such that, $i \in [0, g]$ and $\alpha \in [-.5, .5)$ then α is called a Symbolic Translation.*

The 2-tuple fuzzy linguistic approach is developed from the concept of symbolic translation by representing the linguistic information by means of 2-tuples (s_i, α_i), $s_i \in S$ and $\alpha_i \in [-.5, .5)$:

- s_i represents the linguistic label of the information, and
- α_i is a numerical value expressing the value of the translation from the original result β to the closest index label, i, in the linguistic term set ($s_i \in S$).

This model defines a set of transformation functions between numeric values and 2-tuples.

Definition 2. *[14] Let $S = \{s_0, \ldots, s_g\}$ be a linguistic term set and $\beta \in [0, g]$ a value representing the result of a symbolic aggregation operation, then the 2-tuple that expresses the equivalent information to β is obtained with the following function:*

$$\Delta : [0, g] \longrightarrow S \times [-0.5, 0.5)$$

$$\Delta(\beta) = (s_i, \alpha), \ with \ \begin{cases} s_i & i = round(\beta) \\ \alpha = \beta - i & \alpha \in [-.5, .5) \end{cases}$$

where round(·) is the usual round *operation, s_i has the closest index label to "β" and "α" is the value of the symbolic translation.*

For all Δ there exists Δ^{-1}, defined as $\Delta^{-1}(s_i, \alpha) = i + \alpha$. On the other hand, it is obvious that the conversion of a linguistic term into a linguistic 2-tuple consists of adding a symbolic translation value of 0: $s_i \in S \Longrightarrow (s_i, 0)$.

The 2-tuple linguistic computational model is defined by presenting the comparison of 2-tuples, a negation operator and aggregation operators of 2-tuples.

1. Comparison of 2-tuples. The comparison of linguistic information represented by 2-tuples is carried out according to an ordinary lexicographic order. Let (s_k, α_1) and (s_l, α_2) be two 2-tuples, with each one representing a counting of information:

- If $k < l$ then (s_k, α_1) is smaller than (s_l, α_2).
- If $k = l$ then
  1. if $\alpha_1 = \alpha_2$ then (s_k, α_1) and (s_l, α_2) represent the same information,
  2. if $\alpha_1 < \alpha_2$ then (s_k, α_1) is smaller than (s_l, α_2),
  3. if $\alpha_1 > \alpha_2$ then (s_k, α_1) is bigger than (s_l, α_2).

2. Negation operator of 2-tuples: $Neg((s_i, \alpha)) = \Delta(g - (\Delta^{-1}(s_i, \alpha)))$.

3. Aggregation operators of 2-tuples. The aggregation of information consists of obtaining a value that summarizes a set of values, therefore, the result of the aggregation of a set of 2-tuples must be a 2-tuple. In the literature we can find many aggregation operators which allow us to combine the information according to different criteria. Using functions Δ and Δ^{-1} that transform without loss of information numerical values into linguistic 2-tuples and viceversa, any of the existing aggregation operator can be easily extended for dealing with linguistic 2-tuples. Some examples are:

Definition 3. (Arithmetic Mean). *Let $x = \{(r_1, \alpha_1), \ldots, (r_n, \alpha_n)\}$ be a set of linguistic 2-tuples, the 2-tuple arithmetic mean $\bar{x}^e$ is computed as,*

$$\bar{x}^e[(r_1, \alpha_1), \ldots, (r_n, \alpha_n)] = \Delta\left(\sum_{i=1}^{n} \frac{1}{n} \Delta^{-1}(r_i, \alpha_i)\right) = \Delta\left(\frac{1}{n} \sum_{i=1}^{n} \beta_i\right).$$

Definition 4. (Weighted Average Operator). *Let $x = \{(r_1, \alpha_1), \ldots, (r_n, \alpha_n)\}$ be a set of linguistic 2-tuples and $W = \{w_1, \ldots, w_n\}$ be their associated weights. The 2-tuple weighted average $\bar{x}^w$ is:*

$$\bar{x}^w[(r_1, \alpha_1), \ldots, (r_n, \alpha_n)] = \Delta\left(\frac{\sum_{i=1}^{n} \Delta^{-1}(r_i, \alpha_i) \cdot w_i}{\sum_{i=1}^{n} w_i}\right) = \Delta\left(\frac{\sum_{i=1}^{n} \beta_i \cdot w_i}{\sum_{i=1}^{n} w_i}\right).$$

Definition 5. (Linguistic Weighted Average Operator). *Let* $x = \{(r_1, \alpha_1), \ldots,$ $(r_n, \alpha_n)\}$ *be a set of linguistic 2-tuples and* $W = \{(w_1, \alpha_1^w), \ldots, (w_n, \alpha_n^w)\}$ *be their linguistic 2-tuple associated weights. The 2-tuple linguistic weighted average* $\overline{x}_l^w$ *is:*

$$\overline{x}_l^w[((r_1, \alpha_1), (w_1, \alpha_1^w)) \cdots ((r_n, \alpha_n), (w_n, \alpha_n^w))] = \Delta \left(\frac{\sum_{i=1}^n \beta_i \cdot \beta_{W_i}}{\sum_{i=1}^n \beta_{W_i}} \right),$$

with $\beta_i = \Delta^{-1}(r_i, \alpha_i)$ *and* $\beta_{W_i} = \Delta^{-1}(w_i, \alpha_i^w)$.

We must point out that in [6] we apply the 2-tuple fuzzy linguistic modelling in the design of a Web multi-agent system as a way to overcome the problems of loss of information observed in the Web multi-agent system presented in [5].

3.3 The Multi-Granular Fuzzy Linguistic Modelling

In any fuzzy linguistic approach, an important parameter to determinate is the "granularity of uncertainty", i.e., the cardinality of the linguistic term set S used to express the linguistic information. According to the uncertainty degree that an expert qualifying a phenomenon has on it, the linguistic term set chosen to provide his knowledge will have more or less terms. When different experts have different uncertainty degrees on the phenomenon, then several linguistic term sets with a different granularity of uncertainty are necessary (i.e. multi-granular linguistic information) [12, 15, 20]. The use of different label sets to assess information is also necessary when an expert has to assess different concepts, as for example it happens in information retrieval problems, to evaluate the importance of the query terms and the relevance of the retrieved documents [19]. In such situations, we need tools for the management of multi-granular linguistic information, i.e., we need to define a *multi-granular fuzzy linguistic modelling*. In [12] we define a proposal of multi-granular fuzzy linguistic modelling based on the ordinal fuzzy linguistic modelling and in [15] we define other one based on the 2-tuple fuzzy linguistic modelling. In this paper, we follow that defined in [15] which uses the concept of *Linguistic Hierarchies* to manage the multi-granular linguistic information.

A *linguistic hierarchy* is a set of levels, where each level is a linguistic term set with different granularity from the remaining of levels of the hierarchy [3]. Each level belonging to a linguistic hierarchy is denoted as $l(t, n(t))$, t being a number that indicates the level of the hierarchy and $n(t)$ the granularity of the linguistic term set of the level t.

Usually, linguistic hierarchies deal with linguistic terms whose membership functions are triangular-shaped, symmetrical and uniformly distributed in [0,1]. In addition, the linguistic term sets have an odd value of granularity representing the central label the value of *indifference*.

The levels belonging to a linguistic hierarchy are ordered according to their granularity, i.e., for two consecutive levels t and $t + 1$, $n(t + 1) > n(t)$.

Therefore, each level $t + 1$ provides a linguistic refinement of the previous level t.

A linguistic hierarchy, LH, is defined as the union of all levels t: $LH = \bigcup_t l(t, n(t))$. To build LH we must keep in mind that the hierarchical order is given by the increase of the granularity of the linguistic term sets in each level. Let $S^{n(t)} = \{s_0^{n(t)}, \ldots, s_{n(t)-1}^{n(t)}\}$ be the linguistic term set defined in the level t with $n(t)$ terms, then the building of a linguistic hierarchy must satisfy the following linguistic hierarchy basic rules [15]:

1. To preserve all *former modal points* of the membership functions of each linguistic term from one level to the following one.
2. To make *smooth transactions between successive levels*. The aim is to build a new linguistic term set, $S^{n(t+1)}$. A new linguistic term will be added between each pair of terms belonging to the term set of the previous level t. To carry out this insertion, we shall reduce the support of the linguistic labels in order to keep place for the new one located in the middle of them.

Generically, we can say that the linguistic term set of level $t + 1$, $S^{n(t+1)}$, is obtained from its predecessor level t, $S^{n(t)}$ as: $l(t, n(t)) \rightarrow l(t+1, 2 \cdot n(t) - 1)$. Table 1 shows the granularity needed in each linguistic term set of the level t depending on the value $n(t)$ defined in the first level (3 and 7 respectively).

Table 1. Linguistic Hierarchies

	Level 1	Level 2	Level 3
$l(t, n(t))$	l(1,3)	l(2,5)	l(3,9)
$l(t, n(t))$	l(1,7)	l(2,13)	

A graphical example of a linguistic hierarchy is shown in Fig. 2:

In [15] was demonstrated that the linguistic hierarchies are useful to represent the multi-granular linguistic information and allow to combine multi-granular linguistic information without loss of information. To do this, a family of transformation functions between labels from different levels was defined:

Definition 6. *Let* $LH = \bigcup_t l(t, n(t))$ *be a linguistic hierarchy whose linguistic term sets are denoted as* $S^{n(t)} = \{s_0^{n(t)}, \ldots, s_{n(t)-1}^{n(t)}\}$. *The transformation function between a 2-tuple that belongs to level* t *and another 2-tuple in level* $t' \neq t$ *is defined as:*

$$TF_{t'}^t : l(t, n(t)) \longrightarrow l(t', n(t'))$$

$$TF_{t'}^t\left(s_i^{n(t)}, \alpha^{n(t)}\right) = \Delta\left(\frac{\Delta^{-1}(s_i^{n(t)}, \alpha^{n(t)}) \cdot (n(t') - 1)}{n(t) - 1}\right)$$

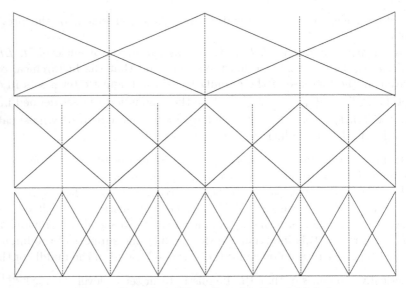

Fig. 2. Linguistic Hierarchy of 3, 5 and 9 labels

As it was pointed out in [15] this family of transformation functions is biyective.

3.4 The Unbalanced Fuzzy Linguistic Modelling

In any problem that uses linguistic information the first goal to satisfy is the choice of the linguistic terms with their semantics, for establishing the label set to be used in the problem. In the literature, we can find two different possibilities for choosing the linguistic terms and their semantics:

- We can assume that all the terms of the label set are equally informative, i.e., symmetrically distributed as it happens in the above fuzzy linguistic modelling.
- We can assume that all the terms of the label set are not equally informative, i.e., not symmetrically distributed. In this case, we need an *unbalanced fuzzy linguistic modelling* [10, 11] to manage the linguistic term sets with different discrimination levels on both sides of the mid term (see Fig. 3). As was known in [10], in information retrieval systems the use of unbalanced linguistic term sets seems more appropriate than the use of symmetrical linguistic term sets, as to express the importance weights in the queries as to represent the relevance degrees of the documents.

To manage unbalanced linguistic term sets we propose a method based on the 2-tuple fuzzy linguistic modelling. Basically, this method consists of representing unbalanced linguistic terms from different levels of an *LH*, carrying out computational operations of unbalanced linguistic information using the 2-tuple computational model. The method consists of the following steps:

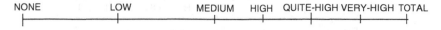

Fig. 3. Unbalanced Linguistic Term Set of 7 Labels

1. Represent the unbalanced linguistic term set S by means of a linguistic hierarchy, LH.

    1.1. Chose a level t^- with an adequate granularity to represent using the 2-tuple representation model the subset of linguistic terms of S on the left of the mid linguistic term.

    1.2. Chose a level t^+ with an adequate granularity to represent using the 2-tuple representation model the subset of linguistic terms of S on the right of the mid linguistic term.

2. Define an unbalanced linguistic computational model.

    2.1. Choose a level $t' \in \{t^-, t^+\}$, such that $n(t') = \max\{n(t^-), n(t^+)\}$.

    2.2. Define the comparison of two 2-tuples $(s_k^{n(t)}, \alpha_1)$, $t \in \{t^-, t^+\}$, and $(s_l^{n(t)}, \alpha_2)$, $t \in \{t^-, t^+\}$, with each one representing a counting of unbalanced information. Its expression is similar to the usual comparison of two 2-tuples but acting on the values $TF_{t'}^t(s_k^{n(t)}, \alpha_1)$ and $TF_{t'}^t(s_l^{n(t)}, \alpha_2)$. We should point out that using the comparison of 2-tuples we can easily define the comparison operators Max and Min.

    2.3. Define the negation operator of unbalanced linguistic information. Let $(s_k^{n(t)}, \alpha)$, $t \in \{t^-, t^+\}$ be an unbalanced 2-tuple then:

    $$\mathcal{NEG}(s_k^{n(t)}, \alpha) = \mathrm{Neg}(TF_{t''}^t(s_k^{n(t)}, \alpha)), t \neq t'', \ t'' \in \{t^-, t^+\} .$$

    2.4. Define aggregation operators of unbalanced linguistic information. This is done using the aggregation processes designed in the 2-tuple computational model but acting on the unbalanced linguistic values transformed by means of $TF_{t'}^t$. Then, once it is obtained a result, it is transformed to the correspondent level t by means of $TF_{t'}^t$ to express the result in the unbalanced linguistic term set.

Assuming the unbalanced linguistic term set shown in Fig. 3 and the linguistic hierarchy shown in Fig. 2, in Fig. 4 we show how to select the different levels to represent the unbalanced linguistic term set.

4 A Model of Multi-Granular Fuzzy Linguistic Multi-Agent Systems Based on Filtering Techniques

In this section we present a new model of Web multi-agent systems that combines both techniques aforementioned to improve their information gathering processes on the Internet, i.e., it is designed using information filtering techniques and assumes a multi-granular fuzzy linguistic modelling.

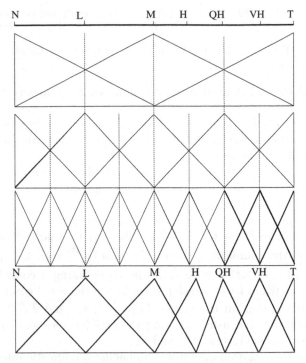

Fig. 4. Unbalanced Linguistic Term Set of 7 Labels

As it is known, a promising direction to improve the effectiveness of search engines concerns the way in which it is possible to "filter" the great amount of information available across the Internet. Then, this new model incorporates in its activity the two more important existing filtering techniques, i.e. content-based filtering and collaborative filtering [26, 28]. On the other hand, our fuzzy linguistic multi-agent models defined in [5, 6] present the following drawback: we assume the use of the same label set to represent the different linguistic information that appears in the communication processes developed in the multi-agent system, as for example, when user weighted queries and the relevance degrees of the retrieved documents are assessed using a same label set, although they represent different concepts. In this model the communication among the agents of different levels and between users and agents is carried out by using different label sets, i.e. working with multi-granular linguistic information, in order to allow a higher flexibility in the processes of communication of the system.

In what follows, we present the architecture of this multi-agent model and its operation.

4.1 Architecture

In [5, 6] were defined two fuzzy linguistic distributed multi-agent models that use linguistic information to carry out the communication processes among the agents. The architecture of these models is hierarchical and is composed of five action levels: *Internet Users, Interface Agents, Task Agents, Information Agents and Information Sources.* The architecture of our new multi-agent model must allow the application of the content-based and collaborative filtering tools. To do that, we incorporate in its architecture two new action levels: the level of the *content-based filtering agents* and the level of *collaborative filtering agent.* Therefore, this model presents a hierarchical architecture that contains seven activity levels: *Internet Users, Interface Agents, Collaborative Filtering Agent, Task Agents, Content-based Filtering Agent, Information Agents and Information Sources.* Furthermore, it works assuming a 2-tuple based multi-granular fuzzy linguistic modelling, that is, it uses different label sets $(S_1, S_2, S_3 \ldots)$ to represent the different concepts to be assessed in its retrieval activity. These label sets S_i are chosen from those label sets that composes a LH, i.e., $S_i \in LH$. For example, we can use the LH shown in Fig. 2. We should point out that the number of different label sets that we can use is limited by the number of levels of LH, and therefore, in many cases different the label sets S_i and S_j can be associated to a same label set of LH but with different interpretations depending on the concept to be modelled.

- **Level 1:** *Internet user,* which expresses his/her information needs by means of a linguistic multi-weighted query. Each term of a user query can be weighted simultaneously by two linguistic weights. The first weight is associated with a classical threshold semantics and the second one with a relative importance semantics. Then, the user makes a query to look for those documents related to the terms $\{t_1, t_2, \ldots, t_m\}$, which are weighted by a linguistic degree of threshold $\{p_1^1, p_2^1, \ldots, p_m^1\}$ with $p_i^1 \in S_1$, and by a linguistic degree of relative importance $\{p_1^2, p_2^2, \ldots, p_m^2\}$ with $p_i^2 \in S_2$. The user also expresses an information need category $\mathcal{A}_i$ chosen from a list of information need categories $\{\mathcal{A}_i, \ldots, \mathcal{A}_l\}$ provided by the system, and the user's identity $\mathcal{ID}$. All this information is given by the user to the *interface agent.*
- **Level 2:** *Interface agent* (one for user), that communicates the user's weighted query, the information need category and the user identity to the collaborative filtering agent, and filters the retrieved documents from collaborative filtering agent to give to the users those that satisfy better their needs. Finally, informs the collaborative filtering agent on set of documents used by user to satisfy his/her information needs DU.
- **Level 3:** *Collaborative filtering agent* (one for interface agent), that communicates the user multi-weighted query to the task agent, receives the more relevant documents chosen by the task agent, retrieves the recommendations on such documents from a collaborative recommendation system using the information need category expressed by the user

$RC^{\mathcal{A}_i} = \{RC_1^{\mathcal{A}_i}, \ldots, RC_v^{\mathcal{A}_i}\}$ $RC_j^{\mathcal{A}_i} \in S_3 \times [-0.5, 0.5)$, filters the documents by recalculating their relevance using these recommendations, and communicates these documents together with their new relevance degrees to the interface agent. Later, it carries out the tasks to update in the collaborative recommendation system the recommendations on the documents used by the user, i.e., it invites user to provide a recommendation rc_y on each chosen document $d_y^U \in DU$ and this recommendation is stored in the collaborative recommendation system together with the recommendations provided by other users that used d_y^U.

- **Level 4:** *Task agent* (one for interface agent, generally), that communicate the terms of user query to the content-based filtering agents, and filters those documents from every content-based filtering agent that fulfills better the query.

- **Level 5:** *Content-based filtering agent* (one for agent information). Each content-based filtering agent communicates the terms of user query to its respective information agent and filters the relevant documents provided by its information agent by recalculating their relevance using the threshold weights. Then, the task agent receives from every content-based filtering agent h a set of documents and their relevance (D^h, RN^h), where every document d_h^h has associated a linguistic degree of relevance expressed in linguistic 2-tuples $rn_j^h \in S_4 \times [-0.5, 0.5)$ $(j = 1, \ldots, \#(D^h))$. It also receives a set of linguistic degrees of satisfaction $C^h = \{c_1^h, c_2^h, \ldots, c_m^h\}$, $c_i^h \in S_5 \times [-0.5, 0.5)$ of this set of documents D^h with regard to every term of the query t_i.

- **Level 6:** *Information agents*, which receive the terms of user query from the content-based filtering agents and look for the documents in the information sources. Then, each content-based filtering agent h receives from its respective information sources h the set of relevant documents that it found through information sources D^h and their relevance R^h, where every document d_j^h has an associated degree of relevance $r_j^h \in S_4 \times [-0.5, 0.5)$ $(j = 1, \ldots, \#(D^h))$.

- **Level 7:** *Information sources*, consisting of all data sources within the Internet, such as databases and information repositories.

This structure is presented in Fig. 5.

4.2 Operation of the Model

The activity of this multi-agent model is composed of two phases:

1. *Retrieval phase*: This first phase coincides with the information gathering process developed by the multi-agent model itself, i.e., this phase begins when a user specifies his/her query and finishes when he/she chooses his/her desired documents among the relevant documents retrieved and provided by the system.

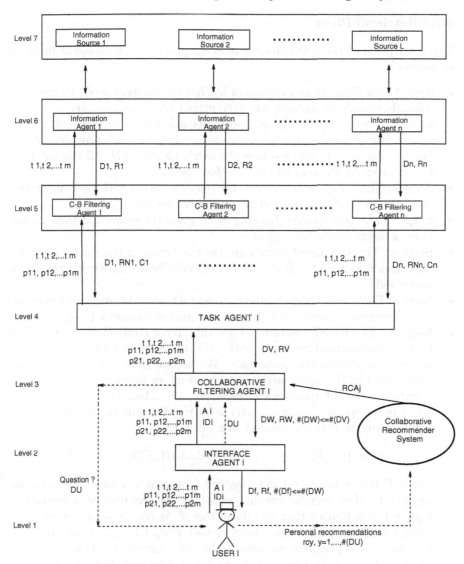

Fig. 5. Structure of Model of Web Multi-Agent System

2. *Feedback phase*: This second phase coincides with the updating process of collaborative recommendations on desired documents existing in the collaborative recommender system, i.e., this phase begins when the *interface agent* informs the documents chosen by the user to the *collaborative filtering agent* and finishes when the recommender system recalculates and updates the recommendations of the desired documents.

In the following subsections, we explain both phases.

4.2.1 Retrieval Phase

The description of the information gathering process of multi-agent model is as follows:

- **Step 1:** An *Internet user* expresses his/her information needs by means of a linguistic multi-weighted query $\{(t_1, p_1^1, p_1^2), (t_2, p_2^1, p_2^2), \ldots, (t_m, p_m^1, p_m^2)\}$, with $p_i^1 \in S_1$ and $p_i^2 \in S_2$, and an information need category $\mathcal{A}_i$ chosen from a list of information need categories $\{\mathcal{A}_1, \ldots, \mathcal{A}_l\}$ provided by the system. The system also requires the user's identity $\mathcal{ID}$. All this information is given by the user to the *interface agent*.
- **Step 2:** The *interface agent* gives the query together with the information need category $(\mathcal{A}_i)$ to the *collaborative filtering agent*.
- **Step 3:** The *collaborative filtering agent* gives the terms and their importance weights to the *task agent*.
- **Step 4:** The *task agent* communicates the terms of the query and their importance weights to all the *content-based filtering agents* to which it is connected.
- **Step 5:** Each *content-based filtering agent* h makes the query to its respective *information agent* h and gives it the terms of the query $\{t_1, t_2, \ldots, t_m\}$.
- **Step 6:** All the *information agents* that have received the query, look for the information that better satisfies it in the *information sources*, and retrieve from them the documents. We assume that the documents are represented in the *information sources* using an index term based representation as in Information Retrieval [17, 18]. Then, there exists a finite set of index terms $T = \{t_1, \ldots, t_l\}$ used to represent the documents and each document d_j is represented as a fuzzy subset

$$d_j = \{(t_1, F(d_j, t_1)), \ldots, (t_l, F(d_j, t_l))\}, F(d_j, t_i) \in [0, 1] \,,$$

where F is any numerical indexing function that weighs index terms according to their significance in describing the content of a document. $F(d_j, t_i) = 0$ implies that the document d_j is not at all about the concept(s) represented by index term t_i and $F(d_j, t_i) = 1$ implies that the document d_j is perfectly represented by the concept(s) indicated by t_i.

- **Step 7:** Each *content-based filtering agent* h receives from its respective *information agent* h a set of documents and their relevances (D^h, R^h) ordered decreasingly by relevance. Every document d_j^h has an associated linguistic degree of relevance $r_j^h \in S_4 \times [-0.5, 0.5)$ which is calculated as

$$r_j^h = \overline{x}^e \left[\Delta\big(g \cdot F(d_j^h, t_1)\big), \ldots, \Delta\big(g \cdot F(d_j^h, t_m)\big) \right] = \Delta\left(g \cdot \sum_{i=1}^m \frac{1}{m} F(d_j^h, t_i) \right),$$

being $g+1$ the cardinality of S_4. Each *content-based filtering agent* h filters documents received from its respective *information agent* h by recalculating their relevance by means of a linguistic matching function

$$e_h : (S_4 \times [-0.5, 0.5)) \times S_1 \rightarrow S_4 \times [-0.5, 0.5) ,$$

which is defined to model the semantics of threshold weights associated with the query terms. This linguistic matching function requires a previous transformation of threshold weights expressed in labels of S_1 that must be transformed in labels of S_4; to make uniform the multi-granular linguistic information, we chose the linguistic term set used to express the relevance degrees. We use the transformation function viewed in definition 6 ($TF_{t'}^t$), to transform the linguistic labels in level S_1 (t) to labels in level S_4 (t'):

$$TF_{S_4}^{S_1}(s_i^{n(S_1)}, \alpha^{n(S_1)}) = \Delta \left(\frac{\Delta^{-1}(s_i^{n(S_1)}, \alpha^{n(S_1)}) \cdot (n(S_4) - 1)}{n(S_1) - 1} \right)$$

obtaining the new linguistic threshold weights $\{p_1^{1'}, p_2^{1'}, \ldots, p_m^{1'}\}$, $p_i^{1'} \in S_4$ for the terms $\{t_1, t_2, \ldots, t_m\}$. Different *content-based filtering agents* can have different threshold matching functions. For example, some linguistic matching functions that we can use are:

1. $e^1(\Delta(g \cdot F(d_j, t_i)), p_i^{1'}) = \begin{cases} (s_g, 0) & \text{if } \Delta(g \cdot F(d_j, t_i)) \geq (p_i^{1'}, 0) \\ (s_0, 0) & \text{otherwise.} \end{cases}$

2. $e^2(\Delta(g \cdot F(d_j, t_i)), p_i^{1'}) = \begin{cases} \Delta(g \cdot F(d_j, t_i)) & \text{if } \Delta(g \cdot F(d_j, t_i)) \geq (p_i^{1'}, 0) \\ (s_0, 0) & \text{otherwise.} \end{cases}$

3. $e^3(\Delta(g \cdot F(d_j, t_i)), p_i^{1'}) = \begin{cases} \Delta(\min\{g, 0.5 + g \cdot F(d_j, t_i)\}) & \text{if } \Delta(g \cdot F(d_j, t_i)) \geq (p_i^{1'}, 0) \\ \Delta(\max\{0, g \cdot F(d_j, t_i) - 0.5\}) & \text{otherwise.} \end{cases}$

Then, each *content-based filtering agent* h calculates a new set of relevance degrees $RN^h = \{rn_j^h, j = 1, \ldots, \#(D^h)\}$ characterizing the documents D^h, which is obtained as

$$rn_j^h = \overline{x}^e [e_h(\Delta(g \cdot F(d_j^h, t_1)), p_1^{1'}), \ldots, e_h(\Delta(g \cdot F(d_j^h, t_m)), p_m^{1'})]$$
$$= \Delta \left(\sum_{i=1}^{m} \frac{1}{m} \Delta^{-1}(e_h(\Delta(g \cdot F(d_j^h, t_i)), p_i^{1'})) \right) .$$

- **Step 8:** The *task agent* receives from every *content-based filtering agent* a set of documents and their new relevance (D^h, RN^h). It also receives a set of linguistic degree of satisfaction $C^h = \{c_1^h, c_2^h, \ldots, c_m^h\}$, $c_i^h \in S_5 \times [-0.5, 0.5)$ of D^h with regard to every term of the query as

$$c_i^h = \overline{x}^e [e_h(\Delta(g \cdot F(d_1^h, t_i)), p_i^{1''}), \ldots, e_h(\Delta(g \cdot F(d_{\#(D^h)}^h, t_i)), p_i^{1''})]$$
$$= \Delta \left(\sum_{j=1}^{\#(D^h)} \frac{1}{\#(D^h)} \Delta^{-1}(e_h(\Delta(g \cdot F(d_j^h, t_i)), p_i^{1''})) \right) .$$

where the $p_i^{1''}$ are the p_i^1 expressed in the set S_5, using the transformation function $TF_{S_5}^{S_1}$ viewed in definition 6.

Then, the *task agent* selects the number of documents to be retrieved from each *content-based filtering agent* h. To do so, it applies the following three steps:

- **Step 8.1:** The *task agent* orders D^h with respect to the new relevance RN.
- **Step 8.2:** The *task agent* aggregates both linguistic information weights, the satisfactions of the terms of the query from every *information agent*, $(c_i^h, \alpha_i^w), c_i^h \in S_5$, and the importance weights that the user assigned to these terms, $(p_i^2, \alpha_i), p_i^2 \in S_2$, using the aggregation process for multi-granular linguistic information presented in [15]:

  1. *Normalization Phase*: the linguistic term set used to express the relevance is chosen to make uniform the multi-granular linguistic information. Then, all the information is expressed in that linguistic term set by means of 2-tuples.

  2. *Aggregation Phase*: through a 2-tuple aggregation operator the information is aggregated. In this paper we use the 2-tuple linguistic weighted average operator, $\bar{x}_l^w$, for combining the satisfactions of the terms of the query and the importance weights.

  Let $\{[(p_1^2, \alpha_1), (c_1^h, \alpha_1^w)], \ldots, [(p_m^2, \alpha_m), (c_m^h, \alpha_m^w)]\}, p_i^2 \in S_2$ and $c_i^h \in S_5$ be the set of pairs of linguistic 2-tuples of importance and satisfaction to be aggregated by the task agent for every information agent h. Then, for combining them first the linguistic values $(p_i^2, \alpha_i), p_i^2 \in S_2$ and $(c_i^h, \alpha_i^w), c_i^h \in S_5$ are transformed in the linguistic term set used to express the relevance degrees, in this case S_4, obtaining their corresponding values $(p_i^{2'}, \alpha_i'), p_i^{2'} \in S_4$ and $(c_i^{h'}, \alpha_i^{w'}), c_i^{h'} \in S_4$. Once the multi-granular information has been unified according to the 2-tuple linguistic weighted average operator definition, the aggregation of the pair associated with every term is obtained as:

$$\lambda^h = \bar{x}_l^w \left([(p_1^{2'}, \alpha_1'), (c_1^{h'}, \alpha_1^{w'})], \ldots, [(p_m^{2'}, \alpha_m'), (c_m^{h'}, \alpha_m^{w'})]\right)$$

- **Step 8.3:** To gather the better documents from *content-based filtering agents*, the *task agent* selects a number of documents $k(D^h)$ from every *content-based filtering agent* h being proportional to its respective degree of satisfaction λ^h:

$$k(D^h) = \text{round} \left(\frac{\sum_{i=1}^n \#(D^i)}{n} \cdot P_s^h\right),$$

where $P_s^h = \frac{\Delta^{-1}(\lambda^h)}{\sum_{i=1}^n \Delta^{-1}(\lambda^h)}$ is the probability of selection of the documents from *content-based filtering agent* h.

- **Step 9:** The *collaborative filtering agent* receives from the *task agent* a list of documents $DV = \{d_1^V, \ldots, d_v^V\}$ ordered with respect to their relevance RV, such that:

1. $r_j^V \geq r_{j+1}^V$,
2. for a given document $d_j^V \in DV$ there exists a h such that $d_j^V \in D^h$ and $r_j^V \in RN^h$, and
3. $\#(DV) = v \leq \sum_{i=1}^n k(D^i)$.

Then, the *collaborative filtering agent* filters the documents provided by the *task agent* using the recommendations on such documents provided by other users in previous searches which are stored in a *collaborative recommender system*. This is done in the following steps:

- **Step 9.1:** The *collaborative filtering agent* asks the *collaborative recommender system* the recommendations existing on DV associated with the information need category $\mathcal{A}_i$ expressed by the user and retrieves them,

$$RC^{\mathcal{A}_i} = \{RC_1^{\mathcal{A}_i}, \ldots, RC_v^{\mathcal{A}_i}\}, \ RC_j^{\mathcal{A}_i} \in S_3 \times [-0.5, 0.5) .$$

- **Step 9.2:** The *collaborative filtering agent* filters the documents by recalculating their relevance using these recommendations $RC^{\mathcal{A}_i}$. Then, for each document $d_j^V \in DV$ a new linguistic relevance degree r_j^{NV} is calculated from r_j^V and $RC_j^{\mathcal{A}_i}$ by means of the 2-tuple weighted operator $\overline{x}^w$ defined in Definition 4:

$$r_j^{NV} = \overline{x}^w \left(r_j^V, TF_{S_4}^{S_3}(RC_j^{\mathcal{A}_i})\right) ,$$

using for example the weighting vector $W = [0.6, 0.4]$.

- **Step 10:** The *interface agent* receives from the *collaborative filtering agent* a list of documents $DW = \{d_1^W, \ldots, d_w^W\}$ ordered with respect to their relevance RW, such that:
1. $r_j^W \geq r_{j+1}^W$,
2. for a given document $d_j^W \in DW$ there exists a i such that $d_j^W = d_i^V$ and $r_j^W = r_i^{NV}$, and
3. $\#(DW) = w \leq v = \#(DV)$.

Then, the *interface agent* filters these documents in order to give to the user only those documents that fulfill better his/her needs, which we call D_f. For example, it can select a fixed number of documents K and to show the K best documents.

4.2.2 Feedback Phase

This phase is related to the activity developed by the *collaborative recommender system* once user has taken some of documents retrieved by the multi-agent system. In the collaborative recommender systems the people collaborate to help one another to perform filtering by recording their reactions to documents they read [21, 28]. In our multi-agent model this feedback activity is developed in the following steps:

- **Step 1:** The *interface agent* gives the user's identity $\mathcal{ID}$ (usually his/her e-mail) together with the set of documents $DU = \{d_1^U, \ldots, d_u^U\}$, $u \leq \#(D_f)$ used by the user to the *collaborative filtering agent*.
- **Step 2:** The *collaborative filtering agent* asks the user his/her opinion or evaluation judgements about DU, for example by means of an e-mail.
- **Step 3:** The *Internet user* communicates his/her linguistic evaluation judgements to the *collaborative recommender system*, rc_y, $y = 1, \ldots, \#(DU)$, $rc_y \in S_3$.
- **Step 4:** The *collaborative recommender system* recalculates the linguistic recommendations of set of documents DU by aggregating again the opinions provided by other users together with those provided by the Internet user. This can be done using the 2-tuple aggregation operator $\overline{x}^e$ given in Definition 3. Then, given a chosen document $d_y^U \in DU$ that receives a recommendation or evaluation judgement rc_y from the Internet user, and supposing that in the collaborative recommender system there exists a set of stored linguistic recommendations $\{rc_1, \ldots, rc_M\}$, $rc_i \in S_3$ associated with d_y^U for the information need category $\mathcal{A}_i$, which were provided by M different users in previous searches, then a new value of recommendation of d_y^U is obtained as

$$RC_y^{\mathcal{A}_i} = \overline{x}^e[(rc_1, 0), \ldots (rc_M, 0), (rc_y, 0)].$$

5 Concluding Remarks

Nowadays Internet users need tools to assist them in his/her processes of information gathering because of the large amount of information available on the Web. We have presented two techniques that could contribute to solve this problem, the information filtering tools and fuzzy linguistic modelling. Then, we have defined a new model of fuzzy linguistic Web multi-agent systems using both techniques. In particular, this new model of Web multi-agent systems is based as on content-based filtering tools as on collaborative filtering tools and on the multi-granular fuzzy linguistic modelling. Some advantages of this model are the following:

- We improve the search and mining processes on the Web and this could increase the users' satisfaction degrees.
- The use of the multi-granular linguistic information allows a higher flexibility and expressiveness in the communication among the agents and between users and agents in the information gathering process.
- The use of the multi-granular linguistic information does not decrease the precision of the system in its results.
- The use of IF techniques allow to filter the information and so, to improve the retrieval process.

References

1. C. Basu, H. Hirsh, W. Cohen, Recommendation as classification: Using social and content-based information in recommendation, *Proc. of the Fifteenth National Conference on Artificial Intelligence,* 1998, pp. 714–720.
2. M. Claypool, A. Gokhale, T. Miranda, Combining content-based and collaborative filters in an online newpaper, *Proc. of the ACM SIGIR Workshop on Recommender Systems-Implementation and Evaluation.*
3. O. Cordón, F. Herrera and I. Zwir. *Linguistic modelling by hierarchical systems of linguistic rules.* IEEE Transactions on Fuzzy Systems, 10 (1) (2001) 2–20.
4. M. Chau, D. Zeng, H. Chen, M. Huang, D. Hendriawan, Design and evaluation of a multi-agent collaborative Web mining system, Decision Support Systems 35 (2003) 167–183.
5. M. Delgado, F. Herrera, E. Herrera-Viedma, M.J. Martín-Bautista, M.A. Vila, Combining linguistic information in a distributed intelligent agent model for information gathering on the Internet, in P.P. Wang, Ed., *Computing with Words,* (John Wiley & Son, 2001) 251–276.
6. M. Delgado, F. Herrera, E. Herrera-Viedma, M.J. Martín-Bautista, L. Martínez, M.A. Vila. A communication model based on the 2-tuple fuzzy linguistic representation for a distributed intelligent agent system on Internet, *Soft Computing,* 6 (2002) 320–328.
7. N. Good, J.B. Shafer, J.A. Konstan, A. Borchers, B.M. Sarwar, J.L. Herlocker, J. Riedl, Combining collaborative filtering with personal agents for better recommendations, *Proc. of the Sixteenth National Conference on Artificial Intelligence,* 1999, 439–446.
8. U. Hanani, B. Shapira, P. Shoval. Information Filtering: Overview of Issues, Research and Systems. *User Modeling and User-Adapted Interaction* 11: 203–259, 2001.
9. F. Herrera, E. Herrera-Viedma, Aggregation operators for linguistic weighted information, *IEEE Trans. on Systems, Man and Cybernetics, Part A: Systems,* 27 (1997) 646–656.
10. F. Herrera, E. Herrera-Viedma, L. Martínez. An Information Retrieval System with Unbalanced Linguistic Information Based on the Linguistic 2-tuple Model. 8th *International Conference on Information Processing and Management of Uncertainty in Knowledge-Bases Systems (IPMU'2002).* Annecy (France) 23–29.
11. F. Herrera, E. Herrera-Viedma, L. Martínez, P.J. Sanchez. A Methodology for Generating the Semantics of Unbalanced Linguistic Term Sets. 9th International Conference on Fuzzy Theory and Technology, Florida, 2003, 151–154, 2003.
12. F. Herrera, E. Herrera-Viedma, L. Martínez. A Fusion Approach for Managing Multi-Granularity Linguistic Term Sets in Decision Making, *Fuzzy Sets and Systems,* 114 (2000) 43–58.
13. F. Herrera, E. Herrera-Viedma, J.L. Verdegay, Direct approach processes in group decision making using linguistic OWA operators, *Fuzzy Sets and Systems,* 79 (1996) 175–190.
14. F. Herrera, L. Martínez, A 2-tuple fuzzy linguistic representation model for computing with words, *IEEE Transactions on Fuzzy Systems,* 8 (6) (2000) 746–752.

15. F. Herrera, L. Martínez, A model based on linguistic 2-tuples for dealing with multigranularity hierarchical linguistic contexts in multiexpert decision-making, *IEEE Transactions on Systems, Man and Cybernetics. Part B: Cybernetics*, 31(2) (2001) 227–234.

16. F. Herrera, Martínez, The 2-tuple linguistic computational model. Advantages of its linguistic description, accuracy and consistency, *Int. J. of Uncertainty , Fuzziness and Knowledge-Based Systems*, 9 (2001) 33–48.

17. E. Herrera-Viedma, Modeling the retrieval process of an information retrieval system using an ordinal fuzzy linguistic approach, *J. of the American Society for Information Science and Technology*, 52(6) (2001) 460–475.

18. E. Herrera-Viedma, An information retrieval system with ordinal linguistic weighted queries based on two weighting elements, *Int. J. of Uncertainty, Fuzziness and Knowledge-Based Systems*, 9 (2001) 77–88.

19. E. Herrera-Viedma, O. Cordón, M. Luque, A.G. López, A.M. Muñoz, A Model of Fuzzy Linguistic IRS Based on Multi-Granular Linguistic Information, *International Journal of Approximate Reasoning*, 34 (3) (2003) 221–239.

20. E. Herrera-Viedma, L. Martínez, F. Mata, F. Chiclana. A Consensus Support System Model for Group Decision-making Problems with Multi-granular Linguistic Preference Relations, *IEEE Trans. on Fuzzy Systems* 2005. To appear.

21. E. Herrera-Viedma, E. Peis, Evaluating the informative quality of documents in SGML-format using fuzzy linguistic techniques based on computing with words, *Information Processing & Management*, 39(2) (2003) 195–213.

22. M. Kobayashi, K. Takeda, Information retrieval on the web, *ACM Computing Surveys*, 32(2) (2000) 148–173.

23. S. Lawrence, C. Giles, Searching the web: General and scientific information access, *IEEE Comm. Magazine*, 37 (1) (1998) 116–122.

24. H. Lieberman, Personal assistants for the Web: A MIT perspective. In M. Klusch (Ed.), *Intelligent Information Agents* (Springer-Verlag, 1999) 279–292.

25. A. Moukas, G. Zacharia, P. Maes, Amalthaea and Histos: Multiagent systems for WWW sites and representation recommendations, in M. Klusch (Ed.), *Intelligent Information Agents* (Springer-Verlag, 1999) 293–322.

26. A. Popescul, L.H. Ungar, D.M. Pennock, S. Lawrence, Probabilistic models for unified collaborative and content-based recommendation in sparce-data environments. In *Proceedings of the Seventeenth Conference on Uncertainty in Artificial Intelligence (UAI)*, San Francisco, (2001) 437–444.

27. L.M. Quiroga, J. Mostafa, An experiment in building profiles in information filtering: the role of context of user relevance feedback, *Information Processing and Management* 38 (2002) 671–694.

28. P. Reisnick, H.R. Varian, Recommender Systems. Special issue of Comm. of the ACM, 40 (3) (1997) 56–59.

29. L.A. Zadeh, The concept of a linguistic variable and its applications to approximate reasoning. Part I, *Information Sciences*, 8 (1975) 199–249, Part II, *Information Sciences*, 8 (1975) 301–357, Part III, *Information Sciences*, 9 (1975) 43–80.

Advanced Simulator Data Mining
for Operators' Performance Assessment

Anthony Spurgin and Gueorgui Petkov

Independent Consultant, 4252 Hortensia St, San Diego 92103, USA
a-jspurgin@cox.net
Technical University of Sofia, Dept. of Thermal and Nuclear Power Engineering,
Block 2, Room 2356, 1797 Sofia, Bulgaria
gip@tu-sofia.bg

1 Introduction

This chapter covers the use of data mining operations associated with power plant simulations for training and other purposes, such as risk assessment. Insights gained from the use of data mining techniques can improve training methods, identify good practices, and specify both crew and individual operator problems with procedures and information displays. It can be used to improve plant safety and availability. The data can be used as a basis for upgrading programs.

It has been found over the years that the dominant effect on plant risk is due to humans rather than equipment. Both equipment and human reliability has improved but the ratio of human to machine reliability has increased. The proper use of human can enhance reliability, but this requires identifying where inadequacies in design, operation, and maintenance lie.

The technique for gauging risk has been the probabilistic risk assessment (PRA) and an important component is the human reliability assessment (HRA). The HRA methodology has been developed in the framework of the PRA methodology in nuclear industry over the last three decades. During this period, both methodologies have been carried out intensively and a great number of approaches have been introduced. The use of the PRA was used initially in the nuclear power field but has moved to be used extensively in many fields, such as oil production, various transportation and space technology (NASA), healthcare, finance, etc.

The HRA is an integral part of the PRA not only as the key to fidelity and acceptability in its issues. It is worth considering on its own. The reason is that the human factor, human-system interaction (HSI) and human performance have been much investigated in terms of productivity, efficiency and humanity

Anthony Spurgin and Gueorgui Petkov: *Advanced Simulator Data Mining for Operators' Performance Assessment*, Studies in Computational Intelligence (SCI) **5**, 487–486 (2005)
www.springerlink.com © Springer-Verlag Berlin Heidelberg 2005

in addition to risk. The HRA worth depends on the rejection of the simplified concept of operator as an automat that is only an element of the machine control loop. The reliability of humans working in a technology field depends on the ability of designers to organize the work place and to match the characteristics of people in order to reduce uncertainty and enhance predictability of the needed operator responses.

The understanding that the human-machine system (HMS) represents a combination of technological and mental processes requires each function of each subsystem (human, machine, technology and organization) to be considered in relation to the larger whole system, rather than on its own. An unbalanced representation of the processes can result in an incorrect interpretation of the dynamics of all system processes and can lead to the incomplete explanation of information, leading to imperfect knowledge and ability to predict the system behavior over the operating domain. This is especially true for the prediction of the dynamic behavior of the system where few concurrent processes are involved and their result could be rare and not observable in short time interval. The approach considered here is for the evaluation of operator performance within the dynamics of plant transient responses. One way to overcome the difficulties of predicting operator performance is to simulate all system processes in parallel and to expose the power plant operators to variations that can occur during various accident scenarios. By computerizing the plausible process models, the simulator allows not only one to monitor and study the processes interaction but also to train the operators to cope with their tasks in realistic conditions. In this chapter we consider that the plant dynamics are modeled sufficiently accurately by power plant simulators for this purpose.

With extension of the PRA applications in different fields, the question is how to use simulators for combined purposes of HRA and training is posed. The requirements for HRA and training are not identical, the HRA is concerned with gathering sufficient data to be able to predict operator performance in terms of reliability and the training requirement is to understand what errors operators make and remove the sources by better training methods. In fact, these needs lead to the definition of data collection processes and the formation of a database for the performance assessment of operators and teams during operation, maintenance and decision-making. In the design of Data Collection System (DCS) there are three basic arguments to consider what is logical, theoretical and practical. This means what is logically required to be collected, what is the theoretical underpinning for the data needs and what can be practically collected?

The attempts to exclude simulator exercises, as a source of data or basis for expert judgments seem to be a dead end. Clearly, experiments on real plants are quite expensive and unthinkable, e.g. even the cost of a simple power plant experiment is approximately equal to the cost of several full-scope simulators (FSS). There are no other better approximations to the responses of operators during real accidents than by the use of simulators.

The expert judgment, by HRA specialists or operators, is in considerable use to overcome the inaccuracies in mental and situational models and simulator exercises' conditions. But it cannot be taken as the basis for data mining: "The use of "expert judgment" is a polite name for "expert guesses", and we do not have data to validate the accuracy of the guesses..." [1]. However, with development and refinement of computerized simulators and especially FSS the importance of expert judgment should eventually be limited. Even current FSS have progressed to the stage of laboratory tests of simulator data mining where the exercises could be extensive, specific and as representative of reality as possible (from "field tests" to "full experiments"). The development of simulation of chemical, biological and high technology processes has been observed and simulators are being acquired. Simulators are used in the air-transport field, but for the most part the use here is really experiential rather than data collection to build reliability models of aircrews. In addition to any logical reasons, the change in theoretical basis of HRA provides good reason to the extensive data mining on simulators. The second-generation HRA methods shift the problem from quantification of the operator behavior and Human Error Probability (HEP) to the determination of the error-forcing context [2], "...any description of human actions must recognize that they occur in context", where the context is a function of time "on a second-by-second basis" [3]. Theoretically, context could be defined as a state of the mind and situation (the technological, mental and environmental processes in their relation). Practically, context may be regarded as a statistical measure of the degree of the HMS state randomness defined by the number of accessible states taking place in the systems' ensemble.

The need for detailed and dynamic determination of operator's performance context leads to the need of continuous monitoring and diagnostics. It forces a need to automatically data mine on the control rooms or simulators and helps reduce the use of experts to make judgments on the effect of context. Currently, most DCS record simulator response data, operator actions, alarms, etc. A few systems also include the observational data made by instructors or similar individuals of the impact of communications, leadership aspects, and stress during transients, etc.

It would appear reasonable that expert opinion will continue to be used to fill gaps in the range of plant conditions covered by simulators, such low power or shut-down conditions. Of course, if an automated context system is designed and is operating, expert opinion may have to be used to compensate for imperfections of the automatic context fixation system by qualitative descriptions, especially if there is no idea how to determine the effect of context as a quantitative measure. Generally, the impossibility of experts to cope with context fixation is summarized as due to: (1) "dead-time" of context registration by an expert; (2) "slack times" of the mental processes; (3) context variety (unsuitable interface and unforeseen circumstances) of specific situation conditions; (4) the context of expert behavior modifies the context of operators' performance.

A practical relationship between the training and HRA can exist to measure the operators' performance. The training needs this measurement to manage operators' training process (individual and crew performances), to improve efficiency of simulator use, to establish explicit standards for simulator training. The HRA needs the operators' reliability database to predict the success or failure of the operators' actions in the PRA. Both are focused on the accident indications, displays, procedures, individual actions and group process for the particular scenario, i.e. accident context on the operators' performance. The training tries to improve individual and crew performances based on current monitored macro-context (possible manifestations) opposite to HRA that considers the systematic (applied across all crews) and individual effects (applies to the activities of crews or individual personnel within a crew) on the micro-context (probable causes). The HRA uses probabilistic concept and more fuzzy psychological factors whereas the training relies on an analysis of engineering factors and conditions evaluation in terms understood by operators. This difference requires maintaining reasonable consistency between training and HRA terms, concepts for context determination and communication and brings up the basic problems that must be solved:

(1) What type of a simulator data mining technology to use?
(2) How to use the simulator data for operators' performance assessment for training needs and for HRA purposes?

2 Data Collection Methods and Analysis Tools

Nuclear power plant (NPP) crews are trained in normal, abnormal and emergency events on FSS. Crews at NPPs are heavily involved in training for all events. So, for example, before a plant is returned to service the crew is exposed to training in power increases on the simulator to ensure the ascent to power goes smoothly. Simulator sessions are held with all the crews at regular intervals during the year. These sessions cover a number of different accidents from mild to severe. Some sessions are training events in which the instructors intervene; other sessions are called re-qualification sessions the crews are expected to restore the NPP to a safe state without help from the instructors.

An FSS is an accurate representation of the NPP technological dynamics. The degree of accuracy of the simulator may vary, but usually the reactor core and associated plant are sufficiently accurately represented to duplicate NPP behavior as far as the crews are concerned. The secondary side of the NPP may not be as accurately simulated as the core. The NPP is both complex and complicated to simulate, but the intent of the designers is to present a representation that is sufficient for training purposes and ensuring that crews take the correct actions in the event of an accident.

Physically a simulator consists of some exact elements and some modeled elements. The control boards, displays, alarms and indicators look and act like real components, but they are driven by signals generated by a series of computers modeling the plant. Some elements may be just duplicates of real components like visual display units (VDUs) and computer display systems. However, the plant variables are derived from models of reactor, steam-lines, pressurizer, steam generators, etc.

The attempt is to try to make the situation as real as possible for the crews. The look and feel should be so real as to make the crews respond as closely as they would during a real plant accident. The arrangement of the control-boards, the type and function of the displays, the lighting, the carpeting, etc., are identical to real control room. It is not totally identical, but it is the closest one the designers can make and it is the best we can do. It is impossible from a safety, availability and cost point of view to carry out anything other that normal plant maneuvers. The results from various reactor test rigs are used to verify/validate dynamic digital models and to ensure that simulated accidents are dynamically similar to the real ones. This data is used to validate the simulator. Often simulators use versions of the dynamic codes as modules within the simulator software.

2.1 Simulator Data Collection Types

One can collect two types of data from simulator operations; these are technical data obtained directly from the simulator itself and the other data derived from the observations of the crew's responses.

The first set of data is generated by the simulator in response to input disturbances and responses by the crew to a disturbance. The simulator data covers "second-by-second" the transient changes to plant variables, i.e. changes in temperature, pressure, flow, etc. Additionally, actions taken by controllers and crews can also be recorded. This data dynamically reflects the response that could be expected from an actual plant; however, DCSs at actual plant are less comprehensive. Often in the simulated control room there are video cameras strategically placed to see and record operator actions.

Video recordings provide a good record of the crew response to the accident scenario. Simulator sessions are run with four or five video cameras and one is faced with a considerable amount of data to process and relate to dynamic data from the simulator itself. The FSS DCS could be extended with sensor modules (mobile equipment for monitoring and presentation of individual operator attention, perception and recognition, e.g. microphones, omni-directional cameras with fish-eye lenses and curved mirrors, motion sensors, etc.) and equipment for systematic management of these sensor modules and interface with the DCS.

It is possible to record by the extended DCS all the transient and behavioral data and store it on a hard disk for later analysis. The recorded data can be recalled by analysts/instructors and transient plots of variables and actions displayed and output. Once the session is over and the next session is started this information is no longer available, so steps have to be taken to download the data to a backup storage unit. The transient data associated with a given session is very valuable for the instructors and crew to critique the response of the crew to a given accident. It could be quite useful to adjust and refine the models of context, operators and crews, if they are systematically constructed. This process is often called a post mortem investigation.

The second set of data relates to observations of the crew and how they respond to an accident. The instructors observe the behavior of the crews and record notes on their observations. Often these notes are not very systematic and therefore are difficult to analyze. Also, the observations of the instructors have to be coordinated with both of the other sets of data. This presents a problem for the instructors, who are trying to decide if the crew is responding well to the accident scenario and where they have difficulties. That is why the second set should be used to confirm or reject the behavioral models assumptions rather than explicit judgment of the crew performance. It is especially true for inexperienced observers and when even semi-automatic observation processing system is not available. Without some form of data processing this job is very difficult to do and the review of the instructors tends to be superficial and arbitrary. Once a given crew has performed well or poorly, the response of the instructors can be biased one way or another for subsequent scenarios undertaken by a crew unless steps are taken. One way that this can be done is to separate observation from evaluation but it means that the observers' data are derived off-line and modulated by observer context in contrast to technical data that are obtained on-line and reflect only simulated context. A well-designed data collection process for both training and HRA purposes will achieve this. Such a system is at the core of the Systematic Approach to Training [4].

2.2 History of Simulator Data Collection Methods

Initially, when there were a limited number of simulators, observations of crew and individual operator performance were made by instructors and recorded by notes made during the sessions. Later, as part of a USNRC funded SROA program [5] to investigate what safety actions could be left to operators and what needed to be automated, a digital computer program was written to collect time data, plant variables and operator actions. Several simulators incorporated such a program, but the Training Departments (TDs) seldom used the program. The SROA project generated a number of Time Reliability Curves (TRCs) and these became very important source of information in HRA development [6, 7].

About the same time, Electricité de France became interested in the effectiveness of their procedures, displays and support aids. Their method of understanding the utility of these items was to use a number of human factors persons to follow and manually record the activities of the operators in responding to simulated accidents [8].

TRCs were generated by [6, 7] and others. The Human Cognitive Reliability (HCR) [7] version of a TRC was generated under an Electric Power Research Institute (EPRI) project. Sub-sequentially, EPRI funded a program to verify/validate the assumptions in a project to collect data at six utility simulators. This project was the Operator Reliability Experiments (ORE) [9]. This project was interesting because both simulator-recorded and observer-recorded data was collected. However, because there were no generally available digital computers most of the data was collected manually. Some simulators had a limited digital program for collecting data from the hard disk, but little post data collection capabilities.

This lack of capability to process data leads EPRI to fund the development of tools to collect and process simulator data. The first of these tools was OPERAS [10]. The staff at the Paks NPP developed a similar capability with a system called COPAS [11]. As more emphasis was placed upon tying plant data to observer data so that better evaluations of operator performance could be made. These later concepts were embodied in various versions of CREDIT [12]. Making detailed observations of operator responses is a very intensive task, especially if the observers record their notes manually. A way out of this quandary is to use barcode readers or pen types of computers. It has been found by experience that barcode readers are both more flexible and easier to use.

2.3 Limitations of Simulators

There are some limits to the utility of simulators for both training and HRA purposes. These are not the usual limits advanced, i.e. that simulators do not reflect the real situation, but real limits. Before discussing these limitations, let's comment on the argument advanced by some people that simulators do not sufficiently duplicate the impact of an accident on the operators. The assumption is that the accident introduces high stress in the operators and the simulator does not have this ability. Careful observations of operators in different countries and many plants have shown this not to be true. During re-qualification scenarios, the operators are under stress to perform since often their job or salary scale is on the line. This means that they take their response to accident scenarios very seriously.

Simulators do have limitations. These limits come from not being able to duplicate plant operations not related to activities in control room. Simulator instructors play the role of plant staff in the field. This is not very realistic, since the field operators may not take the correct actions, whereas the instructors always come back with the correct actions. Another limitation is that most scenarios are run for half to two hours, but in reality the accident and response to it may have to go for days resulting change over of crews.

A simulator is normally designed for operational range of 100% to 20% full power. Low power to shutdown power is not modeled very well and the plant is operated differently under these conditions than under the above power range.

There are questions related to control room manning schedules, instrument ranges, additional personnel required for maintenance and state of the plant all of which affect the relationship of the simulator to actual conditions.

2.4 Analysis of Data

As mentioned above, data consists of technical and observational data. The technical data consists of plant variable responses, i.e. how a variable like primary system pressure changes depending on input disturbances and actions by control systems or operators. Simulators have been used in a number of countries for HRA purposes [15].

Operator actions are of particular interest, specifically the timing of their actions relative to the state of the plant. For any given accident scenario or its deviations the timing of operator actions can vary despite the impact of situation, training and the use of procedures. Of course, for some accident scenarios operators can make mistakes by failing to take correct actions or taking the wrong actions. The interest of TDs and HRA analysts is the reliability of the crews taking actions and the conditions under which the operators take incorrect actions.

TDs are interested in the actions of the crews as well as the actions of individual operators. HRA experts are interested in the ability to gain insights into crew responses, so that they predict human error contributions to the frequency of core or equipment damage depending on the objectives of the PRA. TDs use time plots to check how closely the crews approach limits and how they fail to take the correct actions to terminate or mitigate the accident. They also make observations relative to use of procedures, communications between individual members of crew and overall control of the accident. HRA experts are interested in much the same data, but additionally are interested in the comparison of crew performances. This latter data can be used to form TRCs. Most of the time the crews do not fail, but sometime one or more crews fail to respond within the same time frame, i.e. significantly slower responses. Examination of these non-standard responses yields insights into causes of errors.

It could be concluded that the interests of TDs and HRA experts are not controversial but should be extended for each group.

2.5 Actual Accident Records

Above we have focused on simulator data, since it is the most available and useful and we can gain access to both plant and human response data. The simulated plant data is very detailed yielding every plant variable and every action taken by either operators or by automatic controllers. Also, simulated accidents of various complexities are run very frequently during training sessions and with all of the crews. This provides a lot of data and insights into crew performance. Accidents at plants are infrequent and thank goodness are for the most part not dangerous. But they do provide information on actual accidents and the manner that they can occur. This information should be compared with simulator data. However, it is often the case that only a limited number of plant variables are recorded and the actual cause of the accident may not be recorded or known. In fact, simulators are used to re-construct the features of the accident. Industry experience based upon similar plants can provide useful information to understand plant-induced context effects.

3 Simulator Data and Experiments

This section discusses the steps to be taken in carrying out a set of simu lator experiments and performing an analysis of the results [16]. The steps cover the design of experiments, selection of scenarios, data forms and other methods of collecting data, and data sets resulting from experiments and training sessions. The scenario is defined by the initial state and a set of disturbances (mal-functions or equipment failures, actuated blockages, protections and technological processes) followed by the responses of the crews (operators' actions, commands, conversations and started mental processes) in specific environment (natural, technological, procedural, etc.) and determined stop training study conditions.

3.1 Data Collection Process Steps and Requirements

The steps in the data collection process are:

1. *Design of scenarios.* It is based on the scenario's guidance for simulator training (baseline) and Emergency Operating Procedures (EOPs) related to this scenario. The scenario's designer is working with instructors to understand the basis and details of scenario.

On this step the designer or HRA modeler should answer to the following questions: what to look at, how to observe, identify and interpret sequences of disturbances/symptoms; what procedures contain relevant guidance and how to get to it; how operators have to deal with situation; what observers have to do; what type of objective (list of the recorded parameters) and subjective (observer's data sheets) data to collect.

It is suggested that some sort of activating disturbance such as an instrument failure be included as an introductory event as part of the scenario design. Including such a step goes some way to pull the operators and observers to being more concerned about the accident progression. Recovery from this small test should occur before the second and main disturbance is initiated.

2. *Dry runs.* Use of test runs help to develop observer sheets as well as to see that there are no surprises in the simulator's responses.

The observer sheets should list principal indications and key actions expected to be taken by the operators and should include notes on EOP entry points, transitions and key steps. From the dry runs one can estimate the expected times for both indications and actions. Of course, the times for each of the crews to take actions will vary. The initial alarms and plant states (indications) will be fixed by the disturbances and plant design. There should be some flexibility in the layout of the observer sheets in case the crews' responses are not exactly as expected.

Dry run of scenarios with instructors playing the role of the operators should be carried out to check sequence of things to be observed.

Data sheets should note date, crew identifier, scenario number and observer names.

There should be a number of observers noting what the operators actually do in responding to the accident initiator(s). It is suggested that three observers be used and at least one ought to be an instructor.

3. *Scenario run.* The scenario is then run and the observations are made and recorded.
4. *Crew debriefing.* After the session, a debriefing of the crew is carried out with the operators.

One can follow the usual Institute of Nuclear Plant Operations (INPO) or World Association of Nuclear Operators (WANO) approach and let the shift supervisor lead the review. The objective of the HRA observer group is to listen and to try to understand where the crew says that it had difficulties and where it was easy. Notes should be taken of the comments by crew and shift supervisor. This information can lead to an understanding of the various influences affecting the crews.

Video records are useful on three grounds: to be used during debriefing sessions, as backup to observer notes, and to help when trying to diagnose operator actions.

The records for each observer should be reviewed and the consensus set of times and observations should be formed. If operator actions are recorded by the DCS (which is highly recommended), then this step is unnecessary!

5. *Scenario's context and crew performance inferential statistics.* The steps 2 to 4 should be repeated for each crew.
6. *Analysis and refinement of design and models.* For each scenario the results should be assembled and analyzed. This includes analysis of both simulator-recorded data and observer-recorded data.

3.2 Design of Experiments

The design of experiments covers not only the choice of scenarios but also creating target responses expected of crews. Additionally, if video cameras are used then their placement and keying needs to be covered. If manual methods of data collection are used, then forms for this purpose need to be constructed. However, if barcode readers are used, then the appropriate barcode sheets need to be designed. Sessions purely for requalification training go through a similar preparatory stage. Instructors are less interested in data analysis to determine whether operators are successful or are deemed to fail than HRA analysts and therefore are less inclined to emphasize the data collection aspects! Often instructors are quite happy with their own notes.

We have found that multi-failure scenarios are most useful to test control-room crews. The complete scenario is usually made up of three parts. The first part is a small disturbance such as an instrument failure affecting a control function. The operator response is to place the controller into manual mode and readjust the operating conditions to what they were before the transient. This type of introductory transient helps the crew focus on the realities of controlling the plant. The next part of the scenario is to introduce a small accident like a small line break, which can be isolated without causing a major change in the plant characteristics. The crew is expected to isolate the break and again return to the plant to about the same conditions as before. Then when the crew is relaxing, thinking that the session is about to finish, the third failure is injected. This is usually a major failure calling for heavy involvement of the crew in responding. This might be Steam Generator Tube Rupture (SGTR) or a similar transient. The response to accidents is made more difficult for the crew by safety equipment being out of service or failing to operate when demanded. We have found that these types of scenario do stress the crews and that even following EOPs errors do occur [14].

In HRA experiments the observations are more detailed than those made by instructors during purely training sessions. In addition to observing crew interactions the HRA observers are trying to judge the impact of procedures, instrument arrangements (layout and type), communication protocols, and training on operator performance. The quality of the crews' capabilities are not estimated during the session, but assessed during the examination of the data, including reviewing video recordings after the session. This is to avoid biasing the results. We have noticed that biases do creep into ordinary training sessions since the instructors do not separate data collection from opinion forming! Crew performance can vary from scenario to scenario, very good at one time and not so good at other times. HRA persons are interested in this variation since it may relate to design of procedures, etc., and should be factored into the HEP evaluation.

3.3 Data Collection

Methods of data collection range from hand entries on forms to use of DCSs connected to NPP simulators. In advanced DCSs, observational data is recorded by various methods, such as barcode readers, and entered and processed directly in the DCS [11]. Figure 1 shows a diagrammatic representation of such a system.

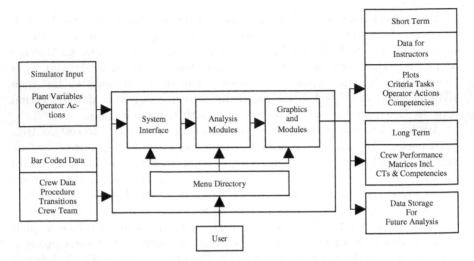

Fig. 1. Data collection system, CREDIT

One such system that we have experience with is the latest version of CREDIT version 3.1 [12]. In this system, simulator data is recorded and selected variable plots are output and operator actions are automatically compared with expected actions and the results are output along with the actual times of the actions. CREDIT version 3.1 uses barcode reader input and then analyzes the observer results and output results in terms of observations versus time responses. Following a simulator session, a debriefing meeting is held with the crew. The collected data, transient plant variable plots, operator action records and observational records are discussed with the crew. If needed, video records are used as back up. Insights gained from these discussions are helpful in determining operators' reasons for certain actions. The barcode process is an effective way of collecting observer data compatible with computer analysis methods, i.e. it forces the observers to have a fixed taxonomy for making observations.

The barcode sheets are organized to collect data or record items such as date of run, personnel, scenario, and chosen mal-functions. Observational data are collected when observed. They cover procedure steps, which operator notices variable changes and who does he communicate with, who decides and takes an action, difficulties with procedure instructions, control room layout of displays and controls, communication protocols, etc. The barcode sheets are organized to be able to cover most situations by the use of partial statements, which can be grouped together to cover the needs of the observers. During a specific observation relating to use of the procedures the observer would swipe the barcode sheet a number of times. The swipes would cover the role of the user, the procedure, the step of the procedure and the difficulty he was having, such as interpretation, readability or where to go next. Associated

Table 1. Manual Data Collected for MSLB/SGTR Accident Scenarios

PWR-1 SERIES-I SCENARIO 2 (MSLB/SGTR) CUMULATIVE TIMES (SECONDS)

Cue/Action Week/Session	Week 1		Week 2		Week 3		Week 4		Week 5	
	AM	PM	AM	PM	AM	PM	AM *	PM *	AM	PM
MSLB on SG 1-1 occurs.	0	0	0	0	0	0	0	0 **	0	0
Containment environment P-250 alarm.	60	58	93	20	60	158	60	60	59	62
Operators diagnose containment activities (MSLB) and manually trip RX (before auto RX trip).	320 +	315 +	320 +	302	289	325 +	310	319 +	289	267 +
Operators identify and isolate faulty SG 1-1.	386	515	365	582	469	559	363	380	342	357
SGTR on SG 1-1 occurs.	902	902	902	902	902	902	902*	902*	903	903
Pzr. level drops rapidly	902	902	902	902	902	902	902	902	903	903
Operators diagnose SGTR and start RCS cooldown (per E-3) by dumping steam to condenser from intact SGs or opening intact SGs 10% dump valves.	2701	2128	1942 ***	2530 ***	1757 ***	1341	2127	1450	1815 ***	N/A
Operators start RCS depressurization (per E-3) to minimize break flow and refill pzr. (e.g., pzr. normal/aux. spray, pzr. PORVs).	1332 ++ 1341 +++	1045 ++	938 ++ 1049 +++	1112 +++	2231 +++	NP	1421	1192	934 ++	N/A

NA = Session ends early.   NP = Not performed.   + = Auto RX trip.   ++ = Use pzr. spray.   +++ = Open pzr. PORVs.
* = TEAM not available.   ** = The MSLB time was assumed to be one minute before P-2***= Use 10% steam dump valves.

with the swipes are times, so the analyst can connect the observations with the plant state. Additionally, the video records associated with the activity can be located easily and used in the post mortem or by HRA analysts. If an observer is interested in the activities of the crew during the accident and needs to review their activities, he can signal this by using the barcode sheet to record this need.

3.4 Results

As mentioned above, data can be collected manually or automatically. A number of organizations have performed simulator sessions for HRA purposes [9, 15] mostly collecting data manually. A typical set of manually collected data from sessions carried out at Pressurized Water Reactor (PWR) NPP are shown in Table 1 and cover ten crews responding to the same accident. Automated recordings would collect the same data and be capable of collecting and store more data, such many more plant variables and other actions taken by the operators. If a DCS is used then observer data could be compiled at the same time by downloading barcode reader recordings. The operators are expected to deal with a number of tasks during a response to an accident. As far as plant safety is concerned, some actions are more important than others,

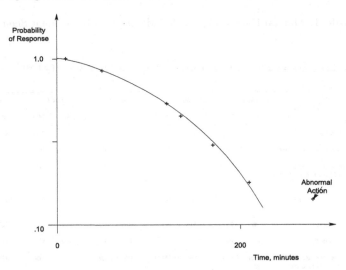

Fig. 2. Typical Time Reliability Curve

for example in responding to a Main Steam Line Break (MSLB) or SGTR on PWR NPP, there are 13 isolation actions, of these only three are really important to plant safety. From a Probabilistic Safety Assessment (PSA) and HRA only these three are important, but as far as TDs are concerned all 13 are important. So there are different criteria for data collection between HRA and Training, however the DCS collects all of the data and the user can apply active filtering of the data and interface management strategies for their requirements be it for HRA or training purposes.

There are important safety actions to be taken by the crews, so in manual data collection the observers have to focus on noting when these actions are taken and at the same time note the impact of influences on operators' responses. This is a very intensive task, what the DCS does is to remove the burden of noting times of activities and the observers can concentrate on impact of procedures, etc., on human activities.

The variability of crew responses for a given activity yields a TRC, such as Fig. 2. There will always some variability in the responses and this reflects human variability even when all crews are successful.

In general, the response curves are smooth curves; however the actions of some crews may fall outside the expected range, see the abnormal action depicted in the figure. Sometimes the crews are successful but their actions are delayed, as for the abnormal action. Sometimes crews are not successful within the time frame of simulator run, or the plant response leads to a different state and the instructors terminate the simulator run. Investigation of the "slow" and unsuccessful results indicates some needed changes to procedures, training or instrumentation. Re-training of the crew(s) is usual step, but it is often not the best solution. The recorded results from the barcode reader download contain information on the use of the procedures, man-machine interface, communications or other observable. Re-reading and interpretation of procedures by the reader and interactions with other members of crew can clearly point to the area of concern. Because of the time tag associated with barcode recordings it is easy to relate problems with time response to human activities.

Having identified a problem, the instructors can revisit other crew responses to see if the problem is a generic or specific issue and decide what is the best way to deal with it. An extreme example of this was seen when 14 out of 15 crews had a problem dealing with a spurious safety injection (SI) signal. After review it was decided that the problem was better addressed by a small change to the order in the procedure together with a logic diagram showing how the logic associated with SI trip operated. Once the changes were incorporated and on a later simulator run all crews were successful. The extra training requirement was minimal, but there was some additional documentation incorporated within the EOPs.

4 Insights from Data, Training and HRA

The data mining practice shows that the quality and applicability of the data obtained on the research and plant simulators strongly depends on the designed DCS, planned experimental effects and methods for data accumulating and processing. The previous two sections give up-to-date tools, forms and procedures for data mining process design, organization and reporting. They are based on numerous experimental trials, experience and efforts to refine them. At the same time, the complaints of a lack of suitable HRA data with psychological realism, scenario's validity, plant-specific relevance and PRA applicability are continuing. The reasons seem to be that the measurements do not imply and model correctly some basic concepts, concerning human error, mental and physical processes, contextual factors that influence operators and crew performances [17]. The question arises as to whether there is an overlapping of the HRA theoretical concepts and research and plant simulator studies.

4.1 Overlapping of Theoretical Concepts and Data Results

The Table 2 represents the overlapping of the HRA concepts and data results. Some answers to the question how to improve the treatment and use of the theoretical concepts and reformulate the HRA methods to connect the human performance modeling, measuring, data mining and evaluation procedures are given. They are illustrated on the base of insights included in the advanced HRA methods: the Holistic Decision Tree (HDT) [18] and Performance Evaluation of Teamwork (PET) [19, 20]. The illustration is limited to the re-worked, reanimated and developed HRA concepts by these methods (indices 1 and 2, respectively).

Table 2. Overlapping of the HRA concepts and data results
Experimental data (E) and Simulator's training data (S), "+" means totally accepted, "−" means totally rejected, •nd "+/−" or "−/+" means partially accepted or rejected, "RW" means re-worked, "PSF" means Performance Shaping Factor

No.	Concept	Insights		Models	
		E	S	HDT	PET
1.	Swain integrates human and machine as a common system.	+	+	+	$+^2$
2.	Human performance depends on context.	+	+	$+^1$	$+^2$
3.	Swain represents the HEP by TRC.		−	−	−
4.	Swain divides the Human Erroneous Actions (HEA) into two parts: cognitive and manual.	+/−	+	+	$+^2$
5.	Swain decomposes a task into sub-tasks and the HEP of the overall task is given by the sum of sub-tasks' HEPs.	−/+	−	−	$-/+^2$
6.	Swain accounts for differences in the environment, under which a task is performed by multipliers (PSFs).	−	−	−	−
7.	Swain associates different circumstances with modifiers (PSFs) to correct the basic TRC		−	−	−
8.	Nominally the PSFs are the same for all crews, yet this does not explain the crew variations		−	$-^1$	$-^2$
9.	Expert judgment can be used for covering uninvestigated scenarios and for Influence Factors (IFs) importance measuring.	−/+	−/+	$+^1$	$-/+^2$
10.	Combination of expert judgment and estimates can be used for HEP evaluation.	−/+	+/−	$+^1$	$-^2$
11.	The HCR model could be verified.		−	−	−
12.	HCR curves represent typical operator responses.		+	$+/-^1$	$+/-^2$
13.	HCR curves represent themselves as a mechanism for generating human error rates.	−	−	−	−
14.	HCR curves represent the variability of humans responding to the same initiator.		+	$+/-^1$	$+/-^2$
15.	Rasmussen's Skill/Rule/Knowledge (SRK) concept is applicable for in HCR curve.		−	−	−
16.	Rasmussen's Step-Ladder Model (SLM) is applicable for decision-making process. Usually two iterative steps are used.		+/−		$+/-^2$
17.	The "violation," "circumvention" or 'masking concepts are applicable.	+	+	$+^1$	$+^2$
18.	Reliability curve is obtained by two curves (coupling/summing up).		+	$-/+^1$	$+^2$
19.	It is not possible to directly measure error probabilities unless the error rate is high!		+	$+^1$	$+/-^2$

Table 2. *continued*

No.	Concept	Insights		Models	
		E	S	HDT	PET
20.	Operator and crew responses depend upon the scenario unfolding (time-line).	+	+	$+^1$	$+^2$
21.	The accident sets up the context for operators, they respond to it and the context determines the potential errors that crews might make.	+	+	$+^1$	$+^2$
22.	The initiator sets up the "second-by-second" context under which the crew operates and drives the displays and alarms, selects the procedures and responses based upon training and knowledge, so all of these are related.	+	+	$+^1$	$+^2$
23.	The accident would affect some items directly (always – displays; sometimes – interpretation of the procedures) and others indirectly.	+	+	$+^1$	$+^2$
24.	A number of IFs together correspond to given HEPs.	+	+	$+^1$	$+^2$
25.	The context controls and influences human performance – holistic approach.	+/−	+	$+^1$	$+^2$
26.	Operator and crew responses depend on their specific scenario's experience.	+	+	+	+
27.	Reason's "pyramid" concept is applicable: Front line operators are at the "sharp end of the pyramid," as opposed to managers who are at the base of the pyramid. Each layer can introduce flaws.		+	+	+
28.	The operator response time appears to fit a lognormal distribution.		+	$+^1$	
29.	The information exchange between control room personnel and local personnel play an important role in the situation awareness.	+	+	$+^1$	$+^2$
30.	The team skill dimensions (supportive behavior, team initiative, leadership, coordination, adaptability) are valuable for successful team performance.	+	+	+	$+^2$
31.	There is a tendency to skip or postpone tasks that human considers to be less important.	+	+	^{1}	^{2}
32.	The operators try to interpret and reason the situation, when it does not follow their images (formed by foresees and expectations rather than just facts).	+	+	$+^1$	$+^2$
33.	Operator makes judgment based on goals, symptoms and tendencies of limited number of parameters or function synthesized by related group of parameters (plant states, abnormal equipment/process status, safety functions...)	+	+	$+^1$	$+^2$

4.2 Advanced HRA Methods for Data Mining

Swain [6] was one of the first to formalize the relationship between machine and human. The principal of PRA/HRA is to represent discrete failure modes covering equipment and human contributions separately. The use of numbers to represent the actions of persons is integral to HRA.

Insights from the HDT Method

The HDT method [18] was developed from results and insights from the EPRI ORE project [9]. An early version was employed in a HRA calculator and was used for latent failures, as a test. Further development of HDT resulted from Paks simulator experiments. HDT was used in the Paks PSA (Full Power) and later used in a number of other PSAs. Development & applications of the HDT method, on review, adopted and reanimated a number of the insights presented in Table 2.

Concept 2. Scenario's "average" context. The HDT is based on the concept that the accident sets up the situation or context. It combines the context in the form of Influence Factors. These IFs together determine the HEP. For a set of accidents the context may vary from one accident to another and this needs to be reflected in the model. The IFs are typically qualities of procedures, training, Man-Machine Interface (MMI), etc. However the quality of an IF may vary according to the accident, e.g. the MMI for a given accident may be good and for another it may be poor.

The HEPs are calculated from the relationship between the IF importance weights, quality descriptors and anchor values. The upper and lower bounds of HEP correspond to the best and worst combination of IF quality descriptors and are derived by a combination of data and judgment. These are the anchor values.

Concept 8. "Average" crew performance. This concept is re-worked in the HDT method to deal with predicting the impact of accident context on the 'average' crew.

Concepts 9 and 10. The expert judgment is widely used by the HDT method for covering and complementing the missing information. The range and specific HEP values can be obtained from simulator experiments and on the base of expert judgment.

Concepts 12 and 14. These concepts are re-worked in that TRCs are a result of crew variability only.

Concept 17. Although violations do occur in practice and are seen in accident reports, the HRA violation concept is not part of the formulation of HDT because it deals with "averaged" crews and violations are not normally an aspect of station behavior. It can be part of HDT if it is observed that the management allows for such deviations from acceptable practices.

Concept 18. It is a re-worked TRC that has been rejected but the idea of recovery is retained!

Concept 19. Simulator data, experiments and actual events recognize the truth of this HRA concept. Normally, the error rate for well-trained crews is low. However, if some aspect has been missed in training, then the failure rate can be high and recorded in simulated accidents.

Concepts 20–25. HDT holistic approach. The effect on the crews is related to the total effect of the combination of the accident scenario, the displays, procedures, training, etc. So these influences should be considered as a whole, i.e. taken together (Holistic aspect). However, it is difficult to separate the effects, so the HDT assumes that they are independent but weighted according to the actual scenario. The IFs are selected based upon a specific scenario. The HEP is a function of the scenario, the IFs and the Quality Values (QVs) are associated with the effect of the scenario on the plant and hence on the crew. So for example, the unit of plant Z exposed to a given accident will have a different HEP to unit of plant X, despite the fact that the plant is identical in most aspects and these differences can be very significant. Plant Z may be well maintained with high-class management and plant X would be the reverse.

Concept 28. This HRA concept is incorporated into HDT formulation.

Concept 29. Communications between MCR crews and plant operators is important for some accident scenarios and the impact included into ET formulation in the PRA.

Concept 31. It is particularly important for training and if consistently observed should be incorporated into HDT.

Concepts 32 and 33. These HRA concepts are important to understand when building a HDT model of crew performance. In the first case, they result in a reduction in error for some accident scenarios, since resourceful crews will bring other skills to accident termination or mitigation. The reverse can occur when crew focus on specific indicators, especially when the accident sequence includes failure of the indications.

Outlines of the HDT Method

We have discussed some aspects in evaluation of the input to the formulation of the HDT method. It is appropriate to look at the model in a little more detail, although a good description is in reference [18]. Context dependent HRA models, such as the HDT method, are the so-called second generation HRA models taking the place of Swain's THERP, HCR and such like HRA methods. Context determines the actions that the operators take along with the consequential errors.

In the HDT model, context is represented by a series of IFs and their associated QVs. Examples of IFs are the Human-System Interface (HSI) and Training. The quality/effectiveness of these IFs can be grouped into categories, such as Efficient or Excellent, Adequate or Good, Supportive or Poor. An accident affects a power plant in a specific way leading to a transient response of the plant, which in turn produces an effect upon the operator via the HSI,

the procedures and resulting from the effects of Training upon the operator and hence determines the operator response to the accident scenario.

The HDT method uses a tree representation to symbolically connect the accident to the IFs and associated QVs to determine the HEP. For a given accident, there is a set of IFs and QVs and they trace a pathway through the tree (somewhat similar to an ET). The end-state of the path leads to a specific HEP, see figure below. The pathway through the tree is shown in blue for illustrative purposes.

In turn, the various displays and indicators reflect the changes in the plant. The quality of the HSI for a specific scenario may vary, for some be well designed and others less so! In Fig. 3, this variability is recognized by the use of supportive and adequate. The effectiveness of the HSI maybe obtained by expert judgment or by test.

The equations (1) and (2) are the mathematical representation of the HDT model. The approach uses anchor values for the lower and higher values of HEP. Values equal to 1.0 E-3 to 1.0 E-4 and 1.0 have been used for the lower and higher HEPs. Other estimates could be used based upon experience with the plant's operational history. HEP formulation takes into account both the IFs and QVs in the S modifier. The IFs are normalized to 1.0 and QFs are relative values, in this case 1, 3, and 9 used in the ISS study. These are for three QVs of Supportive, Adequate and Adverse.

$$\ln(HEP_i) = \ln(HEP_l) + \ln(HEP_h/HEP_l) \left[\frac{S_i - S_l}{S_h - S_l} \right] \tag{1}$$

$$S_i = \sum_{j=1}^{n} (QV_j) I_j \quad \text{with} \quad \sum_{j=1}^{n} I_j = 1 \tag{2}$$

where:

HEP_i = the human error probability of the ith pathway through the HDT
HEP_l = low HEP anchor value
HEP_h = high HEP anchor value
S_l = lowest possible value of S_i.
S_h = highest possible value of S_i.
QV_j = quality descriptor value (i.e. 1, 3 or 9) corresponding to the jth IF
I_j = importance weight of the jth IF
n = number of IFs in the HDT.

The HDT model has been used for a number of PSA studies. The approach to the determination of IFs, QVs, anchor values and verification of HEPs has varied. Use has been made of various domain experts, HRA experts and simulator results. The approaches have varied because of the available tools, experts and time/money. But ultimately, the tool has been useful in providing insights and HEPs for PRA/PSAs.

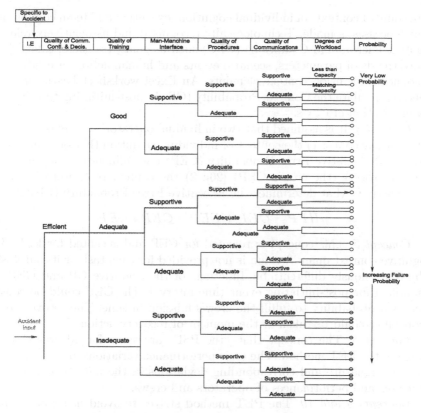

Fig. 3. Portion of Holistic Decision Tree

Insights from the PET Method

Development and application of the PET method adopt and realize the following insights according to the numbers of concepts in Table 2.

Concept 1. The human and machine are represented as a common isolated system for exchanged information in the HMS. It is assumed that the information of an isolated system is conserved and a non-isolated system could be considered as a part of a larger isolated one [19]. The HMS mental and physical processes could be described at each moment by its states.

Concept 2. Dynamic, "second-by-second" context quantification. The context may be regarded as a statistical measure of the degree of the HMS state randomness defined by the number of accessible states taking place in the systems' ensemble. Regardless of the place, moment and agent, the performed human erroneous action (HEA) could be divided into three basic types that determine the reliability of human performance: violation, cognitive (mistake) and executive (slip/lapse) erroneous actions. Based on quantitative definitions of these concepts a PET "second-by-second" macroscopic quantification procedure of contexts of individual cognition, execution and team communication processes is made. Technologically recognised and associatively relevant Context Factors and Conditions (CFC) such as goals, transfers, safety functions, trends of parameters, scenario events and human actions are taken into account as cognition context elements. An Excel worksheet has been developed to calculate the Context Probability (CP) in post-initiating time interval given selection of CFCs.

Concept 4. It is assumed that two individual operator's contexts should be taken into account: 1) the context of individual cognition that determines the individual Cognitive Error Probability (CEP) and influences group decision-making process (the crew's CEP) [26]; 2) the context of operator's sensor-motor activity that determines the Executive Error Probability (EEP).

$$HEP = CEP + EEP - CEP * EEP \tag{3}$$

Concept 5. This concept is rejected for CEP and accepted for EEP. The cognitive context quantification is not provided for each task or its sub-tasks. The continuously differentiable functions of the cognitive CP and CEP are quantified for post-initiating event time interval. The CEP could be considered as a probability to fulfill the crew's mission in time. The recovery error probability could be taken at the moment of recovery action.

Concept 8. The concept that "the PSFs are same for all crews" is rejected in the PET method. The crew performance variations are based on the scenario signature and corresponding deviations in the mental and physical processes and performances of operators and crews.

Concepts 9 and 10. The PET method strives to avoid using any expert judgment. But it is inevitable because of the lack of verified models and proofs of the assumptions. Consequently, the expert judgment could be used for changing and refining of the models. For example, the weighting of the violations and CFCs is not used in the PET method yet. The importance is equal for all CFCs and the assignment of violations to the given CFC is determined by expert judgment (modeler). However, a more conservative variant (without expertise) where violation is assigned to the CFC that gives worst context (highest CEP) is applicable as well.

Concepts 12 and 14. These concepts are re-worked so that crew HEP is a function of the cognition and execution contexts of each individual operator and group processes (communication, information exchange and decision-making).

Concept 16. The PET method re-worked the Rasmussen's SLM framework as a reliability model of individual cognition/decision making process where the non-selective influence of the context is crucial. The identified SLM reliability model of cognition is based on the results of simulator experiments, assuming that the latest model presents the most complete development of the ideas of previous models.

The Combinatorial Context Model (CCM) model and the Violation of Objective Kerbs (VOK) method [19, 20] obtain the CP of a given scenario as a function of time. This CP is used as a probability of connection between sub-processes of cognition in the step-ladder reliability model (SLRM). The non-selective influence assumes equal connection probability between sub-processes. The general view of the obtained function CEP(CP) for different combinations of iterative steps are shown on Fig. 4 in logarithmic scale. The model is constructed and solved by the Analysis of Topological Reliability of Digraphs (ATRD) method [21].

Interpretation of the CEP (CP) curves:

The curves show that in non-severe context $(CP < 0.1)$ the implementation of the cognitive process in more than one iterative step is not important and vice versa in severe context it is crucial.

As the CEP should be decreasing monotonically when the CP is decreasing monotonically, it is obvious that there exists a minimal CP for the operator's response starting. It varies on iterative steps combination (CP, CPP) as follows: for TD&A (0.659, 0.219); •&• (0.584, 0.189); TD&O&A (0.511, 0.107). The initial increasing of the CEP, when CP decreases for more than one step curves, is the time period of the first step. In this period the intention to act increases but the likelihood to respond is small (≈ 0). The last step of the cognitive iterative process must be "Action" but the result for CEP does not depend on the order of steps.

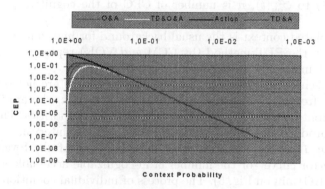

Fig. 4. The general view of dependence between CP & CEP by the ATRD SLRM, where Action (A), Task Definition (TD), Observation (O)

The calculated minimal value of CEP (CP) is limited to the pre-assigned accuracy of the code ATRD SLRM. However, on the base of the implemented PET applications up to now could be concluded that the cases with CP < 0,003 (CEP < 10^{-7}) can be neglected as improbable.

Concept 17. The Reason's concept for violation is re-worked and extended. The extension is based on the quantitative definitions of erroneous actions that follow Reason's qualitative definitions:

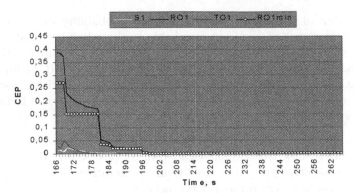

Fig. 5. The CEPs of 1st crew in "Scram" scenario on the FSS-1000, Kozloduy NPP, where Supervisor (S), Reactor Operator (RO) and Turbine Operator (TO)

Errors are "all those occasions in which a planned sequence of mental or physical activities fails to achieve its intended outcome." *Violation is* an "aberrant action" (literally "straying from the path"...)'(*Reason*).

Cognitive error is probable when the $\varphi_{on}(t) \neq \varphi_{sn}(t)$, $n = 1 \ldots N$, where $\varphi_{on}(t)$ and $\varphi_{sn}(t)$ are *objective*, occurred in fact, and *subjective*, considered to have occurred. *Violation* occurs when the objective image of φ_{on} is changed from $\varphi_{on}^1(t)$ to $\varphi_{on}^2(t)$, n is number of CFC of the cognitive process (*PET method*).

The violated context is the usual background for high human error rates. That is why the PET method (by CCM and VOK) represents the violation as the most important contributor for human errors.

This extension means that the dormant conditions are not obligatory to be a result from decisions, actions or inactions of those who are far removed from the front line, such as managers or regulatory authorities. The operators may also produce violations even in the post-accident interval.

Concept 18. The violations determine the slope of the Swain's "slowly reducing error curve" or the number of in-cognizable accessible states of the HMS (see RO1min on Fig. 5). The process of individual cognition determines the slope of the Swain's "rapidly reducing error curve" or the number of unknown accessible states of the HMS. This curve could be represented as remainder of the curves RO1 and RO1min on Fig. 5.

Concept 19. If assumed that the contribution of a given violation to the CEP is constant for a given scenario, it is possible to measure even error probability with Low Error Rate (LER):

$$\text{CEP}_{\text{LER}}(\text{CP}_{\text{NV}}) = \text{CEP}_{\text{LER}}(\text{CP}_{\text{V}}) - [\text{CEP}_{\text{HER}}(\text{CP}_{\text{V}}) - \text{CEP}_{\text{HER}}(\text{CP}_{\text{NV}})]$$

(4)

where indices mean HER – High Error Rate, V – Violated context, NV – Non-Violated context.

Concepts 20–25. PET holistic approach. The HMS is considered by the PET method as a whole. Consequently, the individual cognitive/decision-making is considered as an integrated activity that reveals itself in a context – by analogy to electromagnetic field in induction. The decision-making process includes selective and non-selective influence, but the latter (context influence) is crucial according to the holistic approach. CFCs influence all "control links" of decision-making process. The factors which influence the sub-processes are not included in this PET "holistic approximation" of decision-making process.The context quantification is not provided for individual action. It is necessary for assessing any crucial cognitive error in post-initiator interval to check current situation and to ensure that the outcome could reflect all temporary and permanent influence factors. The CP is a function of time and determines the potential cognitive errors of operators by the SLRM.

Concepts 29 and 30. The context quantification procedure by the CCM and Group Communication Reliability Model (GCRM) of the group interaction gives the opportunity to take into account communication process. The graph GCRM could be extended to include more control room and local operators, but individual CP, CEP and mutual communication probability should be evaluated. The model is solved by the ATRD method.

The PET applications show that the natural communication based on different workable knowledge (different individual CP) in the time of accident is less than 0.05. For that reason, the plant procedures recommend the supervisor to order a number of actions to other operators and to get back their reports. The probability of this initiated communication reaches 0.35 for the "Scram" scenario. Unfortunately, the supervisor obtains this information with a delay, usually after the decision-making process. That is why the impact on team performance of initiated communication is too small because of its inexpedience. As a result, the crew CEP is very close to the supervisor CEP (which is really small for this scenario, Fig. 5).

Concept 31. From the PET standpoint, the measuring of the importance of violations and CFCs (tasks) is very valuable. However, it is better to be evaluated statistically and plant-specifically.

Concept 32. The CCM used in the PET method is based on these concepts and the concept of *"human performance shifts"*, i.e. it assumes that the "context" rate in accident situation is proportional to the deviation in the operator's mental model objective image of past and future from the subjective one. They depend on machine and human, and take into account the total deviation rather than two separate types of deviation.

Concept 33. This concept is the reason to use CFCs as macroscopic parameters. Any CFC depends on specific IFs, & the discovery of operator erroneous (high HEP) should be considered as the starting point of the error investigation, and not the ending point (cause).

Outline of the PET "Scenario Run" Step Algorithm

The PET algorithm for data mining in "scenario run" step includes:

1. Detailed "second-by-second" description of the event by tracing a detailed time-line basically on the simulator-recorded data.
2. Fixation of HMS macroscopic parameters (CFCs) – φ_n. (φ_{sn} and φ_{on}) that are determined in the design of scenario and dry run steps.
3. Specification of initial and boundary conditions. For each situation and for each member the initial φ_{skn} (non-expert) or φ_{skn}^e (expert), and final φ_{okn} or φ_{okn}^v (violated) values of CFCs should be indicated.
4. Calculation of cognition context deviations by the formula:

$$|\varphi_{okn} - \varphi_{skn}| = \Delta\varphi_{kn}, n = 1 \ldots N, k \neq j \qquad (5)$$

5. Calculation of cognition and communication CPs:

$$CP_k(t) = \left[\prod_{n=1}^{N} |\varphi_{okn}(t) - \varphi_{skn}(t)|\right] \bigg/ \left[\prod_{n=1}^{N} |\varphi_{okn}(t) - \varphi_{skn}(t_0)|\right] \quad (6)$$

$$CCP_{kj}(t) = CP_j(t) - CP_k(t), k \neq j \qquad (7)$$

$k, j = 1 \ldots K$, where K is the total number of team members

6. Calculation of individual CEP (by the ATRD SLRM Code).
7. Calculation of team CEP (by the ATRD GCRM Code).

5 Discussion

Simulators are in operation for almost every NPP plant in the world and some investigations have been carried out to examine operator performance at a number of plants [15]. The US Department of Energy started a project to collect HRA data, but the impetus of this work seems to have died. There is a failure on the part of plant managers and others to see the value of this work, despite the knowledge that humans are much more responsible for plant shutdowns and accidents than plant equipment. The estimate from PRA studies is that the human contribution is 70% of core damage risk compared with an equipment contribution of 30%.

There were some theoretical limitations to what was being pursued. In this chapter we tried to explain and compare different approaches to the applicability of expert judged "average" context and "average" crew performance and context description or quantification usability for the HEP evaluation and HRA data mining process.

Practical questions like: Are the simulator data just experiential rather than appropriate for HRA? How far the possibilities of the DCS are spread out and how shall we entrust to expert judgment? How to tie the plant and observer data to fix and treat facts but not to create them? could be overcome by extension and coordination of the HRA and training purposes.

6 Conclusions

It has been shown that data mining can be very valuable to NPP managers, Training managers and instructors, HRA analysts and many others. The analysis of the data can reveal both the strengths and weaknesses in operators and crews. It can reveal the strength of training programs and the quality of trained personnel. Often following the review of accidents the conclusion is reached that the training program is deficient and more time should be spent on training to deal with a specific accident, but this conclusion is wrong. Training is a limited resource and more time should be devoted to understanding what is actually affecting operator performance and then fixing these elements, be it HSI layout, procedures, etc., this is a more effective way to deal with accidents than training. Encouraging data collection and then mining that data for useful information is an intelligent use of corporate funds.

References

1. Moray, N., "Dougherty's Dilemma and the One-sidedness of Human Reliability Analysis," Reliability Engineering and System Safety 29 (1990) 337–344.
2. Barriere, M.T., Bley, D.C., Cooper, S.E., Forester, J., Kolaczkowski, A., Luckas, W.J., Parry, G.W., Ramey-Smith, A.M., Thompson, C., Whitehead, D., Wrethall, J., "Technical Basis and Implementation Guidelines for A Technique for Human Event Analysis (ATHEANA)," NUREG-1624, US Nuclear Regulatory Commission, Washington, D.C., 1998.
3. Hollnagel, E., "Cognitive Reliability and Error Analysis Method – CREAM," Elsevier Science Ltd., London, 1998.
4. Barnes, V., 2001, "The Human Performance Evaluation Process: A Resource for Reviewing the Identification and Resolution of Human Performance Problems," NUREG/CR-6751, US NRC, Washington, DC, USA.
5. Kozinsky, E.J., et al., "Criteria for Safety-Related Operator Actions: Final Report," NUREG/CR-3515, US NRC, Washington, D.C., 1984.
6. Swain, A.D. and Guttman, H.E., 1983, "Handbook of Human Reliability Analysis with Emphasis on Nuclear Power Plant Applications," NUREG/CR-1278, US Nuclear Regulatory Commission, Washington, DC, USA.
7. Hannaman, G.W., Spurgin, A.J. and Lukic, Y., 1984, "Human Cognitive Reliability Model for PRA Analysis," NUS-4531, Draft EPRI Report, Electric Power Research Institute, Palo Alto, California, USA.
8. Villemeur, A., et al., "A Simulator-Based Evaluation of Operator's Behavior by Electricité de France," International Topical Meeting on Advances in Human Factors in Nuclear Power Systems, Knoxville, TN, USA, 1986.
9. Spurgin, A.J. et al., "Operator Reliability Experiments using Power Plant Simulators, Vols. 1,2 &3. EPRI NP-6937, EPRI, Palo Alto, California, 1990.
10. Spurgin, A.J. and Spurgin, J., 1994, "A Data Collection and Analysis System for Use with a Power Plant Simulator," Institute of Mechanical Engineers Seminar, "Achieving Efficiency through Personnel Training – The Nuclear and Safety Regulated Industries, London England.

11. Spurgin, A.J., Bareith A. and Moieni P., 1996 "Computerized Safety Improvement System for Nuclear Power Operator Training," Joint SCIENTECH and VEIKI Report for Brookhaven Laboratory, NY, USA.
12. Spurgin, A.J. and Spurgin, J.P. "CREDIT Vr 3.1 code, Description and Operating Manual", Arizona Public Service contract, Phoenix, Arizona, 2000.
13. Bareith, A. et al., "Human Reliability Analysis and Human Factors Evaluation in Support of Safety Assessment and Improvement at the Paks NPP," 4th International Exchange Forum: Safety Analysis of NPPs of the VVER and RBMK Type, October, Obinsk, Russian Federation, 1999.
14. Holy, J., "NPP Dukovany Data Collection Project," Proceedings of the PSAM5 Conference, Osaka, Japan, 2000.
15. Spurgin, A.J., "Developments in the Use of Simulators for Human Reliability and Human Factors Purposes," IAEA Technical Committee Meeting on Advances in Reliability Analysis and PSA, Szentendre, Hungary, 1994.
16. Spurgin, A.J., Bareith, A., Karsa, Z. "Simulator Data Requirements for HRA Studies," Proceedings of the PSAM7 – ESREL'04 Conference, Springer-Verlag, pp. 1486–1491, 2004.
17. Collier, S., Ludvigsen, J.T., and Svengren, H., "Human Reliability Data from Simulator Experiments: Principles and Context-Sensitive Analysis," Proceedings of the PSAM7 – ESREL'04 Conference, Springer-Verlag, pp. 1480–1485, 2004.
18. Spurgin, A.J., Frank, M.V., "Developments in HRA Technology from Nuclear to Aerospace," Proceedings of the PSAM7 – ESREL'04 Conference, Springer-Verlag, pp. 1748–1753, 2004.
19. Petkov, G., Antao, P. and Guedes Soares, C., "Context Quantification of Individual Performance in Accidents," Proceedings of ESREL'2001, Vol. 3, Torino, Italy, 16–20 September 2001.
20. Petkov, G. Todorov, V., Takov, T., Petrov, V., Stoychev, K., Vladimirov, V., and Chukov, I., "Safety Investigation of Team Performance in Accidents," Journal of Hazardous Materials, ISSN: 0304–3894, Vol. 111, pp. 97–104, 2004.
21. Petkov, G.I., "Development of Techniques and Algorithms for Modeling and Analysis of NPP System Reliability," PhD thesis, MPEI, Russia, 198 p. 1992.

Subject Index